High Performance Computing in Science and Engineering ’23

Thomas Ludwig · Peter Bastian · Michael M. Resch
Editors

High Performance Computing in Science and Engineering ’23

Transactions of the High Performance Computing Center, Stuttgart (HLRS) 2023

Springer

Editors
Thomas Ludwig
Deutsches Klimarechenzentrum GmbH
Hamburg, Germany

Michael M. Resch
High Performance Computing Center
University of Stuttgart
Stuttgart, Baden-Württemberg, Germany

Peter Bastian
Interdisciplinary Center for Scientific Computing (IWR)
Heidelberg University
Heidelberg, Baden-Württemberg, Germany

ISBN 978-3-031-91311-2 ISBN 978-3-031-91312-9 (eBook)
https://doi.org/10.1007/978-3-031-91312-9

The cover figure shows the results of large-scale HPC phase-field simulations combined with concentration diffusion and Navier-Stokes equations for fluid flow. The images illustrate the suspension particle density during freeze-casting, with blue indicating the particle-depleted ice crystals. Depending on the flow conditions and body forces, a regular columnar structure of ice crystals is formed with varying fineness and spacing. See the contribution "Phase-field simulation of sintering processes" by Marco Seiz, Henrik Hierl, Michael Kellner, and Britta Nestler.

This Springer imprint is published by the registered company Springer Nature Switzerland AG
The registered company address is: Gewerbestrasse 11, 6330 Cham, Switzerland

If disposing of this product, please recycle the paper.

Cover Illustration

Details can be found in „Phase-Field Simulation of Sintering Processes" by Brita Nestler, Marco Seiz, Henrik Hierl, and Michael Kellner, Karlsruhe Institute for Technology, Institute for Applied Materials, Kaiserstraße 12, 76131 Karlsruhe, Germany, on pages 185ff.
Freeze casting processes play a crucial role in the development of advanced porous materials with controlled microstructures and tailored properties. They are used for diverse applications such as filtration, catalysis, thermal insulation, and structural components, offering unique advantages in terms of microstructure control, porosity, and functionality.
In freeze casting, a slurry containing a suspension of solid particles in a liquid solvent is cooled below its freezing point. As the solvent freezes, ice crystals form and grow, pushing the solid particles apart and creating pores. After freezing, the solvent is removed by sublimation or freeze drying, leaving behind a regular porous structure.

The figure shows the results of large scale HPC phase-field simulations combined with concentration diffusion and Navier-Stokes equation for fluid flow. The images illustrate the concentration field of the suspension particle density during freeze casting. Depending on the flow conditions and body forces, a regular columnar structure of ice crystals is formed with varying fineness and spacing.

Contents

Molecules, Interfaces, and Solids

Materials Science

Computational Fluid Dynamics

Physics

Simulation relying on HPC remains a major source of insight in physics, be it because an experimental observation is, for principle reasons, impossible, or be it because an accepted fundamental theory is such that an analytic evaluation of the theory is, again for principle reasons, impossible. The physics projects conducted at HLRS relate to both these aspects. This volume contains reports on important simulations in astrophysics, where we try to understand neuron stars and their mergers. Here, time scales prevent us from seeing the full picture in observations, and simulations are mandatory to advance our understanding of these cosmological phenomena. Predicting processes and phenomena to be observed in the large experiments in elementary particle physics need to evaluate quantum chromodynamics, the accepted theory of the strong interaction in various regimes, and since this is impossible analytically, accurate simulation results are required. It is around these two subjects, understanding neutron stars and predicting properties of observables via quantum chromodynamics that the following projects revolve.

The project **microBNS** explores the impact of the nuclear equations of state on the dynamics and the observable features of binary neutron star mergers. They could clarify, for example, which details of the merger are sensitive the equations of state close to saturation density. This and other results are important first steps towards exploring the long-term evolution of the post-merger phase.

The project **GWAnalysis** puts gravitational waves which are generated by neutron star mergers into its focus and thus connects its simulation efforts with possible experimental observations. This is made possible due to the new international network of gravitational wave detectors which will generate more multi-messenger detections. The simulations solve Einstein's field equations in the final stages of a coalescence, and the project has developed a framework to correlate observations with theory using Bayesian methods.

BNSMIC studies microphysical aspects of binary neutron star mergers. A new implementation of a neutrino transport scheme is presented which in particular allows to improve further simulations when it comes to include properties of the weak interactions. In addition to this methodological improvement, the project also explores the postmerger phase by looking at the impact of initial magnetic-field topology on its exponential amplification.

The muon's anomalous magnetic moment has in the past been predicted from QCD theory with a substantial discrepancy compared to the experimentally observed value. The project **GCS-HQCD** improved the simulation methods and result on which the theoretical value of the magnetic moments is based. Several methodological improvements are explored, all aiming at reducing the still relatively large errors due to statistical noise in the simulations. The near-future goal is to reduce these uncertainties to below the percent level.

The muon's magnetic moment is also at the heart of **GS-wprecision**, and it is again by improving the simulation methods and strategies that a much closer agreement between experimental results and theoretical QCD predictions is obtained. One part of the strategy is to use configurations at different lattice spacing, requiring very substantial computational resources as provided by Hawk at HLRS and other installations at other supercomputing centers.

Computations at HLRS in the **stabwf** project present the most important contribution to an international open science project which aims at providing unrestricted access lattice QCD configurations at lattice spacings which allow to simulate at the physical pion mass. While this is mandatory to eliminate systematic errors from the simulations, it is also particularly demanding in terms of HPC resources, and the project introduces a new, stabilized, discretization scheme for the Dirac operator to at least partially remedy this situation.

The PANDA experiment will be one of the key experiments at the Facility for Antiproton and Ion Research (FAIR) at the GSI in Darmstadt. It is important to already today obtain indications on where to search for interesting physics once this experiment is up and running. **RinQCD4PANDA** is a simulation project which based on QCD theory will predict bound states and resonances with a focus on mesons.

These above summaries show that HLRS with Hawk is an indispensable institution for carrying out simulations to understand fundamental physics challenges. I wish you an enjoyable reading when turning to the detailed project reports.

Andreas Frommer
Applied Computer Science
Bergische Universität Wuppertal
Wuppertal, Germany
e-mail: frommer@uni-wuppertal.de

Hadronic Contributions to the Anomalous Magnetic Moment of the Muon from Lattice QCD

G. von Hippel, S. Kuberski, H. B. Meyer, D. Mohler, S. Paul, and H. Wittig

Abstract We report on our ongoing project aimed at providing a precise Standard Model prediction for the hadronic vacuum polarization contribution to the muon's anomalous magnetic moment a_μ, from first principles using lattice QCD. There is currently a discrepancy of $4.2\,\sigma$ between the direct measurement of a_μ and the Standard Model prediction with the hadronic vacuum polarization contribution a_μ^{hvp} evaluated via the traditional data-driven approach based on hadronic cross sections. We present our calculation of the so-called window observable whose value is in tension with hadronic cross section data, which precludes a straightforward interpretation of the new experimental measurement of a_μ. We focus on our efforts to minimize statistical noise in our calculations using deflation techniques and dedicated spectrum calculations to constrain the long-distance regime of the electromagnetic current correlator, which is a crucial step for our goal of providing a refined estimate for a_μ^{hvp} with total uncertainties well below the percent level.

1 Introduction

The anomalous magnetic moment of the muon, $a_\mu \equiv \frac{1}{2}(g-2)_\mu$, parameterizes the fraction of the muon's interaction strength with an external magnetic field due to quantum corrections. It receives contributions from electromagnetism, the weak and the strong interactions where the latter is described by Quantum Chromodynamics (QCD). Thanks to the fact that a_μ can be measured with high precision, a_μ is a

G. von Hippel · H. B. Meyer · S. Paul · H. Wittig
Institut für Kernphysik and PRISMA+ Cluster of Excellence, Universität Mainz, Mainz, Germany

S. Kuberski (✉) · H. B. Meyer · H. Wittig
Helmholtz Institut Mainz, Johannes Gutenberg Universität, Mainz, Germany
e-mail: skubersk@uni-mainz.de

S. Kuberski · H. B. Meyer · D. Mohler · H. Wittig
GSI Helmholtzzentrum für Schwerionenforschung, Darmstadt, Germany

D. Mohler
Institut für Kernphysik, Technische Universität Darmstadt, Darmstadt, Germany

T. Ludwig et al. (eds.), *High Performance Computing in Science and Engineering '23*,
https://doi.org/10.1007/978-3-031-91312-9_1

sensitive probe of the Standard Model (SM) and plays a pivotal role in the quest for new physics which may be able to explain the existence of dark matter or the observed disparity between matter and antimatter. In particular, the observation of a non-zero deficit between experimental measurement [1] and SM prediction would be spectacular signature for the existence of physics beyond the SM. A consensus value for the SM prediction, $a_\mu^{\mathrm{SM}} \equiv a_\mu^{\mathrm{QED}} + a_\mu^{\mathrm{weak}} + a_\mu^{\mathrm{strong}}$, has been reported by the Muon $g-2$ Theory Initiative [2]. By far the biggest share of the theory error is due to the strong interaction, notably the hadronic vacuum polarization (HVP) contribution a_μ^{hvp}. When the latter is evaluated using the traditional data-driven approach based on dispersion integrals and experimentally measured cross sections, one observes a tension of 4.2σ relative to the experimental average. Lattice QCD calculations provide an alternative to the data-driven method but have so far not been able to match the precision, with the exception of the calculation by BMW [3]. Indeed, the BMW result suggests a much smaller tension with experiment. Given the importance of an accurate determination of a_μ for the validity of the SM, it is of paramount importance to provide an independent cross check with comparable or better precision. This is the aim of this project.

In this report we present results for a partial calculation of the full HVP contribution, which is the so-called intermediate window observable defined below. It can be determined with very high statistical precision and provides much better control over systematic uncertainties related to finite-volume effects. Our results corroborate the tension between lattice QCD and the data-driven method based on measured cross sections for $e^+e^- \to$ hadrons, calling for a deeper investigation of hadronic effects.

2 Hadronic Vacuum Polarization Contribution to a_μ

2.1 The Intermediate-Distance Window Contribution

The commonly used approach to compute a_μ^{hvp} on the lattice is based on the time-momentum representation (TMR) [4],

$$a_\mu^{\mathrm{hvp}} = \left(\frac{\alpha}{\pi}\right)^2 \int_0^\infty dt\, G(t)\widetilde{K}(t), \tag{1}$$

where the integration is performed over the Euclidean time variable t, $\widetilde{K}(t)$ is an analytically known kernel function [5], and $G(t)$ is the spatially summed correlator of the electromagnetic current j_μ^γ given by

$$G(t) = -\frac{1}{3}\sum_{k=1}^{3}\sum_{\mathbf{x}} \left\langle j_k^\gamma(x) j_k^\gamma(0)\right\rangle . \tag{2}$$

In the theory with four quark flavours, j_μ^γ is given by

$$j_\mu^\gamma = \frac{2}{3}\bar{u}\gamma_\mu u - \frac{1}{3}\bar{d}\gamma_\mu d - \frac{1}{3}\bar{s}\gamma_\mu s + \frac{2}{3}\bar{c}\gamma_\mu c \equiv j_\mu^3 + \frac{1}{\sqrt{3}} j_\mu^8 + \frac{2}{3} j_\mu^c\,, \tag{3}$$

where j_μ^3 and j_μ^8 denote the isovector and isoscalar components, respectively. An alternative formulation, which is particularly well suited for very large lattices, is based on a covariant integral representation in position space [6]. Its properties have recently been studied [7], but as of now not as part of our project.

The main systematic uncertainty in the computation of a_μ^{hvp} is due to cutoff effects at finite lattice spacing, which are enhanced at small Euclidean time distances $t \lesssim 0.4\,\text{fm}$ [8, 9]. The dominant fraction of the statistical uncertainties stems from the evaluation of the correlation function $G(t)$ in the region of large $t \gtrsim 2\,\text{fm}$, which makes a significant contribution to the integral in Eq. (1) due to the properties of the kernel function $\widetilde{K}(t)$ and an exponentially enhanced signal-to-noise problem in $G(t)$. As suggested in [10], it is useful to divide the integration in Eq. (1) into suitably chosen sub-intervals ("windows"), each enhancing or suppressing different systematic effects. For instance, the intermediate-distance window observable is defined by

$$a_\mu^{\text{win}} = \left(\frac{\alpha}{\pi}\right)^2 \int_0^\infty dt\, G(t)\widetilde{K}(t)\, [\Theta(t, t_0, \Delta) - \Theta(t, t_1, \Delta)], \tag{4}$$

where $\Theta(t, \bar{t}, \Delta) = (1 + \tanh[(t - \bar{t})/\Delta])/2$ is a smoothed step function. If one fixes the parameters as $t_0 = 0.4\,\text{fm}$, $t_1 = 1.0\,\text{fm}$ and $\Delta = 0.15\,\text{fm}$, one finds that a_μ^{win} can be computed with significantly smaller statistical and systematic uncertainties compared to the total a_μ^{hvp}. Therefore, the intermediate window a_μ^{win} has emerged as a benchmark quantity to test the consistency of different lattice QCD calculations, as well as the consistency with data-driven evaluations.

In Fig. 1 we show our results for the isovector contribution $a_\mu^{\text{win,I1}} = 9/10 \cdot a_\mu^{\text{win,ud}}$ to a_μ^{win} for several of our $2+1$-flavour CLS gauge ensembles. Here, $a_\mu^{\text{win,ud}}$ is the contribution of the mass-degenerate up and down quarks from the first term in Eq. (3). This is the dominant contribution to a_μ^{win}. Our results are plotted as function of the squared pion mass, expressed in terms of the quantity $\tilde{y}$, for six different values of the bare inverse gauge coupling β which is in one-to-one correspondence with the lattice spacing. By collecting data in an interval of pion masses from about 420 MeV to 132 MeV and for a large range of lattice spacings, we are able to perform well-constrained extrapolations to the continuum limit and physical quark masses.

In order to facilitate the comparison with other lattice calculations, it is useful to consider individual quark-flavour combinations to a_μ^{win}. In Fig. 2, we present our results for the different flavour contributions that enter a_μ^{win} in isospin-symmetric QCD, as computed in [11], together with the results of other groups. Our final result agrees very well with the recent calculation in Ref. [3], which has similar total uncertainties. The results of different lattice collaborations are in excellent agreement. Considering the large set of different discretizations of the QCD action employed by the various groups, the observed consistency is remarkable.

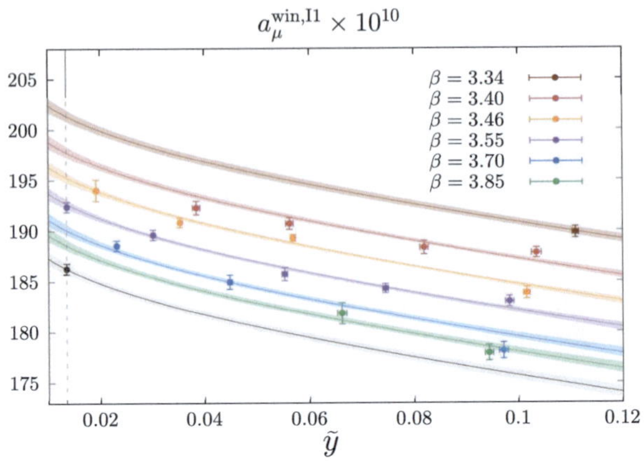

Fig. 1 Typical chiral and continuum extrapolation of the isovector contribution to a_μ^{win}, plotted versus $\tilde{y} = (m_\pi/(4\pi f_\pi))^2$, with f_π the pion decay constant. The black line represents the chiral extrapolation in the continuum limit, and the black point is the result at the physical point. The coloured bands indicate the fit function at finite lattice spacing. Figure taken from [11]

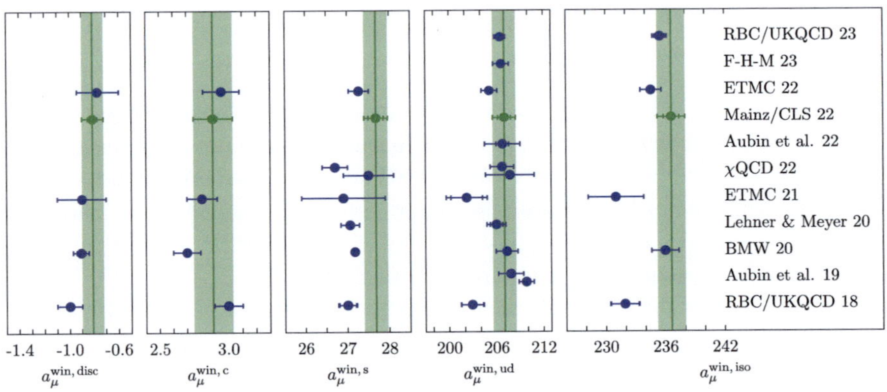

Fig. 2 Comparison of our results for a_μ^{win} (in units of 10^{-10}) with other lattice calculations [3, 10, 12–19] in isosymmetric QCD. The first four panels from the left show compilations of the individual quark-disconnected, charm, strange and light quark contributions. The total result for a_μ^{win} in the isosymmetric theory is shown in the rightmost panel. Our results are represented by green circles and vertical bands

In Fig. 3 we present our results for a_μ^{win} at the physical point as reported in Ref. [11], including an estimate of isospin-breaking effects due to unequal up- and down-quark masses and electromagnetic contributions. Our result is in tension by $3.8\,\sigma$ with a recent data-driven evaluation of a_μ^{win} in Ref. [21]. Note that the four most recent results from lattice QCD shown in Fig. 3 agree very well. Since a_μ^{win} accounts for about one third of the full a_μ^{hvp}, this corroborates the evidence supporting a tension between lattice and data-driven determinations of a_μ^{hvp}.

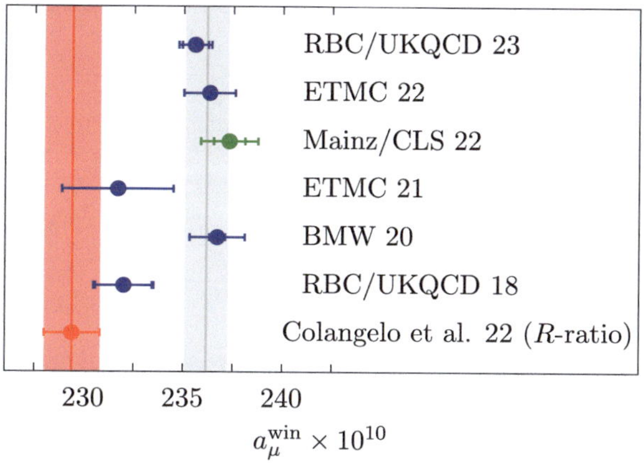

Fig. 3 Comparison of our result for a_μ^{win} including isospin-breaking corrections with the estimates by ETMC [14, 17], BMW [3] and RBC/UKQCD [10, 12]. The grey band denotes a weighted average of the latest lattice QCD results [3, 11, 12, 14], assuming that the data are 100% correlated (see Ref. [20] for further details). The estimate based on the data-driven method of Ref. [21] is shown in red

2.2 *High-Precision Calculation of a_μ^{hvp} via Noise-Reduction Techniques*

The zero-momentum correlation function of the electromagnetic current on the lattice, denoted by $G(t)$ in the previous section, suffers from an exponentially enhanced signal-to-noise problem at large Euclidean distances t. An increase of statistics via stochastic sampling is not the most efficient way to enhance the precision in the long-distance tail of the correlation function, since a reduction of the variance proportional to the number of measurements cannot compete against the exponential loss of information.

We use two complementary approaches to alleviate the signal-to-noise problem in order to achieve our precision goal for a_μ^{hvp}. The first one, which is based on an improved estimator for the correlation function via low-mode averaging (LMA) [22] is discussed in this section. The second one uses an auxiliary computation of finite-volume matrix elements and energies that contribute to the correlation function and is discussed in Sect. 2.3.

When computing a two-point function on the lattice, point-to-all propagators or stochastically estimated timeslice-to-all quark propagators are used in most cases. The computation of exact all-to-all propagators is infeasible because of the sheer number of inversions of the Dirac matrix that would be necessary otherwise. The improved estimator from LMA relies on the fact that the long-distance behaviour of the isovector correlation function is described by the low-energy dynamics of the theory. Therefore, an all-to-all propagator is constructed from a small number of

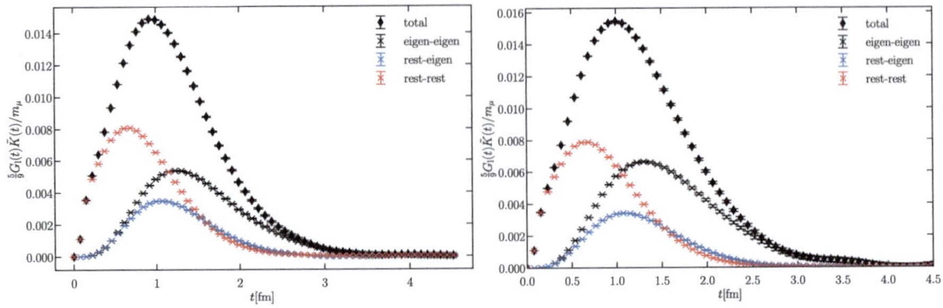

Fig. 4 Contributions to the integrand $G(t)\tilde{K}(t)$ in the isovector channel based on an LMA computation. The total (black diamonds) is the sum of the three contributions denoted by crosses. The black crosses denote the eigen-eigen part that is based on an all-to-all computation in a subspace of the Dirac matrix. Left: Ensemble D450, right: ensemble D452

low eigenmodes of the Dirac operator, where small means a few hundred modes, compared to the $O(10^9)$ modes in total.

Since the long-distance behaviour of $G(t)$ can be described almost completely from these eigenmodes, the all-to-all evaluation of their contribution significantly enhances the attainable precision. At the same time, the remaining contributions to restore the complete correlation function that is defined from all eigenmodes of the Dirac operator, dominate the short distance part where the signal is still relatively good.

In this reporting period, we have used LMA to compute the isovector correlation function to high precision on ensembles D450 and D452 with pion masses slightly above the physical one at a lattice spacing of about 0.075 fm and on ensemble E250 at physical pion mass and a resolution of 0.064 fm (see Fig. 8 for an overview of the ensembles used in this work).

In Fig. 4, we show the integrand $G(t)\tilde{K}(t)$ computed from LMA on the two ensembles D450 and D452. The black diamonds represent the full integrand which is constructed by summing the contributions from the black, blue and red crosses. The black crosses show the contribution from the low modes, which has been evaluated in an all-to-all approach. We have used 608 and 640 modes, respectively, to compute this contribution, denoted by "eigen-eigen" in the figure. One can see from the figure that this part dominates the integrand at large Euclidean time separations. Correspondingly, the signal is well resolved in the relevant region up to 3.5 fm and a_μ^{hvp} can be computed with high precision.

On the left-hand side of Fig. 5 we show the corresponding breakup of the contributions to the integrand for the computation of a_μ^{hvp} on the physical mass ensemble E250. We have used 800 eigenmodes to compute the "eigen-eigen" contribution. This means that we did not scale the number of eigenmodes with the volume, as naively expected. The fact that we still see low-mode dominance in the long-distance tail of the integrand is due to the lighter quark mass on this ensemble, compared to the two boxes D450 and D452 described in the previous paragraph.

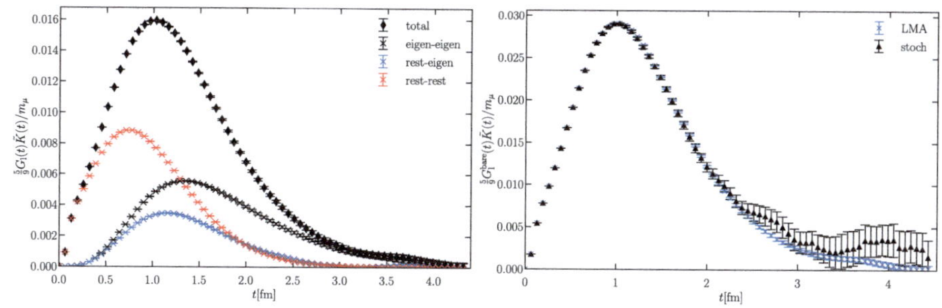

Fig. 5 Left: Contributions to the integrand $G(t)\tilde{K}(t)$ in the isovector channel based on an LMA computation on ensemble E250. The total (black diamonds) is the sum of the three contributions denoted by crosses. The black crosses denote the eigen-eigen part that is based on an all-to-all computation in a subspace of the Dirac matrix. Right: Comparison of the LMA estimator (blue crosses) with our previous data set (black triangles)

On the right-hand side of Fig. 5 we compare the integrand from the LMA estimator with our previous data set, computed from stochastic time slice sources and denoted by "stoch" in the figure. Whereas this is no comparison at equal cost, we stress once more that it would not have been more efficient to increase the number of stochastic sources in the previous calculation to arrive at a similar level of precision in the long-distance tail, due to the exponential loss of signal.

Using the LMA estimator, the variance of the correlation function in the long-distance tail is dominated by the "eigen-eigen" contribution. This means that we have used all available information of the processed gauge configurations to improve our precision. We note that we have computed the correlation function on every second configuration of our Monte Carlo chain for ensemble E250. We currently do not see any autocorrelation between measurements on adjacent configurations. Therefore, we could further improve the signal by doubling statistics. Since we will combine our results with the approach described in Sect. 2.3, we did not yet invest this computing time, as we expect to meet our precision goals with the currently available data set.

2.3 Spectroscopy in the Isovector $\pi\pi$ Channel at Physical Quark Mass

In the long-distance regime, the correlator $G(t)$ is dominated by the isovector two-pion contribution. By performing a detailed study of $I = 1$ $\pi\pi$ scattering at the physical point, one can constrain the correlator $G(t)$ very precisely at large distances. This will be of enormous help for improving the estimation of a_μ^{hvp} at long distances in two ways: first, it reduces the statistical error to better than a percent level on the vector correlator by taking into account its overlap with the $\pi\pi$ states [23] and, second, provides a quantitative estimate of the finite size corrections by computing the time-like pion form factor [4, 24–27].

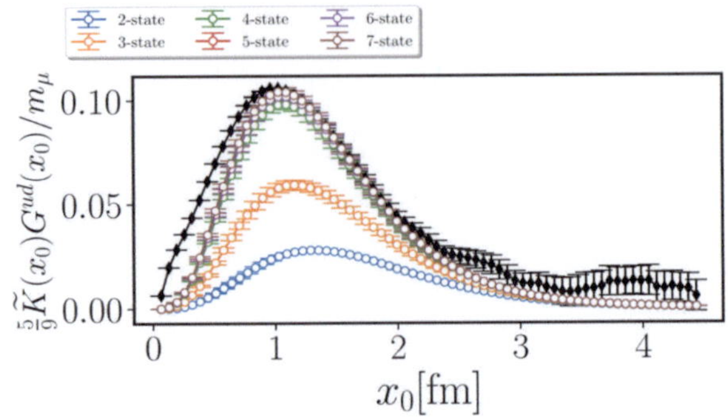

Fig. 6 Preliminary reconstruction of the integrand of the TMR on ensemble E250 from two-pion contributions to the isovector correlation function. The coloured data points denote the accumulation sum of the lowest N contributions, with $N \leq 7$. The black data shows the full correlation function, computed from stochastic time slice sources. Figure from [28]

In this reporting period, we have determined the seven lowest-lying π-π states which have an overlap with the isovector part of the correlator $G(t)$, by solving a Generalized Eigenvalue Problem (GEVP). This calculation is based on measurements on every second configuration of the ensemble E250, matching the set used in Sect. 2.2. The contribution of these π-π states can be used to reconstruct the isovector correlator at large Euclidean time separations where only a few states contribute to its spectral decomposition. Alternatively, they can be subtracted from the isovector correlator after which the improved bounding method can be applied, leading to a more precise estimate for a_μ^{hvp} (see [23] for an extensive discussion of this procedure). We consider all states below the $K\bar{K}$ threshold in our analysis.

In Fig. 6 we show the integrand $G(t)\tilde{K}(t)$ that is used to compute a_μ^{hvp} according to Eq. (1) for the physical mass ensemble E250. The black data points show the computation with stochastic timeslice sources. The coloured data points depict the reconstruction of the integrand, obtained by successively accumulating the contributions from the lowest π-π states. It can be seen that we are able to gradually reconstruct the integrand at large times by adding more and more states. The reconstruction shown in Fig. 6 has been presented at Lattice 2022 [28].

In Fig. 7 we compare the uncertainty on the integrand, computed from three different methods. The black line denotes the statistical uncertainty of our old data set from stochastic timeslice sources that is shown on the right-hand side of Fig. 5. The summation of this data to compute the integral in discretised Euclidean time led to a statistical uncertainty slightly larger than 2%. The blue line, labelled "LMA", shows the significantly reduced uncertainty of the integrand based on the data that has been generated in the current allocation period (see Sect. 2.2). The absolute size of the uncertainties and the slope of the increase towards large Euclidean time separations are significantly decreased.

The red line, labelled "reconstructed", denotes the uncertainty of the integrand that is reconstructed from the finite-volume energies and amplitudes that have been

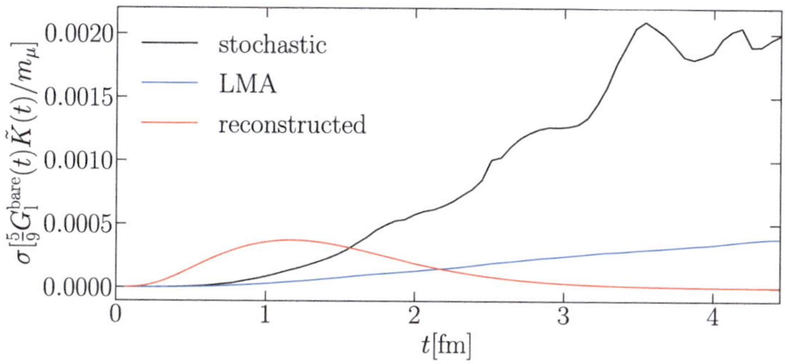

Fig. 7 Statistical uncertainty of the integrand of the TMR from three different estimators. The black line is based on stochastic time slice source, the blue line on LMA and the red line corresponds to the reconstruction from finite-volume energies and amplitudes

computed in this project. Since the uncertainty of the reconstructed correlation function rises only linearly and not exponentially with the Euclidean time t, the signal of this contribution is, by construction, more precise than the data from two-point functions. We will be able to combine the newly generated data from LMA with the reconstructed integrand at large times t and thereby reduce the uncertainty of a_μ^{hvp} even more.

2.4 High-Statistics Calculation of a_μ^{hvp} at Near-Physical Quark Masses

In order to perform high-precision measurements of a_μ^{hvp} for several values of the lattice spacing and a range of light quark masses, we have generated a number of gauge ensembles with close-to-physical pion masses in the scope of this project. The landscape of $2+1$ flavour CLS ensembles that is used in our work is shown in Fig. 8, where ensembles generated in this project are highlighted by a blue circle.

The additional information from these ensembles allows us to tightly constrain the chiral extrapolation to physical quark masses. This is of particular importance since a significant curvature of the dependence of a_μ^{hvp} on m_π can be seen towards physical pion mass [11, 23]. This curvature can be determined more reliably thanks to the two ensembles D450 and D452 at pion masses of 219 and 156 MeV, respectively. Ensemble E250 with $m_\pi = 132$ MeV, on the other hand, tightly constrains the endpoint of the chiral extrapolation to physical pion mass at 135 MeV.

The noise-reduction techniques presented in the previous sections help us to take into account all the information that is present on the gauge ensembles, meaning that our attainable statistics is limited by the length of the Monte Carlo chains. This is not only the case for the isovector contribution to a_μ^{hvp}, which has been discussed in the previous sections, but also for the noisy quark-disconnected contribution. This is

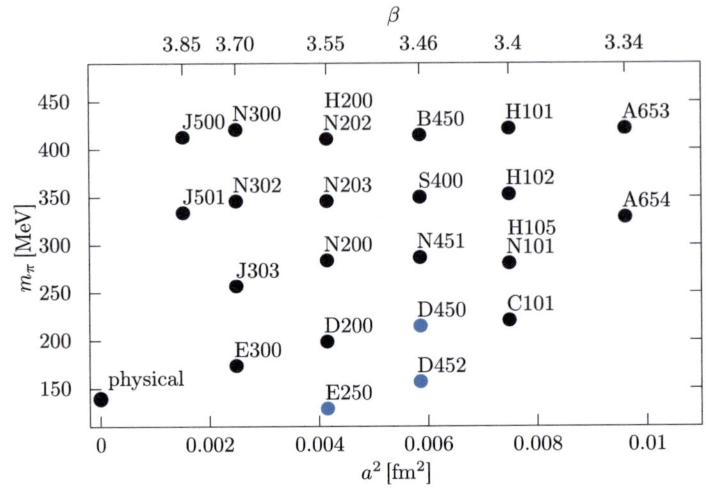

Fig. 8 Overview of $2 + 1$ flavor CLS gauge ensembles used in our computation of a_μ^{hvp} in the plane of squared lattice spacing and pion mass. The sum of the sea quark masses is constant for each value of the bare inverse gauge coupling β. Ensembles marked by blue points have been generated in this project

only possible thanks to the combination of state-of-the-art algorithmic advances for the computation of the contributing diagrams, explained in full detail in [29].

To further improve our precision, we have generated a new Monte Carlo chain comprising 4000 molecular dynamics units (MDU) for ensemble D450, resulting in a threefold increase of the number of gauge configurations. In Fig. 9 we show the Monte Carlo histories of the Hamiltonian deficit at the end of each trajectory of length 2 MDU and of the topological charge. The latter is defined from gauge fields that have been smoothed with the gradient flow [30] and summed over the whole lattice. It has been measured after every second trajectory, i.e. on every gauge configuration that was written to disk. The acceptance rate defined from the Hamiltonian deficit is 0.9051(69).

3 Computational Setup

One of the major computational tasks of our project is the generation of gauge-field ensembles at close to physical light-quark masses. This is challenging because large physical spatial lattice sizes are needed to keep finite-volume effects under control. Furthermore, algorithmic instabilities have been observed in the past at coarse resolutions and close-to physical light-quark masses, as reported in Ref. [31]. In the current reporting period, we finished the production of 1000 gauge configurations (4000 MDU) on ensemble D450 with a pion mass of about 219 MeV at a resolution

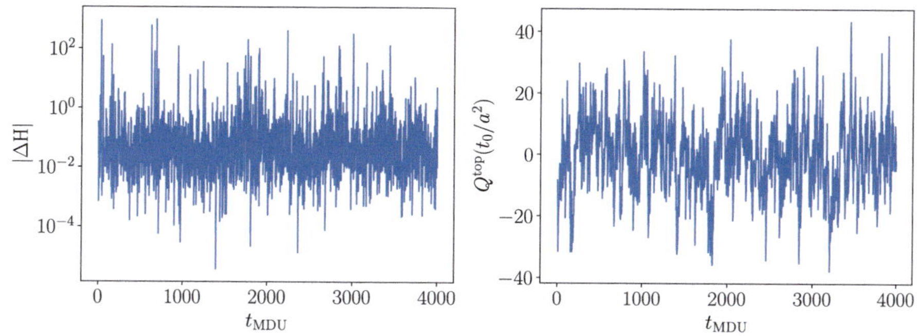

Fig. 9 Monte-Carlo histories of the Hamiltonian deficit ΔH (left) and the total topological charge (right) for the newly generated Monte Carlo chain D450r011

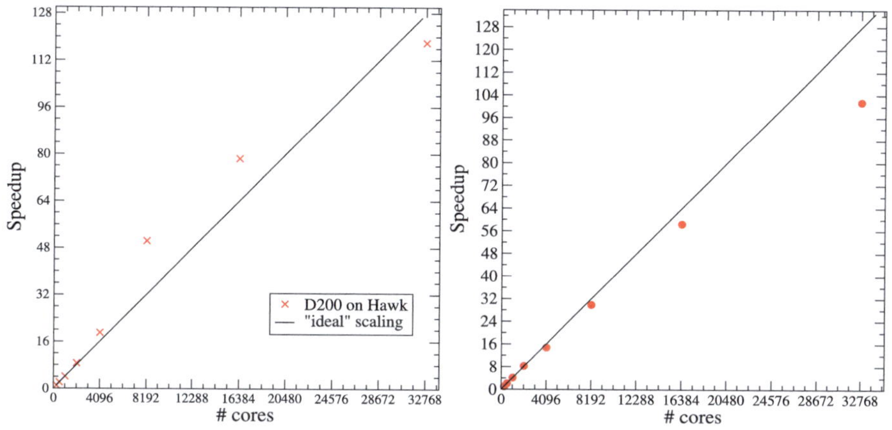

Fig. 10 Left: Strong-scaling behaviour of the openQCD code on Hawk. The plot shows the application of the even-odd preconditioned Wilson-Dirac operator on a 128×64^3 lattice. Speedup factors are defined relative to 256 cores. A clear indication of hyperscaling is seen in this regime. Right: Weak scaling with a local volume of 8^4 on Hawk

of $a \approx 0.076$ fm. The other major computational task is the inversion of the Dirac matrix on these ensembles to compute the observables discussed in this report.

The publicly available openQCD code is used to solve the Dirac equation in all of our projects on Hawk. It exhibits excellent scaling properties over a wide range of problem sizes. Figure 10 shows the strong-scaling behaviour for the system size corresponding to ensemble D200, i.e. for a 128×64^3 lattice, as measured on Hawk (left panel). The timings refer to the application of the even-odd preconditioned O(a) improved Wilson-Dirac operator $\hat{D}_{\rm w}$ to a spinor field, which accounts for the largest fraction of the total computing time. The weak scaling behaviour for a local lattice volume of size 8^4 is shown in the right panel of Fig. 10.

In the current reporting period, we employed three different setups:

A Local lattice of size 8^4 per MPI rank with 8192 MPI ranks on 64 nodes

B Local lattice of size $12 \times 6^2 \times 12$ per MPI rank with 32768 MPI ranks on 256 nodes.
C Local lattice of size $12 \times 6 \times 12^2$ per MPI rank with 16384 MPI ranks on 128 nodes.

For the gauge field generation we used setup A. The spectroscopy runs were performed with setup B on E250 192×96^3. Our runs using LMA on E250 have been performed with setup D. Whenever possible, a hypercube of size 2×4^3 processes was grouped onto a single CPU, to minimise off-CPU communication.

For the current allocation year we received a total of 120 MCore hours. We spent 46% on the spectroscopy, described in Sect. 2.3 and the remaining computer time on the LMA computation of the isovector correlation function that is explained in Sect. 2.2.

4 Summary and Outlook

We have reported on the current status of our precision computation of hadronic contributions to the anomalous magnetic moment of the muon using the *ab initio* formalism of lattice QCD. In the current reporting period, we have focused on the precise computation of a_μ^{hvp} on an ensemble at physical quark mass to tightly constrain the quark mass dependence of the observable. Combined with the highly precise data at slightly larger-than-physical light quark masses on ensembles D450 and D452, we have been able to map out the chiral dependence over a large range of quark masses with much higher precision. We have chosen this approach since it allows us to constrain the continuum extrapolation at unphysically large light quark masses, where the signal-to-noise problem in the computation of the isovector correlation function is milder, and smaller boxes may be used without introducing large finite-size effects that depend on the product $m_\pi L$, where L is the spatial extent of the box. The ability to control the extrapolation to the continuum limit currently limits the attainable precision in the most precise computation of a_μ^{hvp} from lattice QCD [3].

We have found that our computation of the isovector correlation function using low-mode averaging at physical pion mass, despite the additional computational effort, is significantly more efficient than our previously used approach of using stochastic time slice sources. Accordingly, we have been able to significantly reduce the noise in the long-distance tail of the integrand of the time-momentum representation. We are planning to extend our computation on further ensembles at finer resolutions, in order to constrain the continuum extrapolation of our data even more tightly.

Furthermore, we have doubled the statistics for our calculation of $\pi\pi$ scattering at the physical point. In this report, we have outlined the potential to improve the statistical precision of our estimate for a_μ^{hvp} at physical pion mass. We will also use the newly generated data to compute the finite-size corrections in our computation via the time-like pion form factor, which itself is a phenomenologically relevant observable. First results for the phase shifts obtained from the lattice spectrum using Lüscher's quantization condition have been shown in the previous report.

Acknowledgements This work is partly supported by the Deutsche Forschungsgemeinschaft (DFG, German Research Foundation) grant HI 2048/1-1 and by the DFG-funded Collaborative Research Centre SFB 1044 *The low-energy frontier of the Standard Model*. The Mainz $(g-2)_\mu$ project is also supported by the Cluster of Excellence *Precision Physics, Fundamental Interactions, and Structure of Matter* (PRISMA+ EXC 2118/1) funded by DFG within the German Excellence Strategy (Project ID 39083149). The authors gratefully acknowledge the Gauss Centre for Supercomputing e.V. (www.gauss-centre.eu) for supporting this project by providing computing time on the GCS Supercomputer HazelHen and Hawk at Höchstleistungsrechenzentrum Stuttgart (www.hlrs.de) under project GCS-HQCD. Additional calculations were performed on the HPC clusters "Clover" and "HIMster II" at the Helmholtz-Institut Mainz and "MOGON II" at JGU Mainz. D.M. acknowledges funding by the Heisenberg Programme of the Deutsche Forschungsgemeinschaft (DFG, German Research Foundation)—project number 454605793. We are grateful to our colleagues in the CLS initiative for sharing ensembles.

References

1. Abi, B., et al.: Phys. Rev. Lett. **126**(14), 141801 (2021). https://doi.org/10.1103/PhysRevLett.126.141801
2. Aoyama, T., et al.: Phys. Rept. **887**, 1 (2020). https://doi.org/10.1016/j.physrep.2020.07.006
3. Borsanyi, S., et al.: Nature **593**(7857), 51 (2021). https://doi.org/10.1038/s41586-021-03418-1
4. Bernecker, D., Meyer, H.B.: Eur. Phys. J. A **47**, 148 (2011). https://doi.org/10.1140/epja/i2011-11148-6
5. Della Morte, M., Francis, A., Gülpers, V., Herdoíza, G., von Hippel, G., Horch, H., Jäger, B., Meyer, H.B., Nyffeler, A., Wittig, H.: JHEP **10**, 020 (2017). https://doi.org/10.1007/JHEP10(2017)020
6. Meyer, H.B.: Eur. Phys. J. C **77**(9), 616 (2017). https://doi.org/10.1140/epjc/s10052-017-5200-3
7. Chao, E.H., Meyer, H.B., Parrino, J.: Phys. Rev. D **107**(5), 054505 (2023). https://doi.org/10.1103/PhysRevD.107.054505
8. Harris, T., Cè, M., Meyer, H.B., Toniato, A., Török, C.: PoS **LATTICE2021**, 572 (2021). https://doi.org/10.22323/1.396.0572
9. Sommer, R., Chimirri, L., Husung, N.: PoS **LATTICE2022**, 358 (2023). https://doi.org/10.22323/1.430.0358
10. Blum, T., Boyle, P.A., Gülpers, V., Izubuchi, T., Jin, L., Jung, C., Jüttner, A., Lehner, C., Portelli, A., Tsang, J.T.: Phys. Rev. Lett. **121**, 022003 (2018). https://doi.org/10.1103/PhysRevLett.121.022003
11. Cè, M., et al.: arXiv: 2206.06582 (2022)
12. Blum, T., et al.: arXiv: 2301.08696 (2023)
13. Bazavov, A., et al.: Phys. Rev. D **107**(11), 114514 (2023). https://doi.org/10.1103/PhysRevD.107.114514
14. Alexandrou, C., et al.: Phys. Rev. D **107**(7), 074506 (2023). https://doi.org/10.1103/PhysRevD.107.074506
15. Aubin, C., Blum, T., Golterman, M., Peris, S.I.: arXiv: 2204.12256 (2022)
16. Wang, G., Draper, T., Liu, K.F., Yang, Y.B.: arXiv: 2204.01280 (2022)
17. Giusti, D., Simula, S.: PoS **LATTICE2021**, 189 (2022). https://doi.org/10.22323/1.396.0189

18. Lehner, C., Meyer, A.S.: Phys. Rev. D **101**, 074515 (2020). https://doi.org/10.1103/PhysRevD.101.074515
19. Aubin, C., Blum, T., Tu, C., Golterman, M., Jung, C., Peris, S.: Phys. Rev. D **101**(1), 014503 (2020). https://doi.org/10.1103/PhysRevD.101.014503
20. Wittig, H.: In 57th Rencontres de Moriond on Electroweak Interactions and Unified Theories (2023)
21. Colangelo, G., El-Khadra, A.X., Hoferichter, M., Keshavarzi, A., Lehner, C., Stoffer, P., Teubner, T.: arXiv: 2205.12963 (2022)
22. Giusti, L., Hernández, P., Laine, M., Weisz, P., Wittig, H.: JHEP **04**, 013 (2004). https://doi.org/10.1088/1126-6708/2004/04/013
23. Gérardin, A., Cè, M., von Hippel, G., Hörz, B., Meyer, H.B., Mohler, D., Ottnad, K., Wilhelm, J., Wittig, H.: Phys. Rev. D **100**, 014510 (2019). https://doi.org/10.1103/PhysRevD.100.014510
24. Meyer, H.B.: Phys. Rev. Lett. **107**, 072002 (2011). https://doi.org/10.1103/PhysRevLett.107.072002
25. Feng, X., Aoki, S., Hashimoto, S., Kaneko, T.: Phys. Rev. D **91**(5), 054504 (2015). https://doi.org/10.1103/PhysRevD.91.054504
26. Andersen, C., Bulava, J., Hörz, B., Morningstar, C.: Nucl. Phys. B **939**, 145 (2019). https://doi.org/10.1016/j.nuclphysb.2018.12.018
27. Erben, F., Green, J.R., Mohler, D., Wittig, H.: Phys. Rev. D **101**(5), 054504 (2020). https://doi.org/10.1103/PhysRevD.101.054504
28. Paul, S., Hanlon, A.D., Hörz, B., Mohler, D., Morningstar, C., Wittig, H.: PoS **LATTICE2022**, 073 (2023). https://doi.org/10.22323/1.430.0073
29. Cè, M., Gérardin, A., von Hippel, G., Meyer, H.B., Miura, K., Ottnad, K., Risch, A., José, T.S., Wilhelm, J., Wittig, H.: arXiv: 2203.08676 (2022)
30. Lüscher, M.: JHEP **04**, 123 (2013). https://doi.org/10.1007/JHEP04(2013)123
31. Mohler, D., Schaefer, S.: Phys. Rev. D **102**(7), 074506 (2020). https://doi.org/10.1103/PhysRevD.102.074506

Lattice QCD with High Precision

Zoltan Fodor, Fabian Frech, Christian Hölbling, and Balint Toth

Abstract In 2021 an experiment at the Fermi National Accelerator Laboratory (FNAL) near Chicago confirmed a puzzling discrepancy regarding an elementary particle, called muon. The event received significant media attention. On the same day, our ab-initio calculation was published in Nature, challenging previous computations and bringing the theoretical prediction closer to the experimental value: no new forces or particles may be needed to explain the FNAL measurement. Ours is the most precise ab-initio calculation so far, with uncertainties comparable to those of the measurement and of the reference, data-driven computations. These computations were carried out -among other resources- using previous allocations on HAWK.

1 Introduction

The muon is an elementary particle, a short-lived cousin of the electron. For many years the calculation of its magnetic moment has disagreed with its measurement, suggesting that a not-yet-known particle or force perturbs the muon. Such a discovery would have profound consequences on our understanding of Nature.

The experimental measurement at FNAL is performing perfectly and in terms of precision has already outperformed the previous experiment from Brookhaven National Laboratory (BNL). Both the BNL and FNAL results are shown in Fig. 1. Their recently announced combined average [1] is the purple band. In the next few years it is expected, that FNAL will reduce the uncertainty of its current measurement by a factor of four. Already this summer the results of Run2 and Run3

Z. Fodor (✉) · F. Frech · C. Hölbling · B. Toth
Department of Physics, Wuppertal University, Wuppertal, Germany
e-mail: hch@uni-wuppertal.de

C. Hölbling
e-mail: btoth@uni-wuppertal.de

Z. Fodor
Institute for Theoretical Physics, Eötvös University, Budapest, Hungary

Jülich Supercomputing Centre, Forschungszentrum Jülich, Jülich, Germany

T. Ludwig et al. (eds.), *High Performance Computing in Science and Engineering '23*,
https://doi.org/10.1007/978-3-031-91312-9_2

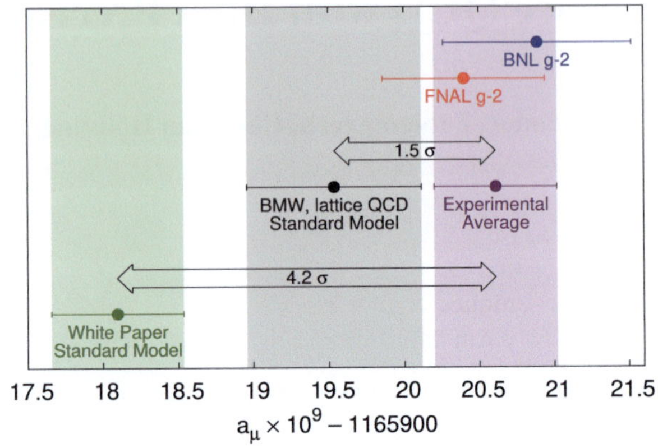

Fig. 1 Current status of theory and experiment for the muon's anomalous magnetic moment. Is there new physics or not?

will be published (the result depicted by the red dot with its error bar contains Run1, one expects that the error will be cut by a factor of two using these two new runs). To fully exploit the potential for new physics searches of this future result, theory computations must reach the same precision.

The gold-standard of the theory computation is given by the White Paper of the g-2 Theory Initiative [2]. Its result is the green band in the figure. It differs by 4.2σ from experiment, which can be considered a very strong signal for physics beyond the Standard Model. The theory result is a sum of many different contributions, most of them are very precisely known. One contribution dominates the total uncertainty (4.0 in units of 10^{-10} out of 4.3, for brevity we will use this "unit" in the rest of this write-up) and there is another one with a subdominant but still sizable error; these two contributions describe the indirect, strong interaction effects on the muon. The calculation of the hadronic light-by-light contribution has made incredible progress in recent years, with different approaches giving consistent results and with a precision that is closing on to the precision of the future FNAL result. It now seems highly unrealistic that this contribution can make up for the 4.2σ discrepancy. The other, more important contribution with large error (4.0 out of 4.3) is the hadronic vacuum polarization, which is obtained in [2] from a phenomenological approach, sometimes called the R-ratio method. It uses a vast amount of data collected from electron-positron annihilation experiments.

In our work [3] we have computed the leading-order hadronic vacuum polarization (LO-HVP) contribution from lattice QCD. We see some tension with the R-ratio method, our result being larger. In Fig. 1 we add to our LO-HVP value the other contributions to the muon's magnetic moment as given in the White Paper. The result is shown by the gray band. This value differs only by 1.5σ from the combined experimental average and is consistent with the new FNAL experiment.

During the last year very important developments happened both on the lattice as well as on the electron-positron experimental side.

Though our results is the only full lattice result, there are already five independent findings (Fig. 2) for the so-called window observable (it was originally introduced by the RBC-UKQCD collaboration; for our result see [3]). It calculates the light connected contribution in the 0.4 to 1.0 fm Euclidean time region. It gives about one third of the total hadronic vacuum polarisation result. This window quantity is far easier to calculate but can be determined using the R-ratio method, too. The two R-ratio determination show our result [3] and an independent calculation, too. These two determinations agree completely, as they should. Not even a one-sigma difference is expected, since they use exactly the same experimental electron-positron data. Note that these R-ratio results are typically obtained by an error-inflation or similar techniques. The reason for that is the internal tensions between the various electron-positron data. The character set "/lat" means that a small contribution from other graphs were subtracted from the R-ratio result to make a direct comparison with the lattice data possible. As it can be seen the lattice results all agree, they do not need any error-inflation. The χ^2 value is around 0.1, which indicates that the errors of the lattice results were—most probably—all overestimated. This merely reflects the fact that the lattice groups are over-cautious with their systematic error estimates. The final conclusions of this plot can be summarized as follows: (a) for the window observables the lattice results all agree and the tension between; (b) independent determinations of the R-ratio perfectly agree (c) most importantly there is a 4.9 to 5.1 sigma tension between the lattice result and that of the R-ratio method.

Another very important new development is coming not from the lattice but from direct measurements of the electron-positron cross section. The CMD collaboration from Novosibirsk, Russia released their latest CMD3 data. The result can be seen in Fig. 3. The energy window is 0.60–0.88 GeV for the two charged pion chanel. This is far the most dominant contribution to the leading order hadronic vacuum polarisation term to the anomalous magnetic moment of the muon in the R-ratio framework. It give more than 50 the most recent CMD3 data is in strong tension with the world average for this quantity. The discrepancy is on the 4.4 sigma level. This is more than the original 4.2 tension between the measured and the White Paper values of the muon's magnetic moment. How this new data with its large tension potential will be included in the White Paper, if at all, is unknown or undecided yet.

These new developments further underlines that the scale determination—the main goal of the present project and report—is needed with unprecedented precision.

2 Project Milestones

This long term project was started nine years ago with the target to compute the leading order hadronic vacuum polarization contribution to the anomalous magnetic moment of the muon, $a_\mu^{\text{LO-HVP}}$. The project had two important milestones.

At the first milestone we were able to reach a relative precision of 2.7% and got the result

$$a_\mu^{\text{LO-HVP}} = 711.0(7.5)(17.3)[18.9] \tag{1}$$

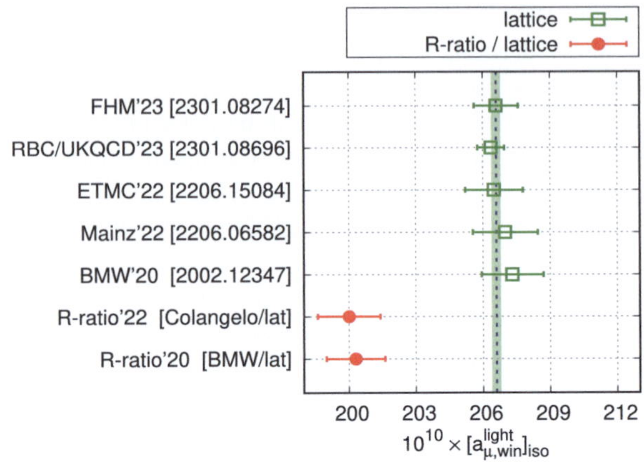

Fig. 2 Current status of lattice (green band) and R-ratio for the muon's anomalous magnetic moment window observable

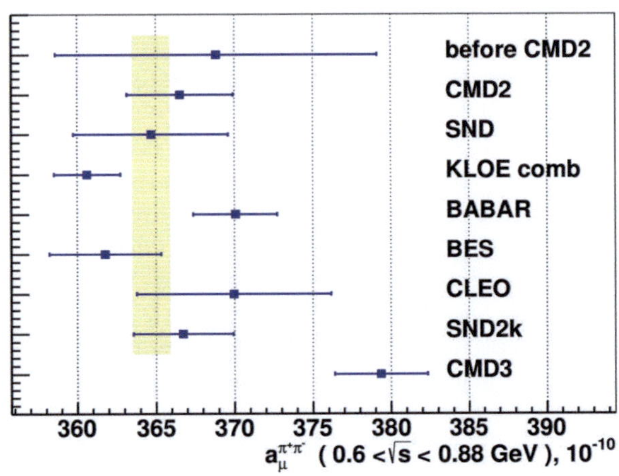

Fig. 3 Current status of the electron-positron input data for the R-ratio method in the 0.60 to 0.88 GeV energy region. The most recent CMD3 result is 4.4σ larger than the previous world average

This was published in Physical Review Letters [4], and the article was then highlighted by the editors as Editors' Suggestion.

At the second milestone we were able to improve the precision by several means and finally obtained the result

$$a_\mu^{\text{LO-HVP}} = 707.5(2.3)(5.0)[5.5] \tag{2}$$

with statistical, systematic and total errors. The relative uncertainty is 0.8%, which is far less than that of other lattice determinations, and is comparable to the errors of R-ratio based computations. The result was published in Nature [3] and become the most cited theory paper in lattice quantum field theory in 2021, 2022 and 2023.

Theory has now important problems ahead. We have to explain the difference between the R-ratio method and our lattice determination. This became even more important because of the high level agreement of the lattice result for the window observable and the 5 sigma tension with the R-ratio method (see discussion above). This is obviously necessary to draw any conclusion about physics beyond the Standard Model. This will be our third milestone. We also have to increase our precision by a factor of four if we want to reach the precision of the future FNAL experiment, this will be the fourth and final milestone of this project.

3 Current Project Targets

To increase the precision from what we currently have in Eq. 2 to the four times smaller target, several problems have to be addressed. The difficulty lies not only in the fact, that higher precision costs more per se, but also that usually with increasing precision additional effects have to be taken into account, that have (justifiably) played a subdominant role in previous determinations. In the following we give a list of problems, on which we are concentrating our efforts.

First and foremost we are preparing a new set of ensembles, with a finer lattice resolution, than we had in [3]. The largest source of the systematic error in Eq. 2 was coming from the continuum extrapolation, 4.2 out of the total systematice error of 5.0. The continuum extrapolation error is obtained as a variation among various approaches to perform the continuum extrapolation. This includes changing the number of lattice spacings in the fit, changing the fit function from linear to quadratic in a^2, multiplying a^2 by powers of the strong coupling constant $\alpha_s(a)$ and applying different improvement procedures. By going to finer lattices we can improve these errors. Our preliminary estimations showed, that we can reduce the continuum extrapolation error by 40% by adding a new ensemble at $a = 0.048$ fm, with $128^3 \times 192$ lattice size.

Beside the hadronic vacuum polarization there are many other important quantities, that we are computing on these fine lattices. For example there are the decay constants of pions and kaons. These can be computed with per-mill level precision, the same precision that we want to reach for the hadronic vacuum polarization itself. They can be considered as benchmark quantities as they provide important an cross-check of our approach to lattice QCD. With the ratio of these two decay constants it is possible to constrain the ratio of two CKM matrix elements V_{ud} and V_{us}. Since these two matrix elements can also be determined by other methods, we get a powerful test to check the consistency of the Standard Model in addition.

A further quantity that we are currently working on is the mass of the Omega baryon (M_Ω). This is required in order to set the length scale in our simulations, an important pre-requisite for the whole project. Its experimental value is known to a very good precision, the challenge is to reach the per-mill level accuracy in our computations. An important target here is a determination of an easily calculable length-scale derived from the gradient flow, called w_0. Though it cannot be measured in experiments, it is very useful to compare results of different lattice collaboration but also to separate the electromagnetic effects from those coming solely from the strong interaction. In this project we go for a 0.1% precise determination of the $M_\Omega \cdot w_0$ product. This requires the computation of Omega propagators on our new fine lattices.

We are also improving the continuum extrapolation procedure by constructing new discretization types for the hadronic vacuum polarization. The discretization artefacts of these can be very different from the one, that we considered in our previous works. Of course they approach the same value in the continuum limit. By combining the different continuum extrapolations that all share the same continuum value we can sizeably reduce systematic error related to the continuum limit. We are currently reanalyzing our coarser lattice spacings with this approach.

4 Preliminary Physics Results

The configurations for the new lattice ensemble at $a = 0.0483$ fm lattice spacing have already been generated in the last two years. We have sufficient statistics to reach the planned error reduction. Figure 4 shows last and recent years progress on the history of the smeared plaquette observable. These runs are using 256 node partitions, the lattice size is $128^3 \times 192$. A single trajectory requires about 13 min on a 256 node HAWK partition. We have already run the computation of simpler quantities, like gradient-flow scale or decay constants, and currently working on the computationally more intensive observables.

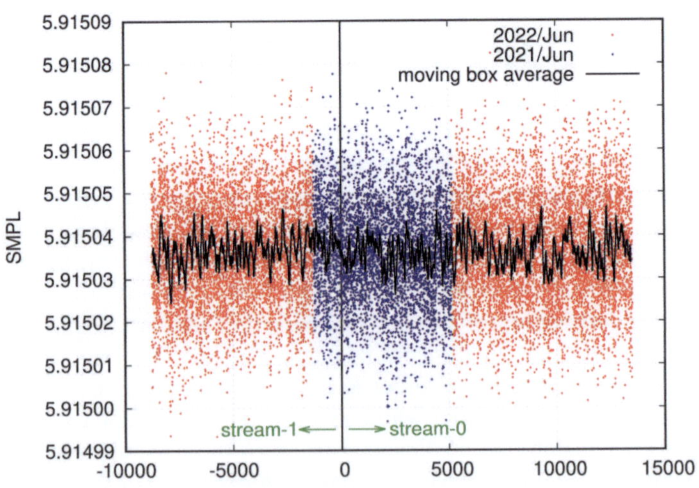

Fig. 4 History of the smeared plaquette on our finest lattice spacing generated on HAWK supercomputer. Blue points show are status in June 2021, red points our statistics a year later. We are running two streams. A moving window average with a width of 100 trajectories is shown with black

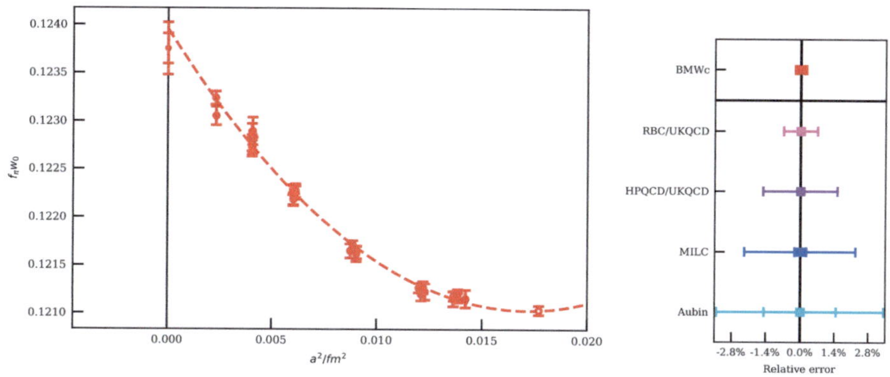

Fig. 5 Left: continuum extrapolation of the pion decay constant, $f_\pi w_0$, an important benchmark observable. The fit includes results from our finest lattice spacing. Right: relative error compared to other lattice determinations

We are now performing the analysis of the decay constants including these fine lattices. Some preliminary results have already been shown at conferences. We are currently doing the final analysis and preparing for publication. The continuum extrapolation of $f_\pi \cdot w_0$ is shown in Fig. 5. A comparison of the relative error to those obtained with other lattice and non-lattice approaches is also shown. We reach a precision of 0.18% without taking into account the error of w_0. This is much better than any previous lattice determination, the previous best was 0.62%. With an expected improvement on the value w_0, we will get close to the error level of the experimental measurements.

In Fig. 6 we show the result of the re-analysis of our coarser lattices with new types of discretizations. These discretizations differ in the size of the lattice artefacts, but they are supposed to give the same continuum limit. The primary goal of this analysis is to increase the confidence in our extrapolation procedure. We also hope to decrease the systematic error by performing a combined continuum limit, where all the different discretizations are forced to approach the same value. The figure shows the so-called window observable, that is part -about $1/3$- of the total $a_\mu^{\mathrm{LO-HVP}}$. This type of analysis is rather expensive. We are currently exploring different ways to compute it on the finest lattice and also for the remaining $2/3$ part.

Our current most important efforts are concentrated on computing the Omega baryon propagators and the total $a_\mu^{\mathrm{LO-HVP}}$ on our finest lattices. We do not have final results on any of those, because the analyses is still going on in a blinded manner by several analysis groups. These two quantities (Omega mass and $a_\mu^{\mathrm{LO-HVP}}$ consume roughly the same amount of computer time as the configuration generation themselves. We try to fit them into the smallest partition possible, so they use 128 and 256 HAWK nodes respectively. The a_μ computation works with thousands of eigenvectors of the Dirac-operator, hence it needs more memory than the Omega measurements. First results were shown in previous year's report. Due to blinding we can't show more and for brevity we do not show last year's plot again.

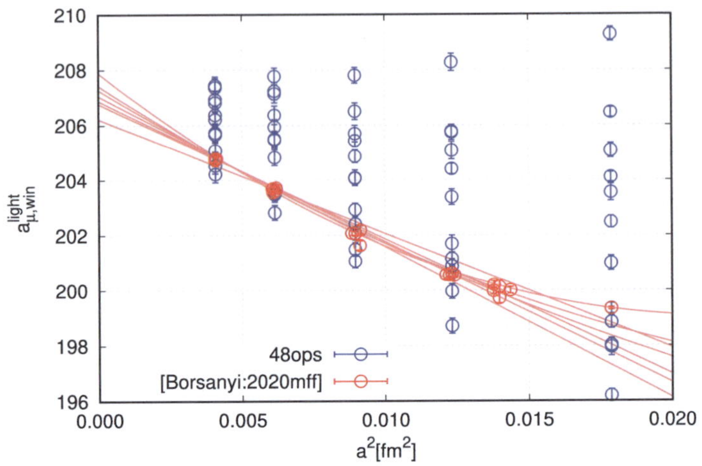

Fig. 6 The window observable, an intermediate distance contribution to $a_{\mu}^{\mathrm{LO-HVP}}$, as a function of lattice spacing. Results are shown from our [3] publication with red, the solid lines are various types of continuum extrapolations. The blue points are the result of the re-analysis of these configurations with new types of discretizations. We expect to improve the continuum limit procedure by including them into the analysis

5 Code Performance

The dynamical fields, both single and double precision, need about 60TB memory. The smallest possible partition is therefore 256 nodes for the configuration generation. We work at fixed lattice size, so only the strong scaling is relevant. In the left panel of Fig. 7 we show the performance of a single iteration of the conjugate gradient solver as a function of the number of HAWK nodes, in a strong scaling setting. This operation is the mostly used function in the application.

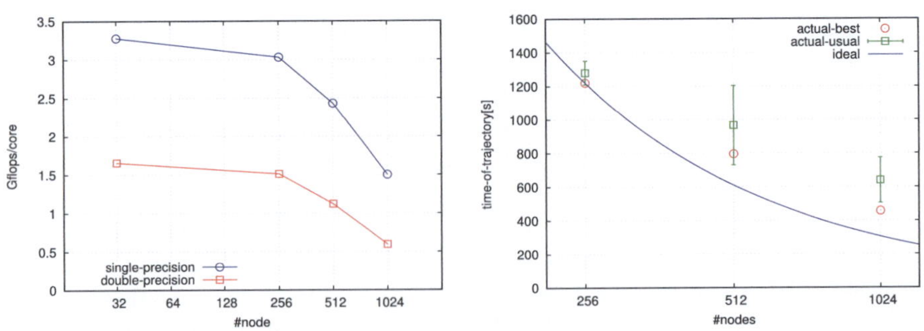

Fig. 7 Code performance as function of the number of nodes. The plots show strong scaling on 128 × 192 lattice. Left: Gflop/core in conjugate gradient iterations. Right: time to solution in seconds, we show the best trajectory with red, average trajectory with green, error bar corresponds to standard deviation

Table 1 Time to solution table

Number of cores	Time to solution [sec]	Ideal time to solution [sec]
256×128	1220	1220
512×128	796	610
1024×128	457	305

The code uses both single and double precision, both performances are indicated in Fig. 7.

In the right panel of Fig. 7 we show the time to solution, which is the total time of a trajectory, starting from 256 nodes. There is a noticeable variation between runs. These have to do with partitions that scatter among more than two racks. On two rack partition, i.e. when the job occupates two racks exclusively, we always reach the ideal timings. The results for the best case scenario are summarized in Table 1.

In ideal case we reach close to 3.0 Gflops/core performance, both for the configuration generation and analysis. Unfortunately there is considerable fluctuation in reality.

6 Publications

As a result of this project we published the following paper (the topic is extremely debated, that is why the various groups wait 3–4 years before coming out with a new result). According to Google Scholar our paper received over 700 citations though it was published only two years ago.

[A] S. Borsanyi, Z. Fodor, J. N. Guenther, C. Hölbling, S. D. Katz, L. Lellouch, T. Lippert, K. Miura, L. Parato and K. K. Szabo, et al. "Leading hadronic contribution to the muon magnetic moment from lattice QCD," Nature **593** (2021) no. 7857, 51-55 doi: https://doi.org/10.1038/s41586-021-03418-1 [arXiv:2002.12347 [hep-lat]].

Other publications:

[B] A. Gerardin et al., Lattice calculation of the π^0, η, η' transition form factors and the hadronic light-by-light contribution to the muon g-2, [arXiv:2305.04570 [hep-lat]].

[C] S. Borsanyi et al., High precision scale setting on the lattice, PoS LATTICE (2022) 371, doi: https://doi.org/10.22323/1.396.0371.

[D] B. Toth et al., Muon g-2: BMW calculation of the hadronic vacuum polarization contribution, PoS LATTICE (2022) 005, doi: https://doi.org/10.22323/1.396.0005.

[E] L. Parato et al., QED and strong isospin corrections in the hadronic vacuum polarization contribution to the anomalous magnetic moment of the muon, PoS LATTICE (2022) 358, doi: https://doi.org/10.22323/1.396.0358.

Last years' invited talks (speaker, conference name, place, date):

- Zoltan Fodor, Quantum Connections, Stockholm, Sweden, June 2023
- Kálmán Szabó, Lattice Gauge Theory Contributions to New Physics Searches, Madrid, Spain, June 2023
- Zoltan Fodor, Muon Physics, Venice, Italy, May 2023
- Zoltan Fodor, Precision Test in Physics, Vienna, Austria, May 2023
- Laurent Lellocuh, BMW prescription, Edinburgh, UK, May 2023
- Zoltan Fodor, Colloquium, Paul Scherer Institute, Switzerland, April 2023
- Zoltan Fodor, Theory Seminar, Duke University, USA, April 2023
- Zoltan Fodor, University Alberta, Lake Louise, Canada, February 2023
- Kálmán Szabó, New Physics Signals Conference, Pisa, Italy, February 2022
- Zoltan Fodor, Hadron Dynamics, Porte Vallarta, Mexico, February 2023
- Zoltan Fodor, Seminar, California State University, LB, February 2023
- Kálmán Szabó, Colloquium, Pisa, Italy, December 2022
- Laurent Lellocuh, Dark Side of the Universe, La Reunion, France, November 2022
- Zoltan Fodor, 25th Workshop HLRS, Stuttgart, Germany, October 2022
- Laurent Lellocuh, Muon g-2 Theory Initiative, Edinburgh, UK, September 2022
- Balint Toth, Flavour changing and conserving processes, Anacapri, Italy, September 2022
- Laurent Lellocuh, Hadron Spectroscopy, Munich, Germany, September 2022
- Balint Toth, International Symposium on Lattice Field Bonn, Germany, August 2022,
- Kálmán Szabó, 14th Conference on Particle and Nuclear Physics, Orlando, US, August 2022
- Finn Stokes, International Symposium on Lattice Field Bonn, Germany, August 2022,
- Zoltan Fodor, PASCOS Conference, Heidelberg, Germany, July 2022
- Laurent Lellocuh, Subnuclear Physics, Erice, Italy, June 2022
- Finn Stokes, SchwingerFest, Los Angeles, USA, June 2022

Acknowledgements The authors gratefully acknowledge the Gauss Centre for Supercomputing e.V. (www.gauss-centre.eu) for funding this project by providing computing time on the GCS Supercomputers at Jülich Supercomputing Centre, at HLRS Stuttgart and LRZ München.

References

1. Abi, B., et al.: Measurement of the positive muon anomalous magnetic moment to 0.46 ppm. Phys. Rev. Lett. **126**(14), 141801 (2021). arXiv: 2104.03281 [hep-ex]
2. Aoyama, T., et al.: The anomalous magnetic moment of the muon in the standard model. Phys. Rept. **887**, 1–166 (2020). arXiv: 2006.04822 [hep-ph]
3. Borsanyi, S., Fodor, Z., Guenther, J.N., Hoelbling, C., Katz, S.D., Lellouch, L., et al.: Leading order hadronic contribution to the muon magnetic moment from lattice QCD. Nature **593**(7857), 51–55 (2021). https://doi.org/10.1038/s41586-021-03418-1[arXiv:2002.12347 [hep-lat]]
4. Borsanyi, S., et al.: Hadronic vacuum polarization contribution to the anomalous magnetic moments of leptons from first principles. Phys. Rev. Lett. **121**(2), 022002 (2018). arXiv: 1711.04980 [hep-lat]

Simulating Binary Neutron Star Mergers

Tim Dietrich, Bernd Brügmann, Edoardo Giangrandi, Henrique Leonhard Gieg, Nina Kunert, Ivan Markin, Vsevolod Nedora, Anna Neuweiler, Henrik Rose, Peter Tsun Ho Pang, Federico Schianchi, Ashwin Shirke, and Maximiliano Ujevic

Abstract To date, about one hundred gravitational-wave events have been detected. Among them, the binary neutron star merger GW170817 was of particular significance since, in addition to gravitational waves, also electromagnetic signatures were observed. As the international network of gravitational-wave detectors has recently restarted, more multi-messenger detections are expected in the coming year. Due to the strong gravitational fields during the final stages of the coalescence, the study of compact binary merger requires numerical-relativity simulations that solve Einstein's Field Equations. These simulations heavily rely on high-performance computing facilities such as HAWK. We use for our research the numerical-relativity code BAM and explore the intricate relation between extreme spacetime and matter beyond the density of atomic nuclei. We further developed a framework that allows us to correlate observations with theoretical models using Bayesian methods. In this way, we can extract valuable physical information from the detected signals and explore the properties of matter on subatomic and cosmic scales.

T. Dietrich (✉) · E. Giangrandi · N. Kunert · I. Markin · V. Nedora · A. Neuweiler · H. Rose · F. Schianchi · A. Shirke
Institute for Physics and Astronomy, University of Potsdam, Potsdam, Germany
e-mail: tim.dietrich@uni-potsdam.de

T. Dietrich
Max Planck Institute for Gravitational Physics Potsdam, Potsdam, Germany

B. Brügmann
Theoretical Physics Institute, University of Jena, Jena, Germany

E. Giangrandi
CFisUC, Department of Physics, University of Coimbra, Rua Larga, 3004-516 Coimbra, Portugal

H. L. Gieg · M. Ujevic
Centro de Ciências Naturais e Humanas, Universidade Federal do ABC, São Paulo, Brazil

P. T. H. Pang
Nikhef, Amsterdam, The Netherlands

Institute for Gravitational and Subatomic Physics (GRASP), Utrecht University, Utrecht, The Netherlands

T. Ludwig et al. (eds.), *High Performance Computing in Science and Engineering '23*,
https://doi.org/10.1007/978-3-031-91312-9_3

1 Introduction

Gravitational waves (GWs) are tiny ripples in spacetime that allow us to study compact binary star systems consisting of black holes and neutron stars. Their first direct observation in 2015 from the merger of a binary black hole system [6] opened a completely new observational window into our Universe and heralded the era of GW astronomy. Since then, about a hundred events have been detected [7], and many more are expected in the coming year as GW detector systems have recently been recommissioned with improved sensitivity. Compact binary mergers containing at least one neutron star, i.e., binary neutron stars (BNS) and black hole–neutron star (BHNS) mergers, are of particular interest because in addition to GWs, electromagnetic counterparts originating from outflowing matter might also be observable. Indeed, we monitored such a multi-messenger event in 2017: The event GW170817 [4] was not only detected via GWs, but also connected to the gamma-ray burst GRB170817A, its non-thermal afterglow, and the kilonova AT2017gfo [2]. This event enabled several studies ranging from confirming general relativity [5], to establishing tighter limits on the behavior at supranuclear densities, e.g., [3, 9, 16, 31], to determining the expansion rate of the Universe using a new, independent method [1]. Moreover, the ejecta of neutron star mergers is thought to be the site for the synthesis of the heaviest elements in our Universe, which require a rapid neutron capture process (r-process), e.g., [15, 35]. The analysis of these signals allows us to learn more about the origin of these elements.

For the study of these systems and the correct interpretation of the detections, we need to match the observational data with theoretical predictions. The latter requires, in the final stages of coalescence, the solution of Einstein's Field Equations together with the equations of general relativistic hydrodynamics due to the strong gravitational fields of the highly dynamic system. For this purpose, our research focuses, firstly, on performing accurate numerical-relativity simulations of the neutron star merger and, secondly, on highly parallel Bayesian parameter estimation tools to match the observations with the theoretical models. This allows us to extract valuable physical information from the detected GW and electromagnetic signals.

We received a total of 152 Mio. CPUhs (75 Mio. CPUhs between January 2021 and January 2022 and 77 Mio. CPUhs since January 2022) for our research studies performed on HAWK. Of these resources, we have approximately 6 Mio. CPUhs remaining to complete our projects. In addition, we were recently granted another 85.05 Mio. CPUhs for follow-up research projects. In the following report, we will summarise a selection of results and publications of our individual projects from the last year and showcase some preliminary results of ongoing studies.

2 Technical Aspects and Computational Details

2.1 The Numerical-Relativity Code BAM

We use our in-house numerical-relativity code BAM [12, 38] for the simulation of BNS and BHNS mergers. Using the methods-of-line approach, BAM solves Einstein's Field Equations and evolves the gravitational field in time, employing finite-difference templates for spatial derivatives. In addition, high-resolution shock-capture methods are applied to handle shocks in the hydrodynamic variables.

BAM uses an adaptive mesh refinement technique to handle different length scales and to ensure sufficient resolution in the strong-field region, while still enabling to have large enough boxes to cover the far-field region to extract gravitational waves. The code employs the Berger-Oliger scheme [11] for local-time stepping. The grid consists of L refinement levels, where the resolution increases by a factor of two for successive refinement levels. To ensure that the compact object is always covered by the highest resolution, the innermost refinement boxes are dynamically moved and adjusted during the simulation.

For parallelization, BAM uses a hybrid OpenMP/MPI strategy in which the refinement boxes are divided into equally sized sub-boxes covered by individual MPI processes. Here, each sub-box contains ghost zones for the communication buffer, which are synchronized after each evolution step. The size of the buffer zones depends on the finite difference stencil used. To ensure load balancing, each MPI processor has a sub-box at each refinement level and operates on the same number of grid points. In addition, each MPI process starts an equal number of OpenMP threads that use shared memory.

We are continuously committed to improving our code. Over the past year, we have focused primarily on improving the physical realism of our simulations. In this regard, we have implemented a new hydrodynamics module handling tabulated equations of state based on nuclear theories to describe the neutron star interior [20]. Furthermore, we have incorporated a sophisticated radiative-hydrodynamics scheme, a first-order multipolar (M1) neutrino transport scheme, to evolve the energy and momentum of neutrinos and are currently also working on including magnetic fields in our simulation. We show in Fig. 1 the scaling behavior for the different code versions: for pure general-relativistic hydrodynamics, for the implementation with magnetic fields, for simulations with tabulated equations of state, and with M1 neutrino transport. The speed of the code decreases, as the complexity of the physical description increases. We typically run BAM on 8192 cores, depending on the resolution of the simulation.

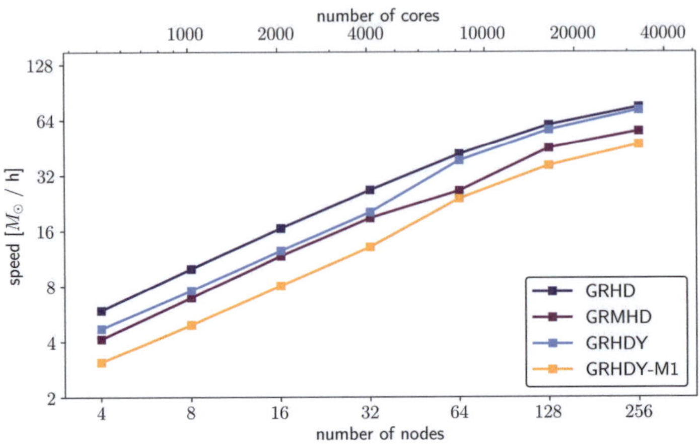

Fig. 1 Strong scaling for the BAM code on HAWK. We show the simulation speed as a function of the number of nodes for simulations with the old BAM version (GRHD), for simulations with magnetic fields (GRMHD), for simulations with tabulated equations of state (GRHDY), and with M1 neutrino transport (GRHDY-M1)

2.2 The Nuclear-Physics and Multi-messenger Astrophysics Framework

For the analysis of the observable signatures, we use our publicly available nuclear-physics and multi-messenger astrophysics NMMA[1] framework that we developed over the last years [30]. Specifically, we use for the individual messengers: the parallel bilby code [10, 36] for GWs, the afterglowpy code [17, 33] for the gamma-ray burst afterglow, and a neural network [8] or Gaussian Process regression [14] for kilonovae using light curve models from the radiative transfer code POSSIS [13].

The simultaneous analysis of GW, gamma-ray burst afterglow, and kilonova data requires considerable computational effort. Therefore, NMMA requires efficient parallelization, which is provided by using the "dynesty" sampler as used in parallel bilby and by implementing MPI parallelization with "mpi4py" and "schwimmbad". The parallelization strategy follows the head/worker model: The "head" estimates the evidence, while the "workers" find new live points to compute the evidence. In general, the accuracy of Bayesian evidence and posterior distribution estimation improves with a higher number of live points. However, due to the nature of parallel prior sampling, perfect ideal scaling cannot be guaranteed, and the theoretical scaling performance [22] is instead approximated by:

$$T(N_{\text{cores}}) = T(N_{\text{core}_1}) \times \frac{\ln(1 + N_{\text{core}_1}/n_{\text{live}})}{\ln(1 + N_{\text{cores}}/n_{\text{live}})}, \tag{1}$$

[1] https://github.com/nuclear-multimessenger-astronomy.

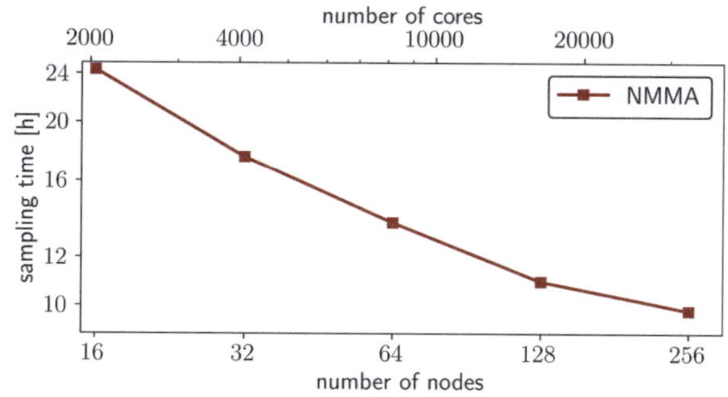

Fig. 2 Scaling results for NMMA on HAWK for a combined analysis of GW and kilonova signals

with the run time $T(N_{\text{cores}})$ for N_{cores} cores, the reference time $T(N_{\text{core}_1})$ for N_{core_1} cores, and the number of live points n_{live}. The scaling results for the analysis of GW and kilonova data using $n_{\text{live}} = 16384$ are presented in Fig. 2. Typically, our jobs run on 8192 cores.

3 Simulating Compact Binaries

Our resources at HAWK allowed us to complete several research projects over the past year, based on simulations of compact binary mergers. These included, for example, an investigation of the mass ratio dependence for prompt-collapse [24], an analysis of the expansion characteristics of the outflowing material to determine how deviations from homologous expansion (typically assumed for light curve modeling) affect the properties of the kilonova [29], and a study of how dark matter in the core of the neutron star might affect the emitted GW signal [18]. In addition, we have performed pioneering numerical-relativity simulations of a merger of a neutron star with a sub-solar-mass black hole and compared the GW signal with existing GW models [25]. Several waveforms from our simulations were published in the second data release of GWs by the Computational Relativity collaboration [21].

The above-mentioned studies accounted for approximately 20 Mio. CPUhs. In the following sections, we discuss a selection of our studies in more detail and present some preliminary results.

3.1 Simulations with Realistic Microphysics

In order to perform a general relativistic hydrodynamics simulation of matter at supranuclear densities, it is necessary to know its equation of state, i.e., the relations that connect different thermodynamic quantities. During the inspiral phase, when neutron stars are cold and their composition does not change, piecewise polytopic fits of pressure over mass density and internal energy are enough to properly capture all the necessary physics. At the merger, however, thermal effects become important, and the composition of nuclear matter can change significantly due to weak interactions with neutrinos. In this regime, piecewise polytropic fits become a poor approximation, and a more accurate description is needed. For this reason, we implemented in [20] a new general relativistic hydrodynamics module where the equation of state is given in the form of a 3D table obtained from sophisticated effective field theory models, e.g., SFHo [37] or DD2 [23]. In this new format, every thermodynamic variable is given as a tabulated function of density, temperature, and electron fraction Y_e.

Our new implementation includes an additional transport equation for Y_e with source terms given by an approximated neutrino leakage scheme. This scheme is a computationally efficient way to model a first approximation for radiative losses (neutrino cooling) but has the systematic issue of neglecting or underestimating the radiation absorption (neutrino heating), leading to a lower Y_e in the ejecta and a more intense cooling with respect to more sophisticated methods.

In [20], we tested our new implementation simulating single star and binary star setups for various tabulated equations of state, comparing the outcome with [19] and obtaining comparable results.

In the second part of this project, we implemented a first-order multipolar radiation transport (M1) [34, 39] scheme to evolve the distribution of the neutrinos in the simulation domain and include the effects of reabsorption (neutrino heating). This scheme is believed to give a more accurate estimate of the ejecta's Y_e and matter's cooling rate through neutrino emission, other than including the effect of neutrino pressure over the fluid.

We tested our new implementation on a set of standard tests widely used in literature, and we employed it for the simulation of four BNS systems with two equations of state and two mass ratios. The results of our tests are in agreement with previous studies performed by other groups and will be published soon, together with a detailed technical explanation of the methods. The effects of neutrino transport using M1 are illustrated in Fig. 3.

Finally, we used the ejecta data obtained from our new implementation to obtain more reliable estimates of nucleosynthesis yields and kilonova light curves using the codes SKYNET and POSSIS, respectively. A crucial advantage of the new implementation is that the electron fraction can now be obtained directly from the simulation, and we no longer have to rely on assumptions based on the entropy of the fluid.

This study was one of our most important scientific contributions to our GWanalysis project last year on HAWK, which is why we devoted most of our resources to it. In total, the above results have required 38.5 Mio. CPUs.

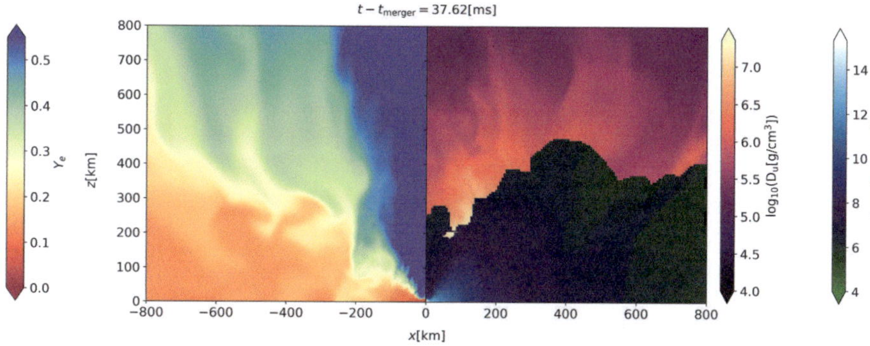

Fig. 3 A snapshot of one of our BNS simulations employing the M1 neutrino transport scheme in the x-z plane at 37.6 ms after the merger. In the left panel, we show the electron fraction Y_e, and in the right one the conserved rest-mass density D. We use two color maps for the gravitationally bound (D_b) and unbound (D_u) mass

3.2 Reverse Phase Transitions in Binary Neutron Stars

The unmatched conditions in the merger of neutron stars, in which we encounter extreme densities, temperatures, fluid velocities, and strong gravitational fields, are perfect laboratories to study and constrain its equation of state and to probe physical principles. In this context, numerical relativity is the best available tool to describe the spacetime and matter fields under these extreme conditions.

Quantum chromodynamics predicts that cold matter undergoes a transition from hadronic matter at low densities to quark matter at very high energies. It is an open question whether isolated neutron stars explore such phase transitions to exotic forms of matter in their cores, which could affect the BNS system dynamics already during the inspiral or during the post-merger phase when matter exists at higher densities. Revealing the presence of a possible phase transition is one of the science goals of next-generation GW detectors. This could be achieved through precise measurements of the tidal deformability of the stars during the inspiral or using the post-merger GWs (where a possible shift in the frequency could be found). Also, electromagnetic signatures might be altered due to different post-merger dynamics.

In [40], we performed new numerical-relativity simulations of BNS systems composed of hybrid neutron stars. The hybrid neutron stars contain a phase transition between the hadronic mantle and the quark-matter core. Our studies revealed that a "reverse phase transition" can happen during the inspiral shortly before the neutron star merger, i.e., the simulated quark core vanishes for the last milliseconds before the collision of the stars. This transition is caused by a decrease in the central mass density due to the strong deformation of the stars. After the collision, a quark core is quickly formed again and, for the particular equations of state that we employed, the remnant collapses promptly to a black hole.

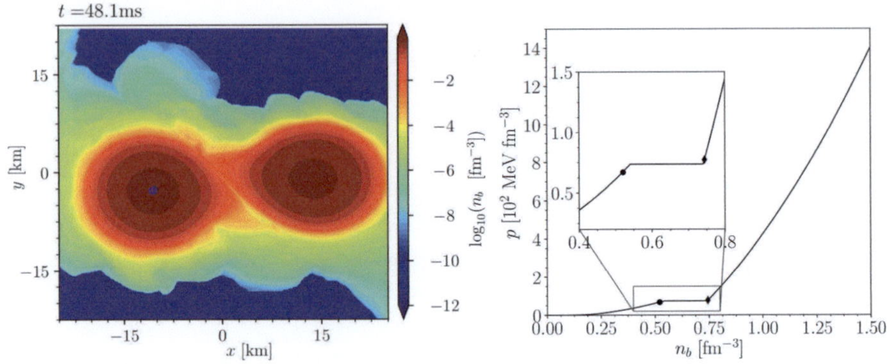

Fig. 4 *Left:* Snapshot of the baryon number density on the equatorial plane at $t = 48.1$ ms of the evolution. The purple circle at the center of the neutron star (left star in the plot) corresponds to a quark matter core. *Right:* Baryon number density-pressure phase diagram of the stars depicting a phase transition from hadronic matter (mantle) to quark matter (core)

Despite the fact that the reverse phase transition is certainly of academic interest, we did not find an imprint on the GW signal or on the mass ejection that is noticeable, given the numerical precision of our simulations. Nevertheless, our simulations verified, in agreement with previous works, that the presence of a phase transition can noticeably soften the equation of state at merger time and can alter the electromagnetic signals from a BNS merger due to different ejecta properties. In Fig. 4, we show a snapshot at $t = 48.1$ ms of one of our BNS simulations together with the equation of state employed. We used for the results above about 6.2 Mio. CPUhs of our resources on HLRS.

3.3 *Modelling Gamma-Ray Burst and Kilonova Afterglows*

One of the key components of the multi-messenger signature of merging compact binaries is the gamma-ray burst afterglow. It encodes information about the environment in which the merger occurs and post-merger conditions, including the nature of the remnant. It is also a key tool for studying the physics of particle acceleration at relativistic shocks. In light of this, accurate and sufficiently fast modeling of gamma-ray burst afterglow is of paramount importance. To this end, we designed a semi-analytical afterglow code `PyBlastAfterglow` [26–28]. The code employs self-consistent, energy-conserving methods to model the dynamical evolution of a trans-relativistic blastwave. It also includes multiple prescriptions for the lateral expansion of the blastwave, for the shock microphysics, and the synchrotron radiation. This allows us to perform systematic uncertainty studies. In addition, we can model the laterally and velocity structured outflows with and compute light curves as well as sky maps. The sky maps are of particular importance as they provide key information regarding the jetted nature of the ejecta and observing angle.

In addition to the gamma-ray burst afterglow, a compact binary merger is expected to produce another non-thermal electromagnetic signature, a kilonova afterglow. This is akin to a gamma-ray burst afterglow, but instead of a highly relativistic collimated ejecta, mildly relativistic quasi-spherically symmetric ejecta is exciting a shock that propagates through the interstellar medium, amplifies magnetic fields, and accelerates electrons. This electromagnetic signature provides a unique window into the moment of the compact binary merger as the fastest ejecta produced at this time also generates the brightest afterglow. In [27], we demonstrate how the parameters of neutron stars can be inferred from late-time X-ray and radio observations. Recently, we have significantly improved the model, by including a more advanced prescription for the electron population behind the shock that consists of thermal and non-thermal electron populations, predicted for these types of shocks. We also included the interaction with the decelerating, laterally expanding gamma-ray burst outflow and a method to compute kilonova afterglow sky maps [26, 28]. For the realization of this project, only minimal computational resources $\lesssim$1 Mio. CPUhs have been consumed.

Currently, we are integrating the code into the NMMA framework as well as adding the model for a magnetar energy injection into both gamma-ray burst and kilonova ejecta. Upon completion, we will have the most comprehensive, in terms of input physics, multi-messenger framework on the market.

4 Multi-messenger Studies

In addition to our research related to numerical-relativity simulations of compact binary systems, we devote a subset of our computational resources to our multi-messenger studies. In the following section, we present one particular study that we recently published.

4.1 Three-Nucleon Interactions in the Einstein Telescope

The equation of state of cold dense matter is fundamental to the understanding of heavy nuclei as well as compact objects like neutron stars. While slightly supranuclear densities can still be studied experimentally in heavy-ion collisions, the density regime in neutron star cores is inaccessible to laboratory experiments. It, therefore, cannot be accurately characterized by phenomenological models of the underlying interactions. Chiral effective field theory offers a systematic ab-initio approach, which is formulated in terms of neutrons and protons but still can be justifiably expanded beyond nuclear saturation density with controlled uncertainties. Implementation of these forces, though, leads to ambiguities that first appear in the parametrization of three-nucleon interactions. These need to be constrained observationally.

To that end, our study [32] analyzed the potential of multi-messenger detections to select among two competing parametrizations through their imprint on the tidal deformability of light neutron stars that can be measured in neutron star mergers' gravitational waveform. As current detections are too limited in number and data quality, we investigated how a population of coalescing systems that are observed by GW detectors of the next generation, e.g., the Einstein Telescope, can inform a preferred choice among these models. We simulated mock data and performed Bayesian inference on binary parameters with our NMMA framework, including known constraints from wider electromagnetic observations.

Our analysis suggests that detections by the Einstein Telescope alone should quickly compile significant evidence even under very conservative assumptions, providing another promising link between GW astronomy and nuclear physics. In total, this study required about 5.5 Mio. CPUhs.

5 Summary

Thanks to our granted project allocation GWanalysis on HAWK, we have been able to perform numerous studies, including simulations of neutron star mergers and analyses of how we can use multi-messenger studies to learn more about nuclear physics. We have also significantly improved our code to account for more realistic physical processes. We were able to publish a total of 7 papers [18, 20, 21, 24, 28, 29, 40] and 4 preprints currently under review [25, 26, 30, 32] in the last year. Including our results from the previous year, this adds up to 21 research articles associated with this allocation at HLRS.

References

1. Abbott, B.P., et al.: A gravitational-wave standard siren measurement of the Hubble constant. Nature **551**(7678), 85–88 (2017). https://doi.org/10.1038/nature24471. arXiv: 1710.05835 [astro-ph.CO]
2. Abbott, B.P., et al.: Estimating the contribution of dynamical ejecta in the kilonova associated with GW170817. Astrophys. J. Lett. **850**(2), L39 (2017). https://doi.org/10.3847/2041-8213/aa9478. arXiv: 1710.05836 [astro-ph.HE]
3. Abbott, B.P., et al.: GW170817: measurements of neutron star radii and equation of state. Phys. Rev. Lett. **121**(16), 161101 (2018). https://doi.org/10.1103/PhysRevLett.121.161101. arXiv: 1805.11581 [gr-qc]
4. Abbott, B.P., et al.: GW170817: observation of gravitational waves from a binary neutron star inspiral. Phys. Rev. Lett. **119**(16), 161101 (2017). https://doi.org/10.1103/PhysRevLett.119.161101. arXiv: 1710.05832 [gr-qc]

5. Abbott, B.P., et al.: Multi-messenger observations of a binary neutron star merger. Astrophys. J. Lett. **848**(2), L12 (2017). https://doi.org/10.3847/2041-8213/aa91c9. arXiv: 1710.05833 [astro-ph.HE]
6. Abbott, B.P., et al.: Properties of the binary black hole merger GW150914. Phys. Rev. Lett. **116**(24), 241102 (2016). https://doi.org/10.1103/PhysRevLett.116.241102. arXiv: 1602.03840 [gr-qc]
7. Abbott, R., et al.: Observation of gravitational waves from two neutron star–black hole coalescences. Astrophys. J. Lett. **915**(1), L5 (2021). https://doi.org/10.3847/2041-8213/ac082e. arXiv: 2106.15163 [astro-ph.HE]
8. Almualla, M., et al.: Using neural networks to perform rapid high-dimensional kilonova parameter inference. In: (Dec. 2021). arXiv: 2112.15470 [astro-ph.HE]
9. Annala, E., et al.: Gravitational-wave constraints on the neutron-star-matter equation of state. Phys. Rev. Lett. **120**(17), 172703 (2018). https://doi.org/10.1103/PhysRevLett.120.172703. arXiv: 1711.02644 [astro-ph.HE]
10. Ashton, G., et al.: BILBY: a user-friendly Bayesian inference library for gravitational-wave astronomy. Astrophys. J. Suppl. **241**(2), 27 (2019). https://doi.org/10.3847/1538-4365/ab06fc. arXiv: 1811.02042 [astro-ph.IM]
11. Berger, M.J., Oliger, J.: Adaptive mesh refinement for hyperbolic partial differential equations. J. Comput. Phys. **53**, 484 (1984). https://doi.org/10.1016/0021-9991(84)90073-1
12. Bruegmann, B., et al.: Calibration of moving puncture simulations. Phys. Rev. D **77**, 024027 (2008). https://doi.org/10.1103/PhysRevD.77.024027. arXiv: gr-qc/0610128
13. Bulla, M.: POSSIS: predicting spectra, light curves and polarization for multi-dimensional models of supernovae and kilonovae. Mon. Not. Roy. Astron. Soc. **489**(4), 5037–5045 (2019). https://doi.org/10.1093/mnras/stz2495. arXiv: 1906.04205 [astro-ph.HE]
14. Coughlin, M.W., et al.: Constraints on the neutron star equation of state from AT2017gfo using radiative transfer simulations. Mon. Not. Roy. Astron. Soc. **480**(3), 3871–3878 (2018). https://doi.org/10.1093/mnras/sty2174. arXiv: 1805.09371 [astro-ph.HE]
15. Cowperthwaite, P.S., et al.: The electromagnetic counterpart of the binary neutron star merger LIGO/Virgo GW170817. II. UV, optical, and near-infrared light curves and comparison to kilonova models. Astrophys. J. Lett. **848**(2), L17 (2017). https://doi.org/10.3847/2041-8213/aa8fc7. arXiv: 1710.05840 [astro-ph.HE]
16. Dietrich, T., et al.: Multimessenger constraints on the neutron-star equation of state and the Hubble constant. Science **370**(6523), 1450–1453 (2020). https://doi.org/10.1126/science.abb4317. arXiv: 2002.11355 [astro-ph.HE]
17. van Eerten, H., Zhang, W., MacFadyen, A.: Off-axis gamma-ray burst afterglow modeling based on a two-dimensional axisymmetric hydrodynamics simulation. Astrophys. J. **722**, 235–247 (2010). https://doi.org/10.1088/0004-637X/722/1/235. arXiv: 1006.5125 [astro-ph.HE]
18. Emma, M., et al.: Numerical simulations of dark matter admixed neutron star binaries. Particles **5**(3), 273–286 (2022). https://doi.org/10.3390/particles5030024. arXiv: 2206.10887 [gr-qc]
19. Foucart, F., et al.: Low mass binary neutron star mergers : gravitational waves and neutrino emission. Phys. Rev. **D93**(4), 044019 (2016). https://doi.org/10.1103/PhysRevD.93.044019. arXiv: 1510.06398 [astro-ph.HE]
20. Gieg, H., et al.: Incorporating a radiative hydrodynamics scheme in the numerical-relativity code BAM. Universe **8**(7), 370 (2022). https://doi.org/10.3390/universe8070370. arXiv: 2206.01337 [gr-qc]
21. Gonzalez, A., et al.: Second release of the CoRe database of binary neutron star merger waveforms. Class. Quant. Grav. **40**(8), 085011 (2023). https://doi.org/10.1088/1361-6382/acc231. arXiv: 2210.16366 [gr-qc]
22. Handley, W.J., Hobson, M.P., Lasenby, A.N.: PolyChord: nested sampling for cosmology. Mon. Not. Roy. Astron. Soc. **450**(1), L61–L65 (2015). https://doi.org/10.1093/mnrasl/slv047. arXiv: 1502.01856 [astro-ph.CO]
23. Hempel, M., Schaffner-Bielich, J.: Statistical model for a complete supernova equation of state. Nucl. Phys. A **837**, 210–254 (2010). https://doi.org/10.1016/j.nuclphysa.2010.02.010. arXiv: 0911.4073 [nucl-th]

24. Kölsch, M., et al.: Investigating the mass-ratio dependence of the prompt-collapse threshold with numerical-relativity simulations. Phys. Rev. D **106**(4), 044026 (2022). https://doi.org/10.1103/PhysRevD.106.044026. arXiv: 2112.11851 [gr-qc]
25. Markin, I., et al.: General-relativistic hydrodynamics simulation of a neutron star - sub-solar-mass black hole merger. In: (Apr. 2023). arXiv: 2304.11642 [gr-qc]
26. Nedora, V., Dietrich, T., Shibata, M.: Synthetic radio images of structured GRB and kilonova afterglows. In: (Feb. 2023). arXiv: 2302.12850 [astro-ph.HE]
27. Nedora, V., et al.: Dynamical ejecta synchrotron emission as a possible contributor to the changing behaviour of GRB170817A afterglow. Mon. Not. Roy. Astron. Soc. **506**(4), 5908–5915 (2021). https://doi.org/10.1093/mnras/stab2004. arXiv: 2104.04537 [astro-ph.HE]
28. Nedora, V., et al.: Modeling kilonova afterglows: effects of the thermal electron population and interaction with GRB outflows. In: (Aug. 2022). https://doi.org/10.1093/mnras/stad175. arXiv: 2208.01558 [astro-ph.HE]
29. Neuweiler, A., et al.: Long-term simulations of dynamical ejecta: homologous expansion and kilonova properties. Phys. Rev. D **107**(2), 023016 (2023). https://doi.org/10.1103/PhysRevD.107.023016. arXiv: 2208.13460 [astro-ph.HE]
30. Pang, P.T.H., et al.: NMMA: a nuclear-physics and multi-messenger astrophysics framework to analyze binary neutron star mergers. In: (May 2022). arXiv: 2205.08513 [astro-ph.HE]
31. Radice, D., et al.: GW170817: joint constraint on the neutron star equation of state from multimessenger observations. Astrophys. J. Lett. **852**(2), L29 (2018). https://doi.org/10.3847/2041-8213/aaa402. arXiv: 1711.03647 [astro-ph.HE]
32. Rose, H., et al.: Revealing the strength of three-nucleon interactions with the Einstein Telescope. In: (Mar. 2023). arXiv: 2303.11201 [astro-ph.HE]
33. Ryan, G., et al.: Gamma-ray burst afterglows in the multimessenger era: numerical models and closure relations. Astrophys. J. **896**(2), 166 (2020). https://doi.org/10.3847/1538-4357/ab93cf. arXiv: 1909.11691 [astro-ph.HE]
34. Shibata, M., et al.: Truncated moment formalism for radiation hydrodynamics in numerical relativity. Prog. Theor. Phys. **125**, 1255–1287 (2011). https://doi.org/10.1143/PTP.125.1255. arXiv: 1104.3937 [astro-ph.HE]
35. Smartt, S.J., et al.: A kilonova as the electromagnetic counterpart to a gravitational-wave source. Nature **551**(7678), 75–79 (2017). https://doi.org/10.1038/nature24303. arXiv: 1710.05841 [astro-ph.HE]
36. Smith, R.J.E., et al.: Massively parallel Bayesian inference for transient gravitational-wave astronomy. Mon. Not. Roy. Astron. Soc. **498**(3), 4492–4502 (2020). https://doi.org/10.1093/mnras/staa2483. arXiv: 1909.11873 [gr-qc]
37. Steiner, A.W., Hempel, M., Fischer, T.: Core-collapse supernova equations of state based on neutron star observations. Astrophys. J. **774**, 17 (2013). https://doi.org/10.1088/0004-637X/774/1/17. arXiv: 1207.2184 [astro-ph.SR]
38. Thierfelder, M., Bernuzzi, S., Bruegmann, B.: Numerical relativity simulations of binary neutron stars. Phys. Rev. D **84**, 044012 (2011). https://doi.org/10.1103/PhysRevD.84.044012. arXiv: 1104.4751 [gr-qc]
39. Thorne, K.S.: Relativistic radiative transfer - moment formalisms. Mon. Not. Roy. Astron. Soc. **194**(2), 439–473 (1981). https://doi.org/10.1093/mnras/194.2.439
40. Ujevic, M., et al.: Reverse phase transitions in binary neutron-star systems with exotic-matter cores. Phys. Rev. D **107**(2), 024025 (2023). https://doi.org/10.1103/PhysRevD.107.024025. arXiv: 2211.04662 [gr-qc]

Neutron Star Mergers: From Nuclei to Cosmic Explosions

Maximilian Jacobi, Federico Guercilena, and Almudena Arcones

Abstract Mergers of binary neutron star systems are among the most fascinating astrophysical sites in our universe. Computational simulations of these events require modeling of the hydrodynamics of nuclear matter at extreme densities and temperatures, strong dynamical gravitational fields, neutrino-matter interactions, and more. Moreover, they are computationally costly requiring the use of highly performant simulation software. In this report, we summarize our recent results in exploring the impact of the nuclear equation of state on the dynamics and observable features of binary neutron star mergers. To this extent, we systematically vary select nuclear matter properties and explore the results in numerical general relativistic hydrodynamical simulations. Our results show that some details of the merger, such as the frequencies of the post-merger gravitational-wave signal and the mass of the shock-heated ejecta, are sensitive to the EOS close to saturation density while others, such as the contraction of the remnant neutron star and the mass of the tidal ejecta, are sensitive to the behavior at higher densities. Furthermore, we show preliminary results of our simulation setup to explore the long-term evolution of the post-merger phase in axisymmetry.

Project acronym and Principal Investigator:
Acid 44163 microBNS
Research field:
Nuclear and Relativistic Astrophysics.

M. Jacobi (✉)
Darmstadt, Germany
e-mail: mjacobi@theorie.ikp.physik.tu-darmstadt.de

F. Guercilena
Trento, Italy

A. Arcones (✉)
Institut für Kernphysik (IKP), Technische Universität Darmstadt, Darmstadt, Germany
e-mail: almudena.arcones@tu-darmstadt.de

T. Ludwig et al. (eds.), *High Performance Computing in Science and Engineering '23*,
https://doi.org/10.1007/978-3-031-91312-9_4

1 Introduction

The allocation "44163 microBNS" on the system HPE APOLLO (Hawk) (formerly on the CRAY XC40 HazelHen system) at HLRS has been awarded to our research group in May 2019 for one year, becoming one of the main computing resources available to us. Multiple extension requests to prolong the allocation time frame were submitted and approved, allowing our operations to continue until December 2023.

The scientific research carried on in our group relates to several broad and intimately related topics in the field of nuclear and relativistic astrophysics. Concerning the research areas explored with the use of the computing resources of HLRS, these topics include:

- the influence of the equation of state (EOS) of dense matter on the dynamics and outcome of binary neutron star (BNS) systems;
- the long-term evolution of BNS merger remnants and accretion disks;
- the study of BNS ejecta and the resulting nucleosynthesis of heavy elements and the associated electromagnetic counterpart signals.

Due to the sheer complexity of the physical models involved in this field of research, no realistic analytical solutions are available. The main tools of our research are therefore large-scale numerical simulations of the physical system of interest. As such, a secondary but significant line of research we follow is the development of better numerical methods to improve the accuracy and/or performance of our results. This makes computing resources such as the ones provided by HLRS a very important asset to us.

2 Current Status and Usage of the Allocation

Over the past three years, we conducted a variety of different simulations on this machine to verify the ability of our codes to fully utilize it. In particular, Hawk's configuration of 2 CPUs with 64 cores per node requires a hybrid MPI/OpenMP approach to exploit it fully. Our codes are based on the EinsteinToolkit framework [14, 17], which implements such a parallelization strategy. They are therefore perfectly suited, in theory, and as we have verified in practice, to run on a machine like Hawk.

As of Tuesday, May 30th, 2023, the total usage of the allocation was $\sim 7.29 \times 10^6$ core-hours on Hawk, or $\sim 29\%$ of the available total of 25.13×10^6 core-hours (converted from the $\sim 16 \times 10^6$ core-hours originally awarded for use on the HazelHen platform). The reason for the rather limited use of resources is the small number of scenarios we have investigated so far in our study. We plan to use the remaining computing time until the end of 2023 (see Sect. 5 for more details on the planned usage). Since our simulation setup and post-processing pipeline have been established and optimized in the previous years new simulations will be run and analyzed comparatively fast.

3 Numerical Methods and Algorithms

To perform large-scale numerical simulations of BNS mergers, we rely on the open-source software platform `EinsteinToolkit` (ET)[1] [14, 17, 38], itself based on the `Cactus Computational Toolkit`[2] [5]. The ET is a collection of computational tools for the solution of large partial differential equation (PDE) systems, geared in particular towards the simulation of general-relativistic astrophysics systems (such as binary black holes (BBHs) or BNSs). It supports large-scale parallelism employing a hybrid `MPI/OpenMP` model and is being routinely run on major computer clusters around the world (see e.g. Bovard et al. [9] for a former example of ET simulations run on the `HazelHen` cluster at HLRS).

To efficiently cover the large range of scales involved in the modeling of BNS, we employ adaptive mesh refinement (AMR) techniques. Our codes leverage the capabilities of the `Carpet` library [32], which implements a vertex- or cell-centered AMR scheme adopting nested and moving grids. This allows us to follow the moving neutron stars with fine-resolution grids while covering with coarser grids a much larger volume needed for the treatment of the ejected matter or gravitational waves (GWs). `Carpet` supports Berger-Oliger mesh refinement [8] with sub-cycling in time.

The solution of the Einstein equations providing the evolution of the spacetime is obtained using standard finite-difference methods where the derivatives are computed with centered stencils, but using upwinded stencils for the terms associated with the advection along the shift vector. Typically, derivatives are computed with a fourth-order accurate scheme, with fifth-order Kreiss-Oliger artificial dissipation added to ensure stability. More specifically, we use the BSSNOK formulation of the Einstein equations [6, 10, 35], as implemented in the public `McLachlan` code [17], part of the ET. We have also experimented with CCZ4 [3, 4, 16] formulation of Einstein's equations, which allows for the propagation and damping of the general-relativistic constraint violations. BSSNOK remains however our standard and preferred choice.

The main piece of code we rely on is the `WhiskyTHC` code [20], which solves the equations of general-relativistic hydrodynamics with nuclear EOSs in Cartesian coordinates on a dynamically evolving, curved spacetime. The equations are solved in conservation form using a finite-volume scheme coupled with the reconstruction of the local primitive variables. `WhiskyTHC` implements a wide range of reconstruction operators, but we employ the fifth-order monotonicity preserving (MP5) scheme, which has proved to be a good compromise between robustness and accuracy [21],

[1] https://einsteintoolkit.org.

[2] https://cactuscode.org.

and has allowed WhiskyTHC to be employed in several different contexts, from BNS mergers [9, 21–24], to core-collapse supernovae [26]. Note that although open-source, free software,[3] WhiskyTHC is not part of the ET; it has however been developed in the context of the ET framework.

The time evolution and the coupling between the hydrodynamic and the spacetime solvers are obtained using the method-of-lines (MoL), in which the PDEs to be solved are first discretized in space with the methods outlined above, and then the resulting system of coupled ordinary differential equations (ODEs) is solved using a standard time integrator. In our case, this integrator is the optimal, strongly stability-preserving third-order Runge-Kutta scheme RK3 [36].

WhiskyTHC is a pure hydrodynamics code, and in particular it does not implement the evolution of electromagnetic fields. It does however include a treatment of neutrino transport and interaction with the fluid. This is of crucial importance for the modeling of BNS (and therefore the achievement of our goals), for various reasons. First of all, neutrinos being produced in the merger remnant and escaping the system work effectively as an energy sink. This implies that the pressure support of the merger remnant is reduced, which can significantly influence the dynamics of the system, including triggering the collapse to a black hole (BH) of the remnant. Secondly, in the absence of the microphysical reactions involving neutrinos, the electron fraction of the matter ejected from the system would be conserved. In practice, the ejecta would remain unrealistically neutron-rich under any conditions. It is instead known that neutrino irradiation triggers weak reactions that leptonize the matter, increasing its electron fraction. It is very important to include such effects in the modeling of the ejecta also in order to provide realistic starting conditions for the subsequent modeling of the ejecta outflow. To this end, WhiskyTHC implements a leakage scheme and the so-called M0 scheme. These methods are described in detail in Galeazzi et al. [13], Radice et al. [25], and references therein. They are both "grey" schemes, i.e. schemes in which the dependence of various quantities on the energy of the neutrinos is not explicitly taken into account: instead, energy-averaged quantities are considered. They account for three distinct neutrino species: electron neutrinos, ν_{e}; electron antineutrinos, $\bar{\nu}_{\mathrm{e}}$; and a species for μ and τ neutrinos and antineutrinos, ν_{x}. Neutrinos are split into two components: a free-streaming one, n_{ν}^{fs}, and a trapped one, n_{ν}^{trap}. The emission of neutrinos from the trapped regime and the subsequent loss of energy is described by a neutrino leakage scheme (NLS) while the neutrino transport and absorption in the free-streaming regime is accounted for by the M0 scheme.

The NLS is based on the method outlined in Ruffert et al. [30], where the local thermodynamical equilibrium chemical potential is used everywhere for all neutrino species while computing opacities as in Rosswog et al. [29]. Table 1 lists the reactions taken into account by this scheme to compute the neutrino production free rates, R_{ν}^{free}, $\nu \in \{\nu_{\mathrm{e}}, \bar{\nu}_{\mathrm{e}}, \nu_{\mathrm{x}}\}$, the free energy release, Q_{ν}^{free}, and the neutrino absorption, $\kappa_{\nu,\mathrm{a}}$, and scattering, $\kappa_{\nu,\mathrm{s}}$, opacities. The scheme distinguishes number density weighted opacities, $\kappa_{\nu,\mathrm{a}}^{0}$ and $\kappa_{\nu,\mathrm{s}}^{0}$, that determine the rate at which neutrinos diffuse out of the

[3] https://personal.science.psu.edu/dur566/whiskythc.html.

Table 1 Weak reactions accounted for in the neutrino transport schemes. The following notation is used: N $\in$ {n, p} denotes a free nucleon, A a nucleus, $\nu \in \{\nu_e, \bar{\nu}_e, \nu_x\}$ a neutrino. The"Reference" column accounts for the corresponding rate implementation

Name	Reaction	Reference
Electron neutrino capture on free neutron	$\nu_e + n \leftrightarrow p + e^-$	[11]
Electron antineutrino capture on free protons	$\bar{\nu}_e + p \leftrightarrow n + e^+$	[11]
Electron-positron annihilation	$e^+ + e^- \rightarrow \nu + \overline{\nu}$	[30]
Plasmon decay	$\gamma + \gamma \rightarrow \nu + \overline{\nu}$	[30]
Scattering off nucleons	$\nu + N \rightarrow \nu + N$	[30]
Scattering off nuclei	$\nu + A \rightarrow \nu + A$	[33]

material, from energy density weighted opacities, $\kappa^1_{\nu,a}$ and $\kappa^1_{\nu,s}$, that determine the rate at which energy is released due to the loss of neutrinos. The neutrino optical depth τ_ν is evolved in time following the scheme presented in [18], which allows the optical depth profile to adapt to the complex geometry of the system. The optical depth is used to define the effective emission rates, which are computed in such a way to provide a smooth interpolation between an estimate of the diffusion rate in optically thick condition and the local production rate in optically thin conditions, based on the optical depth.

The M0 scheme evolves the zeroth moment of the distribution function of free streaming neutrinos and computes their number densities and average energies on a polar grid. This scheme assumes that neutrinos propagate radially at the speed of light, and their evolution is computed on a series of independent radial rays using a first-order, fully-implicit, finite volume method. The mean energies of the free-streaming neutrino component are estimated under the additional assumption of a stationary spacetime and evolved using a fully-implicit upwind first-order finite-difference method. Finally, the coupling with hydrodynamics is handled by interpolating quantities from/to the standard Cartesian AMR grid at every timestep, using trilinear interpolation. The neutrino number and energy rates computed by the combined leakage and M0 schemes appear as sources in the Euler equations for the neutron star (NS) matter. The coupling in this case is handled, at every timestep, by first advancing the hydrodynamic quantities in time disregarding neutrino contributions; neutrino sources are then added to the Euler equations with a semi-implicit first-order method, in an operator split approach.

4 Code Performance and Scaling

The solution of the dynamics of a BNS system requires the storage of the full spacetime metric, the hydro- and thermodynamical degrees of freedom as well as several auxiliary quantities, hence a substantial memory footprint that is typically of a few

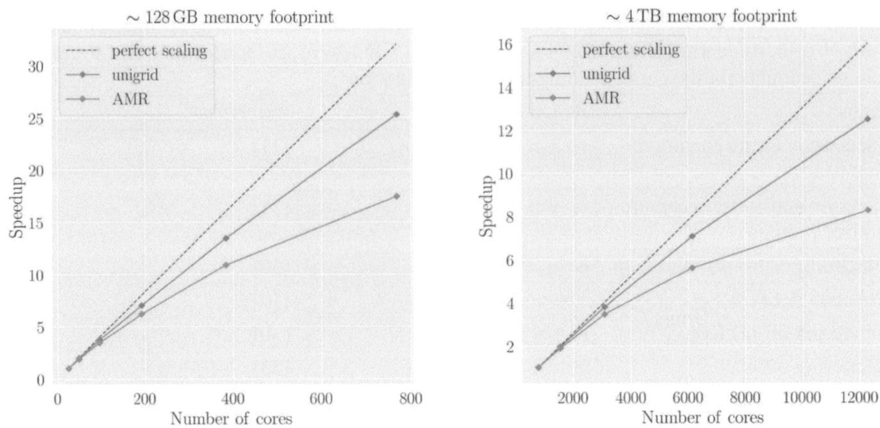

Fig. 1 Scaling plots

hundred GB for realistic (and fully tridimensional) BNS merger simulations. The large number of grid points used, of the order of a few times 10^7, and the large number of equations evolved as well, implies that the simulation running times are quite large, of the order of ~20 days, for which we use the standard checkpointing infrastructure in `Cactus`. In order to demonstrate our code suitability and readiness, we have performed strong scaling tests for representative job sizes of roughly 128 GB and 4 TB of RAM memory footprint. In the first case, the tests are run employing the AMR while the second case uses a simpler unigrid setup. These tests were performed on the HazelHen cluster at HLRS (test allocation `effEOSNS`, ACID 44154). Figure 1 and Table 2 report how the code performs in the strong scaling test.

The code scales quite well in the unigrid setup, in which the only interprocess communication overhead is "horizontal", i.e. due to the necessity of synchronizing ghost zones of adjacent patches of the computational domain. In the AMR case instead there is increased overhead, which is "vertical" in origin, due to the prolongation and restriction operations that are intrinsic to the Berger-Oliger AMR framework. In essence, data must be communicated from coarser refinement levels to finer ones to provide boundary conditions for the latter (prolongation); data also must be communicated from finer levels to coarser ones to ensure the latter do not evolve independently of the former (restriction). When domain patches residing on different refinement levels and being handled by different `MPI` processes (and thus typically different CPUs) communicate in this way, additional overhead arises, limiting the performance in the AMR case.

However, also in this case, the speedup remains essentially linear in the number of cores, and the code performs reasonably up to ~3000 cores. Note that we do not intend, nor need, to run our simulations with more cores than this. Note also that the trade-off for cutting-edge scaling performance is robustness: the ET and the codes based on it implement very well-tested and robust algorithms; we were confident,

Table 2 Scaling tables. Top: memory footprint 128 GB; bottom: memory footprint 4 TB

Cores	Unigrid timing (sec)	Unigrid speedup	AMR timing (sec)	AMR speedup
24	14492	1.0	4864	1.0
48	7406	1.96	2601	1.87
96	3876	3.74	1405	3.46
192	2070	7.00	789	6.16
384	1078	13.44	447	12.67
768	573	25.29	279	17.43
Cores	Unigrid timing (sec)	Unigrid speedup	AMR timing (sec)	AMR speedup
768	4745	1.0	2206	1.0
1536	2432	1.95	1172	1.88
3072	1265	3.75	641	3.44
6144	672	7.06	394	5.60
12288	380	12.49	267	8.26

and have verified in practice, that our runs would not fail on essentially any setup, minimizing the wasting of computational resources.

The simulations we have performed so far allow us to estimate the cost of one high-resolution run as 128 cores $\times$ 8 nodes $\times$ 392 h, i.e. $\sim$400, 000 core-hours. The corresponding cost of a low-resolution run (lower resolution studies are also necessary not only for testing purposes but also for ensuring convergence), which uses a similar setup but runs on a single `Hawk` node instead of eight, would be $\sim$50, 000 CPU-hours.

5 Ongoing Projects

We summarize here the two projects that we are carrying out (and plan on further developing in the coming months) using HLRS's resources:

- (1) **Systematic characterization of EOS effects on BNS post-merger micro- and macrophysics**
  The dependence of the evolution and dynamics of BNS mergers on the EOS employed to describe the microphysics of dense matter is one of the forefront topics of research in modern astrophysics. However, EOSs are complex, and they depend on many degrees of freedom, and the evolution of BNS systems is very non-linear, making it very difficult to isolate the effect of single nuclear physics properties on the outcome of the merger and its properties. Similarly, reversing the process and trying to obtain information on the EOS from the merger outcome suffers from huge uncertainties.

In this project, we are performing high-resolution simulations of BNS mergers, focusing on the analysis of the post-merger dynamics of the remnant and the accretion disk. These simulations employ the EOSs tables presented in Yasin et al. [37], and further ones developed specifically for this project, which allows us to vary one single microphysical nuclear property at a time. By comparing the outcome of various simulations with different input EOSs constructed in this way, we will be able to systematically and unambiguously relate changes in the dynamics of the system to individual nuclear physics inputs, effectively sidestepping the difficulties outlined above.

This analysis, combined with future GW and multimessenger observations of BNS systems, will help constrain the EOS of dense matter. Furthermore, the data produced in the context of this project will serve as initial data for the simulations of the next project we are working on (see below), tying the two together.

- (2) **Study of the long-term evolution of BNS disks and ejecta**

  Many of the most interesting phenomena taking place in BNS systems happen during the late post-merger phase ($\sim$1 s after the merger). These include the ejection of matter by neutrino-driven winds or viscous effects, the r-process nucleosynthesis of heavy elements, and the emission of the electromagnetic signal. Fully tridimensional, relativistic, high-resolution simulations with realistic microphysics such as the ones we are performing in the context of the previous project are necessary to model the inspiral and early-post merger phases, but unfortunately cannot reach such long timescales, being instead limited to $\sim$100 ms after the merger. This is simply due to the sheer amount of computational resources (in terms of computing time, memory requirements, and data storage) that would be needed to push a simulation to cover the late-post merger as well.

  In the post-merger phase, the geometry of BNS systems simplifies and can reasonably be thought of as axisymmetric and therefore two-dimensional. We carry out long-term 2D simulations of the post-merger phase of BNS systems, up to timescales of one second or longer, employing techniques inspired by Fujibayashi et al. [12]. Focusing our analysis on the long-term behavior of the ejected matter, we will be able to study its amount, composition, and thermodynamic state, and use the resulting data to perform detailed nucleosynthesis calculations.

  In order to reach these goals, this project includes also a software development phase. The evolution codes that we rely on have been developed to model the whole inspiral, merger, and post-merger history of a BNS system and are therefore fully tridimensional out of necessity. We have tackled the task of modifying them to be able to leverage the symmetry of the system in the late post-merger, effectively resulting in a 2D code that employs the so-called "cartoon" method as introduced in Alcubierre et al. [2]. This project is closely linked to the one outlined above and complements it because we are going to base the initial data of the 2D runs on the data extracted from the 3D simulations performed as part of the project above.

5.1 Current Status and Planned Developments

5.1.1 Systematic Characterisation of EOS Effects on the Post-merger Micro- and Macrophysics

This project can be conceptually (and temporally) divided into two parts. The first, mostly numerical, consists of performing the production runs of the BNS simulations that we have planned; while the second one includes the data analysis, interpretation, and publishing of the results. The first part is essentially the only one that concerns our usage of HLRS's computational resources in general and the `Hawk` cluster in particular.

The codes we rely on to perform the planned simulations are public domain ones (excluding minor, non-public modifications introduced in the Darmstadt group). As mentioned above they are based on the ET framework [14, 17] and include the `WhiskyTHC` [19, 21] code to handle the discretization of the hydrodynamic equations and microphysics and the `McLachlan` code [17] to solve Einstein's equations and evolve the spacetime. These codes are widely used in the community and are known to scale well on a variety of CPU-based computing clusters (including `Hawk`). Since these codes fully cover our needs "out-of-the-box" (besides the use of our custom-produced EOS tables), we are in a position to easily obtain the data we need.

We completed the production run and analysis of the first set of simulations in 2022. The results form the main part of Maximilian Jacobi's doctoral thesis (April 2023) and has been published (https://doi.org/10.1093/mnras/stad3738).

In 2023 we plan on extending these results with a larger set of follow-up simulations including new EOSs presented in Huth et al. [15]. This set will include more values of the incompressibility and the effective mass and employ a new, improved formulation of the effective mass's density dependence. Furthermore, we plan to vary the symmetry energy and the slope parameter independently.

Numerical setup and overview of the models

We employ a Cartesian-coordinates computational grid made of 7 "box-in-box" refinement levels. The resolution on the finest refinement level is $0.128\,M_\odot \approx 189\,m$ and the grid spans $1024\,M_\odot \approx 1512\,km$ in each direction on the coarsest refinement level. The BNS initial data is generated using the open source `LORENE` library.

So far, we evolved equal-mass BNS configuration, with NS masses of $1.365\,M_\odot$. This setup is consistent with the observations of gravitational waves in GW170817. For the new simulations, we plan to also include asymmetric binaries, for which the NS masses will be chosen in a way to keep this consistency. The evolution is carried out from an initial separation of 45 km (corresponding to 3/4 orbits before the merger) up to about 40 ms after the merger or to the gravitational collapse of the model.

EOS tables

For the first part of this project, we performed 9 high-resolution simulations of merging NSs, employing different EOSs. Six of the models have been first employed

Table 3 Nuclear matter properties (effective mass m^*, binding energy B, incompressibility K, symmetry energy E_{sym}, slope parameter L, and saturation density n_0) for all employed EOSs tables. Also given are the pressure and the dimensionless slope of the pressure at saturation density as defined in the text

EOS	m^* $[m_n]$	B [MeV]	K [MeV]	E_{sym} [MeV]	L [MeV]	n_0 $[\text{fm}^{-3}]$	$P\|_{n=n_0}$ $[\text{MeV fm}^{-3}]$	$\frac{n}{P}\frac{\partial}{\partial n}P\|_{n=n_0}$
LS175†	1.0	16.0	175	29.3	73.7	0.155	3.17	2.77
LS220†	1.0	16.0	220	29.3	73.7	0.155	3.19	3.03
LS255†	1.0	16.0	255	29.3	73.7	0.155	3.19	3.23
$m^*_{0.8}$	0.8	16.0	220	29.3	79.3	0.155	3.39	3.07
m^*_{S}	0.634	16.0	220	29.3	86.5	0.155	3.66	3.13
$(m^*K)_{\text{S}}$	0.634	16.0	281	29.3	86.5	0.155	3.68	3.43
$(m^*K E_{\text{sym}})_{\text{S}}$	0.634	16.0	281	36.9	109.3	0.155	4.45	3.18
SkShen	0.634	16.3	281	36.9	109.4	0.145	5.11	3.18
Shen	0.634	16.3	281	36.9	110.8	0.145	4.98	3.14

in Yasin et al. [37] in the context of core-collapse supernova simulations and two new models were developed specifically for this work. The tables are created with the open-source SROEOS code[4] [31] which is based on a liquid-drop model with Skyrme-type nucleon-nucleon interactions. Their properties are summarized in Table 3.

Following Yasin et al. [37] we take the LS220† EOS as our fiducial model. It employs the same Skyrme parametrization as the LS220 EOS and has an incompressibility parameter $K = 220\,\text{MeV}$ and an effective nucleon mass of $m^* = m_N$, where $m_N = 939.5654\,\text{MeV}$ is the nucleon mass. For each EOS we then systematically vary the nuclear matter properties, one at a time, by tuning the Skyrme parameters used as input for the SROEOS code.

The LS175† and LS255† EOSs models modify the K parameter to 175 and 255 MeV, respectively. In models $m^*_{0.8}$ and m^*_{S} the effective nucleon mass at saturation density m^* is instead changed to $0.8m_N$ and $0.634m_N$, respectively. The lowest used value of $0.634m_N$ is the effective mass of the Shen EOS [34]. In the remaining SROEOS models, we change the nuclear matter properties apart from the effective mass progressively to match the values of the Shen EOS, starting with $K = 281$ MeV ($(m^*K)_{\text{S}}$), followed by the symmetry energy $E_{\text{sym}} = 36.9\,\text{MeV}$ ($(m^*K E_{\text{sym}})_{\text{S}}$), and finally also the value of the saturation density $\rho_0 = 2.43 \times 10^{14}\,\text{g/cm}^3$ and the binding energy per nucleon $B = 16.3\,\text{MeV}$ (SkShen). The SkShen EOS exhibits almost the same nuclear matter properties as the Shen EOS and, therefore, matches it closely around saturation density. At larger densities, however, it predicts higher pressures. Finally, we include the original Shen EOS model for comparison.

Results

In this subsection, we summarize a subset of our results from the first set of simulations.

[4] https://bitbucket.org/andschn/sroeos.

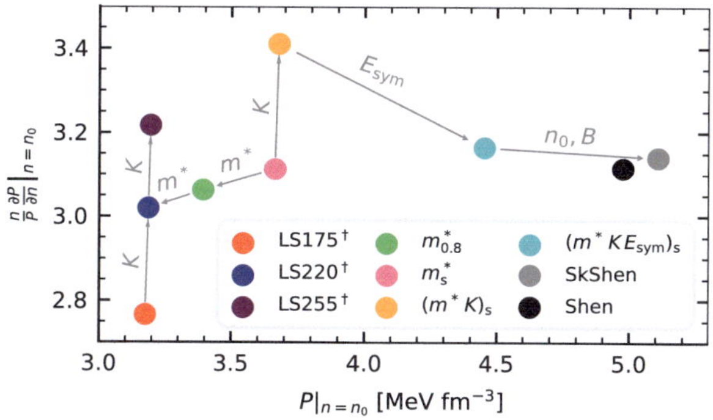

Fig. 2 The pressure and the dimensionless slope of the pressure. Grey arrows show the relevant nuclear matter property changes

EOS models that generally predict lower pressures are often labeled as "soft", while those predicting higher pressures are labeled as "stiff". Our work aims to provide a more precise description of the features of an EOS at different densities and how they impact the dynamics and observable outcome of BNS merger. For this purpose, we characterize the employed EOS models with their pressure at saturation density $P|_{n=n_0}$ and the normalized slope of the pressure with respect to the density $\frac{n}{P}\frac{\partial}{\partial n}P\Big|_{n=n_0}$. These quantities are shown in Fig. 2 and given in Table 3. Two EOSs with similar pressure at saturation density but different slopes predict comparable pressures in the low-density regime but differ at higher densities.

One example of how these density regimes impact different features differently can be seen in Fig. 3 which shows the evolution of the maximum density and the average remnant temperature. The prior determines whether or not a collapse to a BH occurs (as is the case for LS175† and LS220†), while the latter is relevant for the ejection of matter and the mass of the accretion disk. Generally, softer EOSs lead to a more violent merger which produces a more heavily oscillating remnant reaching larger maximum densities and higher temperatures. However, the evolution of the central density naturally depends more on the EOS at high densities. Shocks generated in the oscillating NS only penetrate the lower-density outer layers. The average remnant temperature is therefore more influenced by the lower density EOS.

Merging BNSs eject neutron-rich material which undergoes the so-called r-process in which heavy elements such as gold are synthesized. Furthermore, the decay of the newly synthesized radioactive nuclei powers a thermal transient called kilonova (KN) which is observable from Earth. Such a transient was observed in 2017 for the first time together with the observation of the GW signal GW170817 and was called AT 2017gfo [1]. The correlation of the ejecta mass with the properties of the employed EOSs is a central research topic in the field of BNS mergers since it directly impacts the resulting KN. The most neutron-rich matter is ejected on dynamical timescales (thus called the dynamical ejecta) due to two mechanisms. Tidal forces unbind material from the outward-facing sides of the NSs at the moment

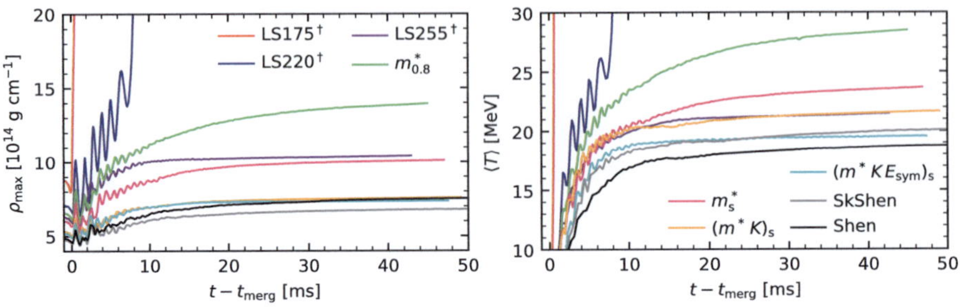

Fig. 3 Maximum density (upper panel) and average remnant temperature (lower panel) versus time

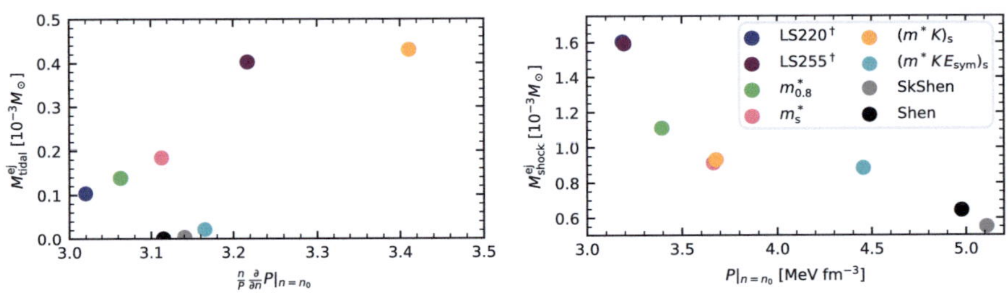

Fig. 4 Masses of the tidal and shock-heated ejecta components as a function of the dimensionless slope of the pressure and the pressure at saturation density, respectively

of merger and shocks created by the collision at the merger "squeeze out" matter from the merger interface. Since the mechanism of ejection is entirely different for the tidal and shock-heated dynamical ejecta components, it is not surprising that they depend differently on the properties of the EOS.

Figure 4 shows the total mass of the shock-heated and tidal ejecta. The mass of the shock-heated dynamical ejecta is correlated with pressure at saturation density but is entirely unaffected by the slope of the pressure at saturation density. As previously mentioned, the matter heated by shocks is close to saturation density so the shock-heated ejection mechanism is sensitive to the low-density part of the EOS. The tidal-ejecta mass, however, does not follow this trend. Instead, we find, that it is correlated to the slope of the pressure at saturation density and is thus significantly larger for LS255† and $(m^*K)_S$ compared to the other models. In the models $(m^*K\,E_{sym})_S$, SkShen, and Shen, neutron-rich matter ejected by tidal forces is overtaken by the shock-heated component early on. This changes the composition of the material such that our definition of the tidal ejecta no longer applies, leading to the small tidal ejecta masses for these 3 models.

We extract the post-merger gravitational wave signal for all of our simulations. The spectrum is initially made up of multiple frequency peaks. Most frequencies decay within ~10 ms, leaving one dominant frequency behind. This frequency is often called f_2 frequency since it is related to the quadrupole ($m = 2$) deformation of the massive NS. Figure 5 shows the Fourier spectra and corresponding time-frequency

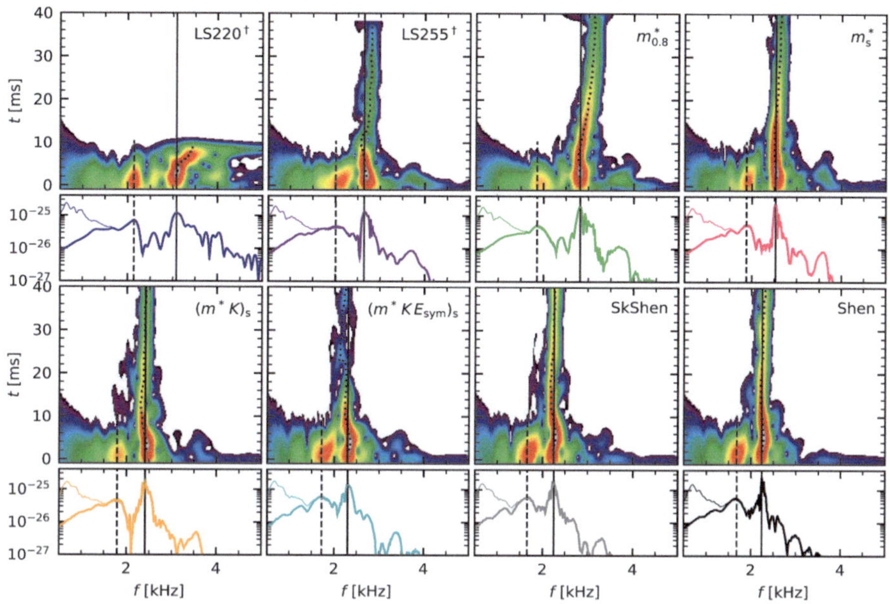

Fig. 5 Post-merger GW Fourier spectra and corresponding time-frequency spectrograms

spectrograms of the post-merger GW emission in 8 of our simulations. Many works have related the f_2 frequency to properties of the equation of state [e.g., 7, 28]. We confirm these relations for our set of simulations. Moreover, we find that the SkShen and Shen models produce a strikingly similar GW signal that matches in both the primary f_2 frequency and the early secondary frequencies. Therefore, we conclude that the post-merger frequencies are determined by the EOS around saturation density where Shen and SkShen are similar and not by the high-density part where they diverge.

Future developments

The results from our simulations have been compiled and analyzed and published (https://doi.org/10.1093/mnras/stad3738). Another follow-up publication focusing on the nucleosynthesis in the ejecta and the resulting electromagnetic transient has also been published (https://doi.org/10.1093/mnras/stae1979). This step only requires a minimal amount of computational resources for post-processing. We extracted the Lagrangian trajectories of all ejected fluid elements for each of the simulations. Based on their density and temperature history, the nucleosynthesis in the ejecta was calculated with the WinNet reaction network [27]. In another post-processing step, we calculated the KN light curve, driven by the radioactive decay of the newly synthesized nuclei. The results are currently being analyzed. By comparing them to KN observations like AT 2017gfo [1], we can uncover insights into the physics of matter at supra-nuclear density and the role of BNS merger in the enrichment of galaxies with heavy elements.

The next step will be to employ new EOS models in BNS merger simulations to investigate the trends suggested by our previous work. This will most likely use all of the remaining core hours of this allocation. We plan on studying the trends observed in our previous work with a larger set of follow-up simulations including more values of the incompressibility, effective mass, and new values of the symmetry energy parameters. Moreover, we plan to extend our study to asymmetric binaries. For this new study, we will include new EOSs from Huth et al. [15]. In this recent work, a new form of Skyrme-functional was introduced which allows for more control of the EOS' properties. Furthermore, it features a novel description of the effective mass informed by chiral effective field theory calculations. This will enable us to study the thermal contributions to the EOS based on ab-initio principles.

5.1.2 Study of the Long-Term Evolution of BNS Disks and Ejecta

As mentioned above the goal of this project is to extend the analysis of BNS systems to the late post-merger phase, about one second after the merger and beyond. Besides the necessity of simply keeping a simulation running until this mark is reached, it is necessary to consider that the matter ejected from the system continues to expand outwards. This implies that in order to capture the dynamics of this matter, the simulation domain has to be much wider than in shorter simulations. A typical value for the radius of the domain, in this case, is $\sim 10,000\,\mathrm{km}$, roughly ten times as large than in typical simulations which stop before $\sim 100\,\mathrm{ms}$ post-merger. Considering that the domain volume scales as the third power of the radius, it is easy to see that the increase in computational resources needed to perform such a simulation is simply not manageable.

In this project, we are circumventing this issue by using the fact that in the late post-merger phase the system approaches an axisymmetric state. Imposing this symmetry, the dimensionality of the simulation domain drops from 3 to 2 dimensions, resulting in a much cheaper setup. Technically this is achieved by formally keeping the code setup as a fully 3D code, but providing specific boundary conditions that enforce axisymmetry, a technique known as the "cartoon method" and first introduced in Alcubierre et al. [2]. Our implementation is in particular inspired by the work of Fujibayashi et al. [12]. One can then use the outcome of a full 3D simulation, such as the ones we are performing in the project above as initial data for an extended run with reduced dimensionality, obtaining the full evolution of the system from inspiral to the late post-merger.

After some initial setbacks, we have now mostly completed the process of converting our codes to this reduced-dimensionality setup and ran a few test simulations. The reduction in computational cost coming from the reduction in the dimensionality of the model is indeed significant. The resulting speed-up, as compared to an analogous 3D simulation, depends on many factors, but we are now able to estimate a cost per high-resolution run of $420\,\mathrm{h} \times 128$ cores (1 node), i.e. $\sim$120,000 core-hours. Note that not only this is a fraction of the cost of a high-resolution run of the previous

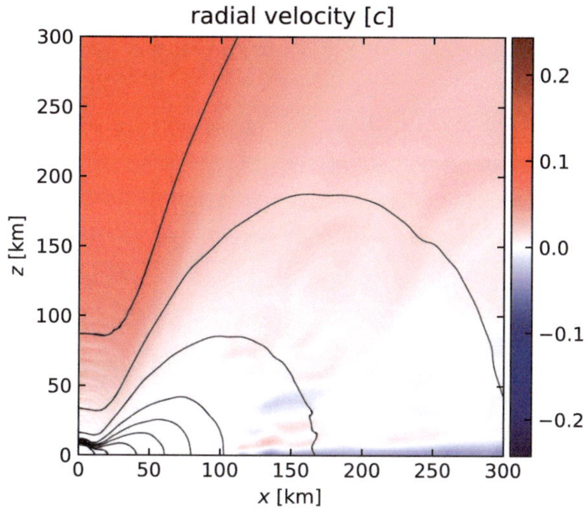

Fig. 6 Radial velocity and density profiles on the meridional plane of a BNS merger remnant+disk system. The contour lines indicate the base-10 logarithm of rest mass density in steps of 1, where the innermost contour corresponds to 10^{14} g/cm^3

project, but it also takes into account the much longer evolution time and much larger domain size.

Figure 6 provides an example of the dynamics of a BNS system modeled with this dimensionality-reducing technique, displaying, in particular, the radial velocity of the fluid and density isocontours. Note that while the figure shows only a region of side 300 km, the simulation covers a domain about 30 times larger.

Unfortunately, some of our simulations crash after hundreds of milliseconds (i.e., about a week of computational time) due to numerical instabilities. In Fig. 7, we show the evolution of several quantities for two of our 2D models. In the model `DD2_2D` the neutrino scheme outlined above is included, while the model `DD2_2D_no`ν does not include a neutrino scheme at all. An instability develops in the center of the model `DD2_2D` after ~390 ms. It can be first seen in the evolution of the maximum density (top-left panel), which suddenly increases and then abruptly drops. At the same time, the minimum of the lapse function (top-right panel) does barely react. Furthermore, the constraint violation (bottom-left panel) stays relatively low, even after 400 ms, and becomes large only later. This points toward a failure caused by the hydrodynamical evolution of the system. Eventually, the conservation of the total baryon mass in the simulation is violated (bottom-right panel).

We are unsure what causes this instability and plan to further investigate it in the future. The fact, that it does appear in the simulation without neutrinos, suggests that the crash is related to the interaction of the neutrino transport scheme and the cartoon method. Once these issues have been addressed, our framework will allow us to extend 3D simulations to multiple seconds. In this way, we will be able to consistently obtain the full mass ejection history of BNS mergers.

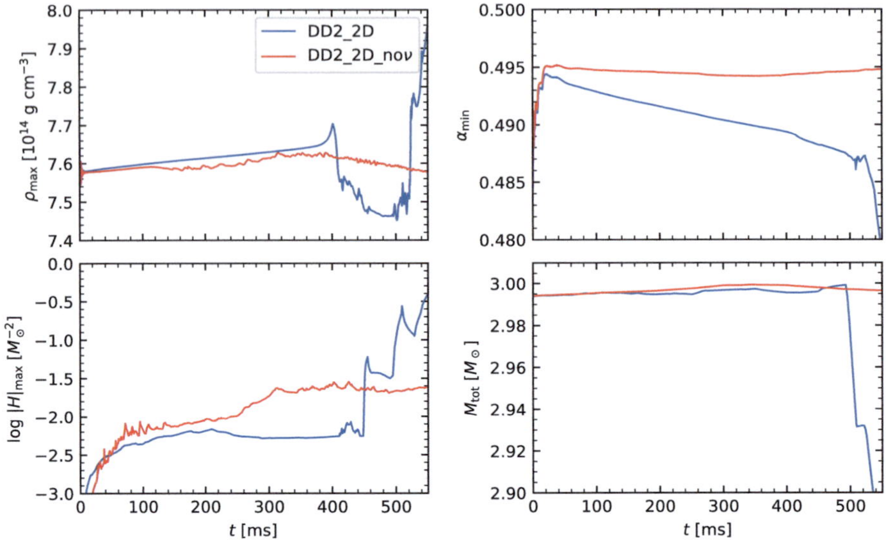

Fig. 7 Time series of several quantities detailing the instability, that arises in DD2. Top left: maximum density, top right: minimum of the lapse function, bottom left: maximum of the absolute value of the Hamiltonian constrain violation H, bottom right: total baryon mass in the simulation domain

6 Conclusions

At the theoretical astrophysics group in Darmstadt, we are pursuing several projects related to the dynamics of BNS mergers. We are specifically focused on the role of the nuclear EOS and the long-term evolution of such systems. Based on the simulations performed on the allocation "44163 microBNS" on the Hawk cluster we have uncovered several correlations between the nuclear EOS and the outcome of BNS mergers. These results make up the main part of the Dissertation of Maximilian Jacobi and will be submitted for publication soon. Furthermore, several follow-up studies are currently being worked on or in planning. We are in the process of finalizing our setup for 2D-axisymmetric long-term simulations of BNS post-mergers with an extended spatial domain. Several simulations have been performed, but some problems remain to be solved. Our setup delivers the expected savings in computational resources and we have confirmed our ability to efficiently make use of the Hawk platform.

Acknowledgements We acknowledge support for the projects mentioned in this report by:
- ERC Starting grant EUROPIUM-667912;
- Deutsche Forschungsgemeinschaft SFB 1245;
- State of Hessen Research Cluster ELEMENTS.

References

1. Abbott, B., et al.: Multi-messenger observations of a binary neutron star merger. Astrophys. J. Lett. **848**(2), L12 (2017). https://doi.org/10.3847/2041-8213/aa91c9. arXiv:1710.05833
2. Alcubierre, M., Brandt, S.R., Brügmann, B., Holz, D., Seidel, E., Takahashi, R., Thornburg, J.: Symmetry without symmetry: numerical simulation of axisymmetric systems using Cartesian grids. Int. J. Mod. Phys. D **10**(3), 273–289 (2001). arXiv:gr-qc/9908012
3. Alic, D., Bona-Casas, C., Bona, C., Rezzolla, L., Palenzuela, C.: Conformal and covariant formulation of the Z4 system with constraint-violation damping. Phys. Rev. D **85**(6), 064040 (2012). https://doi.org/10.1103/PhysRevD.85.064040. arXiv:1106.2254
4. Alic, D., Kastaun, W., Rezzolla, L.: Constraint damping of the conformal and covariant formulation of the Z4 system in simulations of binary neutron stars. Phys. Rev. D **88**(6), 064049 (2013). https://doi.org/10.1103/PhysRevD.88.064049. arXiv:1307.7391
5. Allen, G., Goodale, T., Lanfermann, G., Radke, T., Seidel, E.: The cactus code: a problem solving environment for the grid. In: Proceedings of First Egrid Meeting at ISTHMUS, Poznan, April 2000, http://www.zib.de/visual/projects/TIKSL/Papers/EGrid2000-Cactus.ps (2000)
6. Baumgarte, T.W., Shapiro, S.L., Teukolsky, S.A.: Computing supernova collapse to neutron stars and black holes. Astrophys. J. **443**, 717–734 (1995). https://doi.org/10.1086/175563
7. Bauswein, A., Stergioulas, N., Janka, H.T.: Exploring properties of high-density matter through remnants of neutron-star mergers. EPJA **52**(3) (2016). https://doi.org/10.1140/epja/i2016-16056-7, arXiv:1508.05493
8. Berger, M.J., Oliger, J.: Adaptive mesh refinement for hyperbolic partial differential equations. J. Comput. Phys. **53**, 484–512 (1984)
9. Bovard, L., Martin, D., Guercilena, F., Arcones, A., Rezzolla, L., Korobkin, O.: On r-process nucleosynthesis from matter ejected in binary neutron star mergers. Phys. Rev. D **96**, 124005 (2017). arXiv:1709.09630
10. Brown, J.D.: Covariant formulations of BSSN and the standard gauge. PhysRev D **79**, 104029 (2009). https://doi.org/10.1103/PhysRevD.79.104029. arXiv:0902.3652
11. Bruenn, S.W.: Stellar core collapse: numerical model and infall epoch. Astrophys. J. Suppl. **58**, 771–841 (1985). https://doi.org/10.1086/191056
12. Fujibayashi, S., Sekiguchi, Y., Kiuchi, K., Shibata, M.: Properties of neutrino-driven ejecta from the remnant of a binary neutron star merger: pure radiation hydrodynamics case. Astrophys. J. **846**, 114 (2017). https://doi.org/10.3847/1538-4357/aa8039
13. Galeazzi, F., Kastaun, W., Rezzolla, L., Font, J.A.: Implementation of a simplified approach to radiative transfer in general relativity. PhysRev D **88**, 064009 (2013). https://doi.org/10.1103/PhysRevD.88.064009. arXiv:1306.4953
14. Haas, R., Brandt, S.R., Gabella, W.E., Gracia-Linares, M., Karakaş, B., Matur, R., Alcubierre, M., Alic, D., Allen, G., Ansorg, M., Babiuc-Hamilton, M., Baiotti, L., Benger, W., Bentivegna, E., Bernuzzi, S., Bode, T, Bruegmann, B, Campanelli, M, Cipolletta, F, Corvino, G, Cupp, S, Pietri, R.D., Diener, P., Dimmelmeier, H., Dooley, R., Dorband, N., Elley, M., Khamra, Y.E., Etienne, Z., Faber, J., Font, T., Frieben, J., Giacomazzo, B., Goodale, T., Gundlach, C., Hawke, I., Hawley, S., Hinder, I., Husa, S., Iyer, S., Kellermann, T., Knapp, A., Koppitz, M., Laguna, P., Lanferman, G., Löffler, F., Masso J., Menger L., Merzky A., Miller J.M., Miller M., Moesta P., Montero P., Mundim B., Nerozzi A., Noble S.C., Ott C., Paruchuri R., Pollney D., Radice D., Radke T., Reisswig C., Rezzolla L., Rideout D., Ripeanu, M., Sala, L., Schewtschenko, J.A., Schnetter, E., Schutz, B., Seidel, E., Seidel, E., Shalf, J., Sible, K., Sperhake, U., Stergioulas, N., Suen, W.M., Szilagyi, B., Takahashi, R., Thomas, M., Thornburg, J., Tobias, M., Tonita, A., Walker, P., Wan, M.B., Wardell, B., Witek, H . Zilhão, M., Zink, B., Zlochower, Y.: The einstein toolkit. https://doi.org/10.5281/zenodo.4298887, https://doi.org/10.5281/zenodo.4298887, to find out more, visit http://einsteintoolkit.org (2020)
15. Huth, S., Wellenhofer, C., Schwenk, A.: New equations of state constrained by nuclear physics, observations, and qcd calculations of high-density nuclear matter. Phys. Rev. C **103** (2021). https://doi.org/10.1103/PhysRevC.103.025803

16. Kastaun, W., Galeazzi, F., Alic, D., Rezzolla, L., Font, J.A.: Black hole from merging binary neutron stars: how fast can it spin? Phys. Rev. D **88**(2), 021501 (2013). https://doi.org/10.1103/PhysRevD.88.021501. arXiv:1301.7348
17. Löffler, F., Faber, J., Bentivegna, E., Bode, T., Diener, P., Haas, R., Hinder, I., Mundim, B.C., Ott, C.D., Schnetter, E., Allen, G., Campanelli, M., Laguna, P.: The Einstein toolkit: a community computational infrastructure for relativistic astrophysics. Class. Quantum Grav. **29**(11), 115001 (2012). https://doi.org/10.1088/0264-9381/29/11/115001. arXiv:1111.3344
18. Neilsen, D., Liebling, S.L., Anderson, M., Lehner, L., O'Connor, E., et al.: Magnetized neutron stars with realistic equations of state and neutrino cooling. PhysRev **D89**(10), 104029 (2014). https://doi.org/10.1103/PhysRevD.89.104029. arXiv:1403.3680
19. Radice, D., Rezzolla, L.: THC: a new high-order finite-difference high-resolution shock-capturing code for special-relativistic hydrodynamics. Astron. Astrophys. **547**, A26 (2012). https://doi.org/10.1051/0004-6361/201219735. arXiv:1206.6502
20. Radice, D., Abdikamalov, E., Rezzolla, L., Ott, C.D.: A new spherical harmonics scheme for multi-dimensional radiation transport I. Static matter configurations. J. Comput. Phys. **242**, 648–669 (2013). https://doi.org/10.1016/j.jcp.2013.01.048. arXiv:1209.1634
21. Radice, D., Rezzolla, L., Galeazzi, F.: Beyond second-order convergence in simulations of binary neutron stars in full general-relativity. Mon. Not. R. Astron. Soc. L **437**, L46 L50 (2014). https://doi.org/10.1093/mnrasl/slt137. arXiv:1306.6052
22. Radice, D., Rezzolla, L., Galeazzi, F.: High-order fully general-relativistic hydrodynamics: new approaches and tests. Class. Quantum Grav. **31**(7), 075012 (2014). https://doi.org/10.1088/0264-9381/31/7/075012. arXiv:1312.5004
23. Radice, D., Rezzolla, L., Galeazzi, F.: High-order numerical-relativity simulations of binary neutron stars. In: Pogorelov, N.V., Audit, E., Zank, G.P. (eds.) Numerical Modeling of Space Plasma Flows ASTRONUM-2014, Astronomical Society of the Pacific Conference Series, vol. 498, p. 121 (2015). arXiv:1502.00551
24. Radice, D., Galeazzi, F., Lippuner, J., Roberts, L.F., Ott, C.D., Rezzolla, L.: Dynamical mass ejection from binary neutron star mergers. Mon. Not. R. Astron. Soc. **460**, 3255–3271 (2016). https://doi.org/10.1093/mnras/stw1227. arXiv:1601.02426
25. Radice, D., Galeazzi, F., Lippuner, J., Roberts, L.F., Ott, C.D., Rezzolla, L.: Dynamical mass ejection from binary neutron star mergers. Mon. Not. R. Astron. Soc. **460**(3), 3255–3271 (2016). https://doi.org/10.1093/mnras/stw1227. arXiv:1601.02426
26. Radice, D., Ott, C.D., Abdikamalov, E., Couch, S.M., Haas, R., Schnetter, E.: Neutrino-driven convection in core-collapse supernovae: high-resolution simulations. Astrophys. J. **820**, 76 (2016). https://doi.org/10.3847/0004-637X/820/1/76. arXiv:1510.05022
27. Reichert, M., Winteler, C., Korobkin, O., Arcones, A., Bliss, J., Eichler, M., Frischknecht, U., Fröhlich, C., Hirschi, R., Jacobi, M., Kuske, J., Martínez-Pinedo, G., Martin, D., Mocelj, D., Rauscher, T., Thielemann, F.K.: The nuclear reaction network WinNet. arXiv e-prints arXiv:2305.07048, https://doi.org/10.48550/arXiv.2305.07048 (2023)
28. Rezzolla, L., Takami, K.: Gravitational-wave signal from binary neutron stars: a systematic analysis of the spectral properties. PRD **93**(12), 1–20 (2016). https://doi.org/10.1103/PhysRevD.93.124051. arXiv:1604.00246
29. Rosswog, S., Ramirez-Ruiz, E., Davies, M.B.: High resolution calculations of merging Neutron stars. 3. Gamma-ray bursts. Mon. Not. R. Astron. Soc. **345**, 1077 (2003). https://doi.org/10.1046/j.1365-2966.2003.07032.x, arXiv:astro-ph/0306418
30. Ruffert, M., Rampp, M., Janka, H.T.: Coalescing neutron stars: gravitational waves from polytropic models. arXiv preprint arXiv:astro-ph/9611056 (1996)
31. Schneider, A.S., Roberts, L.F., Ott, C.D.: Open-source nuclear equation of state framework based on the liquid-drop model with Skyrme interaction. Phys. Rev. C **96**(6), 65802 (2017). https://doi.org/10.1103/PhysRevC.96.065802. arXiv:1707.01527
32. Schnetter, E., Hawley, S.H., Hawke, I.: Evolutions in 3D numerical relativity using fixed mesh refinement. Class. Quantum Grav. **21**, 1465–1488 (2004). https://doi.org/10.1088/0264-9381/21/6/014. arXiv:gr-qc/0310042

33. Shapiro, S.L., Teukolsky, S.A.: Black holes, white dwarfs, and neutron stars: the physics of compact objects (1983)
34. Shen, H., Toki, H., Oyamatsu, K., Sumiyoshi, K.: Relativistic equation of state of nuclear matter for supernova and neutron star. Nucl. Phys. A **637**, 435–450 (1998). https://doi.org/10.1016/S0375-9474(98)00236-X. arXiv:nucl-th/9805035
35. Shibata, M., Nakamura, T.: Evolution of three-dimensional gravitational waves: harmonic slicing case. Phys. Rev. D **52**, 5428–5444 (1995). https://doi.org/10.1103/PhysRevD.52.5428
36. Shu, C.W., Osher, S.J.: Efficient implementation of essentially non-oscillatory shock-capturing schemes. J. Comput. Phys. **77**, 439 (1988)
37. Yasin, H., Schäfer, S., Arcones, A., Schwenk, A.: Equation of state effects in core-collapse supernovae. Phys. Rev. Lett. **124**, 092701 (2020). https://doi.org/10.1103/PhysRevLett.124.092701
38. Zilhão, M., Löffler, F.: An introduction to the Einstein toolkit. Int. J. Modern Phys. A **28**, 1340014 (2013). https://doi.org/10.1142/S0217751X13400149. arXiv:1305.5299

Hadron Scattering and Resonances for PANDA Physics

J. Bulava, B. Hörz, R. J. Hudspith, M. F. M. Lutz, D. Mohler, and C. Morningstar

The goal of our project is the determination of bound states and resonances emerging from the theory of the strong interaction, Quantum Chromodynamics (QCD), within the Standard Model of Particle Physics. In the last two decades, many states with properties not well-described by the naive quark model picture of quark-antiquark mesons and three-quark baryons have been observed. A particular focus of our project are mesons with a heavy charm or bottom quark, some of which seem to have a large four-quark component, suggestive of meson-meson bound states. To study these states we employ a discretized version of QCD referred to as Lattice QCD, along with Effective Field Theory (EFT methods).

J. Bulava
Deutsches Elektronen-Synchrotron (DESY), Zeuthen, Germany

B. Hörz
Nuclear Science Division, Lawrence Berkeley National Laboratory, Berkeley, CA, USA

Intel Deutschland GmbH, Feldkirchen, Germany

R. J. Hudspith · M. F. M. Lutz · D. Mohler (✉)
GSI Helmholtzzentrum für Schwerionenforschung, Darmstadt, Germany
e-mail: daniel.mohler@tu-darmstadt.de

D. Mohler
Institut für Kernphysik, Fachbereich Physik, Technische Universität Darmstadt, Darmstadt, Germany

C. Morningstar
Department of Physics, Carnegie Mellon University, Pittsburgh, PA, USA

T. Ludwig et al. (eds.), *High Performance Computing in Science and Engineering '23*,
https://doi.org/10.1007/978-3-031-91312-9_5

1 Introduction

The discovery of the Higgs boson at the Large Hadron Collider in 2012 has completed the Standard Model (SM) of elementary particle physics. While Quantum Chromodynamics (QCD) is firmly established as the theory of the strong-interaction between quarks and gluons within the SM, reliable calculations of QCD at low energies still pose a formidable task. At these energies, the physics is dominated by hadrons, the bound states of the strong interaction. Indeed, low-energy QCD is the last part of the SM that is insufficiently understood, in the sense that we do not understand quantitatively how hadron excitations and nuclei arise from the theory of QCD.

Inspired by the quark model, hadrons are thought of as mesons made from a quark and an antiquark and as baryons composed of three quarks. Recent discoveries by the Belle, BaBar, BESIII and LHCb experiments however revealed many hadrons that do not fit the traditional quark model classification. This is especially striking for heavy mesons containing charm and beauty quarks, where (non-relativistic) potential models were often expected to provide a good description of the spectrum. Instead many unexpected states are seen, including quarkonium-like (in analogy to positronium) states that carry electric charge, and therefore do not fit into the simple picture of quark-antiquark mesons and 3-quark baryons (for a review see [1, 2]), as well as the pentaquark states observed more recently by the LHCb experiment [3, 4]. In the future, new and interesting experimental results regarding mesons and baryon resonances are expected from current and future (BelleII, BESIII, CLAS, COMPASS, GlueX, LHCb, P̄ANDA) experiments, which will likely lead to the discovery of further hadrons with properties not expected from the quark model.

While there is a vigorous experimental program aimed at understanding the properties of those hadrons, our theoretical understanding is largely based on models. In this proposal we focus on two complementary approaches. On the one hand effective field theory (EFT) approaches based on Chiral Lagrangians, and on the other Lattice Quantum Chromodynamics (Lattice QCD). The combination of these methods provides a unique opportunity to determine the spectrum of hadrons. Lattice QCD is an *ab-initio*, non-perturbative approach with quantifiable and systematically-improvable uncertainties needing supercomputer resources. In recent years, the promise held by this approach is finally coming to fruition due to a long standing effort in both the development of novel computational strategies and methods, and due to the general increase in compute power.

This is matched by similar substantial progress on the EFT side, where it was observed that suitable adaptation of the conventional chiral perturbation theory in terms of on-shell meson and baryon masses leads to a significantly larger application domains, that in particular allows for more significant results based on flavor SU(3) chiral Lagrangians [5–9].

Most recently, the importance of left-hand and anomalous-cut contributions in coupled-channel systems was explored in the open-charm sector of QCD [10, 11]. In some cases a strikingly large impact on the choice of quark masses was found.

The proper framework for the study of resonances in Euclidean space is the finite-volume method pioneered by Lüscher [12, 13]. Where it was shown that for elastic scattering in a sufficiently large volume, the energy spectrum $E_n(L)$ in a finite box of spatial extent L is related to the phase shift of the scattering amplitude in infinite volume. Many generalizations (see [14] and references therein) were derived later on, and either increase the information extracted from a single ensemble of gauge configurations, or extend the range of applicability of the method.

We are using the catalog of gauge-field ensembles generated by the Coordinated Lattice Simulation (CLS) consortium. In the current allocation period we focus on the ingredients necessary to suitably enlarge the data basis to tackle coupled-channel scattering. Unlike for elastic scattering, where the quantization condition yields a phase-shift point for each finite-volume energy level, multiple functions of the Mandelstam variable $s = E^2$ need to be constrained from the finite-volume energies through a determinant condition (for a detailed presentation of such quantization conditions see for example [14]). To parameterize the scattering amplitudes in the coupled-channel region, good data-coverage of the relevant energy range is therefore essential. We will achieve this by adding further moderate-sized gauge field ensembles to existing simulation points. Progress regarding this effort is summarized in Sect. 2, where our computational setup is described.

Furthermore expanding the scope of calculations by studying the quark-mass dependence of heavy-light hadrons is absolutely crucial for a meaningful comparison to aforementioned experiments and for future precision calculations. Our long-term goal is to map out the quark-mass dependence of close-to-threshold bound states and resonances with a full quantization of systematic uncertainties from the lattice discretization and from residual volume effects. While our calculations of heavy-light mesons with charm-quarks are progressing, we branched out to heavy-light mesons with bottom quarks in the current allocation period. This resulted in predictions for QCD bound states [15] and some of these results are presented in Sect. 3.1.

2 Computational Setup

The formulation of lattice QCD used in our calculation is the tree-level improved Symanzik gauge action together with the non-perturbatively O(a) improved Wilson fermion action as used by the CLS consortium [16, 17]. In our project we use the deflation-accelerated SAP-GCR solver described by M. Lüscher [18, 19], which is implemented in the `openQCD` code [20].

One of the major computational tasks of our project is the generation of gauge-field ensembles at further lattice volumes at existing points of the CLS library of gauge-field configurations. In addition our calculation relies on the determination of quark propagators and their subsequent contraction to form suitable hadronic correlation functions.

The performance of both parts depends crucially on an efficient solver for the inhomogeneous Dirac equation, i.e.

$$D_{\mathrm{w}}\psi = \eta, \tag{1}$$

where D_{w} is the O(a) improved Wilson-Dirac operator, and η being some source vector. The solution, ψ, to Eq. (1) represents the quark propagator.

Our calculation of the quark propagation employs the stochastic variant (stochastic-LapH) [21] of the method known as distillation [22]. In this framework, quark sources (i.e. the right-hand side of Eq. (1)) are linear combinations of eigenvectors of the three-dimensional gauge-covariant Laplacian with random coefficients. A stochastic estimate of the *perambulator* is obtained from N_η noise sources, and we employ noise partitioning techniques [23, 24] to reduce the variance of the stochastic estimator.

For the calculation of the quark propagators contracted into perambulator matrices, we use version 1.6 of the `openQCD` code [20] developed by M. Lüscher and S. Schaefer. The code (albeit several different versions) is publicly available under version 2 of the GPL license. It has been tested extensively on a variety of platforms, including cluster architectures such as SuperMUC, MOGON-II, Hazel Hen, Hawk, JUWELS, as well as BlueGene/Q machines. Version 1.6 supports the use of AVX and FMA instructions. Parallelisation is implemented via MPI with one MPI process per core for cluster architectures such as MOGON-II, JUWELS, and Hawk.

For the generation of ensembles a newer version of `openQCD` has been used (version 2.4) that allows for hybrid MPI and OpenMP parallelism, which has several other important improvements in the propagator solver routines.

`openQCD` exhibits excellent scaling properties over a wide range of problem sizes. Figure 1 shows the strong-scaling behaviour for the system size corresponding to a CLS "D" ensemble, i.e. for a 128×64^3 lattice, as measured on Hawk (left pane).

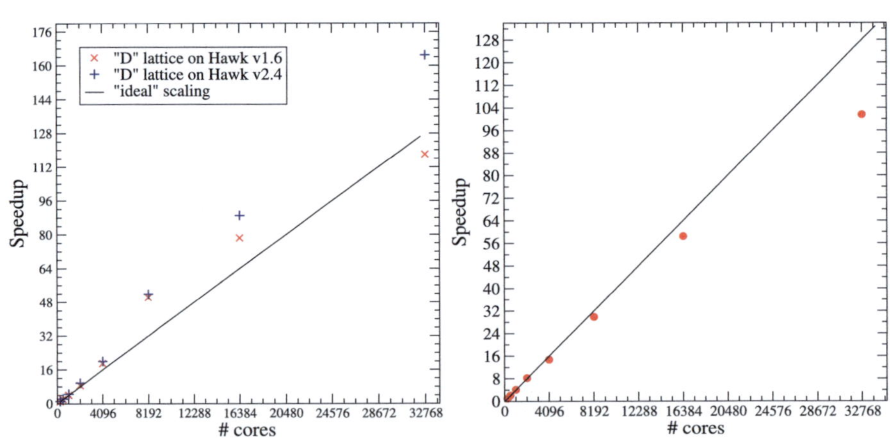

Fig. 1 Lhs pane: Strong-scaling behaviour of `openQCD` on Hawk. The plot shows the application of the even-odd preconditioned Wilson-Dirac operator on a 128×64^3 lattice for `openQCD (v1.6)` (pure MPI) and `openQCD (v2.4)` (mixed MPI/OpenMP). Speedup factors are defined relative to 256 cores with `openQCD (v1.6)`. For our workload, the newer code outperforms the older code consistently and a clear indication of hyperscaling is seen in this regime. Rhs pane: Weak scaling with a local volume of 8^4 on Hawk using `openQCD (v2.4)`

The figure illustrates the speedup for `openQCD (v1.6)` in red, using a baseline of 2 nodes. In addition, the improvements resulting from the mixed MPI/OpenMP implementation in `openQCD (v2.4)` are illustrated by plotting the speedups versus the pure MPI code of version 1.6 in blue. All timings refer to the application of the even-odd preconditioned O(a)-improved Wilson-Dirac operator $\hat{D}_{\mathrm{w}}$ to a spinor field (see (1)) which accounts for the largest fraction of the total propagator computation time. For Hawk, the weak-scaling behaviour for a local lattice volume of size 8^4 is shown in the right pane of Fig. 1.

2.1 Code Development

For charm quarks we need an implementation of the RHQ action from [25], which we developed and used previously in [26]. Originally we had implemented our chosen Relativistic Heavy Quark action (the "Tsukuba" action) solely within an independently-maintained version of `openQCD` based on its release version 1.6 allowing us to work with either heavy quarks or light quarks as different runs. This was a reasonable approach as light- and strange-quark perambulators already existed for several of the ensembles we are interested in, so charm quark perambulators needed to be produced independently, i.e. we recomputed the Laplacian eigenvectors in the stochastic-LapH framework that we use.

However, on the newly-generated "X"-ensembles as well as the new "D"-ensemble produced on HAWK the light-, strange-, and charm-quark perambulators all need to be computed, ideally together. This implies that for efficiency we should reuse the Laplacian eigenvectors for both calculations and hence save a significant amount of computer time. The main difference between the vanilla `openQCD` and the Tsukuba variant we have developed have been a complete rewrite of the Dirac operator (for both the full volume and the SAP subspace in both single and double precision) and a change in the clover implementation.

Unfortunately, due to the structure of `openQCD` having separate pre-computed clover-term inverses in memory was a significant challenge as numerous places in the code require a pointer to the address of that memory depending on the global state of a series of complicated lookup tables. Our implementation required duplication of the majority of the clover and Dirac matrix parts as well as the routines that perform quark inversions, in addition to careful alterations to the flags that control what code path is needed. Testing this code was complicated and highly non-trivial; ensuring that the correct routines were updated when the global state changed was tricky as this is very-much not what the code was originally *designed* for. This part of the project turned out to be a significant software-development undertaking and the merging of this new code with the Chroma-LapH framework was also non-trivial. However, it is now complete and we have a code that can do both light, strange, and heavy quark inversions (or any of them independently) utilising the same Laplacian eigenvectors and gauge fields. We estimate that this will lead to about a 15% saving to all of our future runs, allowing for calculations at greater statistical precision at fixed computational cost.

We have also extended this code for the computation to allow for relativistic heavy bottom quarks and determined the action parameters for all of our lattice spacings. The motivation for doing this is discussed in Sect. 3.1 as part of our ongoing heavy-light scattering studies.

2.2 Workflow and Computational Setup

The gauge fields are generated with `openQCD`. The resulting gauge-field configurations are added to the CLS library with redundant central storage at DESY Zeuthen and at Regensburg University. The gauge fields are also copied to GSI and/or Mainz for further processing.

The workflow for the calculation of perambulator matrices is implemented in the `chroma_laph` software built on top of the well-tested USQCD software stack [27], but using instead the inverter from `openQCD`. As part of a recent project on Hawk, the implementation of the KSRLanczos algorithm in `chroma_laph` has been significantly improved to allow for a much larger Krylov subspace.

For parts of this proposal we moved to the most-recent version of `openQCD`, version 2.4. This, for the first time, allows for a hybrid MPI/OpenMP parallelisation strategy. Benchmarks on Hawk indicate that currently the best setup for maximum parallelism consists of 4 OpenMP threads (with no hyperthreading) on each core complex (sharing a lvl 3 cache) of the AMD Rome platform. Consequently all 128 physical cores can be used by combining the OpenMP threads with 32 MPI ranks per node. While this setup never performs worse than the traditional pure MPI runs, significant speedups are only seen for quite large partitions. We therefore opted mostly for pure MPI runs.

For the gauge-field generation on Hawk we used the following setups:

A Local lattice of size 16×4^3 with an MPI-grid of dimensions 8×10^3 and no OpenMP threading.
B Local lattice of size 64×4^3 with an MPI-grid of dimensions 2×10^3 and a thread-local lattice of size 16×4^3.
C Local lattice of size 8^4 with an MPI-grid of dimensions 16×8^3.

Setups A and B result in a total number of 8000 cores and were used for the generation of ensembles X451 and X150. Setup C results in a total of 8192 cores and was used for the generation of ensemble D251.

For the perambulator runs, setups A and C are also used, while some of the proposed runs for relative strange-quark lines also employ the following setup resulting in 4096 cores:

D Local lattice of size 16×6^3 with an MPI-grid of dimensions 8^4.

2.3 Status of the Gauge-Field Generation

Figure 2 shows an overview of the current library of gauge field configurations by the CLS consortium along the trajectory with constant trace of the quark-mass matrix. The ensembles proposed as part of the current project are highlighted in blue and comprise four new ensembles with a pion mass of roughly 280 MeV at varying lattice spacing.

Table 1 shows an overview of the planned gauge-field generation runs. For the "X" ensembles, the generation of 2000 gauge field configurations, corresponding to 8000 molecular dynamics units (MDU) had been proposed. For X150 and X452 at the coarser two lattice spacings this goal has been achieved through runs on Hawk in the current reporting period. Overall these two ensembles turned out to be roughly 15% more expensive than estimated based on thermalization runs performed on the Mogon II cluster at JGU Mainz.

For ensemble X253 at a finer lattice spacing our original plan was to also produce 2000 gauge configurations corresponding to 8000 MDU on Hawk, while another 1000 gauge-field configurations were to be produced by RQCD. Due to better than anticipated throughput, our RQCD colleagues were able to compensate for the slightly higher than anticipated computational cost of the X150 and X452 ensembles, producing a total of 3005 gauge configurations in a single Monte-Carlo chain.

Similarly 2000 gauge configurations for a larger $T \times L^3 = 128 \times 64^3$ lattice at the finer lattice spacing called D251 were planned, with 1000 gauge configurations

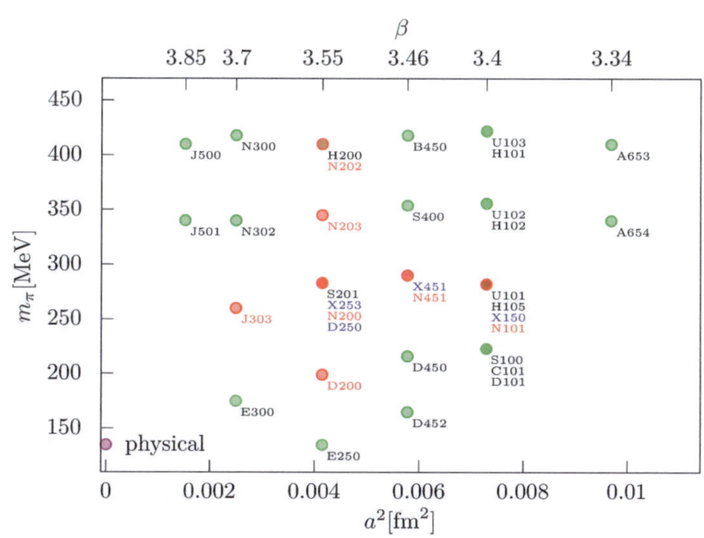

Fig. 2 Overview of the $N_f = 2 + 1$ flavor CLS ensembles with constant trace of the quark-mass matrix. The x-axis indicates the squared lattice-spacing, while the y-axis shows the pion mass. Ensembles that are currently used for the project are highlighted in red, while the new ensembles generated with the current allocation are indicated by blue labels

Table 1 Table of gauge configurations produced on Hawk during the current allocation period and Estimated CPU-time required on Hawk. For a description of the ensembles please refer to the text. All runs have been finalized. The number of cores refers to the number of compute cores used, while the partition size is indicated by the second number. Note that the generation of X150 and X451 at coarser lattice spacing turned out a bit more costly than our initial estimates, while D251 turned out to be cheaper. Ensemble X253 has instead been produced by our CLS partners in Regensburg

Task	Type $T \cdot L^3$	#configs produced/proposed	#cores/partition	Mcore-h used/proposed
X150	$128 \cdot 40^3$	2002/2000	8000	10.6/8.8
X451	$128 \cdot 40^3$	2026/2000	8000	9.0/7.9
X253	$128 \cdot 40^3$	0/2000	8000	0/3.5
D251	$128 \cdot 64^3$	910/1000	8192	8.3/10.0

to be produced as part of this project. In total the runs were stopped after producing 2013 gauge configurations in two chains. Of those, 910 were produced on Hawk.

Figure 3 shows the Hamiltonian deficits $\varDelta H$ as well as the Monte Carlo history of the topological charge for ensembles the ensembles described above. The acceptance rate resulting from the history of $\varDelta H$ is $(91.9 \pm 0.5)\%$ for X150, $(95.8 \pm 0.4)\%$ for X452, $(94.4 \pm 0.3)\%$ for X253, and $(87.0 \pm 0.6)\%$ for D251.

The generation of these chains is now complete and the calculation of physics observables will be started at HLRS on Hawk and on the Virgo II cluster at GSI. As an example Fig. 4 shows the effective masses of the kaon and pion on Ensemble X451. Fits at large time separation lead to $m_\pi = 0.11117(25)$ and $m_K = 0.17806(19)$ which are compatible with the values obtained from the existing ensemble N451 with a larger lattice volume of size $T \times L^3 = 128 \times 48^3$. Furthermore the statistical precision of these results indicates that the generated Monte-Carlo chain is of sufficient length.

2.4 *Status of Perambulator Runs*

In addition to the gauge-field generation runs, quark propagators contracted into so-called perambulator matrices will be calculated as part of our current allocation. Due to our code improvements detailed in Sect. 2.1 progress in these calculations has been a bit slower and until the end of the reporting period in March 2023 no significant resources have been spent on calculating perambulators.

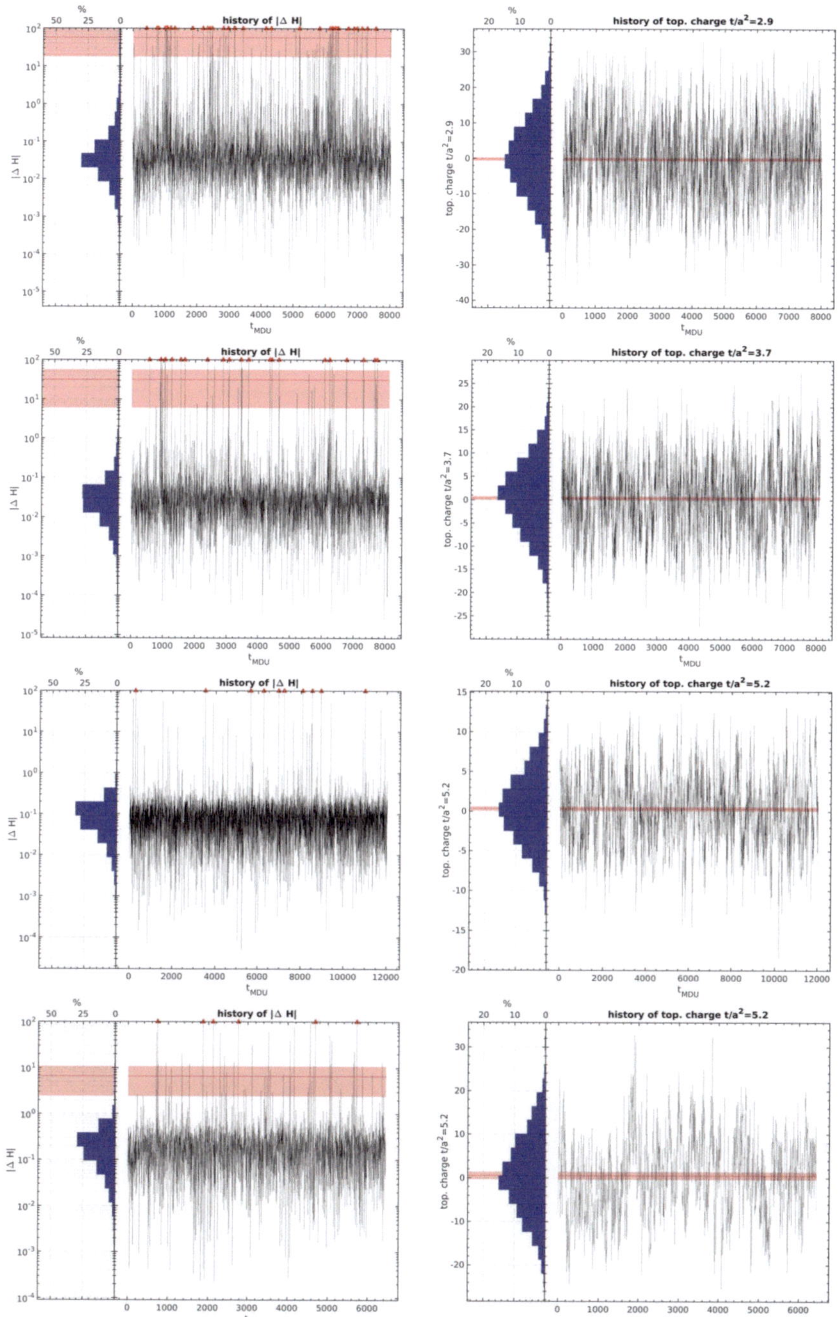

Fig. 3 Monte-Carlo histories of the Hamiltonian deficit ΔH (left) and the total topological charge at finite flow time in the center of the lattice (right) for the ensembles newly produced on Hawk as well as for Ensemble X253. The rows show ensembles X150, X451, X253, and D251 from top to bottom

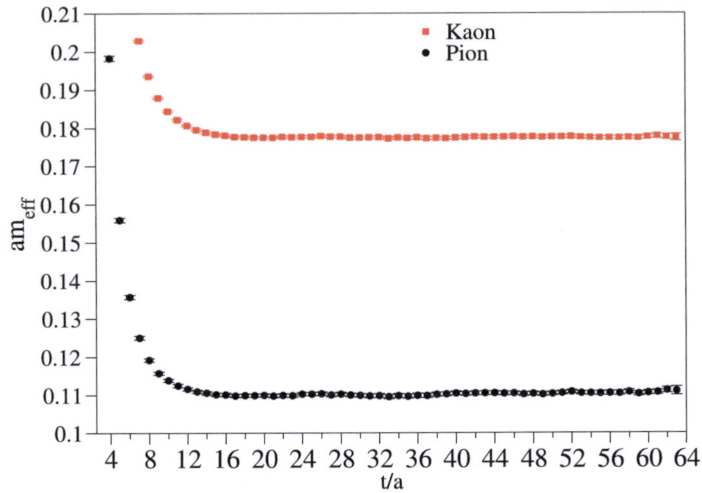

Fig. 4 Pion and kaon effective masses on the newly generated X451 ensemble

3 Spectroscopy and Scattering

While previous studies have created substantial insight into the spectrum of positive-parity heavy-light mesons, precise, modern, and reliable lattice calculations with full control of systematic uncertainties related to the finite lattice spacing and to the light-quark masses used in the simulation are crucial for a meaningful comparison of the heavy-light hadron spectrum to experiment. We report on our efforts to determine QCD bound-states with fully-quantified systematic uncertainties in this section.

3.1 Heavy-Light Mesons with Bottom Quarks

During the current allocation period we decided to extend the scope of our physics objective to also include mesons with a heavy (anti)bottom quark. This work was enabled by our novel neural net based tuning strategy [26] for relativistic heavy-quark actions such as [25, 28]. This strategy has been presented in last year's report.

In [15] we instead tuned lattice Non-Relativistic QCD (NRQCD) nonperturbatively using neural networks to investigate exotic B_s mesons (the $J^P = 0^+$ B_{s0} and $J^P = 1^+$ B_{s1}) and potential doubly-heavy doubly-light tetraquark candidates. Some of our results are shown in Figs. 5 and 6.

Figure 5 gives the chiral and infinite-volume extrapolations for our B_{s0} and B_{s1} mesons. Clearly finite-volume effects are significant on our trajectories but we have enough ensembles with varying spatial extents to control them, likewise the chiral extrapolation is well-controlled. Unfortunately, as this was a lattice-NRQCD cal-

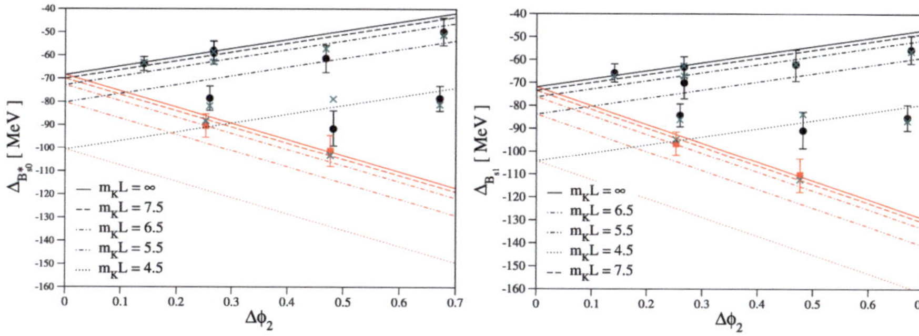

Fig. 5 Chiral and infinite-volume extrapolations for the binding energies of the B_{s0} and B_{s1} mesons with respect to their lowest non-interacting two-meson thresholds, BK and B^*K respectively

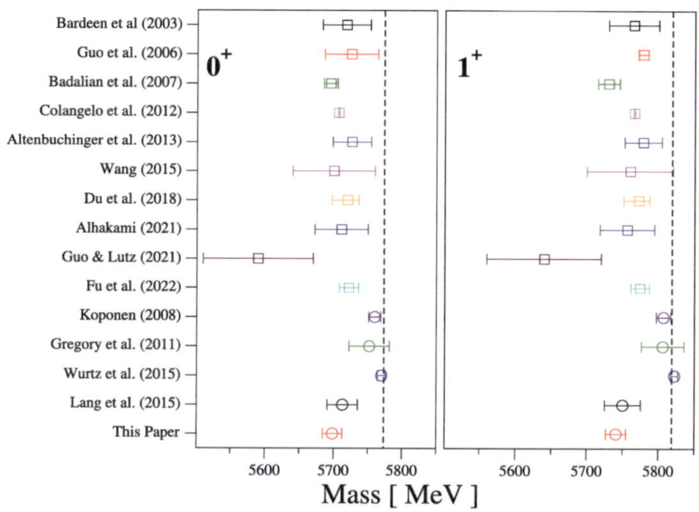

Fig. 6 A comparison of our result for the B_{s0} and B_{s1} mesons compared to other lattice determinations and phenomenological studies

culation no rigorous continuum limit can be taken, nor can the stochastic-LapH framework be used, but as an investigatory tool it is useful.

A comparison of our results to other lattice determinations and phenomenological studies is shown in Fig. 6. Not only do we see consistency with a previous scattering study and several model studies but we obtained a precise determination for these states. There is some tension with earlier lattice studies and future determinations with improved systematics are required to clarify the situation.

The leading uncertainty on our prediction of these states comes from the estimate of discretisation effects. Clearly, an action with relativistic b-quarks that admits a sensible continuum-limit extrapolation is desireable and motivates the extra coding effort discussed in Sect. 2.1 to further investigate these states through the determination of scattering amplitudes with our implementation of a relativistic heavy quark action described in [26].

4 Summary and Outlook

We have reported on the current status of our calculation of the spectrum of heavy-light mesons using Lattice QCD. The extraction of scattering amplitudes through the finite-volume methods first detailed in [12, 13] benefits greatly from using multiple sets of gauge-field configurations with varying spatial volume and otherwise identical parameters. In connection with a similar effort by the RQCD subgroup of the CLS consortium we have therefore embarked on a concerted effort to generate additional gauge ensembles with varying volumes, and we have reported on the progress in this regard.

At the same time our scattering calculations in the D-meson sector are progressing and we hope to have a first physics publication within the next six months. In the meantime we have expanded the scope of our studies to heavy-light mesons and tetraquark states with bottom quarks [15], where we made predictions for QCD bound states not yet observed in experiment. To further reduce the uncertainty estimate from the use of Lattice NRQCD and the lattice discretization for the prediction of these states we hope to employ an RHQ action for b-quarks to study B-meson scattering.

Acknowledgements D.M. acknowledges funding by the Heisenberg Programme of the Deutsche Forschungsgemeinschaft (DFG, German Research Foundation)—project number 454605793; This work was supported in part by: the U.S. National Science Foundation under awards PHY-1913158 and PHY-2209167 (CJM). Calculations for this project were partly performed on the HPC cluster "Mogon II" at JGU Mainz. This research was supported in part by the cluster computing resource provided by the IT Division at the GSI Helmholtzzentrum für Schwerionenforschung, Darmstadt, Germany (HPC cluster Virgo). The simulations were partly performed on the national supercomputer HPE Apollo Hawk at the High Performance Computing Center Stuttgart (HLRS) under acronym RinQCD4PANDA.

References

1. Brambilla, N., et al.: QCD and strongly coupled gauge theories: challenges and perspectives. Eur. Phys. J. C **74**(10), 2981 (2014). https://doi.org/10.1140/epjc/s10052-014-2981-5
2. Lutz, M.F.M., et al.: Resonances in QCD. Nucl. Phys. A **948**, 93 (2016). https://doi.org/10.1016/j.nuclphysa.2016.01.070
3. Aaij, R., et al.: Observation of $J/\psi p$ resonances consistent with pentaquark states in $\Lambda_b^0 \to J/\psi K^- p$ decays. Phys. Rev. Lett. **115**, 072001 (2015). https://doi.org/10.1103/PhysRevLett.115.072001
4. Aaij, R., et al.: Observation of a narrow pentaquark state, $P_c(4312)^+$, and of two-peak structure of the $P_c(4450)^+$. Phys. Rev. Lett. **122**(22), 222001 (2019). https://doi.org/10.1103/PhysRevLett.122.222001

5. Lutz, M.F.M., Heo, Y., Guo, X.Y.: On the convergence of the chiral expansion for the baryon ground-state masses. Nucl. Phys. A **977**, 146 (2018). https://doi.org/10.1016/j.nuclphysa.2018.05.007
6. Guo, X.Y., Lutz, M.F.M.: On light vector mesons and chiral SU(3) extrapolations. Nucl. Phys. A **988**, 48 (2019). https://doi.org/10.1016/j.nuclphysa.2019.02.007
7. Guo, X.Y., Heo, Y., Lutz, M.F.M.: On a first order transition in QCD with up, down and strange quarks. Eur. Phys. J. C **80**(3), 260 (2020). https://doi.org/10.1140/epjc/s10052-020-7818-9
8. Lutz, M.F.M., Sauerwein, U., Timmermans, R.G.E.: On the axial-vector form factor of the nucleon and chiral symmetry. Eur. Phys. J. C **80**(9), 844 (2020). https://doi.org/10.1140/epjc/s10052-020-8417-5
9. Sauerwein, U., Lutz, M.F.M., Timmermans, R.G.E.: Axial-vector form factors of the baryon octet and chiral symmetry. Phys. Rev. D **105**(5), 054005 (2022). https://doi.org/10.1103/PhysRevD.105.054005
10. Korpa, C.L., Lutz, M.F.M., Guo, X.Y., Heo, Y.: Coupled-channel system with anomalous thresholds and unitarity. Phys. Rev. D **107**(3), L031505 (2023). https://doi.org/10.1103/PhysRevD.107.L031505
11. Lutz, M.F.M., Guo, X.Y., Heo, Y., Korpa, C.L.: Coupled-channel dynamics with chiral long-range forces in the open-charm sector of QCD. Phys. Rev. D **106**(11), 114038 (2022). https://doi.org/10.1103/PhysRevD.106.114038
12. Lüscher, M.: Volume dependence of the energy spectrum in massive quantum field theories. 2. Scattering states. Commun. Math. Phys. **105**, 153 (1986). https://doi.org/10.1007/BF01211097
13. Lüscher, M.: Two particle states on a torus and their relation to the scattering matrix. Nucl. Phys. B **354**, 531 (1991). https://doi.org/10.1016/0550-3213(91)90366-6
14. Briceno, R.A., Dudek, J.J., Young, R.D.: Scattering processes and resonances from lattice QCD. Rev. Mod. Phys. **90**(2), 025001 (2018). https://doi.org/10.1103/RevModPhys.90.025001
15. Hudspith, R.J., Mohler, D.: Exotic tetraquark states with two b^- quarks and JP=0+ and 1+ Bs states in a nonperturbatively tuned lattice NRQCD setup. Phys. Rev. D **107**(11), 114510 (2023). https://doi.org/10.1103/PhysRevD.107.114510
16. Bruno, M., et al.: Simulation of QCD with $N_f = 2 + 1$ flavors of non-perturbatively improved Wilson fermions. JHEP **02**, 043 (2015). https://doi.org/10.1007/JHEP02(2015)043
17. Bali, G.S., Collins, S., Georg, P., Jenkins, D., Korcyl, P., Schäfer, A., Scholz, E.E., Simeth, J., Söldner, W., Weishäupl, S.: Scale setting and the light baryon spectrum in $N_f = 2 + 1$ QCD with Wilson fermions. JHEP **05**, 035 (2023). https://doi.org/10.1007/JHEP05(2023)035
18. Lüscher, M.: Local coherence and deflation of the low quark modes in lattice QCD. JHEP **07**, 081 (2007). https://doi.org/10.1088/1126-6708/2007/07/081
19. Lüscher, M.: Deflation acceleration of lattice QCD simulations. JHEP **12**, 011 (2007). https://doi.org/10.1088/1126-6708/2007/12/011
20. Lüscher, M., Schaefer, S.: Lattice QCD with open boundary conditions and twisted-mass reweighting. Comput. Phys. Commun. **184**, 519 (2013). https://doi.org/10.1016/j.cpc.2012.10.003
21. Morningstar, C., Bulava, J., Foley, J., Juge, K.J., Lenkner, D., Peardon, M., Wong, C.H.: Improved stochastic estimation of quark propagation with Laplacian Heaviside smearing in lattice QCD. Phys. Rev. D **83**, 114505 (2011). https://doi.org/10.1103/PhysRevD.83.114505
22. Peardon, M., Bulava, J., Foley, J., Morningstar, C., Dudek, J., Edwards, R.G., Joo, B., Lin, H.W., Richards, D.G., Juge, K.J.: A Novel quark-field creation operator construction for hadronic physics in lattice QCD. Phys. Rev. D **80**, 054506 (2009). https://doi.org/10.1103/PhysRevD.80.054506
23. Wilcox, W.: Noise methods for flavor singlet quantities (1999)
24. Foley, J., Jimmy Juge, K., O'Cais, A., Peardon, M., Ryan, S.M., Skullerud, J.I.: Practical all-to-all propagators for lattice QCD. Comput. Phys. Commun. **172**, 145 (2005). https://doi.org/10.1016/j.cpc.2005.06.008
25. Aoki, S., Kuramashi, Y., Tominaga, S.I.: Relativistic heavy quarks on the lattice. Prog. Theor. Phys. **109**, 383 (2003). https://doi.org/10.1143/PTP.109.383

26. Hudspith, R.J., Mohler, D.: A fully non-perturbative charm-quark tuning using machine learning (2021)
27. Edwards, R.G., Joo, B.: The Chroma software system for lattice QCD. Nucl. Phys. Proc. Suppl. **140**, 832 (2005). https://doi.org/10.1016/j.nuclphysbps.2004.11.254. [,832(2004)]
28. El-Khadra, A.X., Kronfeld, A.S., Mackenzie, P.B.: Massive fermions in lattice gauge theory. Phys. Rev. D **55**, 3933 (1997). https://doi.org/10.1103/PhysRevD.55.3933

QCD Simulations with Stabilised Wilson Fermions, Towards the Physical Pion Mass at a Fine Lattice Spacing

Rocco Francesco Basta, Francesca Cuteri, Anthony Francis, Patrick Fritzsch, Giovanni Pederiva, Antonio Rago, Andrea Shindler, Andre Walker-Loud, and Savvas Zafeiropoulos

Abstract Stabilised Wilson Fermions (SWF) are a new avenue for QCD calculations with Wilson-type fermions. The OpenLat initiative is currently the only collaboration following this new line of research. With resources in the US, Germany, France and Finland OpenLat has been able to produce gauge field ensembles of 2+1 flavours of SWF across a wide range of lattice spacings $a = 0.055, 0.064, 0.077, 0.094, 0.12$ fm, with HLRS providing the most important contribution at the finest resolution of

R. F. Basta · F. Cuteri (✉)
Goethe Universität Frankfurt am Main, Frankfurt am Main, Germany
e-mail: cuteri@itp.uni-frankfurt.de

R. F. Basta
e-mail: basta@itp.uni-frankfurt.de

A. Francis
Institute of Physics, National Yang Ming Chiao Tung University, Hsinchu, Taiwan
e-mail: anthony.francis@cern.ch; afrancis@nycu.edu.tw

P. Fritzsch
School of Mathematics, Trinity College Dublin, Dublin 2, Ireland
e-mail: fritzscp@tcd.ie

G. Pederiva
Jülich Supercomputing Centre, Forschungszentrum Jülich GmbH, Jülich, Germany
e-mail: g.pederiva@fz-juelich.de

A. Rago
Quantum Theory Center and IMADA, University of Southern Denmark, Odense, Denmark
e-mail: rago@qtc.sdu.dk

A. Shindler
Institute for Theoretical Particle Physics and Cosmology, RWTH Aachen University, Aachen, Germany
e-mail: shindler@physik.rwth-aachen.de

A. Walker-Loud
Nuclear Science Division, Lawrence Berkeley National Laboratory, Berkeley, USA
e-mail: walkloud@lbl.gov

S. Zafeiropoulos
Aix Marseille University, Université de Toulon, CNRS, CPT, Marseille, France
e-mail: savvas.zafeiropoulos@cpt.univ-mrs.fr

T. Ludwig et al. (eds.), *High Performance Computing in Science and Engineering '23*,
https://doi.org/10.1007/978-3-031-91312-9_6

$a = 0.055$ fm so far. Matching the physical parameters as well as algorithmic tuning is increasingly challenging with decreasing lattice spacing as one has to switch to open boundary conditions to overcome the well-known topological freezing problem. We report on progress in simulating QCD at $a = 0.055$ fm and decreasing physical pion masses, as well on first observable measurements on such lattices.

1 Introduction

Quantum Chromodynamics (QCD) is the established theory of strong interactions between quarks and gluons—the building blocks of all atomic nuclei—and since its first days it has been capable of explaining an abundance of phenomena that take place in high-energy colliding experiments. However, in the low-energy regime the theory becomes strongly coupled and perturbative techniques cease to apply. The only mathematically rigorous definition and regularisation of the theory is the one discretised on a spacetime lattice which beyond the well-posedness of the theory opens new avenues by rendering the theory amenable to numerical simulations. Numerical *ab initio* simulations of the theory on a lattice provide such a nonperturbative approach and are widely employed. However, the discretisation of the theory is not-unique and any discretisation is valid as long as it respects the basic principles of quantum field theory and yields the correct continuum theory in the appropriate limit. The non-uniqueness of the lattice discretisation implies the existence of different choices, with an impact on the cost and stability of simulations, as well as on the reliability of results when it comes to coarse lattices (large cutoff effects), discretisations with artifacts in their phase diagrams, etc. Improved actions have been designed with the purpose of reducing discretisation effects and improving the approach to the continuum limit by the introduction of counterterms in the discretised action, whose coefficients are tuned to cancel $\mathcal{O}(a)$ terms in the cutoff series expansion, as in the case of Wilson-Clover fermions (WCF) [1].

Lattice QCD simulations, mainly employ the Hybrid Monte Carlo (HMC) algorithm in which a Markov Chain of gauge field configurations is generated by employing a sequence of algorithms involving the generation of auxiliary fields, a *molecular-dynamics* (MD) evolution in the phase space of the gauge field and an *accept-reject step* to make the algorithm exact. These well-known techniques are potentially subject to issues such as integrator instabilities, unbounded fluctuations of the driving forces for the MD, significance losses. Specific fermion actions also come with specific drawbacks that limit the accessible (to stable simulations) parameter space, w.r.t. lattice spacings, volumes and quark masses. With the WCF, for instance, it so happens that: Without chiral symmetry the lowest eigenvalue of the Dirac operator is not protected from taking arbitrarily small values (with more severe consequences, the coarser the lattice spacing and/or the lighter the pion mass); Without automatic $\mathcal{O}(a)$ improvement observables often require rather fine a and entail, therefore, difficulties due to the freezing of topology and critical slowing down.

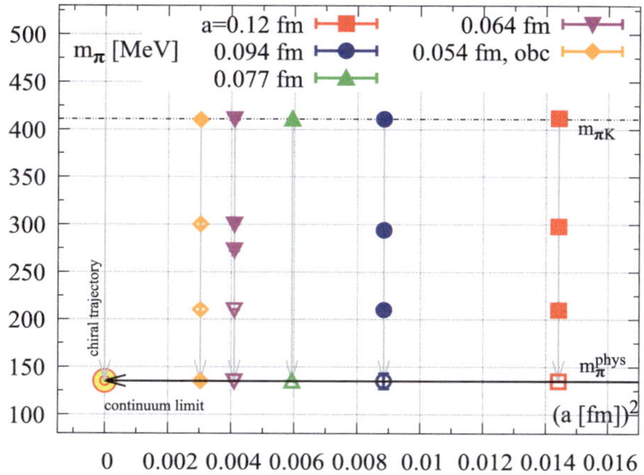

Fig. 1 An overview of all ensembles in a m_π *versus* lattice spacing plane. Included are ensembles in different phases: Thermalisation/tuning (open symbols) and production (solid symbols)

Efforts invested in devising techniques meant to improve stability, while retaining the advantages of the cutoff improvements, can lead to important progress for the efficacy, stability and correctness of simulations of Wilson fermions. Techniques in this direction have been proposed and developed in Ref. [2]: They range from algorithmic improvements like employing stochastic molecular dynamics (SMD) in place of the HMC, and a uniform norm for solver stopping criterion, to an alternative fermion discretisation of the theory: the stabilised Wilson fermion (SWF) discretisation. Such improvements enable stable calculations in parameter regimes in terms of volumes, masses and lattice spacings, which would otherwise be inaccessible.

SWF are proving effective in removing the large volume limitation within the *master-field* framework [3, 4]. The investigations covered in this manuscript aim, instead, at showing continuum scaling and chiral (relative am_q effects) properties of SWF in the traditional volume setup [5]. The opportunity to work at smaller masses on coarser lattice spacings makes SWF attractive for nuclear/nucleon applications, $(g-2)_\mu$, and several other calculations in the FLAG report. The *OPEN LATtice initiative* is enabling the present and further studies through the production and sharing of ever finer ensembles at *ever physical* quark masses (see Fig. 1).

1.1 *The Stabilised Wilson Fermion Toolkit*

Besides established techniques implemented within the open source software package `openQCD-2.0` [6], such as SAP, local deflation, multi-grid, mass-preconditioning, multiple time-scale integrators etc., in the SWF simulation setup a toolkit of measures, some of which designed to improve some aspects of the fermion discretisation and some others meant to improve the algorithmic stability during gauge generation.

1.2 The Exponentiated Clover Action

The SWF toolkit includes a new fermion action resulting from a modification of the WCF action. Within the latter the $\mathcal{O}(a)$-improved Dirac operator takes the form

$$D = \frac{1}{2}\Big[\gamma_\mu\Big(\nabla_\mu^* + \nabla_\mu - a\nabla_\mu^*\nabla_\mu\Big)\Big] + c_{SW}\frac{i}{4}\sigma_{\mu\nu}\hat{F}_{\mu\nu} + m_0 \ , \tag{1}$$

resulting, after even/odd preconditioning, into $\hat{D} = D_{ee} - D_{eo}(D_{oo})^{-1}D_{oe}$, with a diagonal part ($M_0 = 4 + m_0$) given by

$$D_{ee} + D_{oo} = M_0 + c_{SW}\frac{i}{4}\sigma_{\mu\nu}\hat{F}_{\mu\nu} \ . \tag{2}$$

Arbitrarily small eigenvalues can occur as the clover term can saturate the bound $\|\frac{i}{4}\sigma_{\mu\nu}\hat{F}_{\mu\nu}\|_2 \leq 3$ while the clover coefficient c_{SW} is one at tree-level and then grows monotonically with g_0^2. Furthermore, the positive and negative eigenvalues of the clover term are equally distributed. Taken together this makes the above statement more precise as we see that D_{oo} is not protected from arbitrarily small eigenvalues. The probability to encounter arbitrarily small eigenvalues of D_{oo} is larger, the smaller the quark masses, and/or the coarser/larger the lattice.

The exponentiated (bounded from below by construction) clover term [2]

$$D_{ee} + D_{oo} = M_0 + c_{SW}\frac{i}{4}\sigma_{\mu\nu}\hat{F}_{\mu\nu} \ \rightarrow \ M_0 \exp\left[\frac{c_{SW}}{M_0}\frac{i}{4}\sigma_{\mu\nu}\hat{F}_{\mu\nu}\right] \ , \tag{3}$$

was suggested as a remedy to this issue.[1] The SWF action preserves the perturbative Symanzik expansion and the invertibility of the clover term is guaranteed.

1.3 Improving Algorithmic Stability

In the gauge field generation process, an increased stability in the MD evolution, through an effective reduction of unbounded energy violations $|\delta H| \gg 1$, and a reduction of autocorrelation times, can be achieved [7, 8] by resorting to the SMD algorithm as an alternative to the most commonly employed HMC one. Comparatively larger jumps in configuration space can occur with the latter (e.g. due to accumulated integration errors) leading the evolution astray w.r.t the target distribution and therefore requiring short re-thermalisations. An SMD update cycle, with

[1] The Dirac operator can still exhibit exceptionally small eigenvalues, but their source will not be the clover term. In this sense, the SWF only alleviates this problem, for certain parameter regions.

$\pi(x,\mu)$, $\phi(x)$ and $U(x,\mu)$ the momentum, pseudofermion, and gauge fields, respectively,[2] is schematically given by:

1. Random field rotations to refresh $\pi(x,\mu)$, $\phi(x)$ (v and η normal distributed)

$$\pi \to c_1\pi + c_2 v, \qquad\qquad \phi \to c_1\phi + c_2 D^\dagger\eta,$$

   with $c_1^2 + c_2^2 = 1$, $c_1 = e^{-\epsilon\gamma}$, $\epsilon =$ MD integration time, $\gamma =$ friction parameter
2. Short MD evolution
3. Accept/reject-step (exact algorithm)
4. Repeat ⭯

Improved stability is, partly, also a result of using for the solver stopping criterion a volume-independent norm $\|\eta\|_\infty$ that prevents precision losses from local effects:

$$\|\eta - D\tilde{\psi}\|_2 \leq w\|\eta\|_2, \qquad \|\eta\|_2 = \Big(\sum_x (\eta(x), \eta(x))\Big)^{1/2} \propto \sqrt{V}$$

$$\|\eta\|_\infty = \sup_x \|\eta\|_2, \quad V\text{-independent}$$

Finally, quadruple precision—available in `openQCD-2.0`—is adopted to limit the effects of the accumulation of numerical errors for global sums due to the fact that for the global accept/reject step $\delta H \propto \epsilon^P\sqrt{V}$.

2 Computational Setup

We address full $N_f = 2+1$ QCD and to set the scale we use the gradient flow time criterion and [9] the gradient flow time $\sqrt{8t_0} = 0.414(5)$fm.

As a first step, a **tuning of the SWF clover coefficient** is in order. For $N_f = 3$ QCD the tuning, according to the procedure devised in Ref. [10], was performed in the range $\beta \geq 3.8$ in Ref. [2] and was later [5] extended to $\beta \geq 3.685$, (corresponding to a coarse $a = 0.12$ fm lattice spacing that is interesting for nuclear physics studies), only to confirm the validity of the interpolation formula

$$c_{sw}(g_0^2) = \frac{1 - 0.325022g_0^2 - 0.0167274g_0^4}{1 - 0.489157g_0^2}, \qquad \beta = 6/g_0^2 \geq 3.685. \tag{4}$$

[2] Note that SMD coincides with the HMC at fixed ϵ and large γ. Moreover, upon rejection the momentum is reversed and trajectories backtrack with a period $t_{acc} = \delta\tau P_{acc}/(1 - P_{acc})$ [7]. This mandates a high, typically in excess of 98%, acceptance rate.

With Wilson fermions special care needs to be taken when choosing, in bare parameter space, the **trajectory to physical masses and the continuum limit** due to the impact on certain observables of renormalisation and $\mathcal{O}(a)$-improvement effects. We adopted the tuning strategy applied by the QCDSF [11] and CLS [12, 13] collaborations of reducing mass-dependent cutoff effects by working at constant trace of the simulated quark mass matrix, starting at the flavour-symmetric point where all pseudoscalar meson masses are degenerate and

$$\mathrm{tr}[M] = m_u + m_d + m_s = N_f \cdot m_\ell = \mathrm{const} \tag{5}$$

depends on a single mass parameter. As physical input we use the ground state masses of the pion and kaon. Tuning m_ℓ amounts to matching the meson mass combination[3]

$$m_{\pi K}^2 = \frac{2}{3}\left(m_K^2 + m_\pi^2/2\right) \equiv m_{\pi K}^2|_{\mathrm{phys}} \tag{6}$$

to its physical value. With the starting point tuned, the masses between strange and the degenerate light isospin doublet $m_s \neq m_u = m_d$ is split by keeping $\mathrm{tr}[M] = \mathrm{const}$. When the pion mass is decreased to its physical value, the kaon mass also approaches its physical value as a result.

When choosing the input for matching the lattice theory to experiment, we have to subtract small (perturbative) effects arising from strong isospin breaking and electromagnetic effects in physical hadron masses. We follow the procedure outlined in [14], leading to input values (in MeV)

$$m_\pi|_{\mathrm{phys}} = 134.8(3), \quad m_K|_{\mathrm{phys}} = 494.2(3) \quad \Rightarrow \quad m_{\pi K}|_{\mathrm{phys}} = 410.9(2)\,. \tag{7}$$

Once the continuum/chiral trajectory is tuned as described, the generation of large volume gauge ensembles with well defined physical mass and continuum limits is enabled (see Fig. 1).

3 SWF Gauge Generation Workflow

3.1 Validation Observables

We adopt a production protocol for the generation of gauge configurations, that consists in monitoring standard observables and performing sanity checks every $\mathcal{O}(10)$ SMD cycles to verify that the generation is stable and the trajectory to physical masses and the continuum limit is accurately determined. In particular, we look at the:

[3] This definition is motivated by leading order (with higher order contributions known to be reasonably small) chiral perturbation theory, from which it is known that $m_{\pi K}^2 \propto \mathrm{tr}[M]$.

plaquette, topological charge, ΔH distribution, reweighting factors, distributions of the lowest eigenvalue λ of $\sqrt{D^\dagger D}$, and simple spectroscopic observables.

In particular, our checks are meant to ensure that:

1. $\phi_4 = 8t_0(m_K^2 + m_\pi^2/2) = 1.115$ within 0.5%, with an error of max. 1σ.
2. The total reweighting factor fluctuations are mild, and ideally below 5%.
3. The SMD step distance $\delta\tau$ maximises the backtracking period.
4. The distribution of δH matches the one set by the acceptance rate.
5. The distribution of the lowest $\sqrt{D^\dagger D}$ eigenvalue is well-behaved & gapped.
6. The distribution of the lower and upper bounds of the spectral gap for the strange quark are within the input ranges, and the degree of the Zolotarev is sufficiently high, $12(V/2)\delta^2 < 10^{-4}$ [6].
7. There is no significant loss of precision caused by unbalanced contributions to the total action that might drive instabilities in the evolution.
8. The distribution of the flowed topological charge is symmetric around zero with no signs of metastability.

The last two items complement the list we made in earlier references [5] and were very relevant for the thermalisations we run on HAWK: Over the several simulated lattice sizes (see Sect. 3.4) we changed the number, and rearranged the distribution, of poles for the rational approximations based on our criteria Items 6 and 7, and we added one more thermalisation stage on a further intermediate volume size after observing a metastability that prevented the distribution for the topological charge from being symmetric w.r.t. zero over a long range of updates as we demand according to Item 8.

We estimate the distance between two configurations labelled as independent based on the autocorrelation time of the topological charge Q computed via the gradient flow. The boundary conditions are changed to open boundaries (OBC) upon significant increase in the autocorrelation time of Q signalling a possible freezing of topology. We check that there are no visible thermalisation effects in all of the above as well as in the plaquette, $t^2\langle E\rangle$, PP and PA correlators.

3.2 *Tuning and Production*

There are two separate stages in our gauge ensemble generation process: The *tuning* and the *production* stage. In the tuning stage thermalisation is performed and the algorithm's parameters are adjusted over a large number of SMD cycles. Once a stable setup is found, it is run for 100 independent configurations. With all quality criteria listed in Sect. 3.1 fulfilled in this final thermalisation phase, the run is considered ready for production. An overview plot of the existing and planned ensembles is

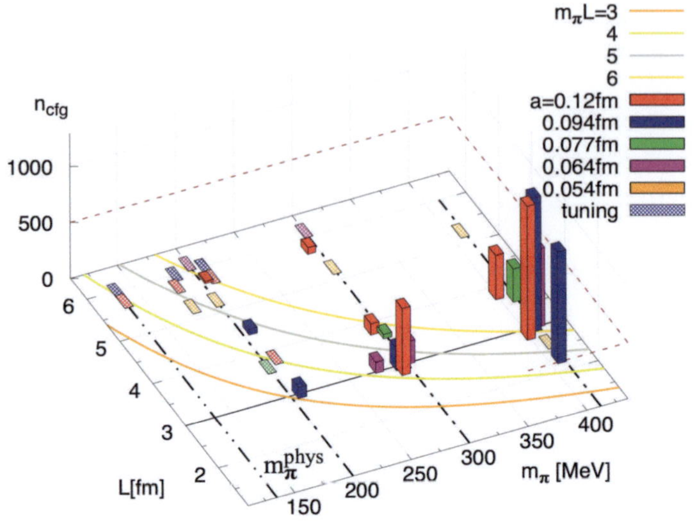

Fig. 2 Starting from the original simulations with stabilised Wilson fermions in Ref. [2], the OPEN LATtice initiative has already been able to make further progress in stable gauge generation towards the continuum limit at the flavour symmetric point ($m_\pi \simeq 410$ MeV) as well as towards the physical point ($m_\pi^{\text{phys.}} = 135$ MeV) at coarse lattice spacing (above 0.08 fm). In our tuning set-up we were able to establish two points at the target physical mass, while also adding more lattice spacings and starting their tuning down to the physical point. The colored solid lines indicate lines of constant $m_\pi L \in [3 - 6]$

given in Fig. 2 in the space of the pion mass, the physical volume and the number of available configurations, and with the lattice spacings distinguished by the colors. Colored bands depicting the regimes $m_\pi L = 3 - 4$, $4 - 5$ and $5 - 6$ are given for reference.

We initially focused on generating ensembles, with periodic/anti-periodic boundary conditions, at lattice spacing $a \in \{0.064, 0.077, 0.094, 0.12\}$ fm. The flavor symmetric point at $m_\pi \sim 412$ MeV has been tuned in all those cases.

The resources that we were granted on HAWK then made possible, among other things, the tuning of the algorithm at the flavor-symmetric point for a target lattice spacing of $a = 0.055$ fm with OBC. The chosen value of the lattice spacing is important because it enables: Heavy-quark physics, better control of cutoff effects for nEDM studies, larger momenta in PDF studies, larger values of the isospin chemical potential before saturation kicks in, etc.[4]

Further ensembles that are being tuned/produced to enable continuum limits at the same values of m_π (see Fig. 1), towards the physical value, at $m_\pi \simeq 300$ MeV and $m_\pi \simeq 210$ MeV. With these ensembles at hand, the next step is to reduce the pion masses.

[4] In all these applications intermediate results at a larger-than-physical pion mass of ~225 MeV are helping, or will help, for continuum extrapolations as well as for chiral extrapolations.

The line at $a = 0.094$ fm is where we reached the lowest available mass so far (aware of the challenges in aiming at the physical pion mass, particularly on coarse lattices [15]).

3.3 Aiming at the Physical Point

With our HAWK contingent we decided to exploit the good scalability of our code on the HAWK platform not only to decrease the value of the lattice spacing down to $a = 0.055$ fm, but also to target our goal to simulate at physical pion mass.[5] In our community it is a well-established fact that simulations below a lattice spacing of about $a = 0.06$ fm suffer the topological freezing problem, i.e., the tunnelling rate between topological sectors breaks down and the algorithm gets stuck in one or a few topological sectors instead of sampling a Gaussian-like distribution. Simulations with open (Neumann) boundary conditions for the gauge fields circumvent to a certain degree this (formerly long-standing) issue and need to be employed to produce correctly sampled gauge ensembles; otherwise one obtains misleading ensemble averages (physics results). As open boundary conditions break physical translation invariance in time-direction, one has to fix the length of the time-direction to the envisaged target. The latter should take into account the physical mass gap of the simulation (pion mass) and an additional 1.5 fm distance from each of the 2 time boundaries (to avoid boundary effects in physical observables). In the past, collaborations chose to set the physical time extent of such simulations w.r.t. the simulated pion mass, requiring a new chain of lattices for each mass, from the symmetric point down to the physical point. Instead, we decided to use the same time-extent for all pion masses such that the target is fixed by the lightest mass, e.g. that of the physical pion. In that case, there is a net gain in effective statistics for many observables at all, but the lightest, masses and only a re-thermalisation in mass (and possibly changing the periodic spatial extent) is required. The latter, however, is part of an optimised thermalisation strategy anyway. The (physical) spatial extent L must be set by requiring $Lm_\pi \geq 4.0$ and $L \gg 3\,\text{fm}$, and T must host an additional 3 fm with open boundary conditions, leading to the requirement $T/a = 192$ at this small lattice spacing when targeting physical pions ($m_\pi = 135\,\text{MeV}$).

With our allocation of **95 Million core-hours**, we managed to scale up the volume in our simulations at the flavor symmetric point up to the target volume for the physical point.[6] Despite the added difficulty of directly scaling the volume up as described, we successfully tuned our simulation algorithms for stability according to our production protocol (see Sect. 3.1) for the finest lattice and the largest volume ever considered within our initiative (details in Sect. 3.4). This represents crucial

[5] Aiming at reaching its production stage in the scope of a future application while extracting physics on the ensembles produced, as an intermediate steps.

[6] Tuning for a new target lattice spacing is always done at the flavor symmetric point first where it is easier done (at any fixed volume) than on other points on the chiral trajectory.

progress towards (1) our plan for a scale setting study that we are working on and plan to publish in the coming months and (2) gauge field generation (at the production stage) in the following two setups:

- Flavour-symmetric point: To identify the bare parameter defining the flavour-symmetric point we made use of a box characterized by $L_s \simeq 3.52$ fm and $L_t \simeq 10.6$ fm in the spatial and temporal directions, respectively, corresponding to a lattice volume of 192×64^3. This setup has the twofold aim of efficiently allowing the parameter tuning and offer a thermalized configuration used to unfold to larger box sizes, that explains the choice of the temporal extent.
- Thermalization towards the physical point: Our long-term strategy encompass a setup with decreasing m_π at fixed tr$[M]$ aiming at $m_\pi^{\text{phys.}} = 135$ MeV and $m_\pi L = 4.8$ on a 192×128^3 volume (corresponding to $L_s \simeq 7.04$ fm and $L_t \simeq 10.6$ fm).

Thermalised starting configurations close to the flavor symmetric point have been made available via the previous allocation on HAWK. Within the scope of the present allocation we aim at generating configurations for $m_\pi = 270$ MeV. Additionally, this setup paves the route towards a simulation of a physical pion.

A physical pion mass ensemble at $a = 0.055$ fm represents a significant step forward in our approach and goes beyond currently available ensembles with WCF [13]. We also want to point out that both in extracting continuum/chiral physics, and in gathering resources for the long term target ensemble, the production costs are being, and will be, shared among projects by several interested parts, at several computing facilities, paving the way for many physics applications within the collaboration, and the wider lattice community (thanks to our FAIR/OPEN science policy that will apply).

3.4 Code Performance and Workflow on HAWK

We have been using on HAWK the openQCD software in its 2.0 version [6], which is released under the GNU public license. The programming language is C and the programming model for parallelisation purposes is pure MPI.

Quoting the author, openQCD *parallelises in 0,1,2,3 or 4 dimensions, depending on what is specified at compilation time. It is highly optimised for machines with current Intel or AMD processors, but should run correctly on any system complying with the ISO C89 (formerly ANSI C) standard, the MPI 1.2 standard and the IEEE 754 standard for floating-point data and arithmetic*. The memory-to-processor bandwidth and the network latencies are the limiting factor for the performance of the program, which is structured to ensure a good average data locality. The machine-specific optimisations include inline AVX/SSE3 assembly kernels for the most computationally intensive part of the code. The code (as typical for lattice QCD applications) requires for most of the cases only next-neighbours node communications ensuring good weak scaling properties. The communication buffers' size scale less than linearly with the

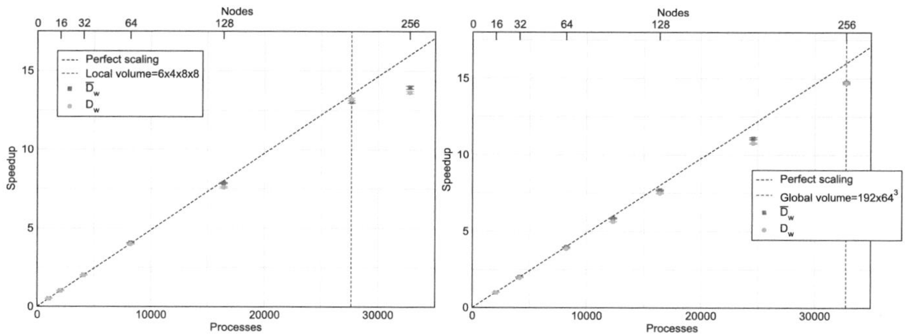

(a) Strong scaling (b) Weak scaling

Fig. 3 Scaling behavior (speedup as function of the number of processes) of openQCD on CPUs at HAWK. For the strong scaling Fig. 3a, the blue line identifies the local lattice size selected for our final run setup. The weak scaling performances Fig. 3b for the local lattice size identified by the strong scaling study. The blue line identifies the volume needed for our intermediate target run (192×64^3)

allocation image (indeed they scale with a power 3/4), while the computational part scales exactly linearly. This ensures a good strong scaling behaviour for large enough local images.

We have run **strong scaling tests** (see Fig. 3a) in a setup with 'maximal network connection' i.e. on a switch with 16 nodes, the minimal setup to obtain a 4-d toroidal topology (2^4), in order to determine what is the optimal local volume per process. Within 16 nodes, we get 16×128 cores at our disposal and we used this as reference point for our scaling test that was extended to use up to 256 nodes.

Each measurement run consisted of 1000 iterations of the application on a random quark field of the exponential $\mathcal{O}(a)$-improved Wilson-Dirac operator D_w and of its even-odd preconditioned counterpart $\hat{D}_w$, in double precision, averaged over 10 stable runs. For the specific case of the strong scaling study, we used a global lattice size of 48×96^3 and scaled the size of the local volume per process accordingly. Our strong scaling analysis of the OpenQCD's performances on HAWK shows a very good scaling performance starting from a local volume of $6 \times 4 \times 8^2$.

We also performed **weak scaling tests** (see Fig. 3b) for the optimal local lattice size based on the results of the strong scaling test in an interconnect-topology-aware manner. In the benchmark phase no physical observable was evaluated, while we estimated the configuration generation cost at the optimal setup identified by the scaling analysis. We ran simulations for the production of our gauge ensemble in a setup with 256 nodes. We let the program generate (MPI I/O) checkpoint configurations that were saved to disk every ~25 SMD cycles.

Two independent streams of simulation are being run, one at the flavour-symmetric point and one at $m_\pi = 270$ MeV. Concurrently to those, stability checks and measurements are being run. While the flavour symmetric point run requires statistics sufficient to comply with our validation requirements, the thermalization towards the ligther pion masses is characterized by short thermalization runs at progressively lower pion mass.

The most computationally-intensive runs performed on HAWK are tuning runs (see Sect. 3.2) for the target gauge ensemble representing the flavor-symmetric counterpart of the physical-pion-mass ensemble.

The typical computational request is of 96 nodes for the generation of the flavour-symmetric runs and 128 for the thermalization runs.

Over the reporting period (April 2022–March 2023), we consumed a total of $\sim$49 Million core-hours i.e. roughly 50% of the allocated budget and we aim at consuming the entirety of our allocation within the time frame of the allocation period. Our projection from the past running months confirms our capability of fully consuming the remaining allocation.

- Within the flavour-symmetric sub-project we managed to generate $\sim$300 thermalized configurations, that have been used to evaluate m_π and t_0 as reported respectively in Figs. 4, 5 and (Fig. 6), leading to the determination of $t_0/a^2 = 7.4156(90)$ and hence $\phi_4 = 1.1477(54)$. We recently reached (for the first time) sufficient statistical accuracy quantifying a mis-tuning w.r.t. the target value of $\phi_4 = 1.12$. We plan to determine a mass-derivative of this observables to identify the target value at the required statistical accuracy. This new set of parameters will be used to generate our final flavour-symmetric ensemble within the remaining allocation period.
- Within the thermalization to lower pion mass subproject we managed to generate two thermalized set of configurations: a first one at $L_\mathrm{s} \simeq 7.04$ fmand $L_\mathrm{t} \simeq 10.6$ fm (192×128^3) for the flavour symmetric point, which has been used as seed for a subsequent thermalization at $m_\pi \sim 270$ MeV, our second set. All the quality indicators defined to validate our runs are well under control (see Sect. 3.1), with the exception of the ϕ_4 mis-tuning that has been address within the other subproject, allowing us to implement a minor thermalization phase at the newly tuned set of bare parameters and start the target production as planned in the scope of the present project.

4 Results

While it is too early to show results for our target parameter setup of $m_\pi^{\mathrm{phys.}} =$ 135 MeV and $a = 0.055$ fm, the OPEN LATtice initiative has presented in Ref. [5] the first results obtained by pooling the resources made available through its members. In particular, with resources available at earlier stages (see Sect. 5) and for significantly less ambitious target parameter setups a large enough statistics (see Fig. 2) over the production phase has been used for extracting physical observables and studying the benefits of SWF w.r.t. continuum scaling and relative am_q effects. Needless to say, the ensemble whose production is being made possible through the resources granted

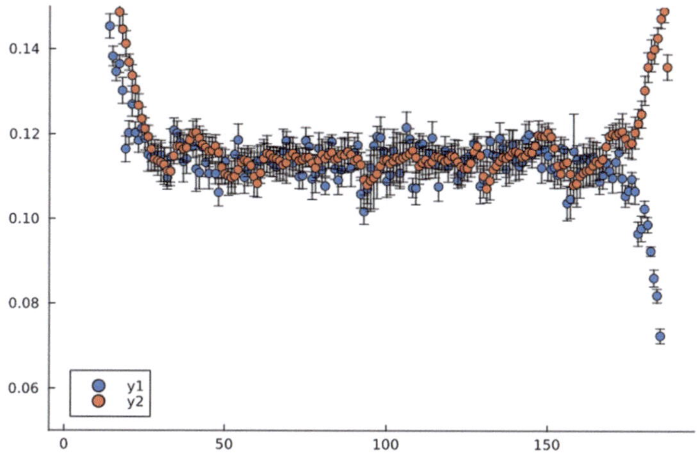

Fig. 4 Determinations of effective masses $am_{\text{eff}}(x_0)$ of the lightest pseudoscalar meson from the A_0P-correlator (blue) and PP-correlator (red). The available statistics allows to extract estimators over a plateau region like $x_0/a \in \{50, 150\}$

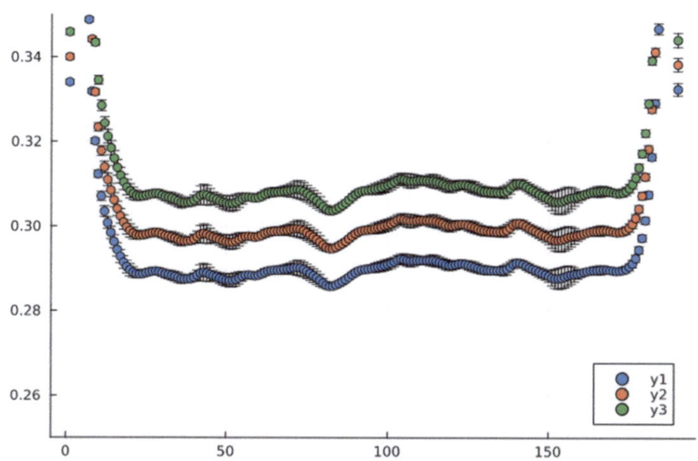

Fig. 5 Gradient-flow observable $t^2E(t, x_0)$ versus Euclidean time x_0 for three different flow times $t = 7.04, 7.36, 7.68$ (blue, orange and green, respectively) used to interpolate to the reference flow-time t_0, implicitly defined by demanding $t^2E(t, x_0)|_{t=t_0} = 0.3$

on HAWK, is crucial both for further corroborating most of the results presented in this section and for enabling new ones. In passing, we note, for instance, that SWF are now also being used for studies in QCD thermodynamics (just like mentioned in the HAWK application), with preliminary results presented at the last lattice conference [16] and currently being extended as finer and finer lattices towards the physical point become available.

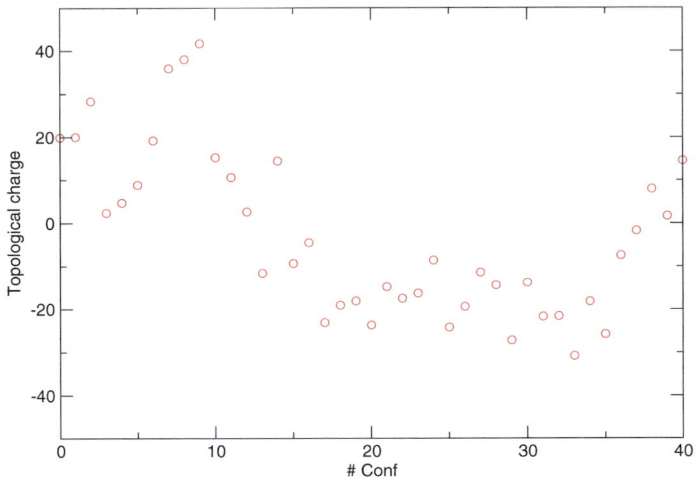

Fig. 6 History of the topological charge for the 192×128^3 flavour symmetric run

4.1 Physics Results Shown Last Summer in the Annual Lattice Conference

In the Lattice 2022 conference we showed [17] our preliminary (at the time) continuum limit results for the light hadron spectrum. We employed a Bayesian framework together with the Akaike information criterion in order to minimize the model dependence and bias of our results. In Ref. [18] the status and plans of our collaboration to generate and share new gauge ensembles using the SWF framework are discussed. In Ref. [19] a novel approach that might alleviate the problem of topology freezing was presented in which some members of our consortium participate. In Ref. [20] the improvement of the axial-vector current in the Schrödinger functional framework was presented. In Ref. [21] results for multinucleon systems were presented and the results using our lattice configurations exhibited very promising continuum scaling properties. In Ref. [16] the first lattice QCD thermodynamics study was performed with our setup and some standard thermodynamical observables were presented. Lastly the Edinburgh group using our lattice configurations (obtained under the open science policy) employed the distillation method in order to pave the way towards the investigation of hadronic D-decays as reported in Ref. [22].

5 Summary and Outlook

SWF build upon recent developments in the generation of gauge fields for use in lattice QCD. The initial studies of [2] show a reduced c_{SW} compared to standard clover simulations, alongside some indication of positive scaling behaviours and benefits in relative am_q effects. In Ref. [5] we also observed good scaling in flow and hadronic observables going towards the continuum. Stable simulations could be run in an extended parameter window. How far SWF do extend such window and what the scaling benefits are is a key motivation for us to study SWF in more depth and to continue their investigation. To make this possible we founded and announced in Ref. [5] a new collaborative effort: the OPEN LATtice initiative. It brings together researchers from different institutes and pools resources at many computing facilities worldwide. Together we want to generate state-of-the-art QCD gauge ensembles for physics applications and share them with the community under the open science philosophy, so that research on SWF can be facilitated and accelerated. In the process we are committed to:

- *Define and uphold quality:* Define standards for control observables, continue to research and improve best practices.
- *Share and maintain repository:* Manage downloads and maintain data integrity, while making all control measurements and data available.
- *Profit from community boosting:* Use resource injections from members and interested/early access parties to expand set of gauges.
- *Grant and enable access:* Ensure that configurations will be made open access with accompanying first publication without further discriminating access control.

We want to exploit the increased stability of the SWF in the coarse regime (especially interesting for nucleon and nuclear physics applications), where lighter pion masses than before seem possible. We also plan on producing ensembles that enable extrapolations from an as broad as possible window in a and m_π to control their systematics. Moreover, our goal is to make all ensembles should available with two volumes in the $m_\pi L \in 4 - 6$ range. The statistics goal for all ensembles is 500 independent configurations in the first, and 1000 in the second, iteration.

Acknowledgements The authors acknowledge support from the HPC computing centres and resources hpc-qcd (CERN), HPE Apollo Hawk (HLRS) under the grant number stabwf/44185, Cori (NERSC), Frontera (TACC) as well as Occigen (CINES), Jean-Zay (IDRIS) and Irène-Joliot-Curie (TGCC) under projects A0080511504 and A0080502271 by GENCI. We acknowledge Martin Lüscher's leading role during the development and implementation of the SWF framework.

References

1. Sheikholeslami, B., Wohlert, R.: Nucl. Phys. B **259**, 572 (1985)
2. Francis, A., Fritzsch, P., Lüscher, M., Rago, A.: Comput. Phys. Commun. **255**, 107355 (2020)
3. Cè, M., Bruno, M., Bulava, J., Francis, A., Fritzsch, P., Green, J.R., Hansen, M.T., Rago, A.: PoS **LATTICE2021**, 383 (2022)
4. Fritzsch, P., Bulava, J., Cè, M., Francis, A., Lüscher, M., Rago, A.: PoS **LATTICE2021**, 465 (2022)
5. Francis, A.S., Cuteri, F., Fritzsch, P., Pederiva, G., Rago, A., Schindler, A., Walker-Loud, A., Zafeiropoulos, S.: PoS **LATTICE2021**, 118 (2022)

6. Lüscher, M.: https://luscher.web.cern.ch/luscher/openQCD/
7. Lüscher, M., Schaefer, S.: JHEP **07**, 036 (2011)
8. Lüscher, M.: EPJ Web Conf. **175**, 01002 (2018)
9. Bruno, M., Korzec, T., Schaefer, S.: Phys. Rev. D **95**(7), 074504 (2017)
10. Lüscher, M., Sint, S., Sommer, R., Weisz, P., Wolff, U.: Nucl. Phys. B **491**, 323 (1997)
11. Bietenholz, W., et al.: Phys. Lett. B **690**, 436 (2010)
12. Bruno, M., et al.: JHEP **02**, 043 (2015)
13. Strassberger, B., et al.: PoS **LATTICE2021**, 135 (2022). https://doi.org/10.22323/1.396.0135
14. Aoki, S., et al.: Eur. Phys. J. C **77**(2), 112 (2017)
15. Mohler, D., Schaefer, S.: Phys. Rev. D **102**(7), 074506 (2020)
16. Basta, R.F., Brandt, B.B., Cuteri, F., Endrődi, G., Francis, A.: PoS **LATTICE2022**, 277 (2023). https://doi.org/10.22323/1.430.0277
17. Cuteri, F., Francis, A., Fritzsch, P., Pederiva, G., Rago, A., Shindler, A., Walker-Loud, A., Zafeiropoulos, S.: PoS **LATTICE2022**, 074 (2023). https://doi.org/10.22323/1.430.0074
18. Cuteri, F., Francis, A.S., Fritzsch, P., Pederiva, G., Rago, A., Shindler, A., Walker-Loud, A., Zafeiropoulos, S.: PoS **LATTICE2022**, 426 (2023). https://doi.org/10.22323/1.430.0426
19. Bruno, M., Cè, M., Francis, A., Green, J.R., Hansen, M., Zafeiropoulos, S.: PoS **LATTICE2022**, 368 (2023). https://doi.org/10.22323/1.430.0368
20. Fritzsch, P., Heitger, J., Kuhlmann, J.T.: PoS **LATTICE2022**, 287 (2023). https://doi.org/10.22323/1.430.0287
21. Green, J.R., Hanlon, A.D., Junnarkar, P.M., Wittig, H.: PoS **LATTICE2022**, 200 (2023). https://doi.org/10.22323/1.430.0200
22. Joswig, F., Erben, F., Hansen, M.T., Lachini, N.P., Portelli, A.: PoS **LATTICE2022**, 063 (2023). https://doi.org/10.22323/1.430.0063

Microphysical Aspects of Binary Neutron Star Mergers

Michail Chabanov, Christian Ecker, Carlo Musolino, Luciano Rezzolla, Samuel Tootle, and Konrad Topolski

Abstract Neutron star mergers represent a unique opportunity to investigate fundamental physics in regimes which are inaccessible in Earth-based laboratories. In this report we present two projects aimed at furthering our understanding of these systems. Firstly, we present a novel implementation of a neutrino transport scheme which will allow us to incorporate weak interactions self-consistently in future simulations. Secondly, we present high resolution simulations aimed at understanding the impact of the initial magnetic-field topology on the exponential amplification in the postmerger phase of neutron star mergers. We also give a brief introduction to the numerical techniques used to model these scenarios and summarise their performance on the HAWK supercomputer system.

1 General-Relativistic Radiative Transport in Binary Mergers

Binary neutron star (BNS) mergers offer us a unique laboratory to explore some of the currently unsolved fundamental questions of astrophysics as well as a way to probe QCD matter in a high density and low temperature regime [1, 2]. They are linked to r-process nucleosynthesis [3] as well as to short gamma-ray bursts [4–6] and, of course, to the emission of gravitational waves (GWs). This makes them the prime candidate system for multi-messenger astrophysics. For this reason a deep theoretical understanding of the properties of these systems has been the subject of many efforts from the astrophysical community for more than a decade. With the advent of advanced third generation and fourth generation gravitational wave detectors [7] we will be able to detect gravitational waves coming from BNS mergers

M. Chabanov (✉) · C. Ecker · C. Musolino · L. Rezzolla · S. Tootle · K. Topolski
Institut für Theoretische Physik, Goethe Universität, Frankfurt am Main, Germany
e-mail: chabanov@itp.uni-frankfurt.de

L. Rezzolla
Frankfurt Institute for Advanced Studies, Frankfurt, Germany

School of Mathematics, Trinity College, Dublin 2, Ireland

T. Ludwig et al. (eds.), *High Performance Computing in Science and Engineering '23*,
https://doi.org/10.1007/978-3-031-91312-9_7

not only during the late inspiral phase, where the dynamics are mostly dictated by gravity with a few corrections coming from the equation of state (EOS) of neutron star matter, but also from the post-merger, where instead the observational signature of the system is expected to be heavily influenced by the EOS. Gravitational waves alone, however, will likely not be enough to provide us with enough information to draw significant conclusions about the state of matter in the cores of neutron stars. For this reason, modeling of BNS mergers has recently started to include more relevant physical effects such as magnetic fields and weak interactions that allow for an accurate investigation of the system's evolution beyond the initial gravity dominated timescale. The inclusion of neutrinos in particular is crucial to be able to model the reprocessing of the initially neutron rich ejected material which leads to the emission of a kilonova [8–10]. Moreover, weak interactions have been postulated to play an important role in the launching of a relativistic jet which is necessary for the emission of a short gamma ray burst (sGRB) [11–13]. Finally, neutrinos play an important role in the compositional and thermal evolution of the disk around the central remnant of the merger, which is a key engine of long-term matter ejection [14–16]. The subject of this project, whose results are available as a preprint [17], was the development of `FIL-M1`, a numerical scheme for the self-consistent evolution of neutrinos in BNS merger simulations. Among all the possible methods of accounting for neutrinos in GRMHD simulations, which range from simple leakage schemes [12, 18, 19] where only local heating/cooling and compositional changes are considered, all the way to full Boltzmann solvers based on MonteCarlo methods [20–22], we chose to implement a grey two moment scheme [23–32], which we believe to be the best compromise between computational affordability and physical consistency currently available. In a grey M1 scheme, the first two moments of the energy averaged radiative transfer equations are evolved with high-resolution shock-capturing methods, and the discretized collisional integral serves as a coupling between the fluid and the radiation. While these equations are formally very similar to the standard GRMHD equations, numerous additional difficulties arise when trying to solve them, and the overcoming of such difficulties was the main focus of this project. Firstly, as for all moment based schemes, an analytic closure has to be given for the first non-evolved moment, which in the case of M1 is the neutrino pressure. This analytic prescription, although it is arbitrary to some degree, influences the solution of the M1 equations significantly and therefore needs to be carefully picked (see e.g. [33] for a comprehensive comparison of available prescriptions). Moreover, the closure requires a nonlinear rootfinding procedure to be evaluated, making it computationally expensive and potentially prone to failure. For these reasons, in `FIL-M1` we implemented multiple closures and made it easy to extend the code to support more. An example of a code test which directly showcases how the closure can impact the solution to the M1 equations is shown in Fig. 1. In this test a spherical region of constant emissivity $1 = Q = \kappa_a$ is set up in vacuum and left to evolve with `FIL-M1` and with three different closures. Since this simple setup is analytically solvable the numerical solution can be compared to the correct one. In particular, one can see that the true solution is only recovered when the closure adopted in `FIL-M1` is a fit to the true closure for this setup which is accurate to within 1% and was given by [33].

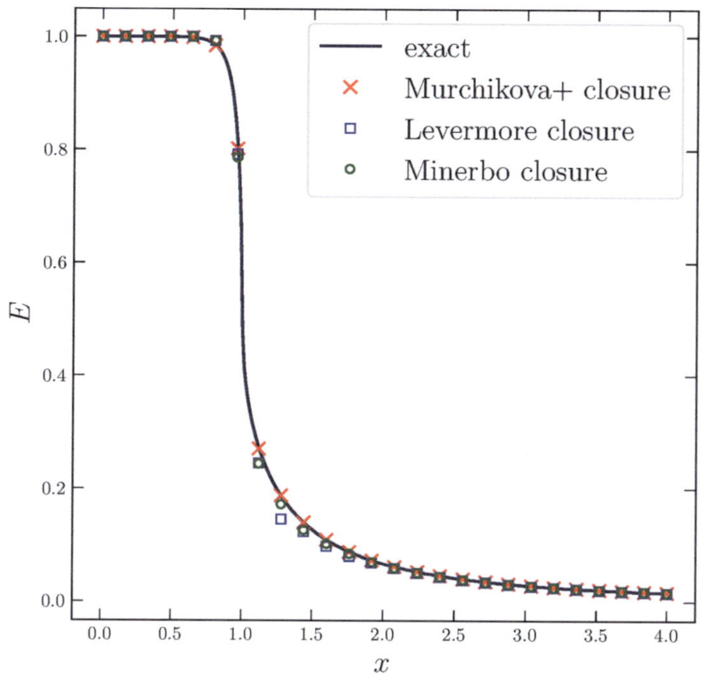

Fig. 1 Radial profiles of the energy density E for a uniform sphere of radiation. The results for different closures (Murchikova+, Levermore and Minerbo) are shown with symbols, while the exact solution is shown with a black solid line

Another delicate aspect of the radiative transfer equations is that they reduce to a set of diffusion equations for the radiation energy density in the asymptotically optically thick regime. In this regime, standard discretization schemes which are suitable for nonlinear hyperbolic PDEs introduce excessive dissipation in the solution therefore achieving a lower convergence order. To overcome this issue in `FIL-M1` we correct the standard HLLE Riemann solver by switching off the diffusive term in the high optical-depth regime (see also [26, 27, 34] for a similar approach). The effectiveness of this correction can be tested by simulating a Gaussian radiation pulse on a uniform medium with a very high scattering opacity. The results of this test can be seen in Fig. 2 where `FIL-M1` can be seen to converge to the analytic result.

The final point which requires special attention when implementing a radiative transfer scheme has to do with the collisional coupling between the radiation and the ordinary fluid. Indeed, in a BNS merger neutrinos can be found in extremely different conditions at different times and in different locations within the system. In particular, in the hot layer of the remnant after merger and in a large portion of the accretion disk, neutrinos are completely coupled to the fluid and their evolution follows essentially the diffusive behaviour described above. On the other hand, in the cold core of the massive remnant and in the low density outskirts of the disk, as well

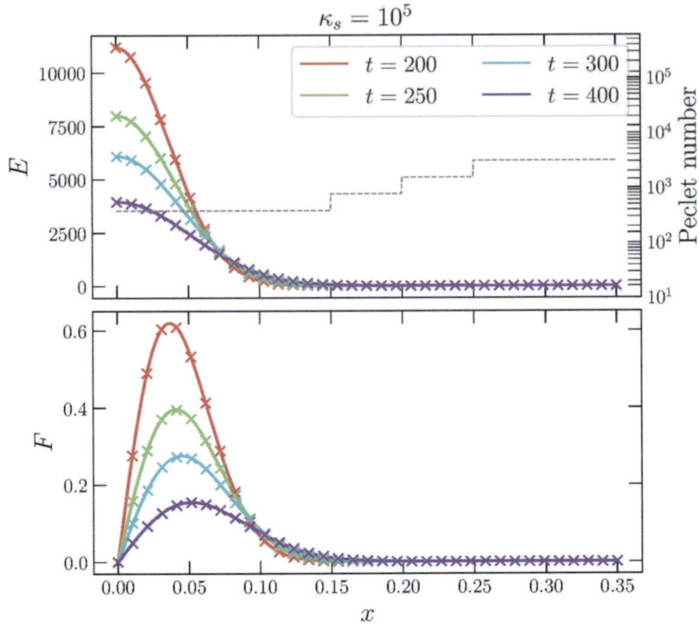

Fig. 2 Radial profiles of the energy density (*Top*) and of the energy flux density (*Bottom*) at different times of a sphere of radiation diffusing in a Minkowski spacetime. The background is flat and static and has a uniform scattering opacity which we set to $\kappa_s^{\mathrm{er}} = 10^5$. The numerical (analytical) solutions are shown with crosses (solid lines) and show a very good agreement between the two. Also shown with a dashed gray line in the top-right panel is the Plecet number $\kappa_s \Delta x$, which is clearly always much larger than unity

as within the ejected material, the mean free path of neutrinos far exceeds the typical lengthscale of the system and they can be considered to be free streaming. Due to these extremely different conditions, the reaction rates involving neutrinos vary by several orders of magnitude across the simulation domain. Moreover, matter in the aftermath of a BNS merger can easily reach mildly relativistic speeds on the order of $0.1\,c$, which also contribute to making the collisional integral (which depends on v_i) numerically larger. Due to this, the M1 equations are well known to be very stiff in BNS merger contexts, and require implicit time integration to be evolved stably and accurately. For this reason and following [30] in `FIL-M1` we evolve the neutrino variables with a two-step implicit integrator and we invert the full set of sources with a globally convergent Newton-Raphson rootfinding algorithm. The system of sources is $4+1$ dimensional. The rootfinding for the implicit sources is perhaps the most delicate aspect of the whole algorithm, since it is very computationally expensive and prone to failure in rapidly moving optically thick media. To test the efficacy of our scheme we performed the test first presented in [30] where a pulse of radiation is initialized on a medium with high scattering opacity which is also uniformly

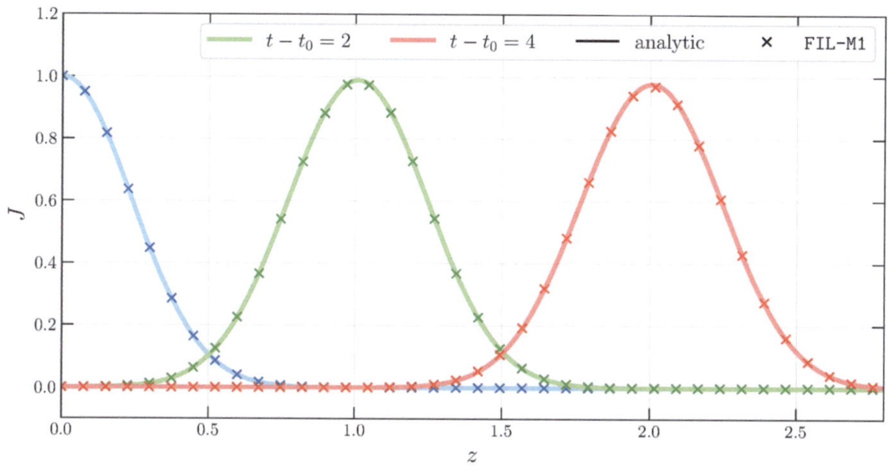

Fig. 3 Radial profiles at different retarded times $t - t_0$ of the energy density in the frame comoving with the fluid J. The numerical (analytical) solutions are shown with crosses (solid lines) and show a very good agreement between the two

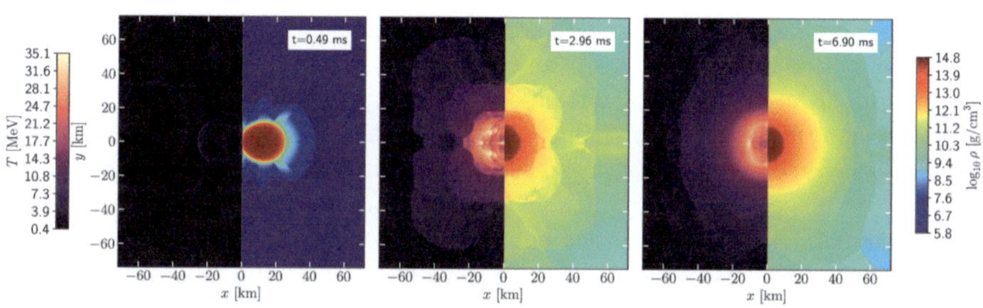

Fig. 4 Representative snapshots of the head-on collision of two equal-mass neutron stars. All panels show the temperature (left half) and the rest-mass density (right side). Note the appearance of a ring-like structure in the temperature of the relaxed post-collision remnant

moving at a velocity $v = 0.5\,c$. The result of this test, where FIL-M1 can be seen to accurately reproduce the analytic solution, is shown in Fig. 3.

Finally, we tested the coupling of the radiation to the full GRMHD system by simulating the head-on collision of two $0.91\,M_{\odot}$ neutron stars described by the DD2 microphysical EOS [35]. We show in Fig. 4 representative snapshots of the temperature and density of this system, which showcase the ability of FIL-M1 to handle realistic systems involving the collisions of neutron stars.

2 Crustal Magnetic-Field Amplification in Binary Mergers

Undoubtedly, strong magnetic fields play an important role in BNS mergers because they lie at the root of many processes which convert large amounts of gravitational binding energy to kinetic energy of ejected material. Hence, they are necessary ingredients in models which aim at explaining and predicting multimessenger observations of BNS mergers [36–42].

One of the most important aspects of magnetic-field evolution in BNS mergers is the strong amplification of magnetic fields which happens directly in the first few milliseconds after the merger. This amplification has already been investigated in a number of studies [43–48] which have shown that the hydrodynamic Kelvin-Helmholtz instability (KHI) is responsible for exponential magnetic-field growth reaching values of up to $\sim 10^{16}$ G and total energies on the order of $\sim 10^{50}$ erg which is ~ 0.1 % of the total rotational energy in the remnant hypermassive neutron star (HMNS).

In this project we set out to investigate whether large magnetic fields are produced universally in long-lived BNS mergers and whether the topology of the magnetic fields can influence or even impede the amplification process. Previous works have shown that the initial magnetic-field topology is quickly destroyed in the violent merger which is subject to strong turbulent flows, see e.g. [49] who investigated dipoles with different strengths, misaligned dipoles, and even a multipolar structure. However, to the best of our knowledge, a topology where magnetic fields are confined to a small region close the neutron-star surface has not been investigated before.

Hence, in this project we simulate the BNS merger where each neutron star only has crustal magnetic fields (the crust-configuration) and contrast it with the standard case in which the cores of the stars are magnetized as well (the full-configuration), see also [50]. A cross-section of the magnetic-field strength in the (x, z) plane shortly before merger for both configurations is shown in Fig. 5. We perform high-resolution, global, and high-order GRMHD simulations with a dynamically evolving spacetime. The simulations employ the high-order high-resolution shock-capturing code `FIL` [19, 51], which is part of the `Einstein Toolkit`, and a temperature-dependent equation of state `TNTYST`. In order to estimate the impact of discretization

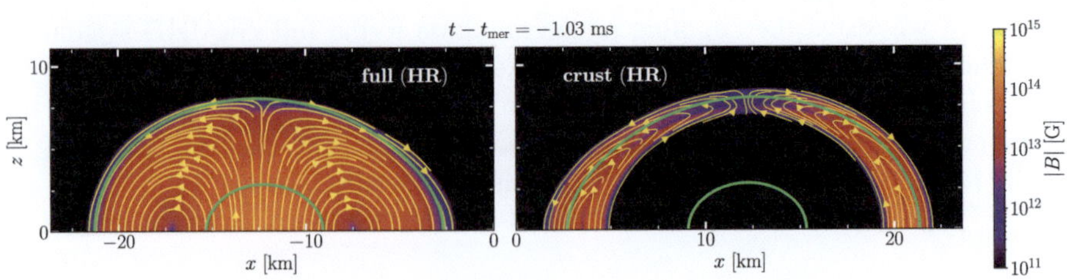

Fig. 5 Distributions on the (x, z) plane of the magnetic-field strength $|B|$ from the HR simulations soon before merger. Magnetic-field lines are shown in yellow, while the left and right panels refer to the full- and crust-configuration, respectively. This figure is reused from [50]

errors on our results we employ two resolutions with $\Delta x \sim 0.047\ M_\odot \approx 70$ m, high-resolution (HR), or of $\Delta x \sim 0.071\ M_\odot \approx 105$ m, low-resolution (LR), where Δx is the grid spacing on the finest refinement level.

We start by describing the overall evolution of the total magnetic energy given by

$$E_B := \int d^3x \sqrt{\gamma}(E^2 + B^2)/8\pi\ , \tag{1}$$

where γ is the determinant of the spatial metric, and $B^2 := B_i B^i$ and $E^2 := E_i E^i$. The spatial vectors B^i and E^i are the magnetic and electric field measured by the Eulerian observer, respectively. The left panel of Fig. 6 shows the evolution of E_B for all simulations. Solid lines denote the HR full-configuration (blue line) and HR crust-configuration (red line). Dashed lines denote the LR full-configuration (blue line) and LR crust-configuration (red line). In the following we will only focus on the results of the HR simulations.

We observe four distinct stages of magnetic-field evolution for both configurations which are denoted by using different shadings for the HR full- and HR crust-configuration in Fig. 6. The first stage, which we name the Kelvin-Helmholtz instability (KHI) driven stage, starts directly when the two neutron stars have their first contact and ends when the total magnetic energy decreases for the first time. For both configurations, this stage is shaped by the generation of turbulent motions in the newly forming HMNS through the KHI. The development of eddies, which drag formerly structured large-scale magnetic-field lines into rotation, leads to exponential amplification of the magnetic field.

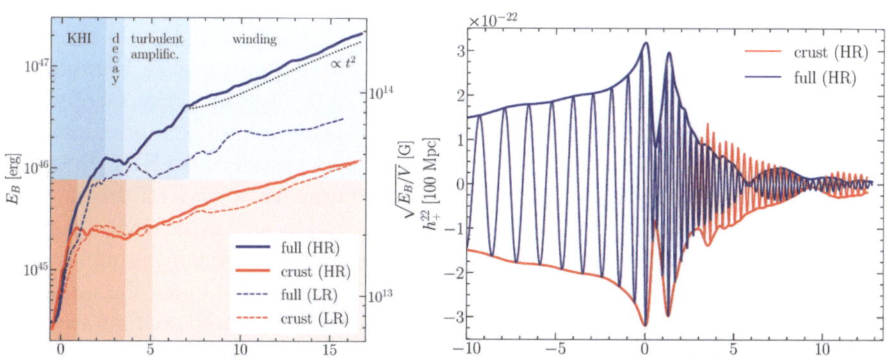

Fig. 6 *Left*: Evolution of the total magnetic energy E_B for the high-resolution (HR) full- (blue solid line) and crust-configuration (red solid line); dashed blue (red) lines refer to the low-resolution (LR) full- and crust-configuration, respectively. Different shadings highlight the four stages of the evolution. The black dotted line shows a quadratic fit to full (HR) in the winding stage and is shifted downwards for better visualization. The quantity V denotes the reference volume of a sphere with a radius of 11 km. *Right*: Gravitational-wave strain in the $\ell = 2, m = 2$ mode of the $+$-polarization extracted at ~ 740 km normalized to a distance of 100 Mpc and for the two configurations. Thick solid lines report the corresponding amplitudes. This figure is reused from [50]

The most significant difference between the full-configuration and the crust-configuration is the early termination of the KHI-driven stage for the crust-configuration. We attribute this behaviour to two physical processes. First, given the magnetic-field topology of the crust-configuration, its amplification is much more sensitive to losses of magnetised material at the stellar surface. Such losses are well studied in the context of dynamical mass ejection [8] which provides an efficient way to remove magnetised material from turbulent regions of the HMNS. Second, after important magnetised material is lost in the crust-configuration it cannot be replenished as in the full-configuration. As magnetised material is removed from the turbulent regions, the amplification stops temporarily until a new process can start.

After the KHI-driven stage, both configurations enter the so-called "decay" stage. At this point, angular momentum redistributes in the HMNS leading to a net fluid expansion which is accelerated by large magnetic-pressure gradients. Hence, magnetic fields are doing work on the fluid, thereby transferring magnetic energy into radial kinetic energy.

In the next stage, the "turbulent-amplification" stage, we observe that turbulent motions can be reinduced in certain regions due to rapid changes in the neutron-star structure. These turbulent motions then lead to magnetic-field amplification.

The last stage is named "winding" stage as in this stage the turbulence is fully developed and the HMNS has reached an approximately axisymmetric structure with a well-developed large-scale velocity profile [52]. For both configurations, magnetic-field amplification in this stage is determined by magnetic winding where a large-scale gradient of the angular velocity leads to the growth of the azimuthal component of the magnetic field [53]. The magnetic energy is growing $\propto t^2$ what we confirm using the black-dotted line in the left panel of Fig. 6.

In order to illustrate the stages we have identified, we show cross-sections of the norm of the magnetic-field $|B| := \sqrt{B_i B^i}$ close to the equatorial plane in Fig. 7 for the full-configuration (crust-configuration) in the top (bottom) panels. From left to right we show snapshots at $t - t_{\rm mer} \in \{-0.12, 0.52, 3.0, 14.98\}$ms, respectively. The first two columns are taken during the KHI-driven stage for both configurations and show the highly magnetized shear layer overturning as a result of the KHI. Note that the second column of the crust-configuration is at a time when the KHI-driven stage has almost ended. It is precisely the magnetised material located at the HMNS surface shown there which will be removed from the turbulent region. The third column shows how in the decay stage the outer density contour has expanded under the influence of the magnetic-pressure gradient. The fourth column shows both configurations at the end of the simulations. A magnetized ring has developed in both configurations which is, however, more strongly pronounced in the full-configuration. The location of the ring coincides with the location of the sharpest increase in angular velocity which provides further evidence for the effect of magnetic winding in this stage.

Finally, we show the emitted gravitational wave signal in the right panel of Fig. 6. While both signals are comparable in amplitude and phase, which is the result of dynamically weak magnetic fields, we find that the signal of the full-configuration

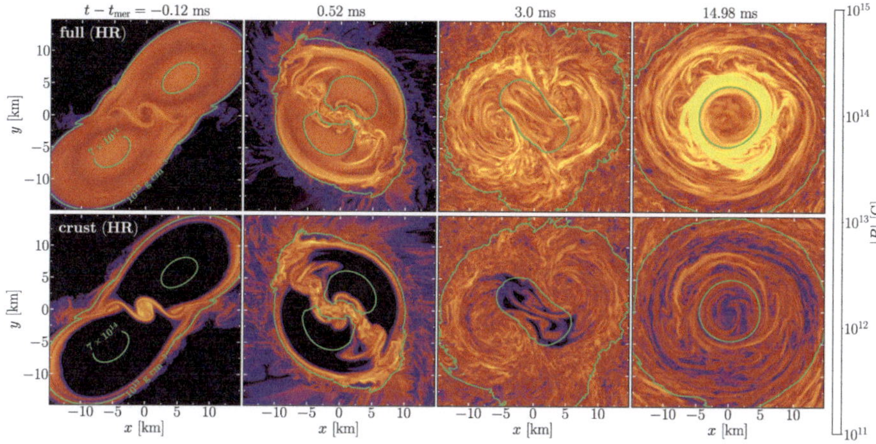

Fig. 7 Distributions on the (x, y) plane of the norm of the magnetic-field strength $|B|$ from the HR simulations at four different times. *Top*: Full-configuration; *bottom*: crust-configuration. This figure is reused from [50]

is stronger damped. The damping of the gravitational wave signal is related to the rapid development of a quasi-axisymmetric structure of the HMNS due to stronger magnetic fields.

3 Code Performance and Computing Resources Used

3.1 Numerical Algorithms

Whether we are interested in accurate BNS mergers in dynamical spacetimes or in the electromagnetic emission and kilonova production from the remnants on secular timescales, our simulations are performed using a variety of numerical techniques and numerical codes, either fully developed by our group - or to whose development the group has contributed. In all cases, the simulations must be able to follow features in the flow that are on sub-horizon scales (i.e. $\simeq$ 150 m), up to the large-scale features associated with the kilonova emission that are a thousand times larger than the horizon (i.e. on scales $\simeq$ 15,000 km). Clearly, only AMR techniques can allow us to cover this large range of scales.

Our internal code `FIL` uses the `Carpet` library as part of the publicly available `Einstein Toolkit`, which implements a vertex-centered adaptive-mesh-refinement scheme adopting nested and moving grids. This allows us to follow the moving neutron stars with fine-resolution grids while covering with coarser grids with a much larger volume which is needed for the treatment of the GWs. `Carpet` supports Berger-Oliger mesh refinement [54] with sub-cycling in time and refluxing. The grid resolution changes by a factor of two between successive levels.

The simulations in projects Sects. 1 and 2 have been carried out using `FIL`, which is a high-order extension of the publicly available general-relativistic MHD code `ILGRMHD` [55]. The basic algorithms in `ILGRMHD` have been developed by the Illinois Relativity Group, but the publicly available version is particularly compact and efficient, providing a speedup factor of almost two when compared with other GRMHD codes developed, for instance, by some of the members in Frankfurt [56]. The original code has a long record of compact-object simulations, including black-hole–neutron-star systems [57], BNS systems [58] and binary black-hole systems with magnetized disks [59] to name just a few of the recent publications including magnetic fields. Among the numerous new developments that the `FIL` code implements, we would like to highlight the fully high-order flux update which employs the so-called ECHO scheme, the capability for finite-temperature dependent EOSs, and an improved primitive-variable recovery [19]. Additionally, as described extensively in project Sect. 1, the new extension of `FIL`, `FIL-M1`, was developed and tested thoroughly as a result of this allocation period paving the way for the inclusion of weak interactions in magnetized BNS merger simulations.

At the same time, merger simulations require the solution of the Einstein equations which ultimately provides a consistent evolution of the spacetime. These equations are solved using standard finite-difference methods where all the derivatives are computed with centered stencils; the only exceptions are the terms associated with the advection along the shift vector for which we use upwinded stencils. Typically, all derivatives are computed with a fourth-order accurate scheme, but sixth and eighth-order are also available, with a fifth-order Kreiss-Oliger artificial dissipation added to ensure stability. More specifically, we use the Z4c formulation of the Einstein equations [60] in the `Antelope` code. Evolution and coupling between the hydrodynamic and the spacetime solvers is done using the method-of-lines (MOL), either with the optimal strong-stability preserving third-order Runge-Kutta scheme, or with a fourth-order one.

3.2 Scaling and Computing Resources Used

In this subsection we want to present the scaling performance of our codes, as well as outline the groups' usage of computing resources that has been provided by the high-performance computing facility HAWK.

We performed strong scaling tests with our code `FIL` using two representative test problems and a fixed combination of 8 MPI ranks per node together with 16 OpenMP threads per MPI rank. In both problems we calculate the first 640 iterations of a BNS head-on collision where the two stars are threaded by strong magnetic fields and whose matter is described by a temperature and composition dependent EOS derived from microphysical calculations. We decided to present the scaling results for a unigrid setup and an AMR setup similar to our production runs. All results can be found in Table 1 and are displayed in Fig. 8. The left panel of Fig. 8 shows the strong scaling results for the unigrid setup. As can be seen, we are able to scale

Table 1 Strong-scaling tests carried out on HAWK. The runtime in seconds is indicated by "time" and the corresponding normalised speedup by "speedup", where the latter is normalised with respect to the lowest core number

`FIL` Unigrid (1 TB)			`FIL` AMR (250 GB)		
# cores	Time	Speedup	# cores	Time	Speedup
640	17661.93	1.00	640	1232.33	1.00
1280	8967.05	1.97	1280	717.34	1.72
2560	4792.36	3.69	2560	426.88	2.89
5120	2592.79	6.81	5120	294.90	4.18
10280	1528.83	11.55	–	–	–

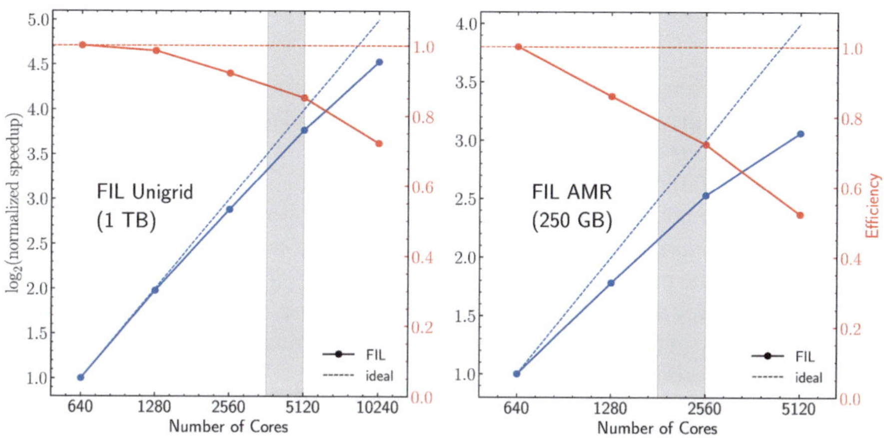

Fig. 8 Strong-scaling tests of Table 1, represented by the normalised speedup as function of core count (blue line with dots, left vertical scale), while the red lines indicate the efficiency (right vertical scale). In the left panel we show the strong scaling performance of `FIL` on a unigrid up to 10240 cores. The right panel shows strong scaling performance of `FIL` employing its full AMR capabilities in a production-like grid setup. The vertically grey shaded regions roughly indicate the number of cores, at which each code is typically used for the proposed simulations

`FIL` up to 10240 cores when using a single grid resolution on the whole domain. Even though we typically make use of the AMR capabilities of `FIL` to save resources and decrease the necessary runtime, the good scaling results for the unigrid setup demonstrate the high degree of parallelization of the `FIL` code.

The right panel of Fig. 8 shows the strong scaling results for the AMR setup. Here, we are able to show moderate scaling up to 2560 cores. The differences between the two setups arise from the overhead stemming from the need to synchronize the solution across different refinement levels covering the same spatial domain which is typical for Berger-Oliger AMR. However, as denoted by the grey shaded region, the

Table 2 Overview over used CPU resources per project

	P1	P2	P3	Total 04/22 → 03/23
Used resources [CPUh]	20M	25M	20M	66M
Average runtime per simulation [h]	244	683	326	342
Average cost per simulation [CPUh]	0.5M	2M	0.7M	0.8M
Average number of cores per simulation	2048	2816	2048	2168
Average number of MPI ranks per simulation	128	176	128	136
Average number of threads per rank per simulation	16	16	16	16

typical number of employed cores per simulation ranges between ~1500 and ~2500 which lies below the regime where the cost of our simulation is heavily influenced by the scaling of our code.

Furthermore, Table 2 displays our usage of computing resources provided by the high-performance computing facility HAWK. In the following we will refer to the project *General-relativistic radiative transport in binary mergers* as `P1`, see also [17] for more details, the project *Crustal magnetic-field amplification in binary merger* as `P2`, see also [50] for additional information. Furthermore, we will refer to resources spent on currently unpublished work as `P3`. The last column of Table 2 denotes our total usage in this allocation period.

Acknowledgements Partial funding comes from the State of Hesse within the Research Cluster ELEMENTS (Project ID 500/10.006), by the ERC Advanced Grant "JETSET: Launching, propagation and emission of relativistic jets from binary mergers and across mass scales" (Grant No. 884631). M. C. and C. E. acknowledge support by the Deutsche Forschungsgemeinschaft (DFG, German Research Foundation) through the CRC-TR 211 "Strong-interaction matter under extreme conditions"—project number 315477589—TRR 211. M. C. acknowledges support by the GSI Helmholtzzentrum für Schwerionenforschung GmbH through the FuE program. S. T., K. T., and L. R. acknowledge the support by the State of Hesse within the Research Cluster ELEMENTS (Project ID 500/10.006).

References

1. Baiotti, L., Rezzolla, L.: Binary neutron-star mergers: a review of Einstein's richest laboratory. Rept. Prog. Phys. **80**(9), 096901 (2017)
2. Paschalidis, V.: General relativistic simulations of compact binary mergers as engines for short gamma-ray bursts. Class. Quantum Gravity **34**(8), 084002 (2017)
3. Metzger, B.D., Martínez-Pinedo, G., Darbha, S., Quataert, E., Arcones, A., Kasen, D., Thomas, R., Nugent, P., Panov, I.V., Zinner, N.T.: Electromagnetic counterparts of compact object mergers powered by the radioactive decay of r-process nuclei. Mon. Not. R. Astron. Soc. **406**, 2650–2662 (2010)

4. Rezzolla, L., Giacomazzo, B., Baiotti, L., Granot, J., Kouveliotou, C., Aloy, M.A.: The missing link: merging neutron stars naturally produce jet-like structures and can power short gamma-ray bursts. Astrophys. J. Lett. **732**, L6 (2011)
5. Palenzuela, C., Lehner, L., Ponce, M., Liebling, S.L., Anderson, M., Neilsen, D., Motl, P.: Electromagnetic and gravitational outputs from binary-neutron-star coalescence. Phys. Rev. Lett. **111**(6), 061105 (2013)
6. Kiuchi, K., Sekiguchi, Y., Kyutoku, K., Shibata, M., Taniguchi, K., Wada, T.: High resolution magnetohydrodynamic simulation of black hole-neutron star merger: mass ejection and short gamma ray bursts. Phys. Rev. D **92**(6), 064034 (2015)
7. Punturo, M., et al.: The Einstein telescope: a third-generation gravitational wave observatory. Class. Quantum Grav. **27**, 194002 (2010)
8. Combi, L., Siegel, D.M.: GRMHD simulations of neutron-star mergers with weak interactions: r-process nucleosynthesis and electromagnetic signatures of dynamical ejecta. Astrophys. J. **944**(1), 28 (2023)
9. Radice, D., Galeazzi, F., Lippuner, J., Roberts, L.F., Ott, C.D., Rezzolla, L.: Dynamical mass ejection from binary neutron star mergers. Mon. Not. R. Astron. Soc. **460**, 3255–3271 (2016)
10. Sekiguchi, Y., Kiuchi, K., Kyutoku, K., Shibata, M., Taniguchi, K.: Dynamical mass ejection from the merger of asymmetric binary neutron stars: radiation-hydrodynamics study in general relativity. Phys. Rev. D **93**(12), 124046 (2016)
11. Eichler, D., Livio, M., Piran, T., Schramm, D.N.: Nucleosynthesis, neutrino bursts and gamma-rays from coalescing neutron stars. Nature **340**, 126–128 (1989)
12. Ruffert, M., Janka, H.-T., Takahashi, K., Schaefer, G.: Coalescing neutron stars - a step towards physical models. II. Neutrino emission, neutron tori, and gamma-ray bursts. Astron. Astrophys. **319**, 122–153 (1997)
13. Just, O., Obergaulinger, M., Janka, H.-T., Bauswein, A., Schwarz, N.: Neutron-star merger ejecta as obstacles to neutrino-powered jets of gamma-ray bursts. Astrophys. J. Lett. **816**, L30 (2016)
14. Dessart, L., Ott, C.D., Burrows, A., Rosswog, S., Livne, E.: Neutrino signatures and the neutrino-driven wind in binary neutron star mergers. Astrophys. J. **690**, 1681–1705 (2009)
15. Perego, A., Rosswog, S., Cabezón, R.M., Korobkin, O., Käppeli, R., Arcones, A., Liebendörfer, M.: Neutrino-driven winds from neutron star merger remnants. Mon. Not. R. Astron. Soc. **443**, 3134–3156 (2014)
16. Fujibayashi, S., Sekiguchi, S., Kiuchi, K., Shibata, M.: Properties of neutrino-driven ejecta from the remnant of binary neutron star merger : purely radiation hydrodynamics case. ArXiv e-prints, March 2017
17. Musolino, C., Rezzolla, L.: A practical guide to a moment approach for neutrino transport in numerical relativity. arXiv e-prints, page arXiv:2304.09168, April 2023
18. Galeazzi, F., Kastaun, W., Rezzolla, L., Font, J.A.: Implementation of a simplified approach to radiative transfer in general relativity. Phys. Rev. D **88**(6), 064009 (2013)
19. Most, E.R., Jens Papenfort, L., Rezzolla, L.: Beyond second-order convergence in simulations of magnetized binary neutron stars with realistic microphysics. Mon. Not. R. Astron. Soc. **490**(3), 3588–3600 (2019)
20. Radice, D., Abdikamalov, E., Rezzolla, L., Ott, C.D.: A new spherical harmonics scheme for multi-dimensional radiation transport I. Static matter configurations. J. Comput. Phys. **242**, 648–669 (2013)
21. Foucart, F., Duez, M.D., Hebert, F., Kidder, L.E., Pfeiffer, H.P., Scheel, M.A.: Monte-carlo neutrino transport in neutron star merger simulations (2020)
22. Roth, N., Anninos, P., Robinson, P.B., Luc Peterson, J., Polak, B., Mangan, T.K., Beyer, K.: General relativistic implicit monte carlo radiation-hydrodynamics. Astrophys. J. **933**(2), 226 (2022)
23. Rezzolla, L., Miller, J.C.: Relativistic radiative transfer for spherical flows. Class. Quantum Grav. **11**, 1815–1832 (1994)
24. Foucart, F., O'Connor, E., Roberts, L., Duez, M.D., Haas, R., Kidder, L.E., Ott, C.D., Pfeiffer, H.P., Scheel, M.A., Szilagyi, B.: Post-merger evolution of a neutron star-black hole binary with neutrino transport. Phys. Rev. D **91**(12), 124021 (2015)

25. Just, O., Obergaulinger, M., Janka, H.T.: A new multidimensional, energy-dependent two-moment transport code for neutrino-hydrodynamics. Mon. Not. R. Astron. Soc. **453**(4), 3386–3413 (2015)
26. Kuroda, T., Takiwaki, T., Kotake, K.: A new multi-energy neutrino radiation-hydrodynamics code in full general relativity and its application to the gravitational collapse of massive stars. Astrophys. J., Supp., **222**(2), 20 (2016)
27. Skinner, M.A., Dolence, J.C., Burrows, A., Radice, D., Vartanyan, D.: FORNAX: a flexible code for multiphysics astrophysical simulations. Astrophys. J. Supp. **241**(1), 7 (2019)
28. Fuksman, J.D.M., Mignone, A.: A radiative transfer module for relativistic magnetohydrodynamics in the PLUTO code. Astrophys. J. Supp. **242**(2), 20 (2019)
29. Weih, L.R., Olivares, H., Rezzolla, L.: Two-moment scheme for general-relativistic radiation hydrodynamics: a systematic description and new applications. Mon. Not. R. Astron. Soc. **495**(2), 2285–2304 (2020)
30. Radice, D., Bernuzzi, S., Perego, A., Haas, R.: A new moment-based general-relativistic neutrino-radiation transport code: methods and first applications to neutron star mergers. Mon. Not. Roy. Astron. Soc. **512**(1), 1499–1521 (2022)
31. Sun, L., Ruiz, M., Shapiro, S.L., Tsokaros, A.: Jet launching from binary neutron star mergers: incorporating neutrino transport and magnetic fields. Phys. Rev. D **105**(10), 104028 (2022)
32. Izquierdo, M.R., Pareschi, L., Miñano, B., Massó, J., Palenzuela, C.: Global high-order numerical schemes for the time evolution of the general relativistic radiation magneto-hydrodynamics equations. arXiv e-prints, page arXiv:2211.00027, Oct 2022
33. Murchikova, E.M., Abdikamalov, E., Urbatsch, T.: Analytic closures for M1 neutrino transport. Mon. Not. R. Astron. Soc. **469**(2), 1725–1737 (2017)
34. Cheong, P.C.-K., Ng, H.H.-Y., Lam, A.T.-L., Li, T.G.F.: General-relativistic radiation transport scheme in Gmunu i: Implementation of two-moment based multi-frequency radiative transfer and code tests (2023)
35. Hempel, M., Schaffner-Bielich, J.: Statistical model for a complete supernova equation of state. Nucl. Phys. A **837**, 210–254 (2010)
36. Abbott, B.P., Abbott, R., Abbott, T.D., Acernese, F., Ackley, K., Adams, C., Adams, T., Addesso, P., Adhikari, R.X., Adya, V.B., et al.: Gw170817: observation of gravitational waves from a binary neutron star inspiral. Phys. Rev. Lett. **119**(16), 161101 (2017)
37. Drout, M.R., Piro, A.L., Shappee, B.J., Kilpatrick, C.D., Simon, J.D., Contreras, C., Coulter, D.A., Foley, R.J., Siebert, M.R., Morrell, N., Boutsia, K., Wan, Z., Whitten, D.D.: Light curves of the neutron star merger GW170817/SSS17a: implications for r-process nucleosynthesis. Science **358**, 1570–1574 (2017)
38. Siegel, D.M., Ciolfi, R., Harte, A.I., Rezzolla, L.: Magnetorotational instability in relativistic hypermassive neutron stars. Phys. Rev. D R **87**(12), 121302 (2013)
39. Kiuchi, K., Kyutoku, K., Sekiguchi, Y., Shibata, M.: Global simulations of strongly magnetized remnant massive neutron stars formed in binary neutron star mergers. Phys. Rev. D **97**(12), 124039 (2018)
40. Fernández, R., Tchekhovskoy, A., Quataert, E., Foucart, F., Kasen, D.: Long-term GRMHD simulations of neutron star merger accretion discs: implications for electromagnetic counterparts. Mon. Not. R. Astron. Soc. **482**(3), 3373–3393 (2019)
41. Ciolfi, R.: Collimated outflows from long-lived binary neutron star merger remnants. Mon. Not. R. Astron. Soc. **495**(1), L66–L70 (2020)
42. Fujibayashi, S., Kiuchi, K., Wanajo, S., Kyutoku, K., Sekiguchi, Y., Shibata, M.: Comprehensive study of mass ejection and nucleosynthesis in binary neutron star mergers leaving short-lived massive neutron stars. Astrophys. J. **942**(1), 39 (2023)
43. Price, D.J., Rosswog, S.: Producing ultrastrong magnetic fields in neutron star mergers. Science **312**, 719–722 (2006)
44. Giacomazzo, B., Rezzolla, L., Baiotti, L.: Accurate evolutions of inspiralling and magnetized neutron stars: Equal-mass binaries. Phys. Rev. D **83**(4), 044014 (2011)
45. Giacomazzo, B., Zrake, J., Duffell, P.C., MacFadyen, A.I., Perna, R.: Producing magnetar magnetic fields in the merger of binary neutron stars. Astrophys. J. **809**(1), 39 (2015)

46. Kiuchi, K., Cerdá-Durán, P., Kyutoku, K., Sekiguchi, Y., Shibata, M.: Efficient magnetic-field amplification due to the Kelvin-Helmholtz instability in binary neutron star mergers. Phys. Rev. D **92**(12), 124034 (2015)
47. Palenzuela, C., Liebling, S.L., Neilsen, D., Lehner, L., Caballero, O.L., O'Connor, E., Anderson, M.: Effects of the microphysical equation of state in the mergers of magnetized neutron stars with neutrino cooling. Phys. Rev. D **92**(4), 044045 (2015)
48. Aguilera-Miret, R., Viganò, D., Carrasco, F., Miñano, B., Palenzuela, C.: Turbulent magnetic-field amplification in the first 10 milliseconds after a binary neutron star merger: comparing high-resolution and large-eddy simulations. Phys. Rev. D **102**(10), 103006 (2020)
49. Aguilera-Miret, R., Viganò, D., Palenzuela, C.: Universality of the turbulent magnetic field in hypermassive neutron stars produced by binary mergers. Astrophys. J. Lett. **926**(2), L31 (2022)
50. Chabanov, M., Tootle, S.D., Most, E.R., Rezzolla, L.: Crustal magnetic fields do not lead to large magnetic-field amplifications in binary neutron star mergers. Astrophys. J. Lett. **945**(1), L14 (2023)
51. Most, E.R., Jens Papenfort, L., Dexheimer, V., Hanauske, M., Stoecker, H., Rezzolla, L.: On the deconfinement phase transition in neutron-star mergers. Eur. Phys. J. A **56**(2), 59 (2020)
52. Hanauske, M., Takami, K., Bovard, L., Rezzolla, L., Font, J.A., Galeazzi, F., Stöcker, H.: Rotational properties of hypermassive neutron stars from binary mergers. Phys. Rev. D **96**(4), 043004 (2017)
53. Shapiro, S.L.: Differential rotation in neutron stars: magnetic braking and viscous damping. Astrophys. J. **544**(1), 397–408 (2000)
54. Berger, M.J., Oliger, J.: Adaptive mesh refinement for hyperbolic partial differential equations. J. Comput. Phys. **53**, 484–512 (1984)
55. Etienne, Z.B., Paschalidis, V., Haas, R., Mösta, P., Shapiro, S.L.: IllinoisGRMHD: an open-source, user-friendly GRMHD code for dynamical spacetimes. Class. Quantum Grav. **32**(17), 175009 (2015)
56. Giacomazzo, B., Rezzolla, L.: WhiskyMHD: a new numerical code for general relativistic magnetohydrodynamics. Class. Quantum Grav. **24**, 235 (2007)
57. Etienne, Z.B., Paschalidis, V., Liu, Y.T., Shapiro, S.L.: Relativistic magnetohydrodynamics in dynamical spacetimes: improved electromagnetic gauge condition for adaptive mesh refinement grids. Phys. Rev. D **85**(2), 024013 (2012)
58. Liu, Y.T., Shapiro, S.L., Etienne, Z.B., Taniguchi, K.: General relativistic simulations of magnetized binary neutron star mergers. Phys. Rev. D **78**, 024012 (2008)
59. Farris, B.D., Gold, R., Paschalidis, V., Etienne, Z.B., Shapiro, S.L.: Binary black-hole mergers in magnetized disks: simulations in full general relativity. Phys. Rev. Lett. **109**(22), 221102 (2012)
60. Bernuzzi, S., Hilditch, D.: Constraint violation in free evolution schemes: comparing BSSNOK with a conformal decomposition of Z4. Phys. Rev. D **81**, 084003 (2010)

Molecules, Interfaces, and Solids

Large-Scale Ab Initio Modeling of Hydrogen Production on Stepped Surfaces of $SrTiO_3$ Perovskite Nanoparticles

Maksim Sokolov, Yuri A. Mastrikov, Eugene A. Kotomin, and Joachim Maier

Abstract As recently was observed, photocatalytic activity toward water splitting of strontium titanate $SrTiO_3$ (STO) is greatly enhanced by creating multifaceted nanoparticles. To understand this effect, we developed a model for two types of surfaces of this nanoparticle, flat and double-stepped. Large-scale density functional theory calculations of water adsorption on these surfaces were performed, to gain insight into water adsorption on single-stepped surface and focuses on proton migration and thermodynamics of hydrogen evolution reaction within the framework of computational hydrogen electrode. We conclude that ridge positions on single- and double-stepped surfaces are nearly identical in terms of adsorption configurations and energetics.

1 Introduction

There is great importance in developing environmentally friendly renewable energy sources, such as wind turbines or photovoltaics. One possible improvement direction is toward the storage of overproduced energy by using hydrogen as an energy vector. Hydrogen itself is produced in hydrogen evolution reaction (HER), which is a cathodic half-reaction of electro- or photocatalytic water splitting. HER reaction equation is $2H^+ + 2\ e^- \rightarrow H_2(g)$, $U^0 = 0$ V versus reversible hydrogen electrode (RHE). There are two different mechanisms of HER: Volmer-Tafel (V-T) mechanism and Volmer-Heyrovsky (V-H) mechanism. The V-T mechanism consists of two Volmer steps, which are the adsorption of two hydrogen atoms to the active site, denoted as * (Eq. 1), followed by recombination of two adjacent hydrogen atoms

M. Sokolov
Theoretical Inorganic Chemistry, University of Duisburg-Essen, Essen, Germany

Y. A. Mastrikov · E. A. Kotomin (✉)
Institute of Solid State Physics, University of Latvia, Riga, Latvia
e-mail: E.Kotomin@fkf.mpg.de

E. A. Kotomin · J. Maier
Max Planck Institute for Solid State Research, Stuttgart, Germany

T. Ludwig et al. (eds.), *High Performance Computing in Science and Engineering '23*,
https://doi.org/10.1007/978-3-031-91312-9_8

and desorption, or Tafel step Eq. 2) to form gaseous H_2. V-H mechanism, however, comprises of a single Volmer step, and the second hydrogen atom is taken from the surrounding media (Heyrovsky step, Eq. 3).

$$H^+ + e^- \rightarrow H^* \qquad \text{Volmer step} \qquad (1)$$

$$H^* + H^* \rightarrow H_2 \qquad \text{Tafel step} \qquad (2)$$

$$H_3O^+ + H^* + e^- \rightarrow H_2 + H_2O \qquad \text{Heyrovsky step} \qquad (3)$$

Strontium titanate $SrTiO_3$ (STO) is a good catalyst for overall water splitting and HER in particular [1]. The catalytic activity of STO toward water splitting can be further improved by creating faceted nanoparticles [2]. Such an improvement is attributed to the preferential deposition of Pt and Co_3O_4 co-catalyst, as well as a lower charge recombination rate. Both of these factors arise due to the anisotropy of these multifaceted nanoparticles.

Water adsorption on STO was studied in detail [3, 4]. In the previous paper, our group investigated the adsorption configurations and energetics of water on two types of surfaces of faceted nanoparticles, namely, flat and stepped $\{1\ 1\ 0\}$ [4]. However, HER on STO stepped surface has not yet been described.

In this paper, we continue our research by investigating water dissociation and proton estimate transition energy barriers for the proton migration and demonstrate that the V-T mechanism can be observed for HER on stepped STO surfaces at low overpotentials. Investigation of the V-H mechanism is out of the scope of this report.

Moreover, we apply the computational hydrogen electrode (CHE) approach of Nørskov and Rossmeisel [5] to estimate the catalytic activity of both flat and stepped $\{1\ 1\ 0\}$ surface toward HER. We use $\Delta\Delta G_{H^*}$ activity descriptor, which is similar to the approach described in [6]. The reason to use $\Delta\Delta G_{H^*}$ over adsorption free energy ΔG_{H^*} is that the latter is very sensitive to exchange-correlation functional and inclusion or neglection of solvation effects. We demonstrate that the flat surface is more active toward HER than stepped, even without considering charge recombination or the presence of co-catalysts. This suggests that the morphology of surfaces also plays a role in improved catalytic activity for HER, which, to the best of our knowledge, was not considered a factor in experimental studies of STO faceted nanoparticles.

2 Computational Details

Gas-phase density functional theory (DFT) calculations were performed with Vienna Ab initio Simulation Package (VASP) [7]. Relaxed rhombohedral $SrTiO_3$ phase ($R\overline{3}c$) with optimized lattice constant a_0 of 5.54 was used. We compare flat TiO_2-terminated $\{001\}$ surface, single-lattice-constant $\frac{a_0}{\sqrt{2}}$ step $\{110\}$ (single-stepped) and double-lattice-constant $\frac{a_0}{\sqrt{2}}$ step $\{110\}$ (double-stepped) surface, shown in Figs. 1, 2 and 3, respectively. Water molecules were placed on both terminations of the slab to neutralize electric dipole moment.

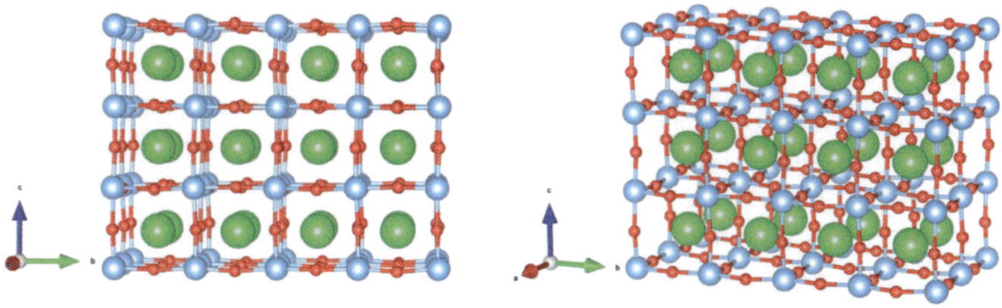

(a) Front view. (b) Isometric view.

Fig. 1 Flat surface cell. Green—Sr atoms, blue—Ti atoms, red—O atoms. Vacuum region is along the c vector. Figure from [4]

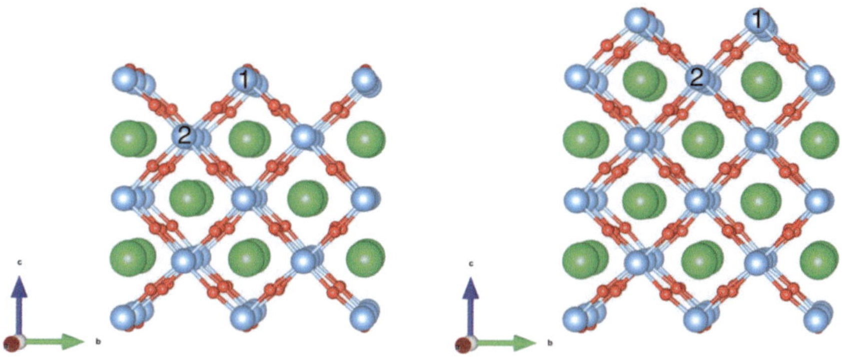

(a) Single step "phase". (b) Single step "antiphase".

Fig. 2 Single-stepped surface cell. 1—ridge adsorption site; 2—gully adsorption site. Green—Sr atoms, blue—Ti atoms, red—O atoms. Vacuum region is along the c vector

Water molecule adsorption energy ΔE_{ads} is calculated by Eq. 4, where $E_{\text{H}_2\text{O}^*}$ is the total energy of configuration with water molecule adsorbed, E_* is the total energy of the corresponding surface without adsorbate and $E_{\text{H}_2\text{O}}$ is the total energy of water molecule. The equation is written for non-symmetric, i.e. single-sided adsorption.

$$\Delta E_{\text{ads}} = E_{\text{H}_2\text{O}^*} - (E_* + E_{\text{H}_2\text{O}}) \tag{4}$$

Water molecule adsorption free energy ΔG_{ads} is given by Eq. 5 (for non-symmetric adsorption), where ZPE stands for zero-point energy, $\Delta\text{ZPE} = \text{ZPE}_{\text{H}_2\text{O}^*} - \text{ZPE}_{\text{H}_2\text{O}} - \text{ZPE}_*$ and $T\Delta S \approx -TS_{\text{H}_2\text{O}}$. ZPE_* and $\text{ZPE}_{\text{H}_2\text{O}^*}$ (for each adsorption site) are calculated via finite differences method as implemented in VASP. $\text{ZPE}_{\text{H}_2\text{O}} = 0.57$ eV and

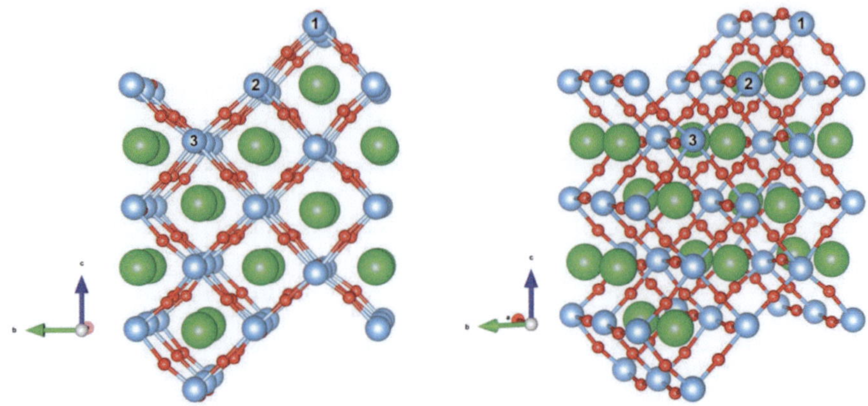

(a) Front view. (b) Isometric view.

Fig. 3 Double-stepped surface cell. 1—ridge adsorption site; 2—slope adsorption site; 3—gully adsorption site. Green—Sr atoms, blue—Ti atoms, red—O atoms. Vacuum region is along the *c* vector. Figure from [4]

$TS_{H_2O} = 0.67$ eV (at 298.15 K and 0.035 bar) [5]. By including the calculated values, we obtain $\Delta G_{ads} = \Delta E_{ads} + 0.53$ eV for ridge and $\Delta G_{ads} = \Delta E_{ads} + 0.79$ eV for slope.

$$\Delta G_{ads} = \Delta E_{ads} + \Delta ZPE - T\Delta S \tag{5}$$

We investigated the impact of water concentrations on the adsorption energies, and the impact is to a vast extend negligible.

To perform thermodynamic simulations of HER, we calculate Gibbs free energy ΔG_{H^*} of Volmer step ($H^+ + e^- \rightarrow H^*$) by applying the CHE approach and using Eq. 6:

$$\Delta G_{H^*} = E_{H^*} - E_* - \frac{1}{2}E_{H_2} + \Delta ZPE - T\Delta S \tag{6}$$

where H^* is hydrogen adsorbed on STO surface. $\Delta ZPE = ZPE_{H*} - \frac{1}{2}ZPE_{H_2}$ with $ZPE_{H_2} = 0.27$ eV and $T\Delta S \approx -\frac{1}{2}TS_{H_2} = -0.20$ eV. ZPE_{H*} is calculated by using finite differences method as implemented in VASP. Please note that the Eq. 6 is written for a single-sided adsorption.

All figures are created in VESTA visualization system.

To estimate the catalytic activity of other adsorption sites, we introduce $\Delta\Delta G_{H^*}$, which is ΔG_{H^*} relative to the ΔG_{H^*} of the flat surface (Eq. 7)

$$\Delta\Delta G_{H^*} = \Delta G_{H^*}(\text{system}) - \Delta G_{H^*}(\text{flat}) \tag{7}$$

3 Results and Discussion

3.1 Water Adsorption

To understand the difference between water adsorption on flat and stepped surfaces, we performed an extensive search of the most favorable water adsorption configuration and considered ~50 initial geometries of a single water molecule per termination [4] where we discussed the most noteworthy ones.

The most important results of the previous study are that on flat surface dissociative adsorption is slightly more preferable and a similar situation is observed on slope of the step. On ridge, only dissociative adsorption is possible, and it is accompanied by spontaneous oxygen vacancy formation. On gully, water adsorption is energetically less favorable.

Additionally, we performed calculations for a single-stepped surface. All energies are at HER conditions of $U = 0$ V versus RHE. Two variations of the single-stepped surface are considered: one where ridge of one side of the slab is directly opposite to ridge on the other side ("phase", Fig. 2a) and the other where ridge is directly opposite to gully ("antiphase", Fig. 2b). On the single-stepped surface (Fig. 2) two modes of dissociative water adsorption are observed. Our results suggest that water adsorption configurations are the same on "phase" and "antiphase" single-stepped surface, so we visualize only "phase" in Fig. 4. In ridge case, water molecule dissociates with spontaneous oxygen vacancy formation ("phase": $\Delta G_{\text{ads}} = -1.04$ eV; "antiphase": $\Delta G_{\text{ads}} = -0.93$ eV). The observed configuration and adsorption energy are almost identical to ridge of double-stepped surface [4].

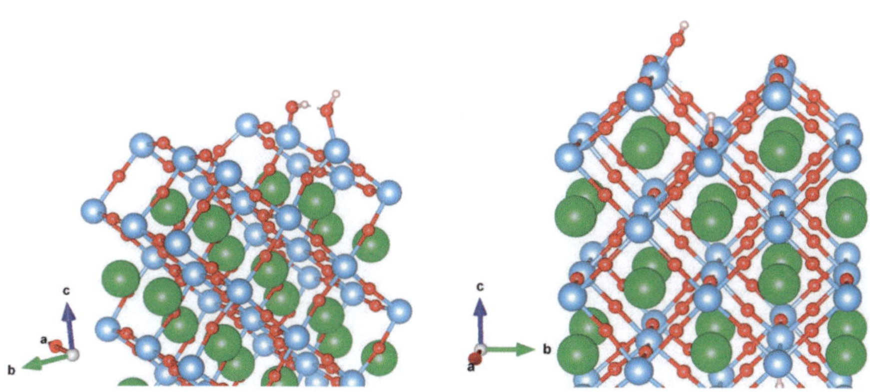

(a) Single-stepped surface, dissociative adsorption on ridge.

(b) Single-stepped surface, dissociative adsorption from ridge to gully.

Fig. 4 Water adsroption on single-stepped surface. Green—Sr atoms, blue—Ti atoms, red—O atoms, white—H atoms

Another case is a less energetically favorable one with OH adsorbate atop ridge and the remaining H is atop gully (“phase”: $\Delta G_{ads} = -0.17$ eV; “antiphase”: $\Delta G_{ads} = -0.14$ eV). The slight difference in energetics can be explained by the shorter distance between adsorbates in “phase” surface.

3.2 Proton Migration

Let us now discuss the possibility of Volmer-Tafel (V-T) mechanism of HER on the STO stepped surface. V-T mechanism requires two adjacent H* adsorbates. Thus, a sufficiently large H* coverage is expected. Also, the V-T mechanism comprises a chemical step (Tafel step), and this step is not affected by the applied electrode potential. Therefore, the free-energy barrier for the Tafel step needs to be sufficiently small so that the V-T mechanism can proceed.

Volmer-Heyrovsky (V-H) mechanism, on the other hand, requires only a single H* adsorbate. Therefore, one may expect that the V-H mechanism may be operative even for a small H* coverage. Given that the V-H mechanism consists of electrochemical steps only that are affected by the applied electrode potential, one could expect that for sufficiently large overpotentials, the V-H mechanism generally excels over the V-T mechanism.

We have demonstrated the low energy barrier for H* migration so that two adjacent H* adsorbates are likely met. Hence, it is suggested that the V-T mechanism is possible for HER over STO stepped surface at low overpotentials. Yet, there is still an opportunity that the V-H mechanism is operative, but the calculation of the transition states for the V-H mechanism is beyond the scope of the current paper.

3.3 Hydrogen Evolution Reaction (HER)

To estimate the catalytic activity of both flat and stepped surfaces we used a standard technique of computational hydrogen electrode approach [5]. We placed a single hydrogen atom atop different oxygen sites, denoted similarly to Ti sites in Fig. 3, with the addition of ridge-slope and slope-gully sites, where H is placed atop O that is between the respective sites. Adsorption atop Ti was not considered, as calculation on flat surface revealed that Ti site to be vastly energetically unfavorable ($\Delta E_{ads} = 2.13$ eV).

As is known from experimental data on nanoparticles [2], HER occurs on the flat part of the nanoparticle, where Pt is deposited as a co-catalyst. Thus we begin by comparing our ΔG_{H^*} value of flat STO surface (−0.13 eV) to that of platinum (−0.09 eV), which is by far the best catalyst for HER [8]. As the values are close to each other, it can be concluded that flat STO surface is active towards HER, according to our calculations. This statement is valid independent of the fact that solvation, overpotential, and kinetic effects may shift the optimum binding energy

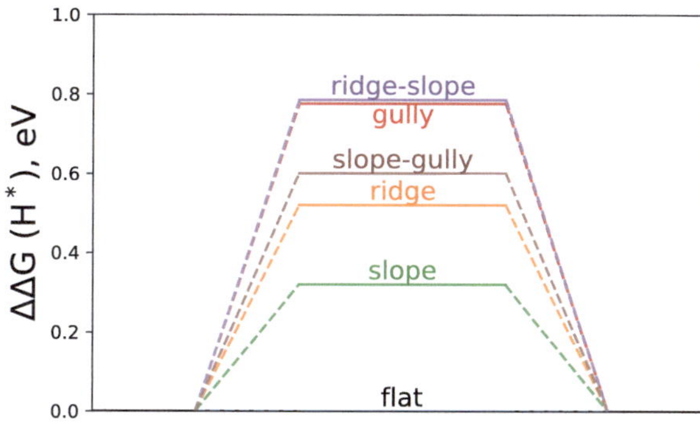

Fig. 5 $\Delta\Delta G_{\mathrm{H}^*}$ values with respect to flat STO surface

of adsorbed hydrogen away from zero, as gas-phase calculations are used here to capture general trends to distinguish between active or inactive sites.

The result is shown in Fig. 5. Ridge, which is the next most active site, has $\Delta\Delta G_{\mathrm{H}^*} = 0.32 \pm 0.2$ eV, and all other sites have $\Delta\Delta G_{\mathrm{H}^*} > 0.32$ eV. Hence it can be concluded that the catalytic activity of STO nanoparticles towards HER is traced to the flat surface when referring to the concept of $\Delta\Delta G_{\mathrm{H}^*}$ as activity descriptor.

3.4 High-Throughput Computing

The STO nanoparticle modeling was performed on the Hawk supercomputer by means of the computer code VASP 5.4. The code is well known for its robust algorithms and at the same time reliable accuracy of the results.

On Hawk the code demonstrated high stability and good scaling. The performance has been measured relative to 512-core reference job. Scaling the job up to 8192 cores, the performance degraded to 20%.

The code reads the input files at the beginning and writes the final state at the end of the execution. No intense I/O operations during the execution are performed.

Each MPI process required ~0.8–1 GB of memory. Output files, depending on the option, 2–4 GB.

Two executables of the VASP code were available—pure MPI and hybrid MPI-OpenMP parallelized. No benefit of the hybrid parallelization on Hawk was observed for the test system. Computational resources required for convergence heavily depend (several orders of magnitude) on the system in question and on the quality of the initial guess of the system's geometry, as well as on the number of particles.

The calculation of hydrogen atom adsorption requires approximately one hour on 4 nodes (512 MPI processes). Calculation of the adsorption of water molecule was taking several hours on 4 nodes (512 MPI processes). Note, that a complete description of the reaction under study required hundreds of calculations of all possible adsorption positions in a combination with various molecule orientations. In addition, necessary for the reaction, the description of water intermediates (O, OH,

H, OOH) required a substantial number of calculations. Large number of calculations failed due to the instability of the initial configuration. Yet, such calculations were necessary for obtaining a complete reaction mechanism.

Taking advantage of good scaling, some heavy tasks were run in 1024 and 2048 MPI processes with the performance drop to not lower than 86%.

The code provides the checkpoints in the form of intermediate geometry as well as electron density. There is a possibility of a manual interruption of the code, with the checkpoints written to the disk. Provided from unfinished calculation, preconverged electron density significantly speeds up the first iteration.

The most computationally demanding jobs were related to the *transition state* calculations. The amount of the resources is **proportional to the number of *Images***—states of the system between the terminating stable states. Technically, each *Image* can be considered a regular calculation. The *transition state* is identified as the Image with the highest energy. Calculation of each *Image* separately would converge in either of two terminating states. Therefore, the method (NEB) implies all intermediate *Images* to be calculated simultaneously, minimizing the forces between them.

Proton migration calculations required several 24-h runs on 44 nodes (4 nodes for each Image, 512 MPI processes per image, 5632 total cores). Running 128 processes per node, we did not encounter any memory limitations. Overall, about 10000 node-hours (10000×128 process-hours) were utilized to calculate various adsorption as well as migration tasks.

4 Conclusions

In this work, we presented the results of our computational study of water dissociation, proton migration process and hydrogen evolution reaction (HER) on flat and stepped surfaces of strontium titanate (STO). CI-NEB calculations suggest that molecular water freely rotates atop slope. It then can dissociate along step virtually without barrier. The dissociated proton can further migrate towards ridge oxygen or gully oxygen, with the latter having a lower barrier. The barriers of proton migration suggest that Volmer-Tafel mechanism of HER in case of incomplete coverage and low overpotentials is possible on slope of the step. Moreover, the barriers on slope of the step are similar to those for flat surface.

Thermodynamic simulations of hydrogen evolution reaction (HER) concluded that flat surface is more catalytically active than stepped surface, which means that not only charge separation and preferential co-catalyst deposition, but surface morphology itself plays a role in enhanced catalytic activity toward HER. The finding of better catalytic activity of the flat surface is consistent with experimental data.

Acknowledgements Calculations have been performed on the computer Hawk at the High Performance Computing Center in Stuttgart (HLRS), within the framework of the project DEFTD 12939.

References

1. Maček Kržmanc, M., Daneu, N., Čontala, A., Santra, S., Kamal, K.M., Likozar, B., Spreitzer, M.: ACS Appl. Mater. Interfaces**13**(1), 370 (2021). https://doi.org/10.1021/acsami.0c16253. PMID: 33351589
2. Mu, L., Zhao, Y., Li, A., Wang, S., Wang, Z., Yang, J., Wang, Y., Liu, T., Chen, R., Zhu, J., Fan, F., Li, R., Li, C.: Energy Environ. Sci. **9**, 2463 (2016). https://doi.org/10.1039/C6EE00526H
3. Cui, M., Liu, T., Li, Q., Yang, J., Jia, Y.: ACS Sustain. Chem. Eng. **7**(18), 15346 (2019). https://doi.org/10.1021/acssuschemeng.9b02672
4. Sokolov, M., Mastrikov, Y.A., Zvejnieks, G., Bocharov, D., Kotomin, E.A., Krasnenko, V.: Catalysts **11**(11) (2021). https://doi.org/10.3390/catal11111326. https://www.mdpi.com/2073-4344/11/11/1326
5. Nørskov, J.K., Rossmeisl, J., Logadottir, A., Lindqvist, L., Kitchin, J.R., Bligaard, T., Jónsson, H.: J. Phys. Chem. B **108**(46), 17886 (2004). https://doi.org/10.1021/jp047349j
6. Yang, T., Zhou, J., Song, T.T., Shen, L., Feng, Y.P., Yang, M.: ACS Energy Lett. **5**(7), 2313 (2020). https://doi.org/10.1021/acsenergylett.0c00957
7. Kresse, G., Hafner, J.: Phys. Rev. B **47**, 558 (1993). https://doi.org/10.1103/PhysRevB.47.558. https://link.aps.org/doi/10.1103/PhysRevB.47.558
8. Hansen, J.N., Prats, H., Toudahl, K.K., Mørch Secher, N., Chan, K., Kibsgaard, J., Chorkendorff, I.: ACS Energy Lett. **6**(4), 1175 (2021). https://doi.org/10.1021/acsenergylett.1c00246

First Principles Investigation of the $LiNbO_3$ and $LiTaO_3$ Lattice Dynamics Under Uniaxial Stress

Mike N. Pionteck, Felix Bernhardt, Kevin Eberheim, Christa Fink, Florian A. Pfeiffer, Nils A. Schäfer, Leonard M. Verhoff, Ferdinand Ziese, and Simone Sanna

Abstract Structural strain severely impacts material properties such as the linear and non-linear optical response. Although μ-Raman spectroscopy is a well-suited technique for the investigation of such effects, it requires the knowledge on the strain dependence of the phonon frequencies. In this work, we model the phonon frequencies in the widely used ferroelectrics lithium niobate and lithium tantalate as a function of uniaxial strain via density functional theory, and compare it with existing μ-Raman spectroscopy results. The majority of phonons shows an increase in frequency under compressive strain, while the opposite is observed for tensile strain. Moreover, for E-type phonons, we observe the lifting of degeneracy already at moderate strain fields (i.e. at $\pm 0.2\%$) along the x and y directions. This work hence allows for the systematic analysis of 3D strains in modern-type bulk and thin-film devices assembled from lithium niobate and tantalate.

1 Introduction

Lithium niobate (LN) and lithium tantalate (LT) are cornerstone materials in nonlinear optics and optoelectronics, with established applications ranging from second-harmonic generation [1, 2], over optical ring resonators, to holographic memory devices [3]. These devices are based on bulk single crystal platforms and have been constantly optimized over the last decades via meticulous analyses of crystal compositional aspects [4–6], degradation effects [7], defect distribution, and the precise imaging of domain structures [8–10]. Single crystalline LN (or LT) thin films on insulating substrates have recently opened up a completely new scenario. This new platform allows for much higher integrability into existing nanoelectronic and

M. N. Pionteck · F. Bernhardt · K. Eberheim · C. Fink · F. Ziese · S. Sanna (✉)
Institut für Theoretische Physik and Center for Materials Research (LaMa), Justus-Liebig-Universität Gießen, Gießen, Germany
e-mail: simone.sanna@theo.physik.uni-giessen.de

F. A. Pfeiffer · N. A. Schäfer · L. M. Verhoff
Institut für Theoretische Physik, Justus-Liebig-Universität Gießen, Gießen, Germany

T. Ludwig et al. (eds.), *High Performance Computing in Science and Engineering '23*,
https://doi.org/10.1007/978-3-031-91312-9_9

nanophotonic structures [11, 12], and outperforms bulk-based devices due to the submicron-scale confinement of optical modes [13–15].

Nevertheless, the transition towards thin-film LN (TFLN) devices emphasizes the importance of crystal properties such as mechanical stress (σ) fields [16, 17], created e.g., due to the bonding process of the TFLN onto the substrate, or occurring as a result of the poling procedure or waveguide fabrication. Accumulated stress does not only increase the chance of critical failure (e.g. due to delamination) [16, 17], but also influences optical key properties such as the refractive index [18, 19], affecting the optical response of a periodically-poled grids [20].

Among the techniques applied for the analysis of LN platforms, Raman spectroscopy can provide information about internal mechanical stress induced, e.g., by lattice mismatch, by external electrical fields [21] or by compositional heterogeneity [22]. In order to quantify occurring stress markers, it is essential to gain fundamental knowledge on the relationship between phonon properties and mechanical stress fields. To date, the only investigations in this regard were conducted via hydrostatic pressure-cells [23, 24], and measure isotropically applied stress. In reality, stress fields in the vicinity of LN domain walls (DW), or mechanical stress in TFLN, are not isotropic [25, 26]. Therefore, it is necessary to theoretically predict and experimentally verify the influence of uniaxial stress on phonon properties, thus providing a fundamental reference for future investigations of mechanical stress fields.

In this work, we provide the theoretical prediction on the evolution of the phonon properties in LN and LT crystals upon uniaxial compressive and tensile stress along the three crystal axes via density functional theory (DFT) and compare our results with existing experimental data.

2 Methodology

The Vienna Ab Initio Simulation Package (VASP) [27] package implementing the DFT offers highly customizable parallelization schemes to exploit the full capability of massively parallel supercomputers such as the HPE Apollo 9000 system (Hawk) at the HLRS. In order to separate the discussion about the computational setup and the discussion about the algorithm performances, we divide the present section in two parts.

2.1 Computational Details

The DFT calculations are performed within the generalized gradient approximation (GGA) [28] in the PBE formulation [29] as implemented in VASP [27]. Thereby, projector augmented waves (PAW) [30] potentials with projectors up to $l = 3$ for Nb and Ta, and $l = 2$ for Li and O have been used, which proved to be accurate enough as demonstrated in a previous work [31]. The electronic wave functions are expanded into a plane-wave basis set up to a kinetic energy of 475 eV. In order to model uniaxial stress, the lattice constants of the rhombohedral unit cell were adjusted according to the expression

$$\boldsymbol{a}_n(\epsilon) = (\boldsymbol{I} + \epsilon)\,\boldsymbol{a}_n(0), \tag{1}$$

where ϵ, $\boldsymbol{I}$, $\boldsymbol{a}_n(0)$, and $\boldsymbol{a}_n(\epsilon)$ denote the strain tensor, the identity matrix, and the n-th unstrained and strained lattice vector, respectively. The strain tensor was then calculated using the elastic compliance tensor S from Ref. [32],

$$\epsilon = S\sigma, \tag{2}$$

and the boundary conditions for uniaxial stress [33]. This means that all components of the stress tensor σ are set to zero, except for the component in the direction in which uniaxial stress is exerted. For instance, if uniaxial stress is to be calculated along the x direction, $\sigma_1 \neq 0 = \sigma_2 = \sigma_3 = \sigma_4$ (Voigt notation) is set as boundary condition. As a result, the strain tensor for uniaxial stress can be calculated using the elastic compliance tensor from Ref. [32] and Eq. 2. Subsequently, the strain tensor and Eq. 1 are used to calculate the lattice vectors under uniaxial stress. By calculating the strained lattice vectors using the compliance tensor, the Poisson ratio and the shear strain are explicitly considered.

The calculations were performed in two ranges for both LN and LT: First, the effects of strain, i.e. compressive as well as tensile strain, on the phonon frequencies have been calculated in fine increments up to values reachable in the experiments. Second, calculations for higher values of strain, but more coarse increments, were performed. For LN and LT this procedure has been performed for lower strains up to 0.10%, as well as for higher values up to 2.4%. Here, increments of 0.02% and 0.4% have been chosen, respectively. The Hellmann-Feynman forces are minimized under a threshold value of 0.005 eV/Å by relaxation of the atomic positions. In this procedure, all lattice vectors calculated from Eq. 1 are fixed. For our calculations of unstrained structures, we have used commonly the accepted experimental lattice constants ($a_R^{LN} = 5.494$ Å, $\alpha^{LN} = 55.867°$ and $a_R^{LT} = 5.474$ Å, $\alpha^{LT} = 56.171°$) [31].

The Γ-centered phonon frequencies and eigenvectors are derived by the frozen-phonon method [34] without symmetry constraints. For the calculation of the Hessian matrix, atomic displacements of 0.015 Å in each Cartesian direction are considered. Since our approach does not take into account the long-range electric fields accompanying the longitudinal-optical (LO) phonons, our calculations are restricted to transversal-optical (TO) phonons only. A $8 \times 8 \times 8$ k-point mesh is used to sample the first Brillouin-zone corresponding to the orthorhombic unit cell, which yields 192 irreducible points.

2.2 Computational Performance

The algorithms implemented in VASP can be parallelized with MPI, OpenMP and a hybrid of MPI and OpenMP. However, our tests show that the code performs best with pure MPI parallelization, i.e. one MPI task for each physical core. In addition, VASP offers several parallelization levels that can be specified by VASP-internal input parameters. The main parallelization levels are determined by KPAR, i.e. the number of **k**-points treated in parallel, and then NPAR, i.e. the number of bands treated in parallel in the matrix diagonalization. According to our tests, these two parameters influence massively the walltime of a calculation. In particular, the parallelization with respect to the **k**-points improves the performance because it divides the problem into independent chunks, whose demand for inter-communication is restricted. Therefore, performance can be maximized by setting KPAR as high as possible, i.e. setting KPAR to the number of **k**-points or the number of nodes, whichever is smaller. In general, the performance is improved when KPAR is a divisor of both the number of **k**-points and the number of nodes. On the other hand, **k**-point parallelization requires a larger amount of memory, since the whole problem must be copied to each **k**-point group. This results in a proportional behaviour between RAM usage and the number of **k**-points treated in parallel. Based on shared memory, NPAR requires less RAM, however, it influences the walltime less than KPAR. NPAR should also be a divisor of bands and number of cores to avoid idle time. The best configuration of KPAR and NPAR depends on the host system. Hence, we needed to test the best configuration for HPE Apollo 9000 (Hawk) and to formulate a rule for setting KPAR and NPAR on Hawk. Then we needed to check the performance of the code with the best setting.

Since the scaling increases with the problem size, we choose a model system representing the $LiNbO_3(0001)$ surface (z-cut), modeled by a large supercell containing 512 atoms, 3072 bands and 32 **k**-points for the performance test. Self-consistent minimization of the electronic energy is performed, which is the main framework of the DFT. The calculations are terminated after self-consistent determination of the electronic ground state. All electronic steps are averaged to the target value T_{LOOP}. This procedure makes the performance test extrapolable to real calculations, which feature a not predetermined number of electronic steps (typically between 30 and 50). Since the first electronic step contains the non-scalable setup of the initial charge, the averaging process yields an upper limit of the electronic step. For the tests, we used 128, 64 and 32 one MPI tasks per node, uniformly distributed among the CPUs. As a result, 64 MPI tasks per node, evenly pinned to every second CPU, turns out to be the most performant configuration.

The left panel of Fig. 1 shows the average walltimes T_{LOOP} of the performance tests with different configurations of KPAR and NPAR as a function of the number of nodes. The green dots denote the average walltime of the best configuration for the respective number of nodes. The tests are in agreement with the experience that setups with higher KPAR require less wall time. However, because of the high

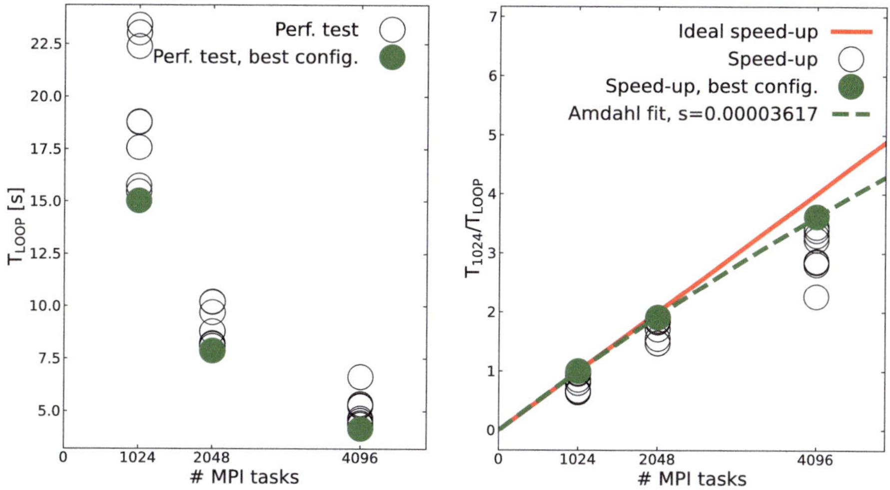

Fig. 1 Results of the performance test; left panel: test results for different parallelization setups, right panel: speed-up of the best guess and Amdahl fits (Amdahl fit for 16, 32 and 64 nodes)

memory demand of the problem, each **k**-point group cannot be arbitrarily scaled. Distributing a single **k**-point pro node is the most performant configuration.

Furthermore, the performance tests yield the rule that NPAR is approximately best set to the square root of the number of MPI tasks employed for each **k**-point group.

Subsequently, we evaluate the scalability of the VASP code on Hawk. As we use 16 nodes at 64 cores at the minimum, we calculate the speed-up as the ratio between the loop walltime for $N = 1024$ and employed cores:

$$S_{1024}(N) = \frac{T_{1024}}{T_{\mathrm{LOOP}}(N)}. \tag{3}$$

The right panel of Fig. 1 shows the speed-up as a function of the number of employed nodes. The green dots again indicate the best configuration of KPAR and NPAR for the respective number of MPI tasks and are fitted to the speed-up by an analytic function. This function is derived from Amdahl's law, which is defined as the speed-up with respect to one MPI task

$$S(N) = \frac{T_1}{T_{\mathrm{LOOP}}(N)} = \frac{1}{s + \frac{p}{N}} \tag{4}$$

where $s, p \in (0, 1)$ with $p = s - 1$ denote the serial and parallel portions of the code, respectively. Thus, the speed-up in (3) can be fitted with the serial part s as parameter in accordance with Amdahl's law

$$S_{1024}(N) = \frac{T_{1024}}{T_{\mathrm{LOOP}}(N)} = \frac{T_{1024}}{T_1}\frac{T_1}{T_{\mathrm{LOOP}}(N)} = \frac{S(N)}{S(1024)} = \frac{s+\frac{1-s}{1024}}{s+\frac{1-s}{N}}. \tag{5}$$

This fit is done for the speed-up of the best configurations at different number of MPI tasks, and plotted as a green line labelled "Amdahl fit" in the right panel of Fig. 1. The obtained serial portion of "Amdahl fit" is at $s = 0.00003617$. This yields a scale efficiency of 87.1% for 4096 cores, i.e. 64 nodes using pinning.

We observe that the speed-up flattens abruptly for more than 64 nodes. This can be explained by our test conditions. Since we use 32 **k**-points for our tests, the parallelization of **k**-points is limited to 32 groups. Consequently, the problem can scale with 16, 32 and 64 nodes. However, for more than 32 nodes, the number of **k**-point groups cannot be increased further, resulting in an abrupt flattening of the speed-up. For calculations featuring a larger number of **k**-points, a linear speedup is expected even for a larger number of cores.

This shows that we are able to scale our code efficiently, if the necessary memory is available and enough **k**-points can be used. Thus, calculations of the optical properties requiring a high number of **k**-points are particularly well suited to exploit the full capability of Hawk. Therefore, we typically use 4096 to 8192 CPUs for our calculations. In total, 65,205,632 core-h were used throughout the reporting period.

3 Results

First, to assess the quality of the models in this work, the calculated phonon frequencies of unstrained stoichiometric LN and LT are compared with previous works [31]. Compared to the experiment, the largest deviation from the experimental values for LN and LT occurs for the $A_1(TO_4)$ mode with 20 and 22 cm^{-1}, respectively. The mean deviation from experimental values for LN and LT is about 8.7 and 8.4 cm^{-1}. Thus, compared with the mean deviation of the theoretical results of Ref. [31, 35] which was about 10.7 and 10.8 cm^{-1} for LN and LT, there is a close agreement with both the theoretical and experimental values. In addition, the typical deviations are also consistent with literature values, such as Refs. [36, 37].

In order to demonstrate the dependence on strain of the phonon frequencies predicted by our calculations, the $E(TO_4)$ mode of LN was chosen [see Fig. 2]. Figure 2a, b show the results of the calculations with lower and higher strain, respectively. The frequency shift of the $E(TO_4)$ mode has an approximately linear dependence on the applied compression. Such a linear dependence is observed in all of our calculations for each mode for both the compressive and tensile strain in stoichiometric LN and LT. This is consistent with the results of LN and LT under hydrostatic pressure by Mendes-Filho et al. [23]. In Fig. 2a the frequency increases up to 6 cm^{-1} and in (b) up to 0.3 cm^{-1}.

Our calculations predict for stoichiometric LN and LT that the degeneracy of the E modes is lifted under strain applied along the x and y direction, due to the reduction

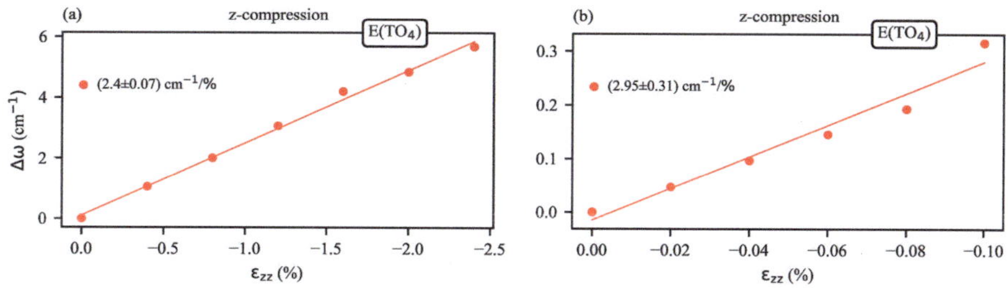

Fig. 2 DFT calculated relative frequency shifts of the E(TO_4) mode of stoichiometric LN as a function of strain in z-direction in **a** steps of 0.4% and **b** steps of 0.02%. The linear fit and the standard deviation of both are plotted and their parameters displayed. The slopes for strain are defined as $(\Delta\omega)/(|\Delta\epsilon|)$ and displayed with the fitting error

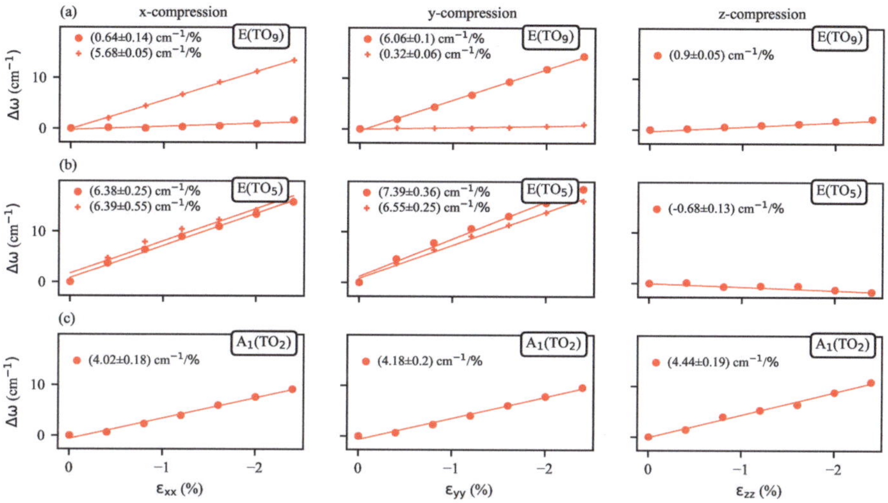

Fig. 3 The DFT calculated relative frequency shifts of **a** E(TO_9) (first row), **b** E(TO_5) (second row) and **c** A_1(TO_2) (third row) modes of stoichiometric LN as a function of strain in steps of 0.4% along the x, y and z direction. The slopes for strain are defined as $(\Delta\omega)/(|\Delta\epsilon|)$ and displayed with the fitting error

of the threefold symmetry. This results in the splitting of the phonon frequency into two branches, which is consistent with previous studies on various 2D materials, that reported a mode splitting under strain due to the reduction of symmetry [38, 39]. Consequently, for strain applied along the x and y direction, all E modes are considered separately. For example, the E(TO_9) mode for compression in x and y directions has a significant splitting [see Fig. 3a]. In contrast, it is noticeable that the E(TO_9) mode remains degenerate for z compression since the considered strain in the z direction of both LN and LT does not lead to any symmetry reduction. This is the case for all E modes of LN and LT under strain.

Table 1 Calculated slopes of transversal A_1 and E modes of stoichiometric LN under compressive and tensile strain, and stoichiometric LT under compressive strain in the x direction, at higher strain in $cm^{-1}/\%$. The slopes for strain are defined as values$(\Delta\omega)/(|\Delta\epsilon|)$

Phonon mode	LN comp. $\Delta\omega[cm^{-1}/\%]$	LN tens. $\Delta\omega[cm^{-1}/\%]$	LT comp. $\Delta\omega[cm^{-1}/\%]$
$A_1(TO_1)$	0.22	−4.49	3.46
$A_1(TO_2)$	4.45	−3.03	5.08
$A_1(TO_3)$	4.82	−7.18	4.69
$A_1(TO_4)$	6.35	−1.03	3.74
$E(TO_1)$	0.48/4.24	−0.61/−2.81	2.25/1.03
$E(TO_2)$	1.74/−0.14	−2.11/−0.11	0.46/3.59
$E(TO_3)$	1.48/3.27	−3.67/−1.05	3.05/2.82
$E(TO_4)$	4.28/-0.11	−4.49/−0.12	2.58/2.90
$E(TO_5)$	6.20/6.94	−8.90/−13.24	10.76/8.61
$E(TO_6)$	12.80/11.03	−7.62/−4.45	13.64/14.27
$E(TO_7)$	4.18/2.40	0.11/0.15	2.00/1.42
$E(TO_8)$	2.87/-0.55	−7.08/0.70	5.30/3.25
$E(TO_9)$	−0.67/7.06	2.98/−7.02	4.77/4.79

Since the slopes for strains in the x and y directions of LN and LT are very similar due to the symmetry of the crystals, Table 1 shows only the slopes for strain in the x direction. With the exception of the $A_1(TO_1)$ and $A_1(TO_4)$ modes, compressive strain leads to an increase in frequency, while tensile strain leads to a decrease in frequency, which is consistent with results from previous research on other materials [40, 41]. Comparing the slopes under the compressive and tensile strain of LN, we find that they hardly differ for the individual modes. This behavior is not surprising, however, it is not universal, as shown for a 2D material by Pak et al. [40]. Although under both compressive and tensile strain the $E(TO_{5/6})$ modes have the largest frequency shifts, the $E(TO_6)$ mode has a larger slope than the $E(TO_5)$ mode under compression. However, this relation turns around under tensile strain as can be seen in Table 1. This applies to strain in the x and y direction for both LN and LT (see, e.g., Ref. [35]).

In addition, the results of LT under compression are listed in Table 1 for the same modes. In direct comparison with the results of LN, they are very similar along both directions, as well as magnitude. This is in agreement with Mendes-Filho et al. [23] and expected since LN and LT are isostructural. Nevertheless, there are some small deviations. In particular, the degeneracy lifting is attenuated compared to LN. For instance, the upper branch of the $E(TO_1)$ mode has a notable lower slope than that of LN. Also, the $E(TO_9)$ mode exhibits a significantly attenuated degeneracy lift for LT under compression than for LN. These deviations can be explained with the help of the eigenvectors. In Fig. 4 the eigenvectors of $A_1(TO_{1-4})$ and $E(TO_{1-9})$ are shown in the rhombohedral unit cell [31, 36]. Such modes with high vibrational contributions from Nb and Ta ions, like the $E(TO_1)$ mode, lead to a slightly different vibrational

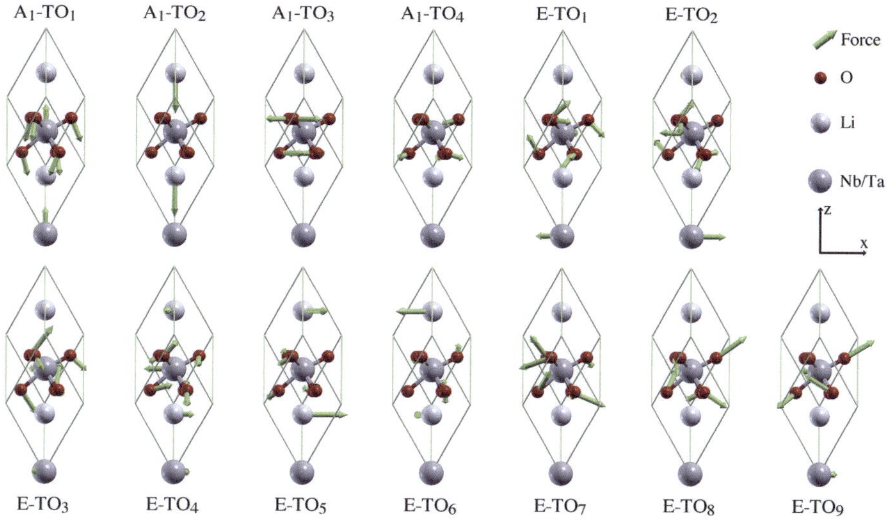

Fig. 4 The eigenvectors of the Raman active vibrations of unstrained stoichiometric LN and LT in the rhombohedral unit cell of the transversal phonon modes with A_1 and E symmetry. Thereby light grey is denoted as Li, dark grey as Nb/Ta, and red as O ion, while the green arrows represent the displacement direction

behavior [31]. The modes that have, in turn, low contributions of Nb and Ta ions with respect to their eigenvectors, are very similar under compression of LN and LT. Since in the E(TO_9) mode the displacement of O is more pronounced than that of Nb/Ta, the reason for the difference between LN and LT may be due to the different nature of the Nb/Ta-O bond, similar to observations in Refs. [31, 36]. The mean Nb-O distance (2.02 Å) is larger than that of Ta-O (2.00 Å). Stronger bonding with the oxygen atoms, and thus larger deformation of the oxygen octahedron, might be the reason for the difference between LN and LT concerning the E(TO_9) mode. Also, slight differences in E(TO_7) and E(TO_8), which are pure distortions of the oxygen octahedron, might be related to the different Nb/Ta-O bonds.

The E($TO_{5/6}$) and A_1(TO_2) modes are characteristic for LN and LT under strain, since these have particularly high slopes, as shown in Table 1 and Fig. 3b and (c). This is also in agreement with Ref. [23]. The E($TO_{5/6}$) are particularly sensitive to strain in the x and y directions. This can be explained by a simple analysis of their eigenvectors. In Fig. 4 it can be seen that these two modes, in contrast to all other modes, are characterized by large displacements of the lithium ions parallel to the xy-plane. Hence, these modes are mostly affected when the distances between the ions parallel to the xy-plane are shortened. The A_1(TO_2) mode has one of the largest slope for strain along the z direction. This can be reasonably explained in terms of the eigenvectors as well. As seen in Fig. 4, the ions of the A_1(TO_2) mode vibrate in z direction. When compression in the z direction shortens the bond length between the ions parallel to their displacement, the frequency is expected to increase. As the

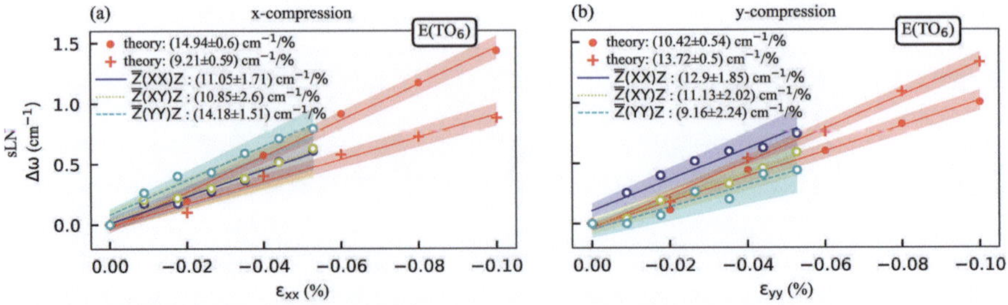

Fig. 5 Comparison of the experimental and theoretical frequency shift (at lower strains): **a**, **b** $E(TO_6)$ phonon mode under compression along the x and y axes of stoichiometric LN. The shaded region around all the curves is the confidence interval of the linear fit

eigenvectors of the $A_1(TO_2)$ mode point mostly in the z direction, it consequently has the largest slope compared to all other modes when compressed in the z direction. All further data and results on LN and LT under compressive and tensile strain are provided in Ref. [35].

As mentioned in the introduction section, the calculations are performed in units of strain and the experiments are performed in units of stress. In order to compare both results we have converted the experimental values of stress into values of strain using the equation given below:

$$\sigma = E\epsilon, \tag{6}$$

where E is Young's Modulus of a sample [42]. The calculated value of Young's Moduli of LN along the x, y, and z axes are 173.6, 173.07, and 201.0 GPa, respectively. For every measured stress, we calculated the respective strain value and plotted the corresponding frequency shift along with the theoretical data as a function of strain [see Fig. 5].

For x compression, the $E(TO_6)$ mode shows lifting of the degeneracy in accordance with the theoretical results. The theoretical results on the degeneracy lifting thus offer a new interpretation of the experimental results. In particular, comparing the data for the $E(TO_6)$ mode with compression in the x direction, it can be seen that the measured data for the $\overline{Z}(YY)Z$ resembles the calculated upper branch, while the measured data for the $\overline{Z}(XY)Z$ and $\overline{Z}(XX)Z$ geometry is similar to the calculated lower branch. For y compression, the same behavior can be observed, with the $\overline{Z}(XY)Z$ and $\overline{Z}(XX)Z$ geometries having the higher slope, respectively, and the $\overline{Z}(YY)Z$ having the lower slope, i.e. the slopes flip for y compression. Consequently, it can be assumed that with the help of different scattering geometries the lifting of the degeneracy by x and y compression can be reproduced experimentally. Based on this interpretation, it can be concluded that the experimental and theoretical data fit very well.

4 Conclusions

Phonon frequencies as a function of uniaxial strain have been calculated for LN and LT and compared with available experimental results. This investigation has shown that all phonon modes are affected by strain. The results are consistent with the roughly linear behavior of the modes and can be reasonably explained by an analysis of the calculated eigenvectors. The linearity of the slopes confirms the previous works under hydrostatic pressure and under external electric fields [21, 23].

Furthermore, calculations have shown that the degeneracy of the E modes is lifted for x and y compression upon symmetry reduction. The splitting has also been observed in available experiments. Based on our investigations, three rules can be formulated to characterize the direction of uniaxial compression of a LN sample using different scattering geometries. This only concerns the E modes. The sample is x compressed if higher frequency shifts are observed for the $\overline{Z}(YY)Z$ geometry than for the $\overline{Z}(XY)Z$ and $\overline{Z}(XX)Z$ geometries when applying the same compressive stress. For y compression, the opposite is the case. In contrast, when no differences are observed between the frequency shift of the E modes for various scattering geometries, it can be concluded that the sample under investigation must be z compressed.

Furthermore, the calculation has shown that the behavior of LT under strain is not significantly different from that of LN. As DWs in LN and LT can be represented in the first approximation as strained lattices, our investigation might be helpful to understand the Raman signal of DWs.

Acknowledgements We gratefully acknowledge financial support from the Deutsche Forschungsgemeinschaft in the research unit FOR5044, (project SA 1948/3-1). The Höchstleistungrechenzentrum Stuttgart (HLRS) is gratefully acknowledged for grants of high-performance computer time.

References

1. Miller, G.D., Batchko, R.G., Tulloch, W.M., Weise, D.R., Fejer, M.M., Byer, R.L.: 42%-efficient single-pass cw second-harmonic generation in periodically poled lithium niobate. Opt. Lett. **22**(24), 1834 (1997)
2. Mouras, R., Fontana, M.D., Bourson, P., Postnikov, A.V.: Lattice site of Mg ion in $LiNbO_3$ crystal determined by Raman spectroscopy. J. Phys.: Condens. Matter **12**(23), 5053–5059 (2000)
3. Yariv, A., Orlov, S.S., Rakuljic, G.A.: Holographic storage dynamics in lithium niobate: theory and experiment. J. Opt. Soc. Am. B **13**(11), 2513 (1996)
4. Iyi, N., Kitamura, K., Izumi, F., Yamamoto, J.K., Hayashi, T., Asano, H., Kimura, S.: Comparative study of defect structures in lithium niobate with different compositions. J. Solid State Chem. **101**(2), 340–352 (1992)
5. Bartasyte, A., Margueron, S., Baron, T., Oliveri, S., Boulet, P.: Toward high-quality epitaxial $LiNbO_3$ and $LiTaO_3$ thin films for acoustic and optical applications. Adv. Mater. Interfaces **4**(8), 2017

6. Li, Y., Schmidt, W.G., Sanna, S.: Defect complexes in congruent $LiNbO_3$ and their optical signatures. Phys. Rev. B **91**(17), 174106 (2015)
7. Furukawa, Y., Kitamura, K., Takekawa, S., Miyamoto, A., Terao, M., Suda, N.: Photorefraction in LiNbO3 as a function of [Li]/[Nb] and MgO concentrations. Appl. Phys. Lett. **77**(16), 2494–2496 (2000)
8. Stone, G., Dierolf, V.: Influence of ferroelectric domain walls on the Raman scattering process in lithium tantalate and niobate. Opt. Lett. **37**(6), 1032 (2012)
9. Nataf, G.F., Guennou, M., Haußmann, A., Barrett, N., Kreisel, J.: Evolution of defect signatures at ferroelectric domain walls in Mg-doped $LiNbO_2$3. Physica Status Solidi (RRL) - Rapid Res. Lett. **10**(3), 222–226 (2016)
10. Rüsing, M., Neufeld, S., Brockmeier, J., Eigner, C., Mackwitz, P., Spychala, K., Silberhorn, C., Schmidt, W.G., Berth, G., Zrenner, A., Sanna, S.: Imaging of 180° ferroelectric domain walls in uniaxial ferroelectrics by confocal Raman spectroscopy: unraveling the contrast mechanism. Phys. Rev. Mater. **2**(10), 103801 (2018)
11. Wang, X., Weigel, P.O., Zhao, J., Ruesing, M., Mookherjea, S.: Achieving beyond-100-GHz large-signal modulation bandwidth in hybrid silicon photonics Mach Zehnder modulators using thin film lithium niobate. APL Photonics **4**(9), 096101 (2019)
12. Rusing, M., Weigel, P.O., Zhao, J., Mookherjea, S.: Toward 3D integrated photonics including Lithium Niobate thin films: a bridge between electronics, radio frequency, and optical technology. IEEE Nanotechnol. Mag. **13**(4), 18–33 (2019)
13. Zhu, D., Shao, L., Yu, M., Cheng, R., Desiatov, B., Xin, C.J., Hu, Y., Holzgrafe, J., Ghosh, S., Shams-Ansari, A., Puma, E., Sinclair, N., Reimer, C., Zhang, M., Lončar, M.: Integrated photonics on thin-film lithium niobate. Adv. Opt. Photonics **13**(2), 242 (2021)
14. Vazimali, M.G., Fathpour, S.: Applications of thin-film lithium niobate in nonlinear integrated photonics. Adv. Photonics **4**(03), 1–18 (2022)
15. Wang, C., Zhang, M., Stern, B., Lipson, M., Lončar, M.: Nanophotonic lithium niobate electro-optic modulators. Optics Express **26**(2), 1547 (2018)
16. Takigawa, R., Tomimatsu, T., Higurashi, E., Asano, T.: Residual stress in Lithium Niobate film layer of LNOI/Si hybrid wafer fabricated using low-temperature bonding method. Micromachines **10**(2), 136 (2019)
17. Prabhakar, K., Patton, R.J., Reano, R.M.: Stress reduction and wafer bow accommodation for the fabrication of thin film lithium niobate on oxidized silicon. J. Vac. Sci. Technol. B **39**(6), 062208 (2021)
18. Friedrich, M., Schmidt, W.G., Schindlmayr, A., Sanna, S.: Optical properties of titanium-doped lithium niobate from time-dependent density-functional theory. Phys. Rev. Mater. **1**(3), 034401 (2017)
19. Zisis, G., Ying, C.Y.J., Ganguly, P., Sones, C.L., Soergel, E., Eason, R.W., Mailis, S.: Enhanced electro-optic response in domain-engineered $LiNbO_3$ channel waveguides. Appl. Phys. Lett. **109**(2), 021101 (2016)
20. Reitzig, S., Rüsing, M., Zhao, J., Kirbus, B., Mookherjea, S., Eng, L.M.: "Seeing Is Believing"—in-depth analysis by co-imaging of periodically-poled X-cut Lithium Niobate thin films. Crystals **11**(3), 288 (2021)
21. Stone, G., Knorr, B., Gopalan, V., Dierolf, V.: Frequency shift of Raman modes due to an applied electric field and domain inversion in $LiNbO_3$. Phys. Rev. B **84**(13), 134303 (2011)
22. Fontana, M.D., Bourson, P.: Microstructure and defects probed by Raman spectroscopy in lithium niobate crystals and devices. Appl. Phys. Rev. **2**(4), 040602 (2015)
23. Mendes-Filho, J., Lemos, V., Cerdelra, F.: Pressure dependence of the Raman spectra of $LiNbO_3$ and $LiTaO_3$. J. Raman Spectrosc. **15**(6), 367–369 (1984)
24. Heyns, A.M., Clark, J.B.: Effect of pressure on the Raman spectra of solids: 1-KC103. J. Raman Spectrosc. **14**(5), 342–346 (1983)
25. Jach, T., Kim, S., Gopalan, V., Durbin, S., Bright, D.: Long-range strains and the effects of applied field at 180° ferroelectric domain walls in lithium niobate. Phys. Rev. B **69**(6), 064113 (2004)

26. Scrymgeour, D.A., Gopalan, V., Itagi, A., Saxena, A., Swart, P.J.: Phenomenological theory of a single domain wall in uniaxial trigonal ferroelectrics: Lithium niobate and lithium tantalate. Phys. Rev. B **71**(18), 184110 (2005)
27. Kresse, G., Furthmüller, J.: Efficient iterative schemes for ab initio total-energy calculations using a plane-wave basis set. Phys. Rev. B **54**(16), 11169–11186 (1996)
28. Perdew, J.P., Chevary, J.A., Vosko, S.H., Jackson, K.A., Pederson, M.R., Singh, D.J., Fiolhais, C.: Erratum: Atoms, molecules, solids, and surfaces: Applications of the generalized gradient approximation for exchange and correlation. Phys. Rev. B **48**(7), 4978–4978 (1993)
29. Perdew, J.P., Burke, K., Ernzerhof, M.: Generalized gradient approximation made simple. Phys. Rev. Lett. **77**(18), 3865–3868 (1996)
30. Blöchl, P.E.: Projector augmented-wave method. Phys. Rev. B **50**(24), 17953–17979 (1994)
31. Sanna, S., Neufeld, S., Rüsing, M., Berth, G., Zrenner, A., Schmidt, W.G.: Raman scattering efficiency in $LiTaO_3$ and $LiNbO_3$ crystals. Phys. Rev. B **91**(22), 224302 (2015)
32. Warner, A.W., Onoe, M., Coquin, G.A.: Determination of elastic and piezoelectric constants for crystals in class (3 m). J. Acoust. Soc. Am. **42**(6), 1223–1231 (1967)
33. The elastic compliance tensor is given in the Cartesian basis which is given by Weis et al. [43]. The applied rhombohedral unit cell is also oriented according to this standard orientation. However, the orientation is not yet unambiguously defined by this convention. In accordance to the standard orientation, the basis vector of the rhombohedral unit cell can point in either the Cartesian +y or -y direction. The former is the general case for the rhombohedral unit cell and the latter for the basis of the elastic compliance tensor. Hence, the basis of the elastic compliance tensor was rotated by 180° about the z axis by a basis transformation according to Ref. [44]. This basis transformation results in the same tensor elements, except from a sign change of $s_{14} \rightarrow -s_{14}$
34. Schmidt, W.G., Bechstedt, F., Srivastava, G.P.: III-V(110) surface dynamics from an ab initio frozen-phonon approach. Phys. Rev. B **52**(3), 2001–2007 (1995)
35. Singh, E., Pionteck, M.N., Reitzig, S., Lange, M., Rüsing, M., Eng, L.M., Sanna, S.: Vibrational properties of $linbo_3$ and $litao_3$ under uniaxial stress. Phys. Rev. Mater. **7**, 024420 (2023)
36. Rüsing, M., Sanna, S., Neufeld, S., Berth, G., Schmidt, W.G., Zrenner, A., Yu, H., Wang, Y., Zhang, H.: Vibrational properties of $LiNbO_{1-x}Ta_xO_3$ mixed crystals. Phys. Rev. B **93**(18), 184305 (2016)
37. Hermet, P., Veithen, M., Ghosez, Ph.: First-principles calculations of the nonlinear optical susceptibilities and Raman scattering spectra of lithium niobate. J. Phys.: Condens. Matter **19**(45), 456202 (2007)
38. Doratotaj, D., Simpson, J.R., Yan, J.-A.: Probing the uniaxial strains in MoS_2 using polarized Raman spectroscopy: a first-principles study. Phys. Rev. B **93**(7), 075401 (2016)
39. Huang, M., Yan, H., Chen, C., Song, D., Heinz, T.F., Hone, J.: Phonon softening and crystallographic orientation of strained graphene studied by Raman spectroscopy. Proc. Natl. Acad. Sci. U. S. A. **106**(18), 7304–7308 (2009)
40. Pak, S., Lee, J., Lee, Y.-W., Jang, A.-R., Ahn, S., Ma, K.Y., Cho, Y., Hong, J., Lee, S., Jeong, H.Y., Im, H., Shin, H.S., Morris, S.M., Cha, S., Sohn, J.I., Kim, J.M.: Strain-mediated interlayer coupling effects on the excitonic behaviors in an epitaxially grown MoS_2/WS_2 van der Waals Heterobilayer. Nano Lett. **17**(9), 5634–5640 (2017)
41. Oliveira, B.G., Zabardasti, A., Goudarziafshar, H., Salehnassaj, M.: The electronic mechanism ruling the dihydrogen bonds and halogen bonds in weakly bound systems of $H_3SiH---HOX$ and $H_3SiH---XOH$ (X = F, Cl, and Br). J. Mol. Model. **21**(4), 77 (2015)
42. Newnham, R.E.: Properties of Materials. Oxford University Press (2004)
43. Weis, R.S., Gaylord, T.K.: Lithium niobate: summary of physical properties and crystal structure. Appl. Phys. Solids Surf. **37**(4), 191–203 (1985)
44. Bond, W.L.: The mathematics of the physical properties of crystals. Bell Syst. Tech. J. **22**(1), 1–72 (1943)

Novel Non-thermal Ablation Mechanics in the Laser Ablation of Silicon

Dominic Klein, Simon Kümmel, and Johannes Roth

Abstract We investigate the non-thermal material dynamics of strongly excited silicon during ultra-fast laser ablation. In contrast to metals, silicon shows strongly excitation-dependent interatomic bonding strengths, which gives rise to a number of unique material dynamics like non-thermal melting, Coulomb explosions and altered carrier heat conduction due to charge carrier confinement. In this study, we report novel non-thermal ablation mechanisms in the ultra-fast single shot laser ablation of silicon and perform large scale massive multi-parallel simulations on experimentally achievable length scales with atomistic resolution. For this, we model the ultra-fast carrier dynamic utilizing the Thermal-Spike-Model coupled to Molecular Dynamics simulations and include the accompanied excitation-dependent non-thermal bonding strength manipulation by application of the excitation-dependent modified Tersoff Potential. Further, we present first results on the systematic construction of the excitation-dependent phase diagram of silicon by thermodynamic integration.

1 Introduction

The non-equilibrium phenomena in highly excited crystalline solids induced by strong laser radiation fields have received much attention in recent years. Despite many theoretical and computational investigations, these ultra-fast processes are still not well understood, especially in the case of covalently bonded semiconductors. Recent publications have reported that covalently bonded semiconductors show bond weakening effects under extreme laser irradiation and can even be driven into

D. Klein · S. Kümmel · J. Roth (✉)
Institut für Funktionelle Materie und Quantentechnologien, Universität Stuttgart, Stuttgart, Germany
e-mail: johannes.roth@fmq.uni-stuttgart.de

D. Klein
e-mail: dominic.klein@fmq.uni-stuttgart.de

S. Kümmel
e-mail: simon.kuemmel@fmq.uni-stuttgart.de

T. Ludwig et al. (eds.), *High Performance Computing in Science and Engineering '23*,
https://doi.org/10.1007/978-3-031-91312-9_10

a state of purely repulsive interatomic bonds. Such ultra-fast ultra-strong excitation gives rise to a lattice dissolution within the order of hundreds of femtoseconds, accompanied with highly unstable states of matter [1–3]. In the context of laser ablation, these non-thermal effects give rise to a complex interplay of competing ultra-fast thermal and non-thermal material dynamics, altering the dominating ablation mechanisms and composition of the ejected material. In this work, we investigate the implications of non-thermal effects on the material dynamics during laser ablation of silicon in two ways. First, we perform massive multi-parallel continuum-atomistic simulations. Here, we apply the so-called Thermal-Spike-Model (TSM) [4] to model the ultra-fast electronic excitation and transport effects within the carrier subsystem and depict the long-time macroscopic material dynamic by the Molecular Dynamics (MD) approach. For the inclusion of non-thermal effects, we implement the electronic excitation-dependent modified Tersoff Potential (MOD*) [3, 5]. Within this simulation framework, we perform ultra-fast single-shot laser ablation simulations on a bulk silicon samples under irradiation of a spatial and temporal Gaussian pulse profile and classify the occurring novel non-thermal ablation mechanisms. Further, we present first investigations on the phase-diagram of strongly excited silicon by thermodynamic integration [6, 7]. Doing so enables us to get more insight into the interplay between the degree of excitation and the structural rearrangements in the material.

2 Simulation Setup

In this work, we model the process of laser ablation on silicon using our in-house simulation package `IMD` [8]. We solve the heat transport equations for the charge carrier density in semiconductors, as well as the charge carrier transport equation by a continuum Finite-Difference (FD) method coupled to MD simulations using the TSM implemented in `IMD`. A detailed description of the method and the applied model can be found in [4, 9]. A simplified sketch of the simulation setup is shown in Fig. 1. The simulation box is spatially divided into CPU cells, distributed over the number of physically available CPUs. Each CPU cell contains the local information of the charge carrier subsystem. On the FD grid, observables of the electronic subsystem, like local carrier temperature T_c or carrier density n_c are calculated within the TSM [4, 9]. For simplicity, only two- or one-dimensional grids are sketched here, while in practice all grids are three dimensional. Each FD cell is further divided into MD cells, which contain the positions and velocities of their associated atoms. The FD and MD subsystems exchange energy by electron-phonon coupling and lead to subsequent heating of the atoms in the embedded MD cell. The force acting on each atom is modeled using the classical modified Tersoff Potential (MOD) [10] and the MOD* [3, 5]. The latter is parametrized from FT-DFT calculations to include non-thermal effects. We choose a spatial and temporal Gaussian laser beam with the surface intensity

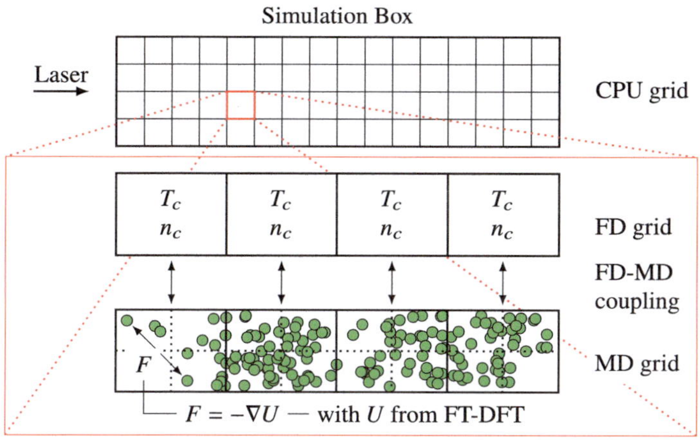

Fig. 1 Schematic presentation of simulation composition in two dimensions. In practice, all grids are three-dimensional. Implementation of periodic boundary conditions and communication schemes are not shown

$$I(t,r) = \underbrace{(1-R)\,\sigma \cdot \frac{4\ln 2}{\pi b^2} \exp\left(-4\ln 2 \frac{r^2}{b^2}\right)}_{\equiv \sigma(r)} \cdot \sqrt{\frac{4\ln 2}{\pi \tau^2}} \exp\left(-4\ln 2 \frac{(t-t_0)^2}{\tau^2}\right) \tag{1}$$

at time t, a spatial full width at half maximum of $b = 500$ nm, a temporal full width at half maximum of $\tau = 100$ fs and a fluence of σ. R is the surface reflectivity and t_0 the time of peak intensity. The optical parametrizations are chosen to reflect a wavelength of $\lambda = 800$ nm and are summarized in [4, 9].

To create the phase diagrams, we use the implementation of the thermodynamic integration in `calphy` 1.0.0 [7] which in our case uses the stable version from 29 September, 2021 of `LAMMPS` [11] to perform the MD part of the thermodynamic integration. For a given pressure and atomic structure, the atomic system is relaxed in an MD simulation, then the temperature is changed over time. From the work done, the temperature-dependent Gibbs free energy can be calculated. The intersection of the Gibbs free energy curves of two structures indicates the phase transition between the two. Repeating this for several different pressures and all pairs of phases, we end up with a complete phase diagram.

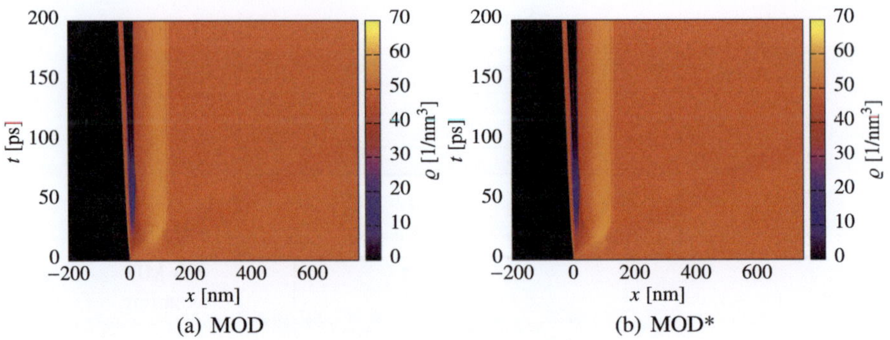

Fig. 2 Comparison of time resolved density histograms for $\sigma = 0.672$ J/cm^2 for the MOD and MOD* interaction potential

3 Results

In Sect. 3.1, we present the novel ablation mechanisms obtained by laser ablation simulations and in Sect. 3.2 we present first data sets on the systematic construction of the phase diagram of strongly excited silicon by thermodynamic integration.

3.1 *Novel Ablation Mechanics and Ablation Depth*

We perform laser ablation simulations on bulk silicon sample Si1D (Sect. 5) applying the established MOD and the excitation-dependent MOD* interaction potential separately. Si1D shows a lateral extension of 22 nm and thus can be seen as a quasi-1D simulation geometry set in the center of the ablation crater. The resulting time resolved density histograms in laser direction x are presented in Fig. 2. The displayed laser fluence $\sigma = 0.627$ J/cm^2 corresponds to roughly 3.5 times the ablation threshold.

The usage of the MOD potential depicts the ablation dynamics being dominated by phase explosion, which is a typical ablation mechanism for metals where non-thermal effects play a minor role in material dynamics [12, 13]. The occurring material responses include a shock wave propagating into the sample by the local speed of sound, a heterogeneous melting front propagating into the sample and the formation of a sub-surface overcritical fluid. The overcritical fluid expands into a mixture of droplets and vapor, ejecting a the surface layer, hence the name phase explosion.

The inclusion of bond weakening non-thermal excitation effects by applying the MOD* potential, however, predict novel ablation mechanisms. We classify the novel ablation mechanics as follows

- Non-thermal evaporation: Non-thermal evaporation is an extreme version of non-thermal melting, where anti-bonding states dominate the interatomic energy landscape. A sufficiently high laser excitation first leads to bond weakening and increase in thermal energy. Like in purely classical theories, the material evaporates, when the thermal energy exceeds the potential energy. However, in the case of non-thermal surface evaporation, the evaporation takes place on the sub-picosecond timescale, leading to an instantaneous ejection of vapor.

- Pre-shockwave non-thermal melting: In laser ablation, we expect the sample surface to rapidly expand due to material heating, thus generating a pressure wave propagating into the sample. With sufficient heating, this results in a heterogeneous melting front. Classically, the melting front propagation speed is restricted to the local speed of sound. However, we observe a non-thermal melting front, propagating with the local speed of light up to the depth, at which internal fluence still meets the conditions for non-thermal melting.
- Non-thermal void formation: Non-thermal void formation is a direct consequence of non-thermal melting. During non-thermal melting, the diamond structure of silicon collapses into a dense amorphous state which is accompanied by an instantaneous drop in local pressure. The melting speed $v_{\text{melt}} = x_{\text{melt}}/\tau_{\text{melt}}$ exceeds the reported equilibrium speed of sound $c_s\,(300\,\text{K}) = 8433$ m/s by three orders of magnitude. As a result, the local material flow is too slow to compensate the resulting vacuum and the material relaxes by the formation of voids.
- Void-induced liquid spallation: Void-induced liquid spallation is another consequence of non-thermal void formation. In the classical sense, spallation is the effect of reflected shockwaves propagating back to the sample surface, interferes with itself upon surface reflection and causes crystalline or liquid material to be ejected. In this new mechanism, the non-thermal voids generate new open surfaces on which the pressure wave gets partially reflected. Additionally, due to the preceding pre-shockwave melting, the material in the laser-affected zone is present in liquid or gaseous form, leading to the emission of liquid droplets.

Despite the drastic deviation from currently published ablation mechanisms in laser irradiated silicon [12, 13], the model is not only in agreement with MD-DFT simulations on non-thermal melting times and non-thermal melting thresholds but also with its prediction of ablation depth. In Fig. 3, we show the resulting ablation depth applying the MOD and MOD* in the range of $0.072\,\text{J/cm}^2 < \sigma < 0.627\,\text{J/cm}^2$, as well as experimental data obtained by single shot ablation experiments under identical laser parameters. The presented data does not only show the significantly better agreement of experimental data and simulation predictions of the MOD* when compared to the MOD but also a nearly 1:1 match to the experimental data underlining the importance of the inclusion of non-thermal effects and validating the novel ablation mechanisms introduced above.

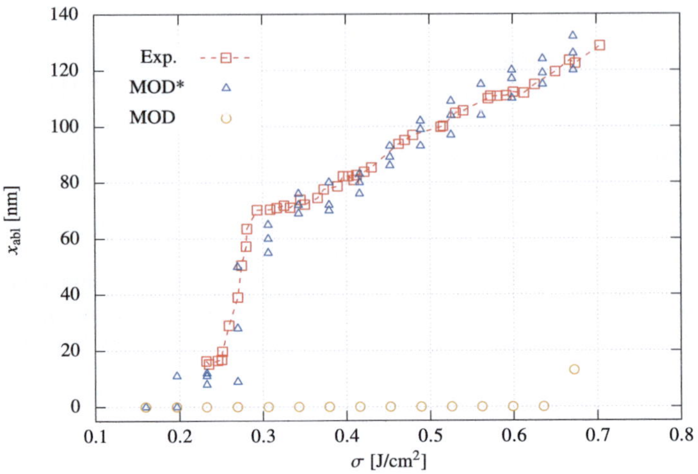

Fig. 3 Ablation depth x_{abl} depending on laser fluence σ. Blue triangles represent data obtained under the MOD* potential and yellow circles represent the data obtained under the MOD potential. Red squares are experimentally measured ablation depths by Zhang [14]

3.2 *Realistic Sample Sizes and Composed Simulations*

The 2020 upgrade from Hazel Hen to Hawk at the High Performance Computing Center Stuttgart and the accompanied substantial increase in performance allow for the atomistic simulation of laser ablation on experimentally achievable laser spot sizes. To our knowledge, this was not possible before and we are the first group to perform atomistic laser ablation simulations at length scales that include the complete crater profile. In Fig. 4, we present snapshots of the spatial density distribution of laser ablation simulations on sample Si2D (Sect. 5) with a lateral extension of over a micrometer, irradiated with a full width at half maximum laser spot size of $b = 0.5\ \mu$m at a wavelength of $\lambda = 800$ nm under usage of the MOD* potential for selected laser fluences at $t = 100$ ps simulation time. Additionally, we show spatial density distribution snapshots obtained by composition of simulation data on the Si1D sample from the previous section. Here, each density distribution along the x-axis corresponds to quasi-1D simulations at peak fluence $\sigma = \sigma(r)$ (see Eq. (1)) at identical simulation time.

Within this simulation framework we observe three notable features:

- Silicon shows three distinct ablation mechanisms. For fluences up to $\sigma = 0.35$ J/cm^2, the ablation mechanism is dominated by conventional thermal ablation, but the morphology of the remaining crater is dominated by the non-thermal melting pool. For fluences of roughly 0.35 J/cm^2 $< \sigma <$ 0.45 J/cm^2, phase explosion becomes the dominating ablation process until it is surpassed non-thermal surface evaporation for fluences $\sigma >$ 0.45 J/cm^2.

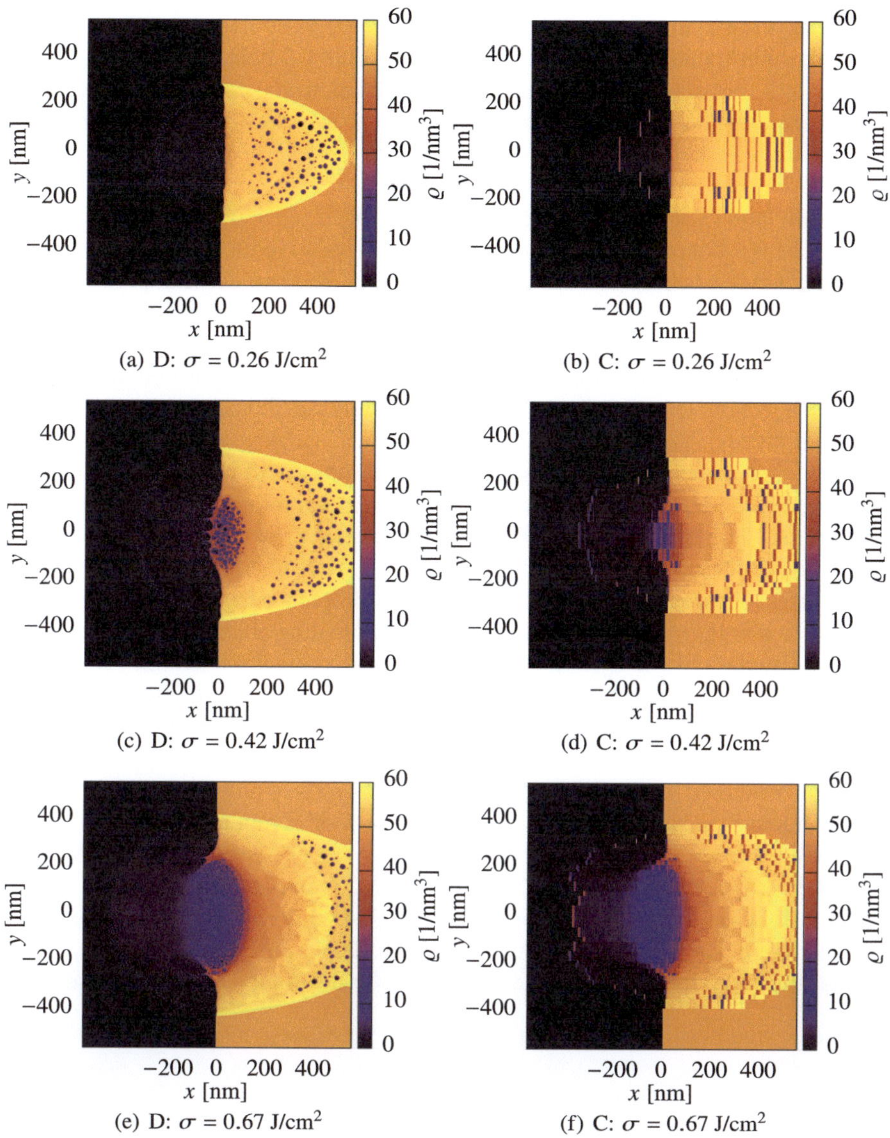

(a) D: $\sigma = 0.26$ J/cm^2

(b) C: $\sigma = 0.26$ J/cm^2

(c) D: $\sigma = 0.42$ J/cm^2

(d) C: $\sigma = 0.42$ J/cm^2

(e) D: $\sigma = 0.67$ J/cm^2

(f) C: $\sigma = 0.67$ J/cm^2

Fig. 4 Snapshots of the spatial density histograms at a simulation time of $t = 100$ ps after peak laser intensity using the MOD* as direct simulations (left D) on sample Si2D and composed from simulations on sample Si1D (right C) for varing laser fluences σ

- Pre-shockwave non-thermal melting occurs at all laser fluences surpassing the non-thermal melting threshold. For regions with $\sigma(r)$ being below the ablation threshold, this non-thermal melting process leads to the formation of an amorphous rim around the ablation crater. The formation of such amorphous structures is currently reported and discussed in literature [15, 16]. Despite the efforts, a conclusive explanation of the underlying mechanisms is still missing. However, our simulation data strongly suggests that the occurrence of the amorphous ring structure is a direct result of non-thermal pre-shockwave melting.
- The accuracy of composed simulations has often been argued and proposed, but not proven since quasi-2D simulations were not feasible in the past. Our composed laser ablation simulations show remarkable accordance depicting the material dynamics of simulations. We observe that the composed simulations yield identical results in terms of crater and non-thermal rim diameters, as well as ablation depth and placement of the ablation crown. However, the lateral material dynamics features are artificially suppressed in the case of composed simulation. Such effects are needed to depict effects, such as nano-bump formation or redeposition of larger clusters slung back by the ablation crown or lateral shockwaves closing non-thermal voids. For fluences above the non-thermal melting threshold, the surface tension of the molt and the ejected droplets become neglectable. This is due to non-thermal evaporation being the dominating ablation mechanic. Composed simulations yield comparable results to the direct simulation setup when only the ejected material and macroscopic crater features are of interest.
  This is especially noteworthy since laser ablation simulation composition shows to be a possibility to save enormous amounts of computational resources. For the shown composed simulations, we used production run simulations at 17 different laser fluences, accumulating to 83,814.13 core-h of computation time. From the obtained data set, all fluences below the chosen peak fluence of $\sigma_p = 0.69$ J/cm^2 can be composed in arbitrary pulse shapes. In comparison, a production run for one fluence on sample Si2D consumes 393.926,42 core-h of computation time for an identical simulation time of $t = 100$ ps. This translates into roughly ≈ 4.7 times the computational resources needed as for the complete Si1D data set.

A more comprehensive and detailed analysis of the effects discussed above can be found in [9].

3.3 The Phase Diagram of Strongly Excited Silicon

To get more insights in the non-thermal effects on the structure, we want to calculate the phase diagrams of silicon at different electron temperatures. This way, we can link the degree of excitation to the structural changes and especially see how the non-thermal melting changes with the degree of excitation.

We started by calculating the melting temperature at zero pressure using the standard MOD [10]. With appropriate parameters and a system containing 4096

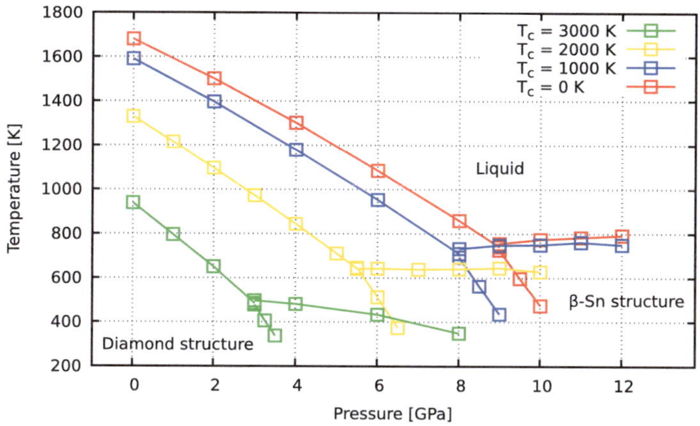

Fig. 5 Temperature-pressure phase diagrams of silicon at different electron temperatures

atoms, we calculate the melting temperature of the diamond structure three times. We find an average melting temperature of 1688.6(12) K which is in excellent agreement with the experimental value of 1687 K [17]. From this first test, we conclude that a system of a few thousand atoms is large enough and decided to use a system containing 4096 atoms in the diamond structure and the liquid phase, as well as 4000 atoms in the β-Sn structure, a high-pressure phase of silicon, to keep the accuracy of the statistics approximately the same for the production runs. When repeated, the melting temperature fluctuates by the order of less than 0.1%.

Since we want to create phase diagrams of silicon at several different electron temperatures, we switch from the standard MOD to the electron temperature-dependent MOD* [3]. These pressure-temperature phase diagrams at different electron temperatures are shown in Fig. 5. The phase transitions are shifted to lower temperatures at higher electron temperatures. This effect is known as bond weakening and is the result of excitation of electrons into antibonding states which are less stable compared to the bonding states. If the electron temperature in silicon is sufficiently high, it undergoes a phase transition to the liquid phase which is a prominent example of a non-thermal phase transition since the lattice temperature might be nearly unchanged for a few femtoseconds [3] after laser-irradiation. This bond weakening can also be seen at high pressures where the β-Sn structure is stable.

We found that the diamond structure is exponentially unstable with the degree of excitation. This also means that non-thermal effect like non-thermal melting become increasingly dominant at high excitations. Unfortunately, it is not possible to find phase transitions into the gaseous state using thermodynamic integration which would have been interesting in the context of laser ablation.

4 Conclusion

In this work, we presented novel non-thermal ablation mechanisms predicted by MD simulations under inclusion of excitation-dependent interatomic interactions. These novel ablation mechanisms show strong implications on the material dynamics of silicon during the ablation process and suggest that silicon shows three distinct ablation regimes dominated by non-thermal melting, phase explosion or non-thermal evaporation. By comparison with experimental data, we have shown that the TSM correctly predicts ablation depth and threshold and thus is a huge improvement on currently existing models. From the electron temperature-dependent phase diagrams, we found the direct connection between the degree of excitation and the instability of the diamond structure in silicon and even the non-thermal melting at sufficiently high excitation. The diamond structure was found to be exponentially unstable with the level of excitement, indicating that non-thermal effects become increasingly dominant in the material dynamics at high excitations.

5 Computational Resources

In Sect. 5.1, we present technical details on the performance of laser ablation simulations using `IMD`. In Sect. 5.2, we present performance reviews of the simulation packages used for the thermodynamic integration of strongly excited silicon. In Sect. 5.3, we summarize the consumed computational resources at the High Performance Computing Center Stuttgart.

5.1 Performance of Laser Ablation Simulations Using IMD

In the current reporting period, we have successfully carried out molecular dynamics simulations of laser ablation on the silicon samples

- Si1D which shows a size of 2225 nm $\times$ 22 nm $\times$ 22 nm and
- Si2D which shows a size of 556 nm $\times$ 17 nm $\times$ 1113 nm

using our in-house simulation package `IMD`. The numbers of atoms, applied processors and achieved performance for a laser ablation production run are shown in Table 1 for each sample.

The used in-house simulation package `IMD` is a typical molecular dynamics simulation code. As such, the main computation load is generated in a single main loop where the classical interactions between atoms are calculated. The program achieves parallelization degrees of the order of 90 and more percent, depending on the simulated experimental setup. The main loop can be fully parallelized with `MPI` due to

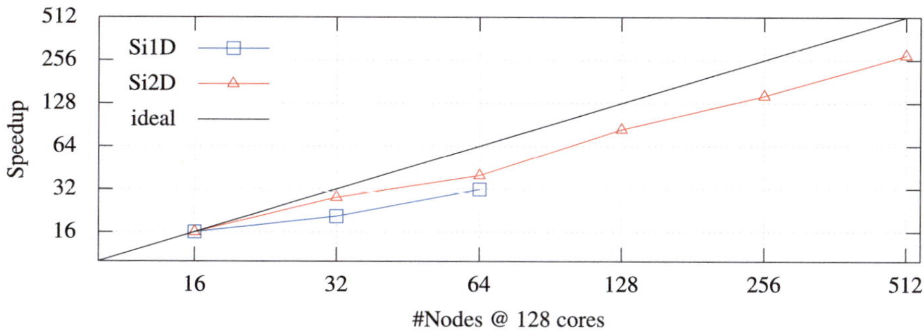

Fig. 6 Scaling behavior of IMD on Hawk at HLRS. This data was obtained with a problem sizes given in Table 1

its short range nature and distributed to any number of processors. The resulting parallelization speedup is only hindered by the communication overhead. The scaling behavior for both samples treated with `IMD` is shown in Fig. 6 and Table 2.

Table 1 Typical resources applied and performance achieved in CPUseconds (cpus per atom n per simulation-step Δt)

Simulation	Wall time	#atoms n	#nodes	#processors	Performance [CPUs/$n\Delta t$]
Si1D	24h	18,121,592	64	8,192	$5.344767 \cdot 10^{-5}$
Si2D	24h	536,870,912	1024	65,536	$5.174720 \cdot 10^{-4}$

Table 2 Scaling behavior in terms of performance (P) and speedup (S) of IMD on Hawk at HLRS. This test was performed with the noted number of particles

# Nodes @	Si1D		Si2D	
128 cores	P [CPUs/$n\Delta t$]	S	P [CPUs/$n\Delta t$]	S
16	$2.688 \cdot 10^{-5}$	16.0	$2.998 \cdot 10^{-5}$	16.0
32	$3.459 \cdot 10^{-5}$	20.6	$5.228 \cdot 10^{-5}$	27.9
64	$5.345 \cdot 10^{-5}$	31.8	$7.463 \cdot 10^{-5}$	39.8
128	–	–	$1.570 \cdot 10^{-4}$	83.8
256	–	–	$2.694 \cdot 10^{-4}$	143.8
512	–	–	$5.175 \cdot 10^{-4}$	276.2

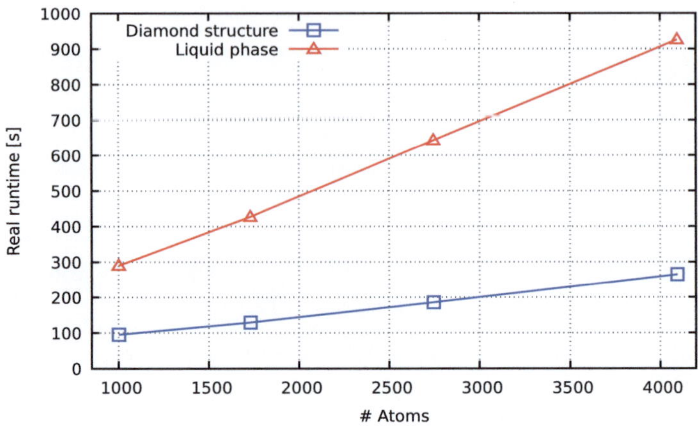

Fig. 7 Scaling of calphy with the number of atoms in a diamond structure and the liquid phase

5.2 *Performance of the Construction of Phase Diagrams*

From calculations with different number of atoms, we find the scaling of the code. As can be seen in Fig. 7, the calculations scale linearly with the number of atoms which can be attributed to the linear scaling of all molecular dynamics simulations with `LAMMPS` [11].

Since calphy uses `LAMMPS` for the molecular dynamics part of the thermodynamic integration which is also the computationally most demanding part, it should be run on several cores. Conveniently, calphy already uses the parallelization schemes implemented in `LAMMPS`. Since we're dealing with comparably small systems in the molecular dynamics context, single-node jobs are an obvious choice.

In order to find the ideal number of cores to use during the production runs, we repeat the calculation of the melting temperature of silicon at zero-pressure with different numbers of cores. The resulting runtime at different number of cores is shown in Fig. 8.

As can be seen in Fig. 8, the best performance is achieved when less than half of a node with 128 cores is used for a single calculation. This allows us to run two calculations at the same time on a single node. A primitive approach would be to always start calculations in pairs. If one calculation is faster than the other, it then has to wait for the slower one to finish before the next pair of calculations is started. Since the runtime of the calculations might vary a lot, we start two calculations simultaneously and track the process IDs of the calculations regularly. This way, if one calculation has finished, we can start a new calculation immediately, ensuring again that only two are running at the same time.

To test this, we calculate the melting temperature of a silicon system containing 4096 atoms three times within on job, two calculations at any given time and do this procedure three times in separate jobs for a good statistics of the runtimes. Each calculation was done on 55 cores, using a total of 110 `MPI` processes per job since

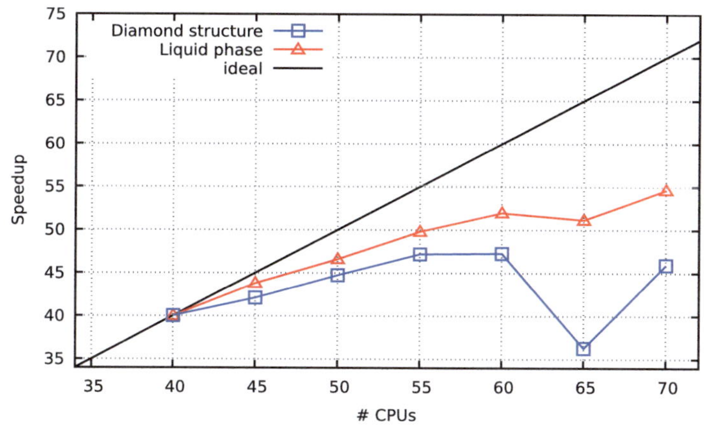

Fig. 8 Speedup of the calculations of the melting temperature of silicon at zero-pressure for different number of cores

we found not using almost all cores of a node to be performing better compared to calculations using all cores. Doing so, we find that each melting temperature calculation now takes an average 427.7(389) s, compared to an average 736.7(291) s when the calculations were done in serial which yields a parallel efficiency of around 86.1% since we can now run two calculation simultaneously.

5.3 Consumed Computing Resources

Since April 2022, a total of 98,248.26 node-hours have been spent on HPE Apollo (Hawk). The used resources are summarized in Table 3. The largest share has been used for the PhD thesis of Dominic Klein [9] studying laser ablation of silicon. Within the course of his master's thesis, Simon Kümmel used 170.86 node-hours for the study of the phase diagram of highly excited silicon. Both studies are subject to the results shown in this work. Azad Gorgis used 22,984.26 node-hours for his master's thesis on the topic of simulations of the 3D printing of metals and the preparation for the project 3DLP-Master (ACID: 44236). Dennis Rapp used 667.70 node-h for test runs and optimization of `IMD`.

Table 3 Consumed computational resources for the project LASMD (ACID:1283) in the time period of April 2022 to May 2023

User	Resources [node-h]	Resources [core-h]	Usage
Dominic Klein	74,425.44	9,526,456.32	Laser ablation
Simon Kümmel	170.86	21,870.08	Phase Diagrams
Azad Gorgis	22,984.26	2,941,985.28	Preparation for 3DLP
Dennis Rapp	667.70	85,465.60	Test runs with `IMD`
Total	98,248.26	12,575,777.28	

References

1. Medvedev, N., Li, Z., Ziaja, B.: Phys. Rev. B - Condens. Matter Mater. Phys. **91**(5) (2015). https://doi.org/10.1103/PhysRevB.91.054113
2. Zier, T., Zijlstra, E.S., Kalitsov, A., Theodonis, I., Garcia, M.E.: Struct. Dyn. **2**(5), 54101 (2015). https://doi.org/10.1063/1.4928686
3. Kiselev, A.: Molecular dynamics simulations of laser ablation in covalent materials. Phd-thesis, Universität Stuttgart (2017). https://doi.org/10.18419/opus-9186
4. Klein, D., Eisfeld, E., Roth, J.: J. Phys. D: Appl. Phys. **54**(1) (2021). https://doi.org/10.1088/1361-6463/abb38e
5. Kiselev, A., Roth, J., Trebin, H.R.: In: High Performance Computing in Science and Engineering '16: Transactions of the High Performance Computing Center Stuttgart (HLRS) 2016 (Springer, Cham, 2017), pp. 189–202. https://doi.org/10.1007/978-3-319-47066-5_14. http://link.springer.com/10.1007/978-3-319-47066-5_14
6. Kirkwood, J.G.: J. Chem. Phys. **3**(5), 300 (2004). https://doi.org/10.1063/1.1749657
7. Menon, S., Lysogorskiy, Y., Rogal, J., Drautz, R.: Phys. Rev. Mater. **5**(10) (2021). https://doi.org/10.1103/PhysRevMaterials.5.103801
8. Roth, J., Eisfeld, E., Klein, D., Hocker, S., Lipp, H., Trebin, H.R.: Eur. Phys. J.: Spec. Top. **227**(14), 1831 (2019). https://doi.org/10.1140/epjst/e2019-800147-7. http://link.springer.com/10.1140/epjst/e2019-800147-7
9. Klein, D.: Laser Ablation of Covalent Materials. Ph.D. thesis, University of Stuttgart (2023). https://doi.org/10.18419/opus-13145
10. Kumagai, T., Izumi, S., Hara, S., Sakai, S.: Comput. Mater. Sci. **39**(2), 457 (2007). https://doi.org/10.1016/j.commatsci.2006.07.013
11. Thompson, A.P., Aktulga, H.M., Berger, R., Bolintineanu, D.S., Brown, W.M., Crozier, P.S., in 't Veld, P.J., Kohlmeyer, A., Moore, S.G., Nguyen, T.D., Shan, R., Stevens, M.J., Tranchida, J., Trott, C., Plimpton, S.J.: Comput. Phys. Commun. **271** (2022). https://doi.org/10.1016/j.cpc.2021.108171
12. Rethfeld, B., Ivanov, D.S., Garcia, M.E., Anisimov, S.I.: J. Phys. D: Appl. Phys. **50**(19), 193001 (2017). https://doi.org/10.1088/1361-6463/50/19/193001. http://stacks.iop.org/0022-3727/50/i=19/a=193001
13. Balling, P., Schou, J.: Rep. Prog. Phys. **76**(3), 36502 (2013). https://doi.org/10.1088/0034-4885/76/3/036502. http://stacks.iop.org/0034-4885/76/i=3/a=036502
14. Zhang, H.: Single-shot femtosecond laser ablation on the nanoscale. Ph.D. thesis, Utrecht University (2013)
15. Bonse, J., Baudach, S., Krüger, J., Kautek, W., Lenzner, M.: Appl. Phys. A: Mater. Sci. Process. **74**(1), 19 (2002). https://doi.org/10.1007/s003390100893, http://link.springer.com/10.1007/s003390100893

16. Werner, K., Gruzdev, V., Talisa, N., Kafka, K., Austin, D., Liebig, C.M., Chowdhury, E.: Sci. Rep. **9**(1), 19993 (2019). https://doi.org/10.1038/s41598-019-56384-0
17. Yamaguchi, K., Itagaki, K.: J. Therm. Anal. Calorim. **69**(3), 1059 (2002). https://doi.org/10.1023/A:1020609517891

Understanding AlInP(001) Surfaces from Massively Parallel Ab-initio Simulations

I. A. Ruiz Alvarado, L. J. Glahn, M. Krenz, A. Bocchini, K. L. Franzke, U. Gerstmann, and W. G. Schmidt

Abstract Total-energy and electronic structure calculations based on density-functional theory are performed in order to determine the atomic structure and electronic properties of $Al_{0.5}In_{0.5}P(001)$ surfaces. It is found that most of the stable surfaces obey the electron counting rule and are characterized by surface atom dimerization. The dimer related surface states are predicted to occur in the vicinity of the bulk band edges. For a very narrow range of preparation conditions, ab initio thermodynamics predicts metal atomic wires formed by surface cations. In case of typical metalorganic vapor-phase epitaxy growth conditions, a surface covered with a monolayer of buckled phosphorus dimers, where half of the phosphorus atoms are hydrogen-saturated, is found to be stable. This structure also reproduces nicely the reflectance anisotropy measured for epitaxially grown AlInP(001) surfaces.

1 Introduction

III-V compound semiconductors are important for various electronic and optoelectronic applications, e.g., high-electron-mobility transistors, light-emitting diodes, photodetectors, electro-optic modulators, and frequency-mixing components. Solar cells made of III-V semiconductors reach the highest efficiencies of any photovoltaic technology so far [1]. The $Al_xIn_{1-x}P$ (AlInP) material system is frequently used as window layer in solar cells, due to its favorable combination of chemical stability, sufficiently wide bandgap, and high quality heteroepitaxial interfaces with many absorbers [2]. In this context, the Fermi level pinning due to AlInP surface states is highly relevant for the device performance [3]. This holds also for the usage of AlInP as intermediate layer to form low resistance ohmic contacts [4].

However, there is little information available on the atomic structure and electronic properties of AlInP surfaces. It has been noted that different growth conditions lead to different degrees of CuPt ordering in the bulk material [5, 6], i.e., alternating group

I. A. Ruiz Alvarado · L. J. Glahn · M. Krenz · A. Bocchini · K. L. Franzke · U. Gerstmann · W. G. Schmidt (✉)
Lehrstuhl für Theoretische Materialphysik, Universität Paderborn, Paderborn, Germany
e-mail: W.G.Schmidt@upb.de

T. Ludwig et al. (eds.), *High Performance Computing in Science and Engineering '23*,
https://doi.org/10.1007/978-3-031-91312-9_11

Fig. 1 $Al_{0.5}In_{0.5}P$ bulk material with cations arranged in CuPt-A and CuPt-B symmetry. Gray, blue, orange and white balls represent Al, In, P, and H atoms, respectively

III layers perpendicular to the $[\bar{1}11]$ or $[1\bar{1}1]$ directions, see Fig. 1. This is likely to be related to the formation of surface dimers, which induce strain in the material [7]. Very recently, some models for clean AlInP(001) surfaces were established, which show that structures different from the ones known from binary III-V(001) surfaces may occur [8]. However, this study was restricted to clean surfaces. Often, AlInP is grown by metal organic vapor phase epitaxy (MOVPE) and chemical beam epitaxy (CBE). Since hydrogen is present in MOVPE and CBE, it can be expected to be present at the AlInP surface as well [9, 10].

Therefore, we here extend the previous study [8] to account for the possibility of surface adsorbed hydrogen. A systematic search for stable $Al_{0.5}In_{0.5}P(001)$ surface structures is performed, where the existence of hydrogen is considered as an additional degree of freedom. The initial geometries for structure optimization are generated by systematically varying the surface stoichiometry of both CuPt-A and CuPt-B type ordered $Al_{0.5}In_{0.5}P$ crystals.

Reflectance anisotropy spectroscopy (RAS), also known as reflectance difference spectroscopy [11], is a powerful tool to characterize surfaces at the atomic level: The measurement of the polarization-dependent reflectivity provides insight in the surface structural and electronic properties [12–15]. RAS is non-destructive and can be used in a wide range of environments. It has been used already to monitor the growth of ternary III-V surfaces and heterostructures [16–20]. The understanding and interpretation of the RAS spectra, however, requires numerical simulations. Therefore, RAS spectra are calculated here for the surface structures most relevant in the surface phase diagram.

2 Theory and Implementation

In detail, density-functional theory (DFT) calculations are performed using the Vienna ab initio simulation package (VASP) [21]. The generalized gradient approximation (GGA) with the PBE functional [22] is used to model the electron exchange and correlation interaction. The electron-ion interaction is described by the projector-augmented wave (PAW) scheme [23, 24]. The electronic wave functions are expanded into plane waves up to a kinetic energy cutoff of 400 eV. The Brillouin zone integration is performed using Γ-centered in an equivalent $8 \times 8 \times 1$ mesh in the full (1×1) surface Brillouin zone. The (001) surfaces are modeled by supercells containing 13 and 14 atomic layers for cation and anion terminated surfaces, respectively, and a vacuum region of ~13.5 Å. The slab bottom dangling bonds are saturated with fractionally charged H atoms. The electric field resulting from the inequivalence of the two surfaces is taken into account by a dipole correction to the electrostatic potential. The atoms are structurally relaxed until they experience forces smaller than 0.02 eV/Å. The calculations are performed at the equilibrium lattice parameter of $Al_{0.5}In_{0.5}P$ calculated to be 5.751 Å.

In order to compare the various clean and hydrogen covered surfaces energetically, the chemical potentials μ_{A_i} of the respective surface constituents need to be taken into account. The surface ground state is determined by the minimum of the thermodynamic potential

$$\Omega = U - TS - \sum_i \mu_{A_i} n_{A_i}, \tag{1}$$

where U is the total energy of the system. In solids, the entropy term, TS, contributes very little to the difference in Ω under usual experimental conditions [25] and is neglected in the following. The chemical potentials μ_{A_i} for A_i = In, Al, and P are restricted by their bulk values

$$\mu_{A_i} \leq \mu_{A_i,\text{bulk}}. \tag{2}$$

Additionally, the stability of $Al_{0.5}In_{0.5}P$ requires

$$\begin{aligned} \mu_{\text{Al}} + \mu_{\text{In}} + 2\mu_{\text{P}} &= \mu_{\text{AlInP}_2,\text{bulk}} \\ &= \mu_{\text{Al,bulk}} + \mu_{\text{In,bulk}} + 2\mu_{\text{P,bulk}} \\ &\quad - \Delta H_{f,\text{AlInP}_2}. \end{aligned} \tag{3}$$

This allows for formulating the formation energy in dependence on $\Delta\mu_{\text{In}}$ and $\Delta\mu_{\text{Al}}$, i.e., the variations of the In and Al chemical potentials with respect to their bulk values. For the heat of formation $\Delta H_{f,\text{AlInP}_2}$ we calculate a value of −1.49 eV.

The hydrogen chemical potential provides an additional and independent degree of freedom. In the approximation of a two-atomic ideal gas, it is written in dependence on partial pressure p and temperature T as

$$\mu_{\text{H}}(p, T) = \frac{k_B T}{2} \left[\ln\{\frac{p\lambda^3}{k_B T}\} - \ln Z_{\text{rot}} - \ln Z_{\text{vib}} \right] - \frac{1}{2} U_{\text{H}_2}, \tag{4}$$

where k_B is the Boltzmann constant, λ the de Broglie thermal wavelength of the H_2 molecule,

$$\lambda = \sqrt{\frac{2\pi\hbar^2}{m k_B T}}, \tag{5}$$

and Z_{rot} and Z_{vib} are its rotational and vibrational partition functions, respectively. By increasing the temperature, the hydrogen chemical potential is lowered, i.e. less energy is gained by taking a H atom out of the reservoir and attaching it to the surface. A variation of the hydrogen chemical potential with respect to its molecular value at zero temperature of $\Delta\mu_{\mathrm{H}} \sim -0.6$ eV corresponds, e.g., to the growth conditions in Ref. [26].

DFT calculations typically underestimate the band gap due to the inaccurate treatment of the electronic self-energy. On the DFT level of theory we calculate a band gap of 1.26 eV for CuPt-ordered AlInP bulk, severely underestimating the experimental value of 2.31 eV [27]. Using self-consistent GW calculations [28], we obtain a value much closer to experiment, 2.33 eV. In addition to self-energy effects, optical spectra are affected as well by local-field and electron-hole attraction effects. They are included here by solving the Bethe-Salpeter Equation (BSE) [29–31]. For numerical reasons, we use a model dielectric function in the BSE [32, 33] and approximate the GW quasiparticle shifts by a scissors operator. This methodology leads to an AlInP bulk dielectric function in very good agreement with experiment, as shown in Fig. 2.

Following Refs. [13, 35–37], the reflectance anisotropy is obtained as

$$\frac{\Delta r}{r}(\omega) = \frac{4\omega}{c}\Im\left[\frac{2\pi\Delta\alpha^{hs}(\omega)}{\varepsilon_b(\omega)-1}\right], \tag{6}$$

where $\varepsilon_b(\omega)$ is the AlInP bulk dielectric function and $\Delta\alpha^{hs}(\omega)$ is the difference of the diagonal tensor elements of the half-slab polarizability corresponding to the respective polarization directions. The calculation of $\Delta\alpha^{hs}$ using a supercell approach contains necessarily not only the surface optical anisotropy, but also contributions from the hydrogen-passivated bottom layer. Fortunately, these contributions are small and occur only for photon energies above 3.5 eV. We get rid of these artifacts by averaging the half-slab polarizabilities for slabs terminated with hydrogen bonds oriented along [110] and [$\bar{1}$10].

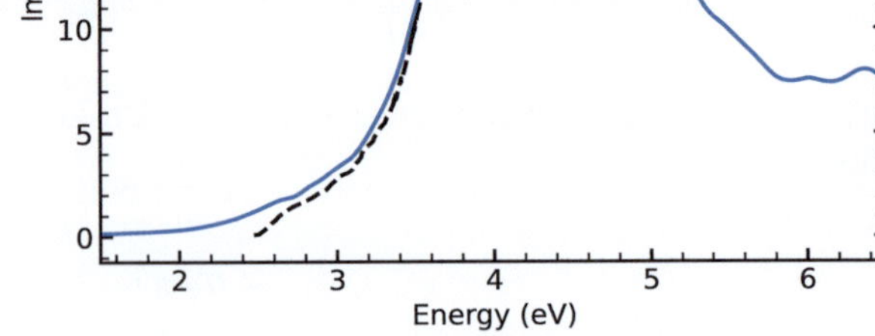

Fig. 2 Imaginary part of the dielectric function of AlInP calculated within the BSE and from experiment [34]

3 Computational Performance

Highly customizable parallelization schemes are implemented in VASP. In particular, parallelization (and data distribution) over bands, parallelization (and data distribution) over plane wave coefficients, as well as parallelization over k-points (no data distribution) can to be used at the same time on massively parallel systems such as the HPE Apollo (Hawk), in order to obtain high computational efficiency.

The performance of Hawk in combination with the available parallelization routines has been tested and optimized with VASP and a test system consisting of 1728 electrons distributed over 1280 orbitals of different symmetry.

The first step of the parallelization procedure is the distribution of the workload related to each orbital on a certain number of cores. It turns out that a number of cores corresponding approximately either to the square root of all available cores or to the total number of cores per node works best for the HRLS Hawk. In addition, this method significantly improves the stability due to reduced memory requirements. Numerical results for the test configuration are shown in Fig. 3 (blue line). This setup results in an improvement up to roughly 2048 cores.

Within the described approach for parallel computing, it is possible to steer the data distribution mode. In particular, the plane wise data distribution in real space can be activated. This allows for a much reduced communication during the Fourier transforms (FFTs). Unfortunately, the resulting load balancing is worsened. Therefore, the suitability of the plane wise data distribution must be tested for the particular computational architecture and in dependence of the number of processors. The results of our tests without plane wise data distribution in real space are shown in Fig. 3 (orange line). As expected, advantages of the plane wise data distribution occur for a relatively small numbers of processors, but are outweighted by load-balancing problems for calculations that employ 4096 cores.

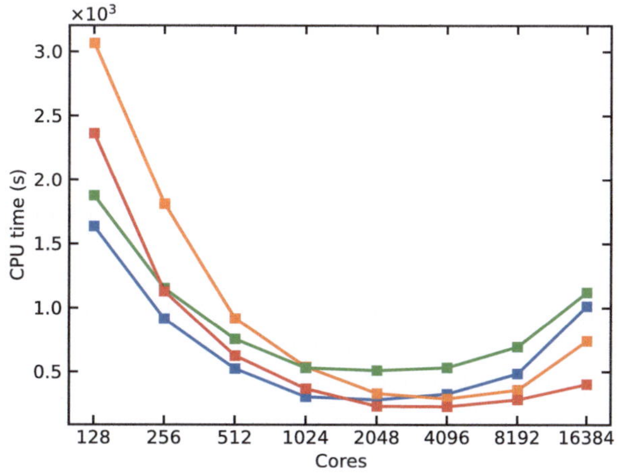

Fig. 3 CPU time on the HRLS Hawk for the self consistent calculation of the electronic structure of a test system consisting of 1728 electrons distributed over 1280 orbitals. See text for details

The computational techniques for the modeling of atomic system discussed here rely on mathematical libraries, in particular the Linear Algebra Package LAPACK or its distributed-memory implementation ScaLAPACK. The calculations discussed above have been performed with ScaLAPACK. This speeds up the calculations by up to a factor of two compared to LAPACK calculations, shown by the green line in Fig. 3. Even if only 128 cores are used, there is a noticeable speed up achieved by using the scalable linear algebra package.

With VASP it is possible to use additionally a parallelization over the k points used to sample the Brillouin zone in the reciprocal space calculations. Thereby it is possible to specify the number of k points that are to be treated in parallel. Within the group of cores that share the work for an individual k point also the electronic states and/or plane wave coefficients are treated in parallel. The results of the corresponding tests are shown in Fig. 3 (red line). It can be seen that the k point parallelization leads to an additional saving of computer time, provided more than 1024 cores are used. The speed up with respect to calculations without k point parallelization amounts up to a factor two. Moreover, this approach allows for the extension of the improved performance up to 4096 cores.

4 Results

4.1 Surface Phase Diagram

Figure 4 contains the calculated phase diagram of the $Al_{0.5}In_{0.5}P(001)$ surfaces structures considered here. Only 10 out of the more than 80 candidate structures considered in our work turn out to be energetically relevant. Schematic top views of these surfaces are compiled in Fig. 5. At hydrogen-deficient conditions ($\Delta\mu_{\mathrm{H}} \sim -1$ eV), P-dimer structures known from clean III-V(001) surfaces [38, 39] dominate the phase diagram. This concerns the c-4×4 structure for anion-rich surfaces, and the β2–2×4 surface for intermediate preparation conditions. The occurrence of [110] oriented P dimers in the P-rich c-4×4 structure and of [$\bar{1}$10] oriented dimers in 2×4 surface reconstructions is in accord with MOVPE-grown AlInP surfaces in N_2 atmosphere [5]. Hetero-dimer structures featuring either In-P or Al-P dimers in the topmost layer occur for cation-rich preparation conditions. They differ from the mixed-dimer structure known from InP(001) by the staggered arrangement of the second-layer cation dimers, due to the size mismatch between Al and In.

For a broad range of intermediate values of the hydrogen chemical potential, $\Delta\mu_{\mathrm{H}} \sim -0.2 \cdots -0.8$ eV, the 2D-2H surface is found to be stable. It corresponds to a 2×2 reconstructed surface formed by P dimers, see Fig. 5. Hydrogen is adsorbed in an alternating sequence on these buckled P dimers. Similar structures were observed for the (001) surface of InP, GaP, and $In_xGa_{1-x}P$ prepared by gas phase epitaxy [9, 10, 40]. The P dimers of the 2×2-2D-2H surface break up at extremely H-rich conditions, $\Delta\mu_{\mathrm{H}} \sim 0$ eV. In this case, the surface P atoms are alter-

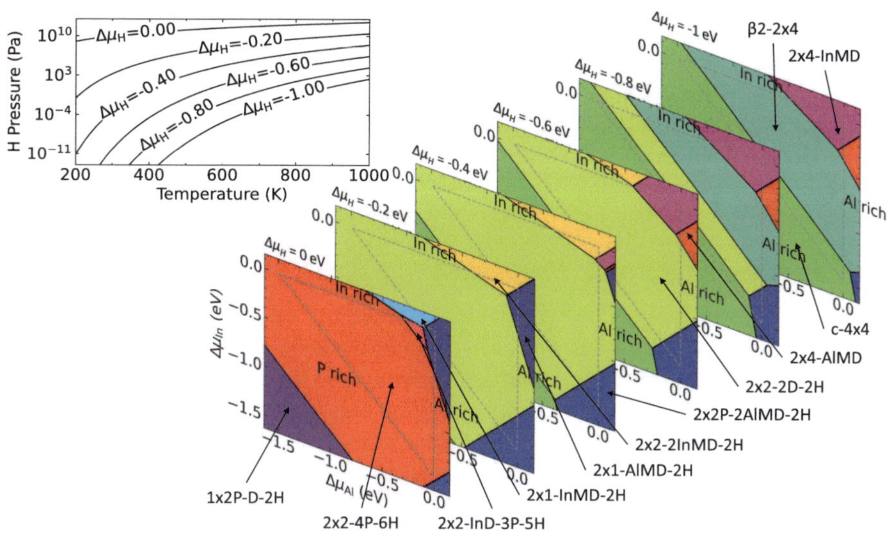

Fig. 4 Energetically favored $Al_{0.5}In_{0.5}P(001)$ surface structures, see Fig. 5, in dependence on the Al, In and H chemical potentials. The thermodynamically allowed range of the chemical potentials according to Eqs. 2 and 3 is indicated by dashed lines. The inset shows the dependence of the hydrogen chemical potential on temperature and partial pressure according to Eq. 4

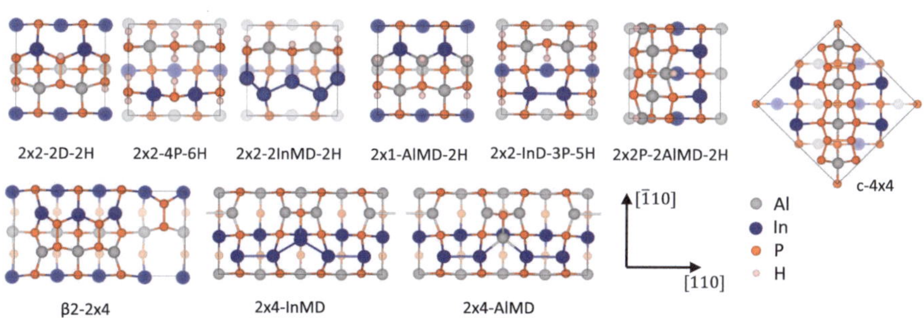

Fig. 5 Top view of stable relaxed clean as well as hydrogenated $Al_{0.5}In_{0.5}P(001)$ surface structures identified in the present work

natingly decorated by one and two hydrogens, forming the 2×2-4P-6H structure. The hydrogen-covered surface P atoms are replaced by H-decorated In-P and Al-P hetero-dimers, if the surface is prepared in hydrogen and cation-rich conditions.

Zunger et al. [41] suggested that the creation of a subsurface selectivity for occupation by a small cation under the strained anion dimer rows, and occupation by a large cation underneath the opening between dimer rows is the main thermodynamic driving force for CuPt ordering in GaInP alloys. The present calculations support this picture for AlInP. All stable P-dimer structures identified here are characterized by P dimers that form above Al, while In occupies the corresponding subsurface positions between the P-dimer rows, see Fig. 5. No such correlation is found for the hetero-dimer structures. Mixed anion-cation dimers form on top of In, e.g., in the 2×2-2InMD-2H structure, as well as above Al, e.g., in the 2×1-AlMD-2H structure. This agrees with the observation that the degree of atomic ordering of MOVPE-grown AlInP depends sensitively on the preparation conditions [5].

4.2 Surface Electronic Properties

Binary III-V surface reconstructions can be understood in terms of a simple electron counting model (ECM) [42]: A surface structure satisfies this model if all cation dangling bonds are empty and all anion dangling bonds are full, given the number of available electrons. This model may be extended to hydrogen-covered surfaces [43] and is satisfied by the majority of the stable ternary surface structures identified here. In fact, the 2×1-AlMD-2H structure is the only exception to the ECM. It is stable in a very narrow range of preparation conditions that are both cation and hydrogen rich.

Not only is the 2×1-AlMD-2H structure the only stable structure that does not comply with the ECM. It is also the only structure that gives rise to a metallic band structure, see Fig. 6. It features one-dimensional Al-In atomic wires along the [110] direction, which are partially H decorated. The Al-In metal bonds result in a quasi-one-dimensional electron band, which disperses along the atomic wire direction and pins the Fermi energy at mid-gap position. Similar electronic properties are calculated for the 2×2-2InMD-2H structure. Here, an In-In atomic wire extends along the [110] direction and gives rise to two strongly dispersive electron bands. These two bands, one occupied and one empty, are separated by a small band gap of about 0.2 eV and pin the Fermi energy at mid-gap position. However, these two structures are stable

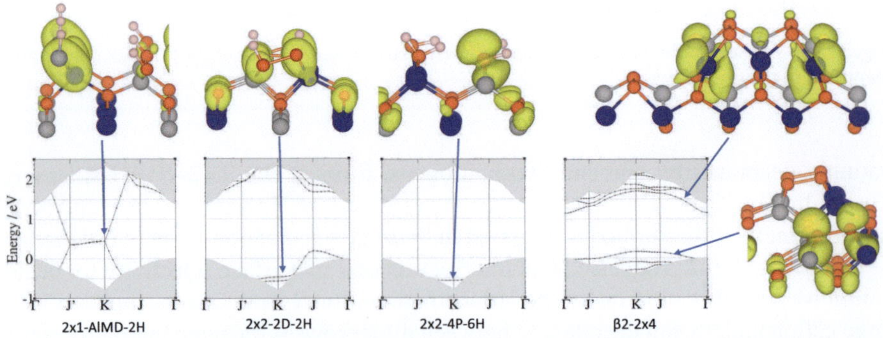

Fig. 6 Band structures of the thermodynamically most relevant surface structures as well as of the 2×1-AlMD-2H surface. Grey regions indicate the projected bulk band structure. ΓJ and ΓJ' are parallel to [$\bar{1}$10] and [110], respectively. The orbital character of state (at the K point) is indicated for prominent surface states. See Fig. 5 for notation of atoms

only in a very small window of preparation conditions, where the cation-rich surface is exposed to hydrogen.

The $\beta2$–2×4, the 2×2-4P-6H, and, in particular the 2×2-2D-2H structures are far more prominent in the calculated surface phase diagram. They correspond to P-rich surfaces without hydrogen, in extremely hydrogen-rich, and at intermediate conditions, respectively. In case of the 2×2-2D-2H structure, an occupied bound surface state is observed that extends slightly above the bulk valence band maximum (VBM), see Fig. 6. This state is mainly related to the dangling bonds on the up atom of the phosphorus dimer. Exposing the P-rich AlInP(001) surface to very hydrogen-rich conditions leads to the 2×2-4P-6H structure and removes essentially all surface states from the band gap region. In fact, the only remaining surface state occurs at about 0.2 eV below the bulk VBM and corresponds to dangling bonds at under-coordinated surface P atoms. In case of the hydrogen-free $\beta2$–2×4 surface, an occupied surface state is observed at about 0.2 eV above the bulk VBM. This only weakly dispersive feature corresponds to an anti-bonding π^* combination of p_z orbitals localized at the third-layer P dimer. Additionally, an empty surface state extends slightly into the region of the bulk band gap. It is related to empty dangling bonds located at the three-fold coordinated second-layer cations. It is mainly In localized, but also has some Al contribution.

Generally, with the exception of the 2×1-AlMD-2H and 2×2-2InMD-2H structures discussed above, it is found that all stable surfaces are characterized by occupied surface states close to the bulk VBM that are derived from surface anions or anion-cation-bonds. The unoccupied surface states occur close to the bulk conduction band minimum and are cation related.

4.3 Surface Optical Properties

In Fig. 7 RAS spectra simulated for energetically relevant surface structure models are shown. Obviously, the calculated spectra are highly surface specific. This demonstrates that RAS is suitable to discriminate between the various stable AlInP surface structures and may be used for growth monitoring. Among the structures considered here, best agreement with experiment [44] is found for the 2D-2H surface model. It is the only model that reproduces the pronounced negative feature at 2.5 eV. Given that the 2D-2H surface is also the most dominating structure in the calculated surface phase diagram and that it is consistent with the measured low-energy electron diffraction and X-ray photoelectron spectroscopy data [26], it can be concluded that the AlInP(001) surface produced in the MOVPE environment is composed of a complete layer of phosphorus dimers. Half of the P dangling bonds on the dimers are hydrogen saturated, and the other half are filled with lone pairs of electrons.

While the surface with a 2D-2H reconstruction is stable for a wide range of growth conditions, there occur also other structures in the calculated AlInP(001) surface phase diagram [26]. The 4P-6H surface, see Fig. 5, stable for very H-rich conditions, features six hydrogens bonded to the topmost four P atoms in the surface

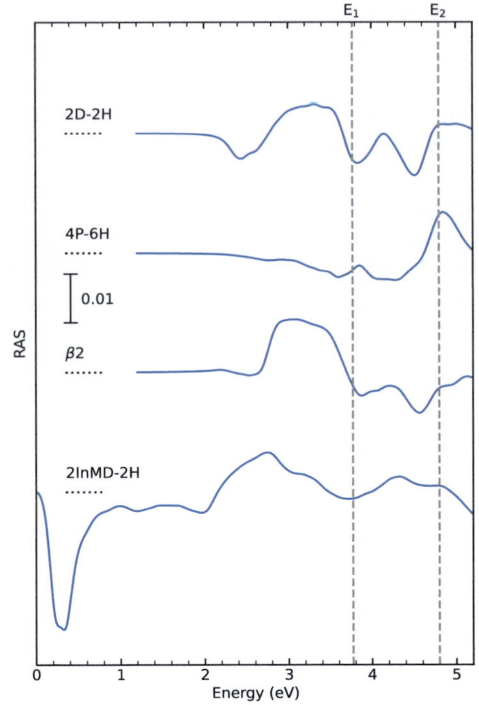

Fig. 7 RAS spectra calculated for energetically relevant surface structure models, see also Fig. 5. The 2InMD-2H spectrum is calculated on the DFT level of theory

unit cell. The hydrogen termination removes essentially all surface states from the AlInP band gap region, see Fig. 6. Accordingly, no significant RAS features appear for low photon energies, see Fig. 7. The optical anisotropies at around 3.8 and 4.8 eV coincide with bulk critical point energies and are likely to be related to surface-modified bulk electronic states [45].

The $\beta 2$ surface, known from various binary III-V compounds, is stable for hydrogen-poor conditions. It is characterized by occupied surface states around the bulk valence band maximum that are due to antibonding π^* combinations of p_z orbitals localized at the P dimer atoms. Electronic transitions from these dimer states into unoccupied surface resonances as well as transition from occupied surface resonances into empty sp^2 orbitals at threefold coordinated surface cations cause a series of positive RAS features between 2.5 and 4.5 eV, see Fig. 7.

Another very interesting structure is formed for a small range of H-rich and In-rich conditions. It consists of In atomic wires that form along the [110] direction. The band structure of this 2InMD-2H surface features two bands, one occupied, one empty, that strongly disperse in wire direction, but have nearly no dispersion perpendicular to the wire direction. These two bands are separated by only about 0.3 eV along the $J' - K$ direction of the surface Brillouin zone and pin the Fermi energy at midgap position. Transitions between these two bands give rise to an extremely strong optical anistropy for low photon energies, see Fig. 7. Similar findings are reported for other surface atomic wire systems [46].

5 Conclusions

In summary, both ab initio thermodynamics as well as the calculated surface optical anisotropy strongly suggest that InAlP(001) surfaces produced in the MOVPE environment are terminated by a complete layer of phosphorus dimers. Half of the P dangling bonds on the dimers are hydrogen saturated, and the other half are filled with lone pairs of electrons. These lone pairs form a bound surface state slightly above the valence band maximum. The calculations suggest the existence of further structures, depending on the surface preparation conditions. P-rich surfaces are characterized by P dimers, while Al-P and In-P hetero-dimers form for more cation-rich surface preparation conditions. Most stable surface structures obey the electron counting rule. Dimer-related surface states are found close to the bulk valence band maximum. Exposure of the surface to extreme hydrogen-rich conditions is predicted to quench the dimerization and to remove all surface states from the region of the bulk band gap. For cation- and hydrogen-rich preparation conditions, metal atomic wires may form on the surface. They give rise to quasi-one-dimensional surface bands that pin the Fermi energy in the mid gap region. These metal atomic wires are predicted to give rise to very strong optical anisotropies.

Acknowledgements The Höchstleistungs-Rechenzentrum Stuttgart (HLRS) is acknowledged for grants of high-performance computer time. Financial support by DFG (PAK981) is gratefully acknowledged.

References

1. Kalogirouo, S.A. (ed.): McEvoy's Handbook of Photovoltaics: Fundamentals and Applications. Academic Press (2017)
2. Kim, C.W., Park, G.Y., Shin, J.C., Kim, H.J.: Appl. Sci. **12**(2) (2022)
3. Lepkowski, D.L., Kasher, T., Boyer, J.T., Chmielewski, D.J., Grassman, T.J., Ringel, S.A.: IEEE J. Photovolt. **10**(3), 758 (2020)
4. Iwata, K., Asahi, H., Ogura, T., Sumino, J., Gonda, S., Ohki, A., Kawaguchi, Y., Matsuoka, T.: In: Seventh International Conference on Indium Phosphide and Related Materials, pp. 183–186 (1995)
5. Jinghua, Z., Xiaohong, T., Jinghua, T.: CrystEngComm **11**(6), 1068 (2009)
6. Yuan, W., Hall, D.C.: J. Appl. Phys. **113**(10), 103515 (2013)
7. Zhang, S.B., Froyen, S., Zunger, A.: Appl. Phys. Lett. **67**(21), 3141 (1995)
8. Ruiz Alvarado, I.A., Karmo, M., Runge, E., Schmidt, W.G.: ACS Omega **6**(9), 6297 (2021)
9. Schmidt, W.G., Hahn, P.H., Bechstedt, F., Esser, N., Vogt, P., Wange, A., Richter, W.: Phys. Rev. Lett. **90**, 126101 (2003)
10. Hahn, P.H., Schmidt, W.G., Bechstedt, F., Pulci, O., Del Sole, R.: Phys. Rev. B **68**, 033311 (2003)
11. Aspnes, D.E., Studna, A.A.: Phys. Rev. Lett. **54**, 1956 (1985)
12. Weightman, P., Martin, D.S., Cole, R.J., Farrel, T.: Rep. Prog. Phys. **68**, 1251 (2005)
13. Schmidt, W.G.: Phys. Status Solidi (b) **242**, 2751 (2005)
14. May, M.M., Lewerenz, H.J., Hannappel, T.: J. Phys. Chem. C **118**(33), 19032 (2014)

15. Lastras-Martínez, A., Ortega-Gallegos, J., Guevara-Macías, L.E., Nuñez-Olvera, O., Balderas-Navarro, R.E., Lastras-Martínez, L.F., Lastras-Montaño, L.A., Lastras-Montaño, M.A.: APL Mater. **2**(3), 032107 (2014)
16. Zorn, M., Jönsson, J., Krost, A., Richter, W., Zettler, J.T., Ploska, K., Reinhardt, F.: J. Cryst. Growth **145**(1–4), 53 (1994)
17. Haberland, K., Bhattacharya, A., Zorn, M., Weyers, M., Zettler, J.T., Richter, W.: J. Electron. Mater. **29**(4), 468 (2000)
18. Krahmer, C., Philippens, M., Schubert, M., Streubel, K.: Phys. Status Solidi c **3**(3), 655 (2006)
19. Krahmer, C., Philippens, M., Schubert, M., Streubel, K.: J. Cryst. Growth **298**, 18 (2007)
20. Supplie, O., Heinisch, A., Sugiyama, M., Hannappel, T.: In: 2018 IEEE 7th World Conference on Photovoltaic Energy Conversion (WCPEC) (A Joint Conference of 45th IEEE PVSC, 28th PVSEC & 34th EU PVSEC), pp. 3923–3926. IEEE, Waikoloa Village, HI (2018)
21. Kresse, G., Furthmüller, J.: Comp. Mat. Sci. **6**, 15 (1996)
22. Perdew, J.P., Burke, K., Ernzerhof, M.: Phys. Rev. Lett. **77**, 3865 (1996)
23. Blöchl, P.E.: Phys. Rev. B **50**, 17953 (1994)
24. Kresse, G., Joubert, D.: Phys. Rev. B **59**, 1758 (1999)
25. Wippermann, S., Schmidt, W.G.: Phys. Rev. Lett. **105**, 126102 (2010)
26. Glahn, L.J., Ruiz Alvarado, I.A., Neufeld, S., Zare Pour, M.A., Paszuk, A., Ostheimer, D., Shekarabi, S., Romanyuk, O., Moritz, D.C., Hofmann, J.P., Jaegermann, W., Hannappel, T., Schmidt, W.G.: Phys. Status Solidi (b) **259**(11), 2200308 (2022)
27. Cheong, J.S., Ng, J.S., Krysa, A.B., Ong, J.S.L., David, J.P.R.: J. Phys. D Appl. Phys. **48**(40), 405101 (2015)
28. Shishkin, M., Marsman, M., Kresse, G.: Phys. Rev. Lett. **99**(24), 246403 (2007)
29. Albrecht, S., Reining, L., Del Sole, R., Onida, G.: Phys. Rev. Lett. **80**(20), 4510 (1998)
30. Rohlfing, M., Louie, S.G.: Phys. Rev. Lett. **81**(11), 2312 (1998)
31. Schmidt, W.G., Glutsch, S., Hahn, P.H., Bechstedt, F.: Phys. Rev. B **67**, 085307 (2003)
32. Bechstedt, F., Del Sole, R., Cappellini, G., Reining, L.: Solid State Commun. **84**(7), 765 (1992)
33. Liu, P., Kim, B., Chen, X.Q., Sarma, D.D., Kresse, G., Franchini, C.: Phys. Rev. Mater. **2**(7), 075003 (2018)
34. Ochoa-Martínez, E., Barrutia, L., Ochoa, M., Barrigón, E., García, I., Rey-Stolle, I., Algora, C., Basa, P., Kronome, G., Gabás, M.: Sol. Energy Mater. Sol. Cells **174**, 388 (2018)
35. del Sole, R.: Solid State Commun. **37**(7), 537 (1981)
36. Del Sole, R., Fiorino, E.: Phys. Rev. B **29**(8), 4631 (1984)
37. Manghi, F., Del Sole, R., Selloni, A., Molinari, E.: Phys. Rev. B **41**(14), 9935 (1990)
38. Schmidt, W.G.: Appl. Phys. A **65**, 581 (1997)
39. Schmidt, W.G.: Appl. Phys. A **75**, 89 (2002)
40. Cheng, S.F., Sun, Y., Law, D.C., Visbeck, S.B., Hicks, R.F.: Surf. Sci. **600**(14), 2924 (2006)
41. Bernard, J.E., Froyen, S., Zunger, A.: Phys. Rev. B **44**, 11178 (1991)
42. Pashley, M.D.: Phys. Rev. B **40**, 10481 (1989)
43. Karmo, M., Ruiz Alvarado, I.A., Schmidt, W.G., Runge, E.: ACS Omega **7**(6), 5064 (2022)
44. Ruiz Alvarado, I.A., Zare Pour, M.A., Hannappel, T., Schmidt, W.G.: Phys. Rev. B (submitted)
45. Schmidt, W.G., Bernholc, J.: Phys. Rev. B **61**, 7604 (2000)
46. Chandola, S., Hinrichs, K., Gensch, M., Esser, N., Wippermann, S., Schmidt, W.G., Bechstedt, F., Fleischer, K., McGilp, J.F.: Phys. Rev. Lett. **102**(22), 226805 (2009)

Molecular Dynamics Simulation of Laser-Based Powder Bed Fusion

Kevin Vietz, Jonas Schmid, and Johannes Roth

Abstract Laser-based powder bed fusion describes an additive manufacturing process in which a metal or alloy powder is printed layer by layer. For each layer, metal powder is distributed on the underlying sample, before the powder is fused onto the previous layer by melting it with a laser. In order to give predictions on the experimental side in terms of defects structures and the mixing of alloys, we focus on molecular dynamics (MD) simulation. This allows us to describe the highly non-equilibrium dynamics on an atomistic scale.

1 Introduction

An introduction to the process of laser-based powder bed fusion, also known as selective laser melting (SLM), has been given already in a previous HLRS report [1]. This method is a type of additive manufacturing. A laser beam is scanned across a powder, melting and fusing it at desired locations and thereby generating a product. The unmelting powder is removed and the product is polished.

Here, we will report on a few selected topics from the whole process: the SLM simulation of binary alloys, the study of resolidification after SLM, and performance improvements.

K. Vietz · J. Schmid · J. Roth (✉)
Institute for Functional Matter and Quantum Technologies, University of Stuttgart, Stuttgart, Germany
e-mail: johannes.roth@fmq.uni-stuttgart.de

K. Vietz
e-mail: kevin.vietz@fmq.uni-stuttgart.de

J. Schmid
e-mail: jonas.schmid@fmq.uni-stuttgart.de

T. Ludwig et al. (eds.), *High Performance Computing in Science and Engineering '23*,
https://doi.org/10.1007/978-3-031-91312-9_12

2 Theoretical Background

The dynamics of the powder, as well as the interaction between powder and laser is modeled by molecular dynamics (MD) simulation using the `IMD` [3]. The use of MD simulations allows a structural analysis for melting as well as the nucleation and the mixing of alloys on an atomistic base.

The interacting potentials between the atoms are depicted by the embedded atom method [2]. The total interacting energy of the system

$$E_{\text{tot}} = \sum_i \left[F(\rho_i) + \frac{1}{2} \sum_{i \neq j} \phi(r_{ij}) \right] \tag{1}$$

contains of the electrostatic core-core repulsion $\phi(r_{ij})$, as well as an embedding function F depending on the host electron density ρ_i.

The laser power P_0 is described by a Gaussian power profile. The laser velocity is set constant, while also the total laser power is approached constant due to the short time scale of the simulation.

The laser absorption is modeled by an energy re-scaling on an atomistic base. The radiant power inside the metal at a depth z is calculated according to the Lambert-Beer law given by

$$P(z) = P_0 \, \mathrm{e}^{-\alpha z} \tag{2}$$

depending on the material specific absorption coefficient α.

3 Laser-Based Powder Bed Fusion of TiAl

We investigate the applicability of binary alloys for additive manufacturing by performing simulations of laser-bed powder bed fusion with titanium aluminide and studying the material dynamics. Therefore, we compare coated powder grains with pre-alloyed grains and investigate the influence of the laser scanning speed. To speedup our simulations, some performance improvements are shown as well as resources and the scalability of our computations.

3.1 *Suitability of Coated Powder Grains*

We perform selective laser simulations of coated powder grains, as well as pre-alloyed titanium-aluminide powder grains. The schematic illustrations of the initial grain structures are shown in Fig. 1. We construct the initial powder grains with a diameter

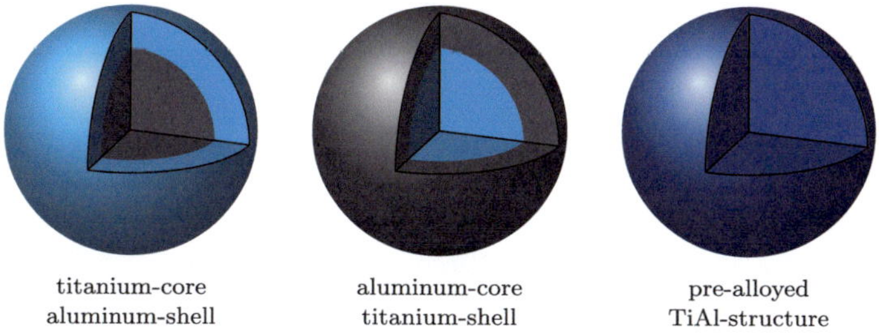

Fig. 1 Schematic illustration of the investigated powder grain structures

of 400 Å and place them on a fixed ground of atoms. The initial titanium:aluminum ratio is set to 50:50 which leads to about 2 000 000 atoms per powder grain.

The time evolution during selective laser simulations for the different structures is shown in Fig. 2. Due to the chosen laser parameters, we observe that all grains burst and melt completely. But despite infusing enough energy to melt the grains completely we observe an inhomogeneous mixing for the coated structures in the long time limit.

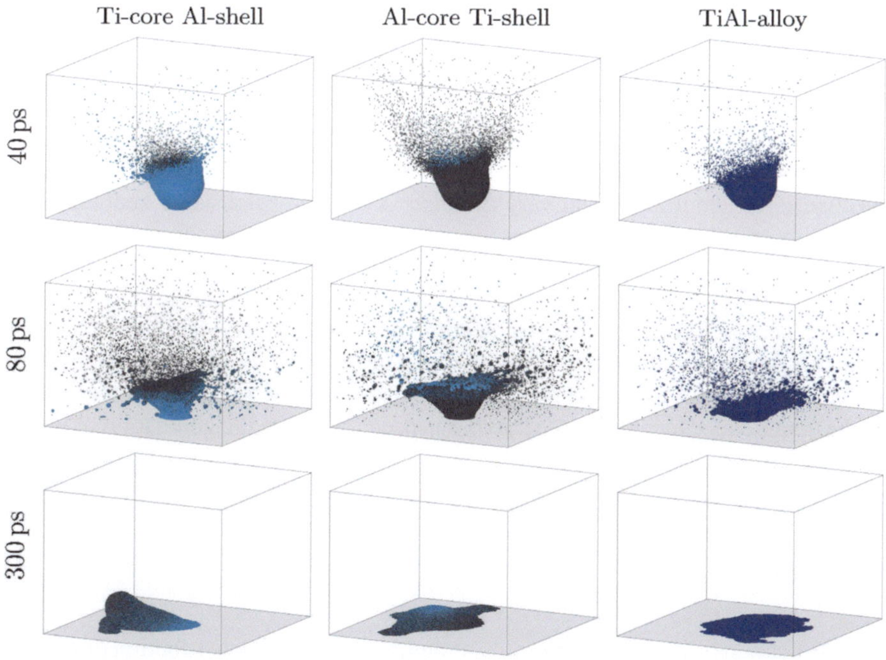

Fig. 2 Time evolution of selective laser simulations for coated grains in comparison to pre-alloyed titanium-aluminide grains

Fig. 3 Snapshot after 80 ps of different laser scanning speeds. From left to right the used velocities are $v = 0.1$, $v = 0.3$ and $v = 0.5$ in IMD-units

For the investigation of the distribution ratios, we perform a cluster analysis and only consider atoms within the biggest leftover cluster. We then find titanium:aluminum ratios of 58:42 for the aluminum coated titanium core, 52:48 for the titanium coated aluminum core and 55:45 for the pre-alloyed grain simulation. So we are left with less aluminum in all simulations, but the more aluminum atoms are protected by surrounding titanium atoms the less aluminum is vaporized.

3.2 *Laser Scanning Speed Comparison*

Because of the different scaling between simulated and real powder grains and especially the scaling in time, it is by far not trivial to choose a reasonable set of laser parameters to depict the material dynamics in a good manner. Therefore, we perform selective laser simulations and vary the laser scanning speed while keeping other parameters constant. The simulations are performed with the pre-alloyed TiAl-grains and velocities are given in IMD-units, which corresponds to $\text{Å}/10.18\,\text{fs}$.

Figure 3 shows a snapshot during the selective laser simulations for different laser scanning speeds. While the infusing energy varies greatly, we obtain very different behaviors during the simulation. The grain on the left shows a strong explosion, while the grain on the right side is not fully melted. Also, a kind of splashing fountain effect can be obtained for increasing the scanning speed, results in the generation of bigger detached droplets.

4 Resolidification

Resolidification has been studied in a simulation of a bimodal distribution of spheres as shown in Fig. 4. The initial sample is composed of aligned spheres covered with argon gas (Fig. 5). After the laser has passed the sample, the material solidifies (Fig. 6). Aluminum recovers its energetically favorable crystalline fcc structure, some grains fuse coherently while others crystallize in random orientation and are

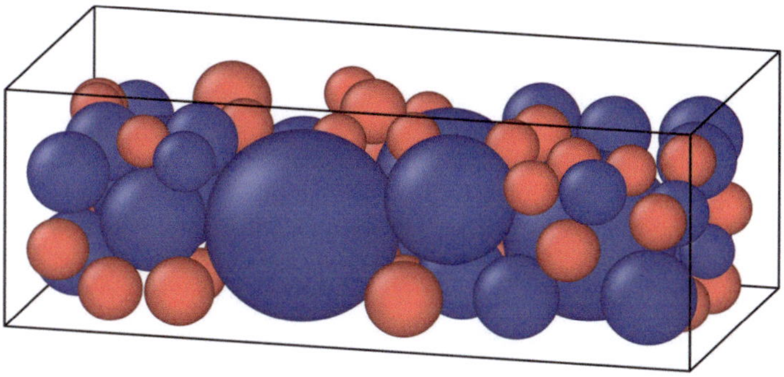

Fig. 4 Bimodal distribution of spheres with Gaussian size distribution

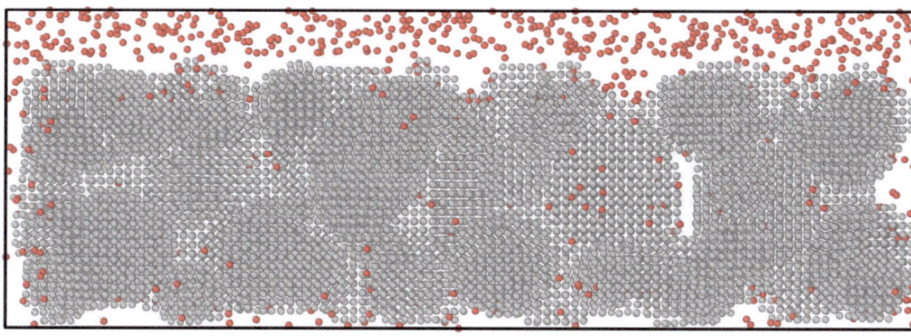

Fig. 5 Crystal structure of the aluminum spheres (grey) covered by argon (red)

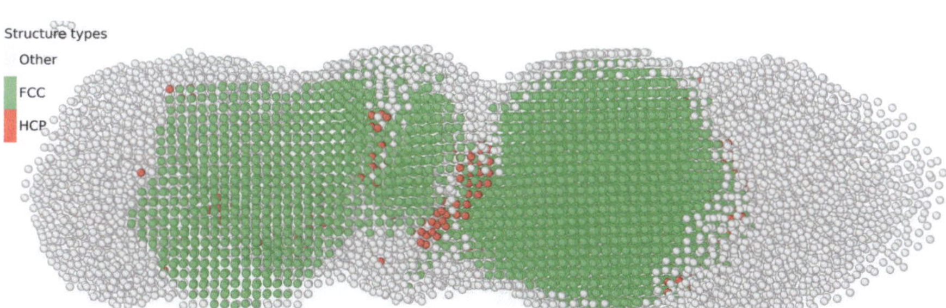

Fig. 6 Cross section through the sample after resolidification. Colors show different crystal structure types and grain boundaries. Liquid atoms or atoms with undefined crystal structure are colored gray, red stacking faults (hcp) ar observed between grains with genuine aluminum crystal structure (fcc), colored green. At the left side two grains have fused coherently. In the center, two grains in random orientation are separated by a stacking fault

separated by stacking faults, indicated by the hcp crystal structure. Studying the distribution of stresses in the resolidified sample would be the next step but currently the samples are too small for reliable results.

The argon gas still plays a minor role: if gaps persist, they are not closed and the gas escapes. Thus, the geometry and the size of the sample has to be modified to increase the probability of gas pockets.

5 Summary and Conclusions

Our simulations have demonstrated that it is possible to achieve meaningful results for laser-based powder bed fusion (SLM) with molecular dynamics (MD) simulations. A drawback are the long simulation times, short real time scale of the simulations and the small samples sizes. These problems can be overcome by optimizations of the simulation parameters like the adaption of the simulation timestep for different parts of the simulation, as well as the simulation code. Nonetheless, existing supercomputers are large enough for bigger simulations. A huge advantage of MD simulations based on ab-initio-interactions is the fact that much less assumptions have to be made compared to continuum descriptions used in finite element methods. All basic material properties can be derived directly from the simulations and compared to experiments. The MD simulations permit a structural analysis of melting as well as the nucleation and the mixing of alloys on an atomistic level. The simulations have already resulted in reproducing qualitatively basic problems of SLM like inclusions, lack of fusion and boding defects, or balling. Cracks and pores are expected to follow. The observation of keyhole effects might be more difficult. It is expected that larger simulations can be used to produce quantitative results which can directly be compared to experiments.

6 Performance Analysis

6.1 Performance Improvements

By performing the simulations with `IMD`, we have two major problems which arise and slow down our computation greatly. The first problem occurs due to the fast dynamics of the system. Some atoms collect so much energy from the laser that they leave the simulation box. Since we cannot apply periodic boundaries in this case, the atoms are assigned to the corresponding ceiling-CPU. Due to this behavior, the slowdown could be immense if those CPU's collects to much atoms. Therefore, we simply remove atoms which leave the simulation box.

Another problem arises from the strong inhomogeneity of the system. As the most part of the system doesn't contain atoms in the first place, some CPU's have hardly anything to calculate. But this is only the case if we distribute the system as a unit grid to the CPU's. We bypass this problem by using adaptive load-balancing. The functionality of an adaptive load-balancer is shown in the schematic illustration in Fig. 7 for two dimensions. Firstly, the system is rasterized onto a Cartesian grid whose cells are called MD-cells. Afterwards, the center of mass is calculated and corners of domains are attracted by domains with high load. Finally, the MD-cells are assigned to the domains.

In Fig. 8, the performance of a simulation using adaptive load-balancing is compared to a regular unbalanced load. We performed short simulations with 10 000 steps and 778 642 atoms distributed in two powder grains. We observed a speedup of 3.5 for those simulations.

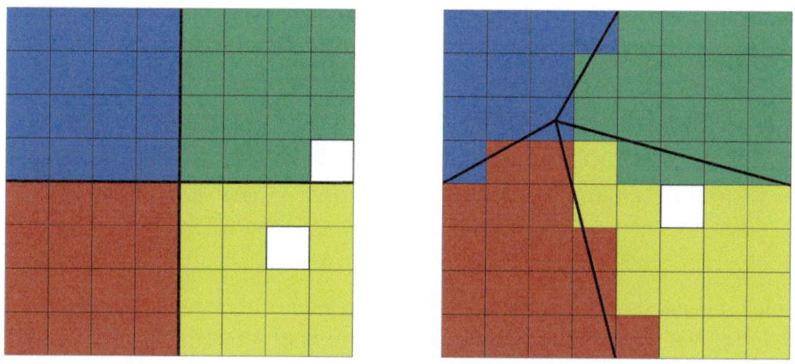

Fig. 7 Schematic illustration of adaptive load-balance. The different colors represents different CPU's whereas the fine grid are the individual MD-cells. On the left a regular unbalanced load and on the right a balanced load with moved domain corners

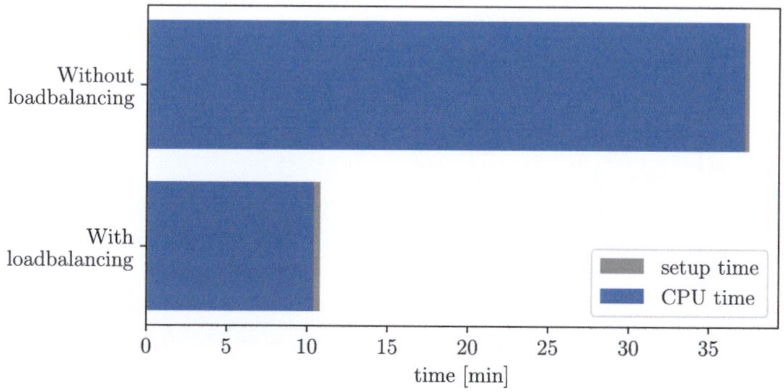

Fig. 8 Benchmark of the adaptive load-balancing performance in comparison to regular processor distribution

6.2 Computational Resources

In the current reporting period, we applied molecular dynamics simulations to study laser-based powder bed fusion of binary alloys. We used the simulation package `IMD` which is capable of balancing the load for inhomogeneous atom distributions. The typical resources used for our simulations are listed together with the number atoms in Table 1. Since removing atoms during simulations as described in Sect. 6.1, the initial number of atoms at time $t = 0$ is given. To distinguish between pre-alloyed grains, we write coated grains with an underscore between coating and core material in the order from outside to inside.

We obtain some inconsistencies in relation to the computation time, especially the simulation Ti_Al 2 seems to do not scale well and consumes approximately twice as much CPU time. We observed that in some cases the load-balancer may get stuck in a local minimum. This then leads to an increasing local imbalance.

The scaling behavior of simulation TiAl 1 is shown in Fig. 9 and Table 2. Due to the parallelization capabilities of `IMD`, we can distribute our simulations to any number of processors. But the inhomogeneity of the system impedes the parallelization speedup because the more processors we use, the more processors got empty space allocated.

Table 1 Typical resources for the selective laser simulations of a single powder grain. As atoms are removed during the simulation, the number of atoms is given at $t = 0$

Simulation	Steps	Atoms	Nodes	CPU's	CPU time [s]
Al_Ti 1	100 001	3 857 316	16	2048	11 234
Al_Ti 2	100 000	3 704 817	16	2048	11 613
Ti_Al 1	100 001	3 857 966	16	2048	10 399
Ti_Al 2	100 000	3 759 746	16	2048	19 582
AlTi 1	100 001	3 985 507	16	2048	17 170

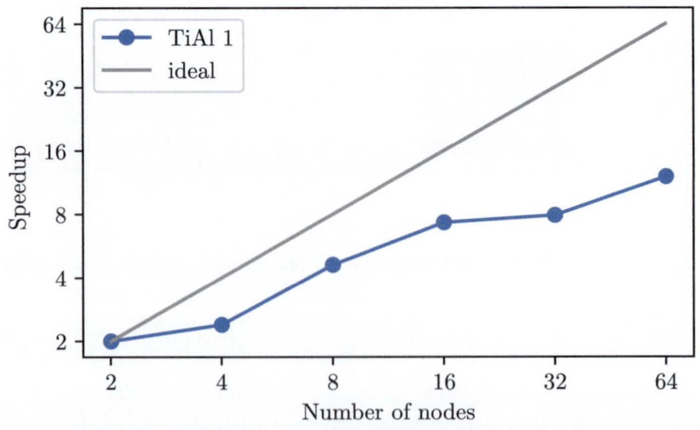

Fig. 9 Scaling behavior of simulation TiAl 1 using `IMD` on HLRS's Hawk supercomputer

Table 2 Scaling behavior of simulation TiAl 1 using `IMD` on HLRS's Hawk supercomputer

Nodes	CPU's	Performance [10^{-5} s]	Speedup
2	256	2.409 946	2.0
4	512	2.883 283	2.4
8	1024	5.556 496	4.6
16	2048	8.812 321	7.3
32	4096	9.537 388	7.9
64	8192	14.525 28	12.1

References

1. Oelschlaeger, F., Klein, D., Müller, S.: J. Roth (2023); Molecular dynamics simulation of selective laser melting, high performance in science and engineering. In: Nagel, W.E., Kröner, D.B., Resch, M.M. (eds.), vol. 21, pp. 65–80. Springer, Berlin (2023)
2. Daw, M.S., Baskes, M.I.: Embedded-atom method: derivation and application to impurities, surfaces, and other defects in metals. Phys. Rev. B **29**, 6443 (1984)
3. Roth, J.: IMD: a typical massively parallel molecular dynamics code for classical simulations - structure, applications, latest developments, sustained simulation performance. In: Resch, M.M., Bez, W., Focht, E., Kobayashi, H., Kovalenko, Y. (eds.), pp. 63–76. Springer, Berlin (2013)

Materials Science

Production of Defects in Two-Dimensional Materials Under Ion and Electron Irradiation: Insights from Advanced First-Principles Calculations

Silvan Kretschmer, Mahdi Ghorbani-Asl, Rico Friedrich, Xuemei Zhang, Tom Barnowsky, Francis Davies, Mitisha Jain, Prosun Santra, and Arkady V. Krasheninnikov

Abstract The main goal of this research project carried out with the extensive use of HLRS computational resources is to get fundamental insights into the formation of defects in two-dimensional (2D) materials under electron and ion irradiation using atomistic simulations. The effects of native and irradiation-induced defects on properties of the materials and the interaction mechanism of ions (including highly charged ions) with 2D materials have been studied as well. Motivated by applications in energy storage and electronic structure tuning, the role of defects in Li intercalation into heterostructures based on 2D materials have been also investigated. Another research direction is the search for new 2D materials and assessment of their properties. Putting aside the previous papers in peer-refereed scientific journals reporting results obtained with the help of HLRS resources (30+ publications), 14 papers, see Refs. [1–14], have been published since the submission of the last annual report. The obtained results have also been disseminated through popular articles at different internet resources, including the HLRS webpage.

1 Introduction

In this project, by combining first-principles calculations carried out at the density functional theory (DFT) level of theory with analytical potential molecular dynamics simulations, we studied the formation of defects in 2D materials under electron

S. Kretschmer (✉) · M. Ghorbani-Asl · R. Friedrich · X. Zhang · T. Barnowsky · F. Davies · M. Jain · P. Santra · A. V. Krasheninnikov
Institute of Ion Beam Physics and Materials Research, Helmholtz-Zentrum Dresden-Rossendorf, Dresden, Germany
e-mail: s.kretschmer@hzdr.de

A. V. Krasheninnikov
e-mail: a.krasheninnikov@hzdr.de

T. Ludwig et al. (eds.), *High Performance Computing in Science and Engineering '23*,
https://doi.org/10.1007/978-3-031-91312-9_13

and ion irradiation. The influence of defects on the electronic, optical and catalytic properties of 2D materials has been studied. The role of defects in Li intercalation into heterostructures based on 2D materials have been also investigated, as this topic is important in the context of energy storage and electronic structure tuning. Other related topic, e.g., alkali metal atom storage in organic crystals, have been addressed as well. The highlights from our research are presented below.

2 Results

2.1 Ion-Beam Mediated Modification of Two-Dimensional Materials

Ion beams are an established tool to image (in case of He) and pattern 2D materials. At the same time, the detailed interaction of—especially highly charged—ions with the 2D target systems is not fully understood. Together with our experimental collaborators from University Duisburg-Essen, we investigated the velocity distribution of sputtered Mo atoms by Xe ions of different charge states—thus varying the potential energy of the ions—impinging on 2D MoS_2 supported by a Au(111) surface [1]. From the comparison to our empirical potential molecular dynamics simulations, see Fig. 1, we observe that for increasing charge state of the ion (with increasing potential energy) the amount of emitted slow neutral projectiles increases, opposed to the impact of single charged ions with the same kinetic energy where the velocity

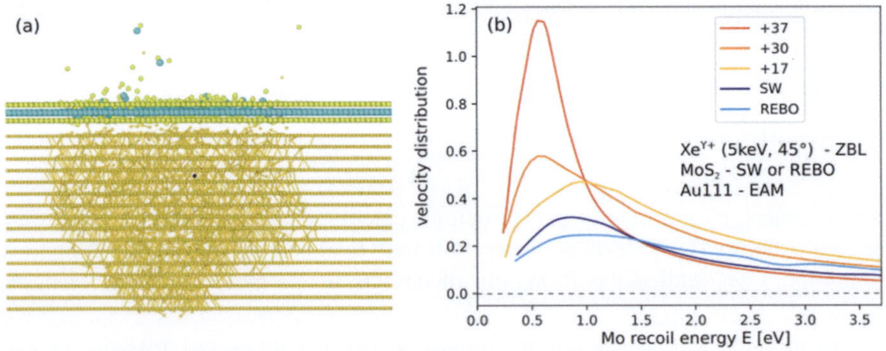

Fig. 1 **a** Snapshot of LAMMPS atomistic simulations of irradiation of 2D-MoS_2 on Au(111) substrate with Xe ions. Xe atom is represented as a dark blue ball, Mo–cyan, S– yellow, and Au orange. **b** Velocity distribution of sputtered Mo atoms obtained using the Stillinger-Weber (SW) and Reactive-bond-order (REBO) potential for (effectively neutral) projectiles. The results are compared to the experimental data for ions with charge states +17, +30, +37. Note the systematic trend in the Mo velocity distribution attributed to the gradual transition from the kinetic to potential energy dominated energy transfer. Image adapted from Ref. [1]

distribution is systematically shifted to faster recoils. From that, and the absence of ultra-slow recoils, it can be concluded that electron-phonon coupling rather than a Coulomb explosion dominates the energy transfer from the highly charged ion (HCI) to the target system. Further investigations of the neutralization behaviour of energetic ions transmitting 2D materials in collaboration with TU Wien have been carried out, and a manuscript is currently under review.

Detailed understanding of the interaction mechanism of ions with (supported) 2D materials is mandatory for employing ion beams for material patterning and modification. In a preliminary study for the directed growth of hBN dots on monolayer and bilayer graphene (MLG/BLG) supported by silicon carbide carried out together with our experimental collaborators at Ferdinand-Braun-institute and Paul-Drude-institute, Berlin, we modelled the effect of He and Ne ions on the defect production in the graphene layers [2]. Figure 2 shows the results for the helium ion irradiation in the SiC-surface layer graphene (zero layer—ZLG) in comparison to MLG, BLG and free-standing graphene (FSG). We find a slight increase of the number of single vacancies in MLG and BLG as compared to the free-standing graphene at higher ion

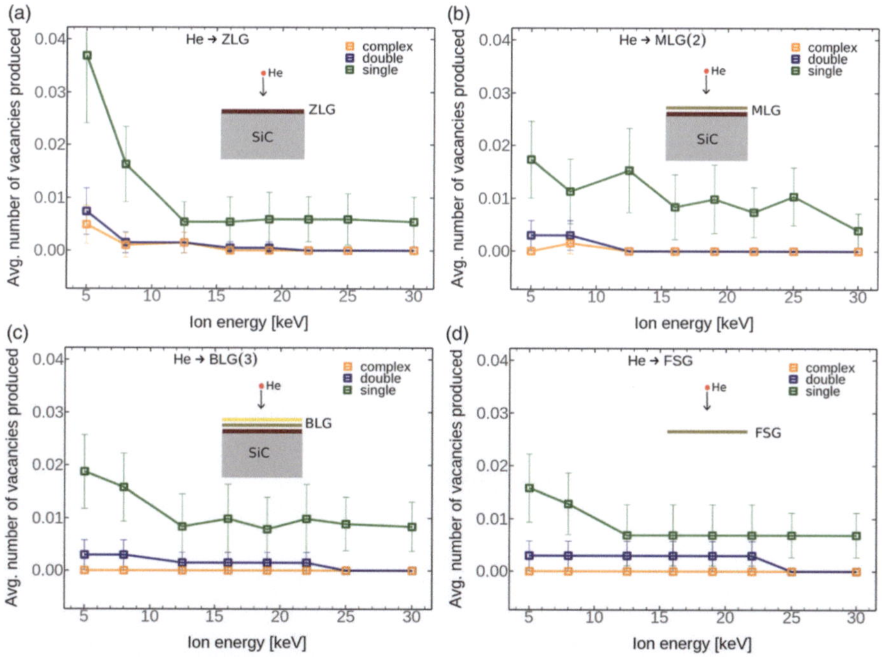

Fig. 2 Defect production by He ion irradiation in graphene on SiC. The average number of single vacancies (single), divacancies (double) and complex defects is shown as evaluated by empirical potential MD using the LAMMPS package for the graphene buffer layer (ZLG, **a**), MLG (**b**), BLG (**c**) and free-standing graphene (**d**)—model system illustrated by schematics. Image adapted from Ref. [2]

energies, which can be attributed to the collision cascade in the substrate. No significant amount of complex or extended defects (which would qualify as nucleation sites for hBN growth rather than isolated point defects) are observed. Further investigations considering the cumulative effect of higher dose irradiation are currently under way.

2.2 *Electron-Beam Induced Structural Transformations in Few-Layer $MnPS_3$*

Recently, new types of layered materials based on transition-metal phosphorus trisulfides (TMPTs) have attracted considerable amount of attention. In particular, manganese phosphorus trisulfide ($MnPS_3$) is an exciting candidate for multifunctional electronics/spintronics applications due to its semiconductor character and inherent magnetic properties that remain even when the material is reduced down to a few layers.

Here, we investigated in detail the effects of electron irradiation and thermal annealing on few-layer $MnPS_3$. We demonstrated that the appropriate electron fluence causes the transformation of the host material into a cubic α-$MnS_{1-x}P_x$ phase embedded into the parent host material. The [111] direction of the crystals was found to be parallel to the incidence direction of the electrons onto the parent $MnPS_3$. This phase transformation of $MnPS_3$ can be locally controlled by tuning the size of the electron beam down to a nanometer scale. Additionally, we estimated that the newly identified γ-MnS type phases alloy with P to create γ-$MnS_{1-x}P_x$ structures. Other orientations of the ultrathin $MnS_{1-x}P_x$ phases also emerge in the γ-MnS type when the host material has completely degraded (Fig. 3a, b). To get more insights into the effects of phosphorus alloying on the electronic structure of MnS, the electronic structures of the mixed compounds were analyzed. The P dopants in MnS push the Fermi level to the valence band because P has one less electron than S, and it is known as an acceptor dopant (Fig. 3c). The electron density plot also indicates that the defect-induced states have a highly localized character on the P atom (Fig. 3d). Overall, our results showed that electron beam irradiation and thermal annealing can be utilized to grow phases with distinct properties starting from freestanding quasi-2D materials.

2.3 *A Van Der Waals Heterostructure with an Electronically Textured Moiré Pattern*

The creation of moiré structures as a result of long-range coincidence lattices created by combining materials with various lattice constants or rotation (twist) angles is one of the most fascinating aspects of van der Waals (vdW) heterostructures. As

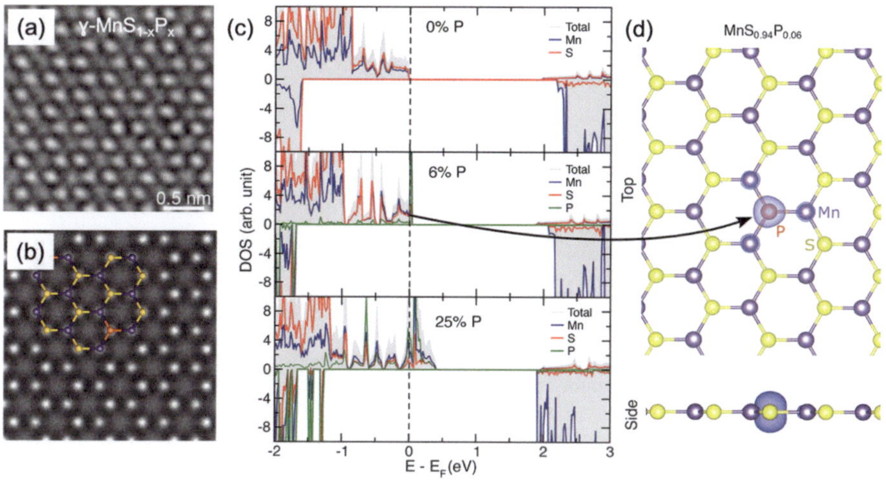

Fig. 3 **a** High-resolution TEM images obtained at electron acceleration voltage of 80 kV and **b** and simulated HRTEM images of $MnS_{1-x}P_x$ crystals. **c** Projected density of states of the γ [0001]-$MnS_{1-x}P_x$ monolayer with the spin-up and spin-down components corresponding to positive and negative values. The Fermi level is shifted to zero. The P-associated states are clearly seen at the edge of the valence band. **d** Partial charge densities at the Fermi level for the $MnS_{0.94}P_{0.06}$ monolayer. The image adapted from Ref. [3]

a result, these moiré patterns in vdW heterostructures provide an additional degree of engineering of materials properties, which has been used to illustrate novel materials design for prospective applications. We have already shown that the electronic structures of Pt-dichalcogenides are significantly impacted by interlayer interaction. It is anticipated that the lateral modifications of the interlayer contacts in vertical heterostructures based on these materials would give rise to the opportunity to texture the electronic structure. Using scanning tunneling spectroscopy experiments carried out by our coworkers and first-principles simulations, we investigated the effects of local variation of the interlayer atom coordination on the electronic structure of $PtSe_2$/$PtTe_2$ heterostructures. It was found that varying interactions of the chalcogen p-orbitals between the two materials cause interlayer states that vary within the moiré structure. The local density of states (LDOS) below the Fermi level is larger in region (ii) as compared to the other sites. The variations in LDOS near the band edge are consistent with its experimental topographic STM plot (Fig. 4a), suggesting that region (ii) has stronger tunneling contrast than the other two moiré regions. According to our density functional theory calculations, local changes in atom coordination in the moiré unit cell lead to changes in the charge transfer from $PtSe_2$ to $PtTe_2$, which in turn affects the interface dipole value (Fig. 4b).

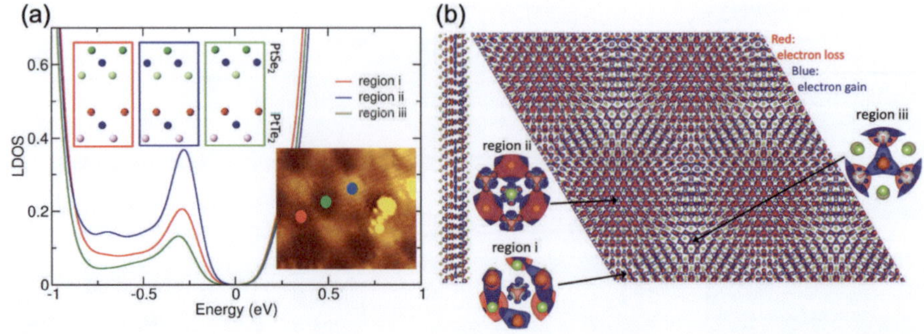

Fig. 4 **a** Local density of states for different regions in the moiré structures of $PtSe_2/PtTe_2$ heterostructure. The LDOS is shown for $PtSe_2$. Inset shows STM images of the different regions shown with the same color-coded spectra. Charge transfer between $PtSe_2$ and $PtTe_2$ layers is calculated as the difference between the charges in the heterostructure and the isolated monolayers. The image is adapted from Ref. [4]

2.4 *Quasi-2D FCC Lithium Crystals Inside Defective Bi-Layer Graphene*

In-situ TEM investigations have discovered quasi-2D crystals within bi-layer graphene [5]. It was also known that Li crystals contain various impurities. We carefully examined the interactions of isolated Li atoms with vacancy-type defects in graphene using first-principles calculations. It was found that the presence of defects in graphene significantly strengthens the bonding between Li and graphene as compared to the pristine system (Fig. 5a, b). In particular, the Li-graphene interaction is enhanced for large vacancies, e.g., the tetra vacancy. The results demonstrated that quasi-2D Li crystals enclosed between graphene sheets nucleate at the defects. These results provide an explanation for the experimentally observed dominating nucleation and development of q-2D FCC Li crystals between bilayer graphene. The charge density difference maps indicated that the region between Li ions and the graphene is primarily depleted in electron density, which reflects charge transfer from Li to the graphene lattice (Fig. 4c, b). In addition, we investigated the behavior of impurities, such as O, N, S, and F, most likely to be found in the enclosed Li crystals. It was found that all impurity atoms take up octahedral interstitial sites and interact strongly with atoms in Li crystals, preventing the delithiation process.

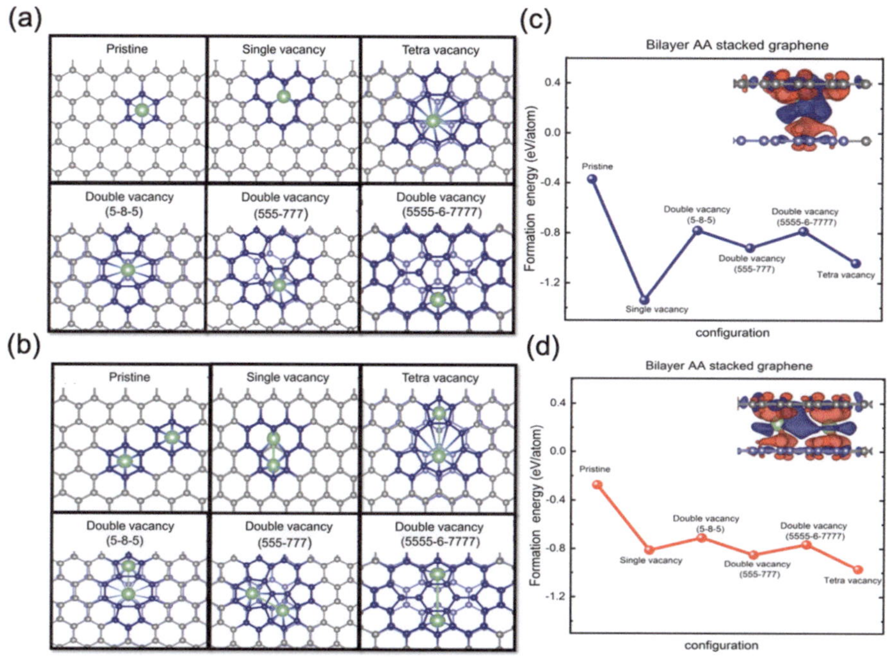

Fig. 5 Atomic structures of bilayer graphene with **a** one lithium and **b** two lithium atoms adsorbed on pristine area, single vacancy, double vacancy (5-8-5), reconstructed double vacancy (555-777), reconstructed double vacancy (5555-6-7777), and tetra vacancy. Formation energies of atomic configuration with **c** one lithium and **d** two lithium atoms adsorbed on pristine and defective bilayer graphene. The image is adapted from Ref. [6]

2.5 Hydrogenic Spin-Valley States of the Bromine Impurities in 2H-MoTe$_2$

For quantum technologies, creating semiconductors doped with specific chemical elements that enable the encoding of a qubit within spin states is of utmost importance. Strong spin-orbit coupling in transition metal dichalcogenides semiconductors results in locked spin-valley states with long coherence times. In this work, we examined the 2H-MoTe$_2$ with substituted bromine Br_{Te} dopant. Our coworkers used electron spin resonance (ESR) to detect the Br_{Te} spin signal and relate the spin lifetime to the electronic parameters determined through transport measurements and angular resolved photoemission spectroscopy. According to observations of electron spin resonance, Br_{Te} dopant has a spin with a long-lived nanoseconds coherence time. By combining first-principles calculations with scanning tunneling microscopy and spectroscopy, we found that the spatial modulations of the hydrogenic states related to the dopant levels are due to their hybridization to the Q-valleys of the conduction

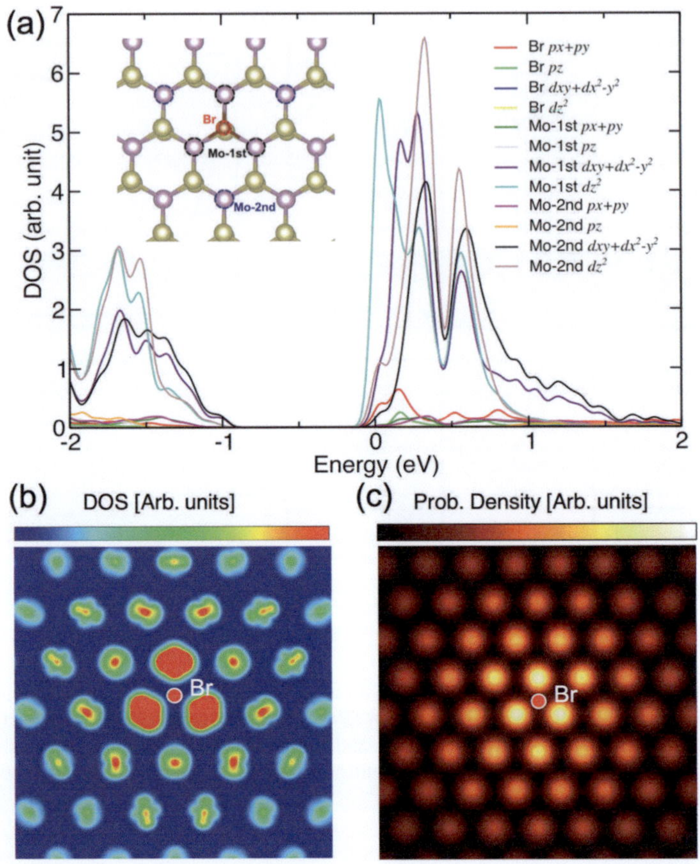

Fig. 6 **a** Partial density of states (PDOS) for Br_{Te} dopant in a $MoTe_2$ monolayer. The PDOS is shown as a function of energy for the p-orbitals and d-orbitals of the Br atom and the $p-$ and $d-$orbitals of the 1st and 2nd neighbors Mo atoms surrounding the Br atom. **b** Map of the local DOS integrated between $E_F + 0.1$ V and $E_F + 0.3$ V, showing a modulation of the DOS around the dopant. **c** This modulation is a consequence of the hybridization of the p orbital of the dopant with the Bloch states at the K-point of the conduction band. Both images are plotted on the same scale. The Br atom is indicated as a red dot at the center of the images. The image is adapted from Ref. [7]

band. In particular, our DFT calculations showed that all three orbital components $p_{x,y,z}$ of the anion site have a large contribution to the Q-valleys in the conduction band, see Fig. 6a. The spatial modulation of the local DOS indicates that the p orbitals hybridize to the Bloch states at the K-point of the conduction band, which is compatible with the Bloch model, Fig. 6b, c.

2.6 Computational Exploration of Novel Two-Dimensional Materials

We have outlined a new group of 2D non-van der Waals (vdW) materials with ultra low exfoliation energies [8] by generalizing our previous investigations [9]. This study is highlighted in the GCS Inside magazine summer 2023 issue and was enabled by an extended screening of the AFLOW materials database according to the structural prototype of previously realized experimental systems taking into account, in addition to oxides, sulfides and chlorides. The computational resources were employed for the structural relaxations as well as to calculate the electronic and to carefully investigate the magnetic properties. An open question is, however, whether a universal search criterion for non-vdW 2D materials can be derived. This can pave the way to guide experiments in fully leveraging the space of this novel nanoscale materials class. Another key question is how the properties of these predicted candidates can be controlled. The ambition is thus to utilize high-throughput calculations to screen further materials and to investigate the chemical functionalization of these systems by surface passivation with terminal groups such as H, OH, and H_2O. The optimum design of magnetic features is also an open question.

In this context, we have carried out a large part of the required calculations. We have identified eight additional non-vdW 2D systems in part exhibiting ultra low exfoliation energies. By accounting for long-range dispersive interactions in the so called SCAN+rVV10 scheme (see Fig. 7a and b), we could prove that the exfoliation energies of the new candidates are comparable to the ones of traditional 2D compounds and are ultimately as small as the one of graphene. $SbTlO_3$ (see Fig. 7c) appears to be a corner case at the intersection between vdW and non-vdW materials

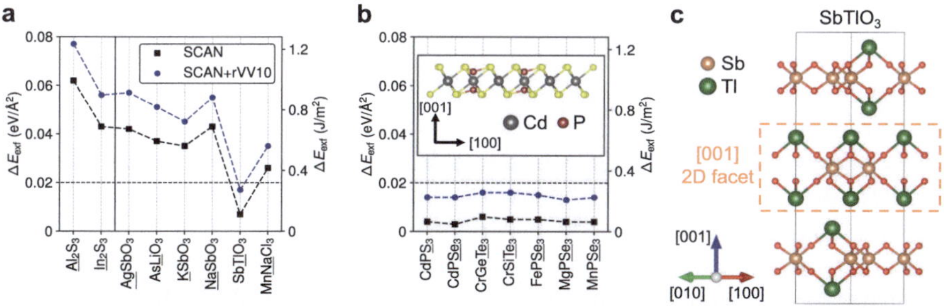

Fig. 7 Comparison of the calculated exfoliation energies from SCAN and SCAN+rVV10 for **a** the eight non-vdW 2D candidates and **b** seven vdW 2D systems with the same structure. Inset in panel (**b**): side view of $CdPS_3$. The vertical black line in panel (**a**) separates binaries from ternaries. As a reference, the exfoliation energy of graphene [10] is indicated by the dashed horizontal black lines. For the ternaries, the data for the slabs with the energetically favorable termination are plotted and for all systems the terminating element is underlined at the bottom axis. The dashed lines connecting the data points are visual guides. c) Atomic structure of $SbTlO_3$. The exfoliable [001] facet (monolayer) is indicated in the orange dashed box. The black line denotes the conventional unit cell

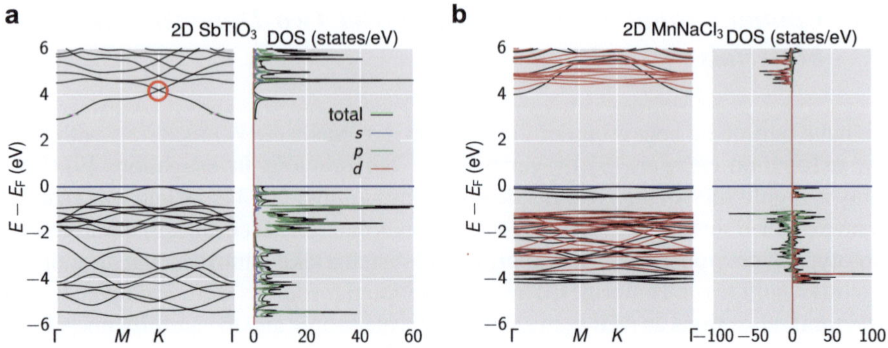

Fig. 8 Band structures and densities of states. Band structure and density of states (DOS) of **a** 2D $SbTlO_3$ and **b** 2D $MnNaCl_3$. The energies are aligned at the respective Fermi energy E_F. For 2D $SbTlO_3$, a linear band crossing at the K-point is highlighted by the circle. For the spin polarized band structure in (**b**), majority spin bands (positive DOS) are indicated in black while minority spin bands (negative DOS) are in red

spaces as evident from a detailed analysis of binding energy contributions and structural features. When computing the electronic band structures of the candidates with the smallest exfoliation energies, $SbTlO_3$ and $MnNaCl_3$, (see Fig. 8) we found an appealing linear band crossing as well as a strongly spin polarized 2D ferromagnet.

With respect to the chemical functionalization, the largest portion of the comprehensive calculations for hydrogen passivation has been completed. Full and a partial coverage of all 35 candidates were carefully investigated including structural relaxation, computing of the corresponding adsorption energies as well as re-relaxation of randomized geometries and phonon calculations to validate the dynamic stability. From this large scale analysis, it turned out, that only for four candidates (two for partial H, two for full H coverage), the hydrogen passivation is energetically favourable compared to the pristine system and isolated molecules *and* the resulting systems are dynamically stable. For the 2D sheets that passed all energy and stability checks, band structures and densities of states were computed. For magnetic candidates, the lowest energy magnetic state was retrieved from a combinatorial search checking for all ferromagnetic and antiferromagnetic alignments of the spins for five different sizes of the induced moments. The electronic and in particular magnetic properties of the systems exhibit appealing modifications which might be useful to control the magnetic state of these systems. In a second step, also the passivation of all new 2D materials by OH groups was attempted which turned out to be computationally demanding due to the many candidates exhibiting energetic favorability. These calculations are thus still ongoing and we expect to obtain the results during the next year.

3 Implementation of the Project and Technical Data

3.1 CPU Time Used

The group has used 23 million CPU hours in the time frame from April 2022 to March 2023. Note that almost half of CPU time is invested in on-going research whose results still need to be published.

Depending on the simulation project, the number of nodes (processors) used varied form 1 to 20 (128–2048 processors). Scaling of the standard DFT codes VASP and GPAW, along with the empirical potential code LAMMPS is presented below.

3.2 Codes Employed and Their Scaling

Three different codes were used to carry out the simulations, GPAW [11], VASP [12, 13] and LAMMPS [14]. All these codes are widely used (thousand of users) in atomistic simulations on various platforms with massive parallel architecture, and a good scaling behavior has been demonstrated. The scaling behavior of LAMMPS, VASP and GPAW codes are presented in Fig. 9.

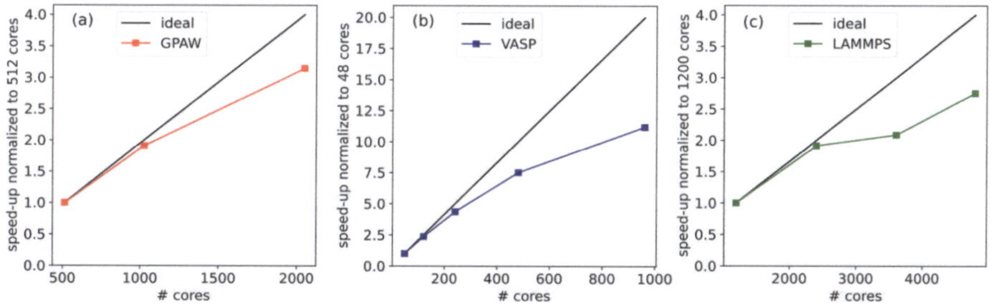

Fig. 9 Scaling plots of the employed codes. **a** Scaling behavior of GPAW when running Ehrenfest dynamics simulations on CPUs at HLRS. Data obtained for the impact simulation of a proton (20 keV) on graphene (96 atoms) with a timestep of 5 as and grid spacing of 0.1 Å(240 bands). **b** Scaling behavior of the VASP code when running molecular dynamics simulations on CPUs at HLRS. Data obtained for a system composed of 108 atoms, cut-off energy of 400 eV and 5 k-points. **c** Scaling behavior of the LAMMPS code when running ion irradiation simulation of graphene deposited on substrate. The speedup is normalized so that at the first data point (1200 processing elements)

References

1. Skopinski, L., Kretschmer, S., Ernst, P., Herder, M., Madauß, L., Breuer, L., Krasheninnikov, A.V., Schleberger, M.: Velocity distributions of particles sputtered from supported two-dimensional mos_2 during highly charged ion irradiation. Phys. Rev. B **107**, 075418 (2023). Feb
2. Jain, M., Kretschmer, S., Höflich, K., Lopes, J.M.J., Krasheninnikov, A.V.: Atomistic simulations of defects production under ion irradiation in epitaxial graphene on sic. Phys. Status Solidi (RRL)—Rapid Res. Lett. **17**, 2200292 (2022)
3. Storm, A., Köster, J., Ghorbani-Asl, M., Kretschmer, S., Gorelik, T.E., Kinyanjui, M.K., Krasheninnikov, A.V., Kaiser, U.: Electron-beam- and thermal-annealing-induced structural transformations in few-layer $MnPS_3$. ACS Nano **17**(5), 4250–4260 (2023)
4. Li, J., Ghorbani-Asl, M., Lasek, K., Pathirage, V., Krasheninnikov, A.V., Batzill, M.: A van der Waals heterostructure with an electronically textured Moiré pattern: $Ptse_2$/$ptte_2$. ACS Nano **17**(6), 5913–5920 (2023)
5. Kühne, M., Börrnert, F., Fecher, S., Ghorbani-Asl, M., Biskupek, J., Samuelis, D., Krasheninnikov, A.V., Kaiser, U., Smet, J.H.: Reversible superdense ordering of lithium between two graphene sheets. Nature **564**(7735), 234–239 (2018)
6. Zhang, X., Ghorbani-Asl, M., Zhang, Y., Krasheninnikov, A.V.: Quasi-2D FCC lithium crystals inside defective bi-layer graphene: insights from first-principles calculations. Mater. Today Energy **34**, 101293 (2023)
7. Sheina, V., Lang, G., Stolyarov, V., Marchenkov, V., Naumov, S., Perevalova, A., Girard, J.-C., Rodary, G., David, C., Sop, L.R., Pierucci, D., Ouerghi, A., Cantin, J.-L., Leridon, B., Ghorbani-Asl, M., Krasheninnikov, A.V., Aubin, H.: Hydrogenic spin-valley states of the bromine donor in 2H-$MoTE_2$. Commun. Phys. **6**, 135 (2023)
8. Barnowsky, T., Krasheninnikov, A.V., Friedrich, R.: A new group of 2D non-van der Waals materials with ultra low exfoliation energies. Adv. Electron. Mater. **9**(4), 2201112 (2023)
9. Friedrich, R., Ghorbani-Asl, M., Curtarolo, S., Krasheninnikov, A.V.: Data-driven quest for two-dimensional non-van der Waals materials. Nano Lett. **22**(3), 989–997 (2022)
10. Zacharia, R., Ulbricht, H., Hertel, T.: Interlayer cohesive energy of graphite from thermal desorption of polyaromatic hydrocarbons. Phys. Rev. B **69**(15), 155406 (2004)
11. Enkovaara, J., Rostgaard, C., Mortensen, J.J., Chen, J., Dułak, M., Ferrighi, L., Gavnholt, J., Glinsvad, C., Haikola, V., Hansen, H.A., Kristoffersen, H.H., Kuisma, M., Larsen, A.H., Lehtovaara, L., Ljungberg, M., Lopez-Acevedo, O., Moses, P.G., Ojanen, J., Olsen, T., Petzold, V., Romero, N.A., Stausholm-Møller, J., Strange, M., Tritsaris, G.A., Vanin, M., Walter, M., Hammer, B., Häkkinen, H., Madsen, G.K.H., Nieminen, R.M., Nørskov, J.K., Puska, M., Rantala, T.T., Schiøtz, J., Thygesen, K.S., Jacobsen, K.W.: Electronic structure calculations with GPAW: a real-space implementation of the projector augmented-wave method. J. Phys. Condens. Matter **22**, 253202 (2010)
12. Kresse, G., Hafner, J.: Ab initio molecular dynamics for liquid metals. Phys. Rev. B **47**, 558–561 (1993)
13. Kresse, G., Furthmüller, J.: Efficiency of ab-initio total energy calculations for metals and semiconductors using a plane-wave basis set. Comput. Mater. Sci. **6**, 15–50 (1996)
14. Plimpton, S.: Fast parallel algorithms for short-range molecular dynamics. J. Comput. Phys. **117**, 1–19 (1995)

Phase-Field Simulation of Sintering Processes

Marco Seiz, Henrik Hierl, Michael Kellner, and Britta Nestler

Abstract Predicting the microstructure of materials is of paramount importance, as many properties of interest depend directly on the microstructure. Thus the overarching goal of the project is to predict the microstructure with the phase-field method, which allows for direct 4D (3D+time) observation of the microstructure. The simulations shed light on the mechanisms of microstructure evolution and allow the prediction of the evolution under a change of processing conditions as well as changes of the material itself. The processes of solidification and sintering are investigated in this project. Alloy solidification, specifically the coupled growth of dendrites and eutectics, forms one part of the work on the solidifcation, with the freeze-casting process as a preprocessing step for sintering forming the other. The main topic of sintering is investigated by employing not only phase-field methods but also molecular dynamics simulations. By combining results from both methods, it was possible to construct the first phase-field model which correctly accounts for the densification evolution of large three-dimensional packings. This model is then used to investigate the competing processes of grain growth and densification. Furthermore, the colloidal structure produced during freeze-casting is also computationally sintered

M. Seiz (✉) · H. Hierl · M. Kellner · B. Nestler
Karlsruhe Institute of Technology (KIT) - Institute for Applied Materials (IAM), Karlsruhe, Germany
e-mail: marco.seiz@kit.edu; marco@kit.ac.jp

H. Hierl
e-mail: henrik.hierl@kit.edu

M. Kellner
e-mail: michael.kellner@partner.kit.edu

B. Nestler
e-mail: britta.nestler@kit.edu

M. Seiz
Faculty of Mechanical Engineering, Kyoto Institute of Technology, Kyoto, Japan

B. Nestler
University of Applied Sciences Karlsruhe - Institute of Digital Materials Science, Karlsruhe, Germany

T. Ludwig et al. (eds.), *High Performance Computing in Science and Engineering '23*,
https://doi.org/10.1007/978-3-031-91312-9_14

with the method and thus represents the first simulation series of the freeze-casting process starting from a suspension and finishing with a porous, sintered structure.

1 Scientific Work Accomplished and Results Obtained

1.1 Coupled Dendritic-Eutectic Growth

The coupled growth of dendrites and eutectics is investigated by coupling the classical phase-field model of alloy solidification with a heuristic nucleation mechanism. The goal is to determine whether the coupled growth, which usually occurs in uncontrolled solidification, will influence the microstructure significantly. The Al-rich side of the Al-Cu system is used as a model system, since many prior experimental investigations show coupled structures. Within the considered composition range, it forms the fcc α-Al phase strongly rejecting Cu, the intermetallic phase $\theta - Al_2Cu$ as well as a liquid melt. The remaining details should be taken from the publication [1].

First, two-dimensional simulations are carried out, with the simulations employing between 128 and 1024 cores. The growth parameters are the velocity v, the temperature gradient G and the concentration c_0 of Cu in the melt. The parameter range is chosen based on a pre-study such that the transition region between a dendritic-eutectic and a eutectic structure is investigated. The transition region is observed to depend on all three growth parameters. In the simulations it is generally observed that the relationship between velocity and front undercooling does not significantly change w.r.t. the pure growth forms. Typically, the formation of secondary arms is suppressed if coupled growth occurs, as the eutectic effectively overgrows the arms before they can fully form. This depends on the distance between the eutectic front and the dendrite tip, as complete solidification simulations could show, see Fig. 1 A

Fig. 1 Intermediate state of a complete directional solidification simulation. The initial state only had a single circular, periodic α-Al seed (black), which then grew into the melt (grey) resulting in multiple dendrites with secondary arms. The secondary arms showed coarsening as well as dissolution depending on the local environment. The eutectic (white+black) grows in the interdendritic channels

significant amount of time is spent in primary solidifcation, during which secondary dendrite arms developed fully and even partial dissolution of secondary arms took place. After the eutectic started nucleating, it grew in the interdendritic channels while also nucleating in the pockets between secondary dendrite arms, suggesting that eutectics of different dominant orientation would form around dendrites. The spacing of the eutectic during coupled growth is generally slightly refined w.r.t. a pure eutectic structure, with this likely being a combination of the far-field being changed by the dendrite as well as structural effects. Close to the dendrite, the θ phase grows more thickly due to the increased Cu concentration around the dendrite. This also causes the eutectic front to be curved close to the dendrite.

In two dimensions the eutectic can only form lamellar patterns. In order to test whether the pattern is changed by coupled dendritic-eutectic growth, a number of 3D simulations are carried out, employing 4096 cores. It is observed that while the eutectic pattern changes from a matrix-fiber to a lamellar structure (Fig. 2), it is not entirely due to growing coupled to a dendrite. Rather, with one phase within the eutectic being anisotropic, this forces the entire eutectic into a different pattern, at least for the present solidification conditions.

1.2 *Freeze-Casting*

The freeze-casting process is modelled by resolving the suspended particles not individually, but as a concentration field. This allows the use of the classical phase-field model of alloy solidification. The details of how to homogenize over the particles in order to reproduce the freeze-casting process are given in [2]. Here we only reproduce the results. The simulations are generally run on 128–512 cores.

The goal was to relate the microstructural lengths L—tip distance λ, the ice trunk diameter d_{ice} and suspension channel diameter d_{sus}—to the process parameters of velocity v, temperature gradient G and solids loading c_0 as well as the particle size r. This allows the determination of which parameters need to be recorded in experiments in order to make full comparisons possible. These parameters are varied and for each particle size, a scaling law of the form $L = Ac_0^n v^m G^o$ for each length is determined. The magnitude and sign of these exponents allows the determinationof how relevant each parameter is and what qualitative influence (sign) it should have on the corresponding length. Furthermore, the simulation model allows future exploration of data-driven design proposals to develop novel tailored freeze-casting materials.

Generally it is observed that the tip distance λ is only weakly dependent on the solids loading c_0 regardless of the particle size, in contrast to the diameters being strongly dependent on it. This is due to the tip distance roughly being the sum of diameters, d_{sus} rising with c_0 and d_{ice} decreasing with c_0. Since usually both diameters are of interest, the solids loading should always be recorded.

The velocity always plays a significant role and thus should be recorded. In contrast, the temperature gradient is somewhat less significant, but still changes all lengths. Interestingly, for larger particle sizes it is observed that d_{sus} becomes less

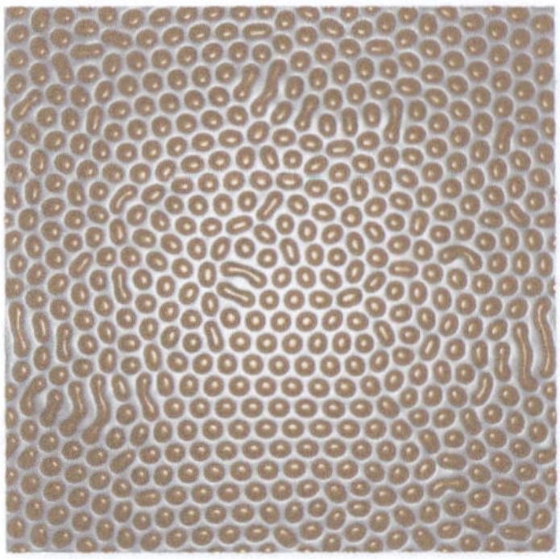

(a) Matrix-fiber structure resulting from isotropic α-Al and isotropic θ seeds.

(b) Lamellar structure resulting from anisotropic α-Al and isotropic θ seeds.

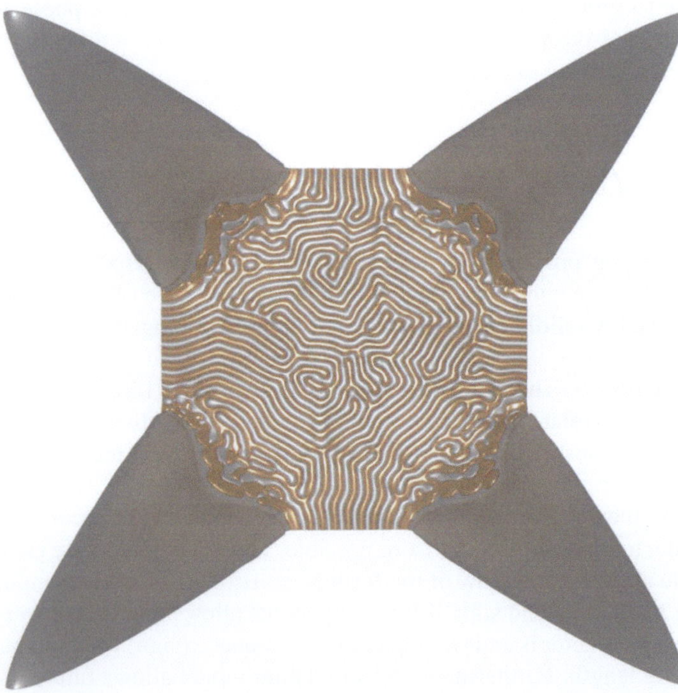

(c) Scale-bridging solidification microstructure containing large-scale dendrites with fine lamellar eutectic being observed in the interdendritic space. This structure results if an additional dendritic seed is added to the initial conditions of fig. 2a. The initial isotropic α-Al is replaced over time by the anisotropic variant.

Fig. 2 Final states of 3D simulations, showing the distribution of solid phases (α-Al = silver, θ = orange) in the entire domain. It is quite clear that any anisotropy in the eutectic leads to a lamellar structure for the present solidification conditions. Thus coupled growth need not induce a morphological change dependent on G. This is likely due to the liquidus curve describing the equilibrium concentration of the suspension becoming more and more step-like as the particle radius is increased.

The effect of fluid flow on the microstructure evolution is qualitatively investigated by a small number of simulations. For this, a simulation which reached steady-state is continued with a body force acting (anti-)-parallel to the growth direction and simultaneously solving the coupled phase-field, concentration and Navier-Stokes equations. The effect of the suspended particles having a higher density than water is taken into account with a solids loading c dependent density, which induces the particles (solids loading) to preferentially move in the direction of the body force. The resulting concentration fields are shown in Fig. 3. Generally a body force acting antiparallel to the growth direction causes a coarsening of the microstructure and suppresses side branches, with parallel forces showing the inverse effect but to a lesser degree. It should be noted that for "large" parallel body forces, plumes of particles from the dendrite tips dominate the evolution; it is likely that the simulation domain was too small to properly account for these long-range effects.

A small number of 3D simulations were run initially on 2048 cores. These showed that indeed the 3D structure approximates ice lamellas, which allows the reduction of the problem to two dimensions, since the spacing between these lamellas approximates the tip distance λ. One of these will be used as input to the later sintering simulations, which will also show the resulting structure.

1.3 Solid-State Sintering

The problem of solid-state sintering is investigated by both molecular dynamics (MD) and phase-field (PF) simulations. MD is employed in order to determine an appropriate model for the calculation of velocities necessary for densification during sintering. By constructing appropriate geometries, it is possible to establish a direct link between the number of absorbed vacancies on grain boundaries (GBs) and how grains move in response to this. A model which accounts for the observed qualitative rules consists of a specification of the displacement jump $\Delta u_{\alpha\beta}$ across each $\alpha\beta$ grain boundary and a matrix equation linking the displacement jumps to the grain displacements u:

$$\Delta u_{\alpha\beta} = \frac{\Omega}{A_{\alpha\beta}} \Delta N_{\alpha\beta} \tag{1}$$

$$\underbrace{\mathrm{C}}_{B\times N} \underbrace{\mathrm{u}}_{N} = \underbrace{\Delta\mathrm{u}}_{B} \tag{2}$$

which has the known displacements jumps Δu across all grain boundaries B and the unknown grain displacements u of the N grains. The displacement jump of each grain boundary $\Delta u_{\alpha\beta}$ depends on the atomic volume Ω, the grain boundary area $A_{\alpha\beta}$

(a) $g = -g_0$ (b) $g = -\frac{g_0}{10}$ (c) $g = 0$ (d) $g = \frac{g_0}{1000}$ (e) $g = g_0$

Fig. 3 Solids loading field of a simulation study of dendritic morphologies for various body forces g. Blue indicates the particle-rejecting ice crystal, purple-red the suspension of medium solids loading and yellow the suspension of the highest solids loading. The simulations represent full-field multiphysics computations of the coupled phase-field, diffusion and Navier-Stokes equations together

and the number of absorbed vacancies $\Delta N_{\alpha\beta}$. The matrix C describes the contacts of individual particles, with $B \geq N$. Each of the B rows has only two non-zero entries (1, –1), which describe the orientation of the contacting particles relative to each other in that dimension. This system is, in general, overdetermined and hence is solved in a least-squares sense. It should be noted that this requires tracking the area and how many vacancies are absorbed for each grain boundary individually. This can pose memory and communication challenges, which are addressed in Sect. 2.

Based on the improved model, simulations are conducted and compared to previous simulations of sintering in large scale three-dimensional green bodies with an initially monodisperse distribution of grain sizes. The simulation domains are of size N^3, $N \in \{200, 400, 800\}$ voxels, which is also used as a label. The previous simulations are based on [3] (DO) and an adaption of [3, 4] (ADV-μ) together with corrections accounting for model problems observed in [5, 6]. Depending on the

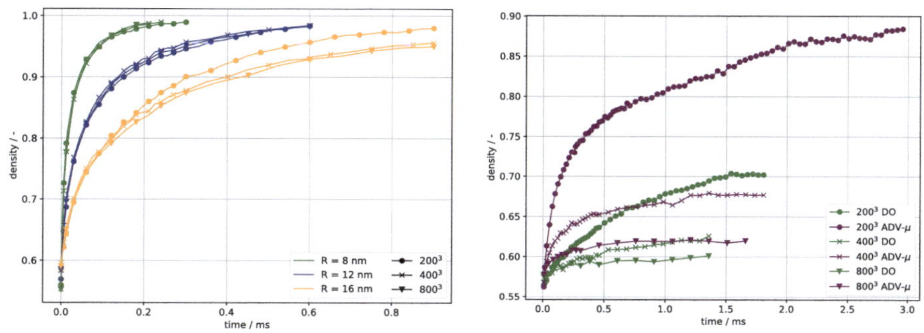

(a) MD-inspired model

(b) literature models ($R = 12\,\mathrm{nm}$)

Fig. 4 Comparison of densification for the present MD-inspired model and phase-field models of sintering based on literature. The latter clearly have a strong dependence on domain size, which the MD-inspired model eliminates once a RVE is reached

investigated size of the domain, between 128 and 8192 cores are employed. The calculation time varied, with more slowly densifying simulations being run for longer; the shortest time was 2 h, with the longest around 70 h for the later simulations. The density evolution of these simulations is shown in Fig. 4. It is quite obvious that the previous simulations did not accurately account for the green body size, as there is a huge influence of it on their densification. The present model however solves this problem entirely and has the correct qualitative dependence of sintering rate on particle size, as also shown in Fig. 4. Furthermore, the representative volume element (RVE) is likely mainly determined by how many particles are considered at once. The simulation 200^3 $R = 16\,\mathrm{nm}$ clearly shows a deviation from its bigger companions, but at the same domain size the simulations with smaller particles do not. Thus a certain minimum number of particles should be contained within a three-dimensional packing, with that threshold likely lying between 98 particles (200^3 $R = 16\,\mathrm{nm}$) and 262 particles (200^3 $R = 12\,\mathrm{nm}$). For polydisperse packings this threshold is likely much larger, but this will be left to further research.

Given that the densification is now accurately modelled, several topics of interest within the sintering process are investigated:

- coupled grain growth and densification
- evolution of closed porosity

The simulations described so far had grain growth kinetically suppressed by choosing a small GB mobility $\tau_{gb} = 100\tau_{gv,0}$, with the mobility $M \sim \frac{1}{\tau}$. By setting the mobility to be the same as that of the grain-vapor interface $\tau_{gb} = 1\tau_{gv,0}$, coupled grain growth and densification can be investigated. The effect of coupled grain growth and density is visualized by looking at the isolated porosity and whether it gets entrapped by grains or stays attached to GBs. In the latter case, vacancy annihilation will eventually remove these pores, given that they are not gas-filled. Whether a pore gets entrapped depends on the speed at which the grain is moving and how

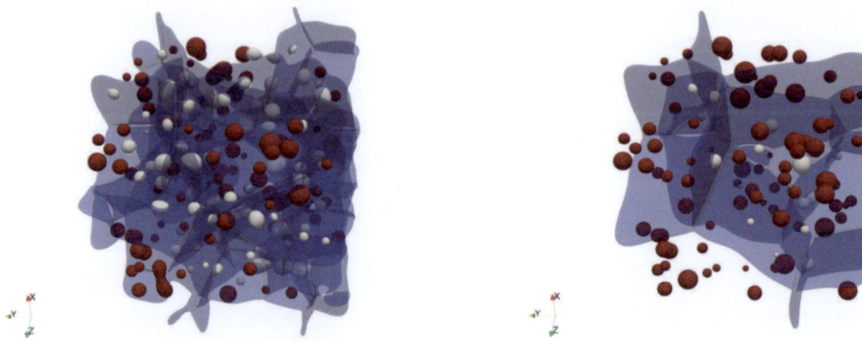

(a) $t = 0.6075\,\text{ms}$ (b) $t = 1.8075\,\text{ms}$

Fig. 5 A 400^3 voxel domain at two different simulation times with grain growth not suppressed. The grain boundaries (blue) get reduced due to grain growth, with isolated pores (grey) forming independently. Some of these get entrapped (red) due to grains moving faster than the pores

fast the pore can move, and thus grain growth obviously plays a role. Figure 5 shows a simulation with the GBs shown as transparent blue, the isolated pores as grey and entrapped pores as red. The entrapped pores never appear if grain growth is kinetically suppressed, but as observed in the simulations, once grain growth happens they are effectively unavoidable. Since these pores are no longer attached to a grain boundary, their further shrinkage is strongly inhibited. Thus grain growth is detrimental to the full densification of a body. It should be noted that these are the first simulations ever to start from a green body and ending in a dense sintered body, with macroscopic densification, pore closure, grain growth and pore entrapment all occurring within the same simulation.

As further variables of interest the values of the grain boundary D_{gb} and surface diffusivities D_s are identified and thus these are varied relative to their base values, with the resulting evolution of the density and average grain size shown in Fig. 6. For each of these variations, a simulation with kinetically suppressed grain growth is run (triangle) and one where grain growth is not suppressed (circles). The colors within the plot uniquely identify a parameter variation. For the base case as well as reduced surface diffusion, a change of mobility has little influence on the density evolution up to close to 95% density. This is likely due to the timescale of densification being much faster than that of grain growth for the present set of parameters: By the time the grain size has doubled, 95% density has already been reached. Afterwards, a significant difference in the density evolution is observed, with the simulation without grain growth achieving a higher end density. The surface diffusion has little effect on the densification; a reduction seems to increase the densification slightly though. This is likely a combination of a physical effect (less neck growth without shrinkage) and a model defect since the surface chemical potential may no longer be relaxed adequately.

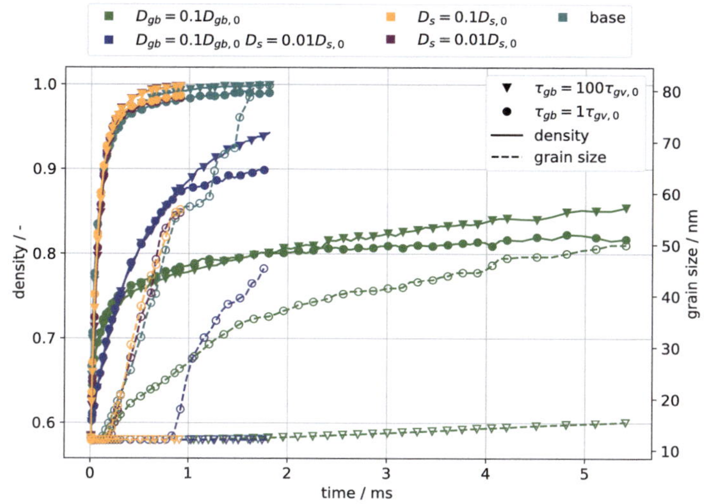

Fig. 6 The influence of grain growth on densification for various diffusion coefficients is shown. The color of the line indicates the (changed) parameters and the symbol whether grain growth is kinetically suppressed. For quick densification, surface diffusion has little influence on either. A reduced grain boundary diffusion slows down both processes. With densification happening slowly enough, a reduced surface diffusion increases densification

Changing the grain boundary diffusivity has a massive effect on the densification, as it directly controls the concentration of vacancies on grain boundaries. Furthermore, grain growth is also influenced by the grain boundary diffusivity. This is likely due to grain motion inducing grain growth [7], with smaller D_{gb} also leading to less grain motion. Due to the decrease of densification rate, the inhibitory effect of grain growth on densification becomes clear at around 80% density.

Interestingly, reducing the surface diffusion for the simulation with reduced grain boundary diffusion increases densification significantly and adds an inhibitory period to grain growth. The aforementioned effect of slower neck growth is most likely the cause of this, as it now proceeds sufficiently slow to be used for vacancy annihilation and thus densification. The inhibitory period is likely also related to this, since grain growth requires a curvature differential across the neck. If the sintering neck is small, the curvature differential of the individual grains is not transmitted fully between the grains, leading to almost no grain growth. Once the neck is of sufficient size to transmit the curvature differential, grain growth of speed comparable to the base case is observed.

Based on these observations a few conclusions can be made: Changes to the material or the process which influence the grain boundary mobility, the surface diffusion as well as the grain boundary are all relevant. Doping is one way of changing these quantities without significantly affecting the properties of the finished material. Based on the present investigation, the ideal dopant would be one which segregates to both the surface and grain boundary, increases the grain boundary diffusion while reducing the grain boundary mobility and the surface diffusion. While such a dopant is unlikely to exist, even one which reduces both grain boundary diffusion and mobility, such as doping Y_2O_3 with La or Gd [8], can be effective in inhibiting grain growth until high density is reached. In this sense, the base case as well as the simulations with reduced surface diffusion represent ideally doped systems. Another factor, not considered here, is the decrease of grain boundary energy via doping. This reduces the driving force and hence speed of grain growth, while at the same time increasing the dihedral angle which favors densification.

Furthermore, any changes to the process which lead to more time spent in a state region where grain boundary diffusion occurs much faster than grain growth will also help with this goal. In experiments it is generally observed that sintering at high heating rates produces denser materials with a lower grain size, whereas significant grain growth occurs for low heating rates. Based on the present simulation results, this can be interpreted to be the difference of timescales between densification and grain growth. High heating rate sintering processes would likely correspond to the base simulation of Fig. 6: Most of the densification takes place before the grain size has doubled and thus a dense material at low grain size can be achieved. If the heating is not controlled well, it will likely take the path of the simulation without kinetic suppression of grain growth and thus result in large grains and entrapped porosity. The present simulation approach can be employed to estimate when to stop heating the sample and thus improve materials properties. In contrast, conventional sintering is more akin to the simulation with lowered grain boundary diffusion, in which the timescales of densification and grain growth are similar and thus multiple doublings of the grain size are observed before high density is reached—with the concomitant increase of entrapped porosity which cannot be eliminated on processing timescales.

Finally, as an application of the model, the freeze-cast structure of the earlier section is simulatively sintered. For this, the regions of high particulate density are resolved with spheres, with the remaining regions effectively only containing ice which is transformed into the surrounding atmosphere. This roughly mimicks an evaporation process, which is part of the process chain to a freeze-cast product. In order to obtain densification in reasonable times, the particles are resolved at a physical resolution of 8 nm instead of the particle size during freeze-casting (250 nm). After this preprocessing step, the sintering model is applied as for the previous simulations. The microstructure evolution over time is shown in Fig. 7. The copper-like material shows the grains with the grain boundaries implied by the shadows with an etching-like effect. As expected the structure densifies, but keeps a large amount of open porosity which is desired in freeze-casting of porous materials. At simulation end no isolated porosity remains. Both the distance between the bridges as well as their thickness shrinks by about 16–19%, similar to experimental observations [9]. The shrinkage in the freezing direction (upwards) is lower at around 10 to 13%. In contrast to experiments by [10], the freezing direction shrinks *less* than directions normal to it, suggesting that further work ought to be done to investigate the anisotropic shrinkage.

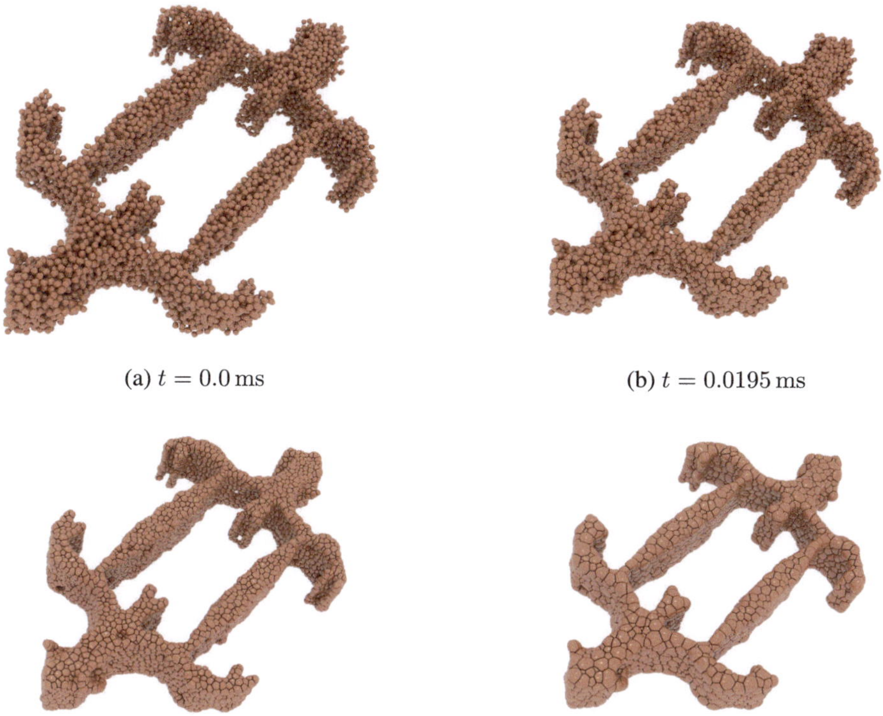

(a) $t = 0.0\,\text{ms}$ (b) $t = 0.0195\,\text{ms}$

(c) $t = 0.2175\,\text{ms}$ (d) $t = 1.2075\,\text{ms}$

Fig. 7 Microstructural evolution of a freeze-cast structure during the sintering process. The grain structure is shown as a reflective copper-like material with the black lines indicating grain boundaries

2 Realization of the Project

The simulations concerning solidification all employ the PACE3D framework in its general formulation. Thus they are based on finite differences defined on a regular Cartesian grid of nodal values, with a conservative flux-differences approach of order two being used to resolved divergences. Laplacians are similarly resolved with central differences. Time integration generally employs the explicit Euler method. MPI is employed for parallelization by decomposing the global domain into MPI process local subdomains, with ghost exchanges between adjacent processes.

The simulations of dendritic and eutectic growth under directional solidification conditions are conducted with a moving window technique, as a steady-state needs to be reached. The domain is shifted in the direction antiparallel to the growth direction whenever the interface crosses a cutoff. This allows the simulation of a quasi-infinite domain in one direction with only a finite domain. However, this requires a parallel reduction operation in each time step in order to determine when to shift the domain.

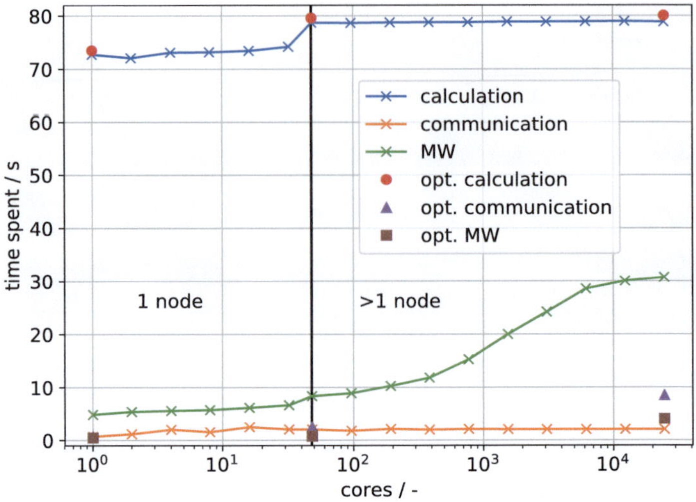

Fig. 8 Time expenditure of different program parts from 1 core up to 24 576 cores. Without the present optimization, the time spent waiting for the moving window technique grows excessively

A new criterion was developed based on the actual solidification speed which introduces an asynchronous test of the cutoff crossing and executes the domain shift in a delayed manner. This was developed in tandem with a project at the LRZ and thus Fig. 8 shows results obtained there. This shifts some of the load-imbalance onto the communication time, but the total execution time is reduced significantly.

LAMMPS [11] is employed to conduct small-scale molecular dynamics (MD) simulations, with the scaling and performance given in the reference.

The phase-field simulations concerning sintering employ a high-performance and massively parallel implementation based on [12]. This implementation is extended with advective terms to account for the macroscopic densification of the green body. In order to reduce numerical diffusion, a WENO-5 scheme [13] is employed for the advection discretization. Due to this new discretization the single-core performance is investigated anew. The floating-point operations (FLOP) within the compute regions are counted and related to the execution time of these regions, measured with lightweight timers. This results in a single-core performance of 9.4 GFLOP/s and given the single-core peak performance of the AMD EPYC 7742 of 36 GFLOP/s, about 26.1% of the theoretical peak performance is reached.

The improved model for calculating the displacement poses several challenges:

- a $O(nN^2)$ scaling with the phase number N and process count n
- dynamic generation of systems of equations with a complicated communication pattern.

First, the $O(nN^2)$ scaling results from the fact that several properties of the system need to be stored on a dual interaction $A_{\alpha\beta}$ basis of the phase-fields of which there are N. The number of MPI processes n enters, as each worker naïvely stores its own

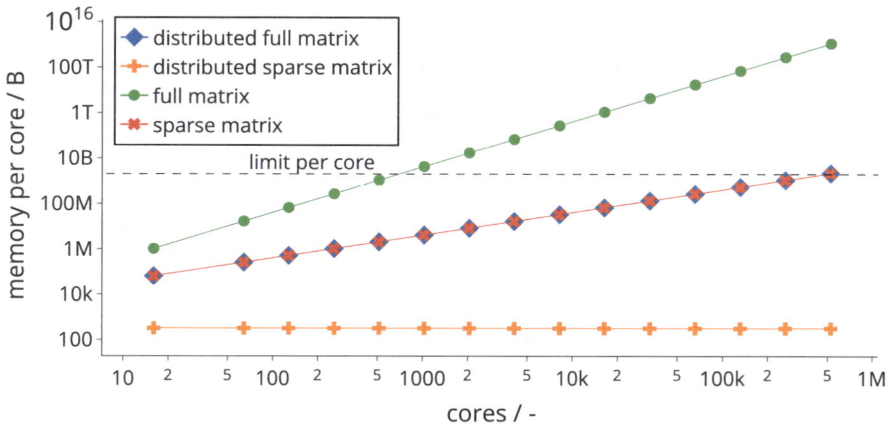

Fig. 9 Memory consumption per process for all the four variants within the test setup if weak scaling is conducted. If each process holds the full matrix, the memory requirements quickly become too stringent. If either the sparsity of the matrix is utilized or the matrix is distributed across the processes, the memory requirements grow linearly. Utilizing both optimizations yields a constant memory consumption per core. Note that the communication requirements also scale with the necessary memory

private copy of the dual interactions. However, most of the entries will be zero, as only actually interacting phase-fields will have non-zero entries. Thus storing only a sparse representation can already reduce the necessary storage and communication significantly by effectively achieving $O(nN)$ scaling. Distributing the matrix itself across the processes removes the factor n out of the scaling and thus achieves constant cost in a weak-scaling sense, as shown in Fig. 9. In this figure several variants are shown: The naïvely stored full matrix, only storing non-zero entries (sparse) as well as distributed versions of either. Updates of the sparse distributed matrix are achieved by employing MPI one-sided communication.

The dual interactions $A_{\alpha\beta}$ serve as input for the dynamic generation of a system of Eq. (2) whose solution describes the instantaneous displacement of the grains. The least squares problem could theoretically be solved e.g. with an ADMM approach [14], but this requires a collective reduction over N scalars in each iteration step. Based on early tests and the packings to be considered, a single reduction of that kind takes on the order of 10 ms even on a single compute node. The local sparse matrix-vector operations will be dominated by memory accesses, as the subproblems will be small. Assuming favourably for a full solution that each memory access takes 100 ns (RAM), the time of completion for one iteration will be on the order of 10 μs. Thus almost all time would be spent in waiting on communication. This part of the calculation cannot be interleaved with the other computation kernels, as it depends on results from these; thus solving the problem fully would massively reduce the scalability of the solver. An alternative would be gathering the transpose of the matrix and solve the normal equations e.g. with a CGLS [15] algorithm only requiring single scalar reductions and irregular point-to-point communication. However, gathering the transpose of a distributed matrix is quite time-consuming on its own and thus also unfavourable for parallelization. Furthermore, the transpose matrix cannot be cached in general, as it is a result of the dynamic evolution of the microstructure.

In order to avoid these problems, "typical" solutions of the problem are sought and then used to massively simplify the system of equations, while still recovering the salient points. For this, random particle packings are employed as input from which the system is constructed and the solution is serially calculated in a least-squares sense with the LSQR method [16]. The right hand-sides are assumed to be composed of a purely random part, realized with a normal distribution, plus a dependence on the spatial position of either adjacent grain. It is observed that the shape of the solution strongly correlates with the spatial distribution of right-hand sides, specifically linear functions are generated from spatially uniformly distributed right-hand sides, quadratic functions from spatially linearly distributed right-hand sides and so on. Given that the present phase-field model does not account for different grain boundary types and their absorption activity, assuming a spatially uniformly distributed right-hand side is justified. Thus the problem can be reduced in the following way: Replace the unknown displacements u with a model function $u(x) = m(x - x_0)$ with only m unknown, x being the center of mass of individual particles and x_0 being the total center of mass:

$$\underbrace{\mathrm{C}}_{C\in\mathbb{R}^{BxN}} u = \underbrace{\mathrm{C}(x - x_0)}_{D\in\mathbb{R}^{B}} m = \Delta u \tag{3}$$

$$Dm = \Delta u \tag{4}$$

$$\underbrace{D^T D}_{q\in\mathbb{R}} m = \underbrace{D^T \Delta u}_{p\in\mathbb{R}} \tag{5}$$

$$m = \frac{p}{q} \tag{6}$$

which only requires a parallel reduction operation for the values p, q per dimension, with each worker calculating the necessary matrix/vector products of their submatrices in parallel. Defining the mean absolute relative error as $E = \frac{1}{n}\sum_n |\frac{x_{LSQR}-x_{fit}}{x_{LSQR}}|$ yields an observed error of about 10%. The magnitude and shape of the solution are preserved quite well as shown in Fig. 10.

The scaling of this implementation is tested by observing complete runtimes of actual runs with the same block size per worker but different loads due to calculation skips, including I/O. This resulted in an observed parallel efficiency of 73% when scaling from one node (128 cores) to 64 nodes (8192 cores), much below the efficiency observed in [12] for the explicit solver. This originates from the one-sided communication as follows: Since the time to update the globally distributed $A_{\alpha\beta}$ depends on how the origin and target are connected, it will naturally induce some load imbalances. Furthermore, in order for $A_{\alpha\beta}$ to be in a consistent state across processes, every process needs to have finished `MPI_Win_Flush_all` and thus an

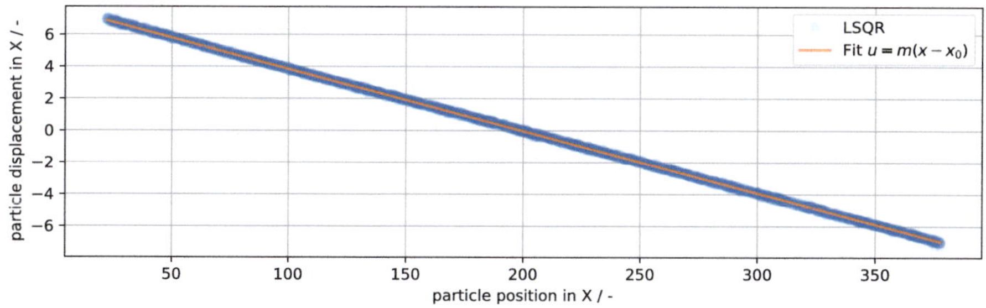

Fig. 10 Comparison of serially solving the system with the LSQR method and the simplified fit solution

asynchronous barrier is started after this call. The wait on this asynchronous barrier, due to the aforementioned load imbalances, almost entirely accounts for the observed drop in efficiency.

The improved phase-field model of sintering massively reduced the number of time steps necessary to reach at least 90% density and thus the computing time that was initially applied for has not been completely used. Furthermore, the new model also showed convergence to a representative volume element at a lower size than expected and thus smaller domain sizes could be employed. This also opened up the possibility of investigating different aspects of solidification. The goal of these is to allow the generation of a green body by sampling a solidified microstructure with spheres, approximating the local solid fraction by doing so. This allowed the complete simulation of the freeze-casting process starting from a suspension and ending in a porous sintered body for the first time ever.

References

1. Seiz, M., Kellner, M., Nestler, B.: Simulation of dendritic-eutectic growth with the phase-field method (2023) (arXiv)
2. Seiz, M., Nestler, B.: Modelling and simulation of the freeze casting process with the phase-field method. Comput. Mater. Sci. **193**, 110410 (2021)
3. Hötzer, J., Seiz, M., Kellner, M., Rheinheimer, W., Nestler, B.: Phase-field simulation of solid state sintering. Acta Mater. **164**, 184–195 (2019)
4. Wang, Y.U.: Computer modeling and simulation of solid-state sintering: a phase field approach. Acta Mater. **54**(4), 953–961 (2006)
5. Seiz, M.: Effect of rigid body motion in phase-field models of solid-state sintering, Comput. Mater. Sci. **215**. Publisher, Elsevier B.V (2022)
6. Yang, Q., Gao, Y., Kirshtein, A., Zhen, Q., Liu, C.: An interfacially consistent phase-field model for solid-state sintering without artificial void generation. SSRN Electron. (preprint) (2022)
7. Mcfadden, G.B., Boettinger, W.J., Mishin, Y.: Effect of vacancy creation and annihilation on grain boundary motion, 1–36 (2019)
8. Kindelmann, M., Ran, K., Rheinheimer, W., Morita, K., Mayer, J., Bram, M., Guillon, O.: Segregation-controlled densification and grain growth in rare earth-doped Y_2O_3. J. Am. Ceram. Soc. **104**(10), 4946–4959 (2021)

9. Wilke, S.K., Mack, J.B., Kenel, C., Dunand, D.C.: Evolution of directionally freeze-cast Fe2O3 and Fe2O3+ NiO green bodies during reduction and sintering to create lamellar Fe and Fe-20Ni foams. J. Alloy. Compd. **889**, 161707 (2021)
10. Lichtner, A., Roussel, D., Röhrens, D., Jauffres, D., Villanova, J., Martin, C.L., Bordia, R.K.: Anisotropic sintering behavior of freeze-cast ceramics by optical dilatometry and discrete-element simulations. Acta Mater. **155**, 343–349 (2018)
11. Thompson, A.P., Aktulga, H.M., Berger, R., Bolintineanu, D.S., Brown, W.M., Crozier, P.S., In't Veld, P.J., Kohlmeyer, A., Moore, S.G., Nguyen, T.D., Shan, R, Stevens, M.J., Tranchida, J., Trott, C., Plimpton, S.J.: LAMMPS - a flexible simulation tool for particle-based materials modeling at the atomic, meso, and continuum scales. Comput. Phys. Commun. **271**, 108171 (2022)
12. Hierl, H., Hötzer, J., Seiz, M., Reiter, A., Nestler, B.: Extreme scale phase-field simulation of sintering processes. In: 2019 IEEE/ACM 10th Workshop on Latest Advances in Scalable Algorithms for Large-Scale Systems (ScalA), Denver, CO, USA, 18 Nov 2019, pp. 25–32. Institute of Electrical and Electronics Engineers (IEEE) (2020), 35.14.01; LK 01
13. Shu, C.-W.: High order weighted essentially nonoscillatory schemes for convection dominated problems. SIAM Rev. **51**(1), 82–126 (2009)
14. Boyd, S.: Distributed optimization and statistical learning via the alternating direction method of multipliers. Found. Trends® Mach. Learn. **3**(1), 1–122 (2010)
15. Ji, H., Li, Y.: Block conjugate gradient algorithms for least squares problems. J. Comput. Appl. Math. **317**, 203–217 (2017)
16. Paige, C.C., Saunders, M.A.: LSQR: an algorithm for sparse linear equations and sparse least squares. ACM Trans. Math. Softw. **8**(1), 43–71 (1982)

Computational Fluid Dynamics

The following chapter presents selected contributions from the field of computational fluid dynamics (CFD) during the reporting period. All computations were conducted on the HPE Apollo supercomputer "HAWK" at HLRS. Following the trend from the previous years, this field of application benefits strongly from the available supercomputing resources. Over 60% of computing time spent on HAWK is allocated to CFD. Many of the computing projects combine the expertise of at least two groups located at different research institutions across the country; some also include international partners.

In this period, 19 individual annual reports were submitted and evaluated in a peer-review process for scientific excellence and efficient use of the computing resources. The voluntary work of the reviewers is highly appreciated. Based on their recommendations, twelve have been selected for publication in this book. The spectrum of the presented projects ranges from fundamental research into the intricate process of individual droplets hitting a film-covered surface or the soot oxidation processes in mixing jets to applications in the field of turbomachinery components and the noise generated at wind energy turbine trailing edges. The same breadth is also found in the numerical methods preferred by each group to solve the underlying equations—from classical finite volume schemes for the incompressible Navier-Stokes equations to Lattice-Boltzmann methods and to specialized high-order accurate finite element or finite difference formulations for compressible flows; from statistical formulation like the RANS equations to scale-resolving approaches like LES and DNS. Also the influence of data-driven approaches becomes more and more apparent. Most of the codes that drive these projects are established research codes developed and maintained at universities, but also commercial software packages are employed. As becomes obvious from the following detailed reports, the commercial solvers employed currently on HAWK cannot take full advantage of the provided resources, while the groups developing specialized research codes report better parallelization efficiency, but also a considerable amount of work efforts spent on the design, implementation and validation of HPC-ready algorithms.

In the following, the individual project contributions are introduced briefly. The full papers follow after this introduction. In their paper "Towards Exascale CFD Simulations Using the Discontinuous Galerkin Solver FLEXI" the authors from the Institute of Aerodynamics and Gas Dynamics, University of Stuttgart, demonstrate the exascale readiness of the code FLEXI which solves the compressible Navier-Stokes equations using a high-order discontinuous Galerkin (dG) discretization in space and an explicit Runge–Kutta method in time. Because of their high computational intensity these methods require less memory bandwidth per core and excellent strong scaling up to the half a million cores on HAWK is demonstrated. Another innovative aspect of this project is the efficient incorporation of trained machine learning models from TensorFlow into the Fortran code FLEXI.

The contribution "Computational Aspects for DNS of Compressible Turbulent Boundary Layers on up to Full Machine Size" by authors from the Institute of Aerodynamics and Gas Dynamics, University of Stuttgart, contributes research on direct numerical simulation of the turbulent boundary layer over a flat plate using the code NS3D. This code features 6th compact finite differences in space combined with a classical 4th order explicit Runge–Kutta scheme in time and a 10th order filter for stabilization. Impressive computations on up to 24 billion grid points and half a million cores have been performed at excellent scaling. Moreover, the authors have developed a highly interesting framework to investigate performance differences of the compute nodes in detail.

The report on "HPC-Simulations of Unsteady Turbomachinery Flows" by the members of the Institute of Thermal Turbomachinery and Machinery Laboratory of the University of Stuttgart reports on high-fidelity CFD models for the prediction of the aerodynamic excitation for three vaneless radial turbines. A number of simulations at different operating points by means of URANS and hybrid RANS/LES simulations are conducted in order to assess the excitation maps for resonance effects. The usage of a HPC system made it possible to achieve a detailed insight in the forcing behavior over a wide range of operating conditions. The authors report that the scale-resolving simulations showed a more accurate prediction of the diffuser flow field in terms of pressure recovery and velocities when compared against experimental data. All simulations were all performed on HPE Apollo Hawk system at the HLRS Stuttgart using ANSYS CFX in combination with the "Open MPIHAWK" parallelization. The reported scaling results show an acceptable scaling up to O(10) nodes and underline the need for more suitable simulation tools for HPC applications.

The contribution "Towards Large Scale Blunt Trailing Edge Noise Simulations" by the authors from the German Aerospace Center (DLR), Institute of Software Methods for Product Virtualization, addresses the numerical simulation of trailing edge noise on wind turbines. Of particular interest to them is blunt trailing edge noise, caused by finite thickness of the trailing edge which produces a von Kármán-type quasi-periodic vortex shedding. This induces an additional noise component onto the broadband spectrum. The authors perform LES of a corresponding wind tunnel experiment with the Lattice-Boltzmann solver musubi. Musubi is shown to scale on up to 512 nodes. The authors introduce a non-blocking reduction from MPI-3 for global fault checking as a novel feature and present initial results of a comparison of a very large eddy simulation with wind tunnel data.

The paper "Soot Oxidation in a Turbulent Temporally Evolving Non-premixed Air/ethylene Jet" by the group from the Institute for Combustion Technology, RWTH Aachen University, presents a direct numerical simulation of a turbulent temporally evolving non-premixed air/ethylene jet specifically designed for soot oxidation on approximately 1.5 billion grid points. This constitutes currently the largest available data set on soot formation and consumption in a turbulent environment. In addition, it is the first large-scale simulation to cover the later stages of soot evolution, and it provides a detailed description of this process with a particular focus on oxidation. In addition, an assessment of existing subfilter models to account for soot-gas phase interaction, particularly relevant for soot oxidation, is performed, and possible improvements are discussed. The well-established high-order finite difference code CIAO is used in this computational campaign on HAWK on up to 512 nodes. The authors show very good scaling on HAWK and other machines and identify the solution of the elliptic pressure equation, the transport of Lagrangian particle trajectories for soot and the chemistry solver as the most challenging operations for HPC.

The paper "CFD of Transient Operations of a Pump Turbine" from the Institute of Fluid Mechanics and Hydraulic Machinery, University of Stuttgart establishes a workflow on HAWK that was developed to enable the simulation of transient operations of a pump-turbine. This workflow requires a combination of mesh deformation and remeshing as well as the usage of a porous domain to represent the motion of the guide vanes as well as 3D-1D coupling. The simulation chain was validated by comparing a load increase from 6 to 15 degrees of guide vane angle against experimental results. The simulations were performed on HPE Apollo Hawk system at the HLRS Stuttgart using ANSYS CFX in combination with the "Open MPIHAWK" parallelization.

Authors from the same institution present their results on a scale resolving hybrid RANS-LES simulation of a Kaplan turbine with over 400 million degrees of freedom. The simulation with ANSYS CFX was run with good efficiency on 4096 cores of HAWK. The schemes use an automatic selection of a wall-resolved or wall-modeled formulation. The results of the RANS-LES simulation indicate that most of the simulation domain is running in LES mode and that the requirements on the spatial grid for LES are met. The results are compared against a steady-state RANS simulation at 20 million DOF. Initial comparisons with experimental results indicate a better match of the unsteady results to PIV measurements. The authors report the setup, scaling and results in their contribution titled "Highly Resolved Hybrid RANS-LES Simulations of Different Operating Points of a Model Kaplan Turbine".

In the contribution "Modeling of Coupled Turbulent Channel-Porous Media Flow through a Deep Autoencoder-Echo State Network Framework" from the researchers at the Cluster of Excellence SimTech and the Institute of Aerospace Thermodynamics, University of Stuttgart as well as the Institute of Thermodynamics and Fluid Mechanics, Technische Universität Ilmenau, the authors propose a machine learning model, namely a combined Convolutional Deep Autoencoder–Echo State Network, for the analysis and forecasting of the dynamics and low-order statistics in coupled turbulent channel-porous media flows. Coupled porous and free-flow systems are particularly relevant in the suppression of flow noise through porous liners or transpiration cooling systems. The authors conduct fully resolved direct numerical simulations using the well-established HPC-capable, h/p adaptive spectral element code Nektar++ to generate the data for training. Their investigations reveal up to 70% strong scaling on up to 512 nodes of HAWK. With the generated flow data, the autoencoder is trained to represent a dense low-order representation of the flow features, which are advanced through an echo state network. This enables effective modeling and prediction of dominant flow behaviors, particularly within the laminar regime of the porous domain.

As in the previous paper, the authors of the report titled "CFD-Driven Surrogate-Based Optimization" from the Chair of Fluid Mechanics and Institute of Aerodynamics, RWTH Aachen, focus on the development of data-driven surrogate models. They set out to investigate the potential of spanwise traveling transversal surface waves as an active drag reduction strategy for turbulent boundary layer flows by the means of wall-resolved Large Eddy Simulation for a weakly and fully compressible case. The simulations are carried out with the body-fitted, structured grid Finite Volume-solver "m-AIA" on HAWK. The code is well-established as an HPC scheme, as also shown by the excellent strong scaling results for the full machine. The actuation is realized by prescribed characteristics of the waves (amplitude, wave length, time interval) and realized by an moving mesh in arbitrary Lagrangian-Eulerian formulation. From the simulation data, two surrogate models based on support vector regression and Gaussian process regression are generated and exploited in the optimization of the actuation. Via Bayesian optimization, prior mean information is incorporated into the process. The presented initial results show the considerable potential of this approach.

The authors of the report on "Optimization of a Non-standard Pump from Galvanic Industry with the Island Model" from the Institute of Fluid Mechanics and Hydraulic Machinery at the University of Stuttgart present a fully automated optimization toolchain on HAWK for parameter sweeps in engineering design. They apply their approach to the evolutionary optimization of a non-standard pump from galvanic industry, parallelized by the island model. This model is a version of the Estimation of Distribution Algorithms and maps individual population to processors (as islands), such that each island evolves independently and only occasionally exchanges results and evolution processes with other processors. The aim of this optimization is to improve the efficiency the pump at a specified operating point. Each geometry is parameterized by nine degrees of freedom, and the commercial code Ansys CFX version 20.2 with its known HPC limitations is used, so each run is limited to 7 nodes of HAWK. The authors discuss load distribution between the simulations and optimization runs. Their results indicate that an increase in efficiency of close to 50% is possible over the baseline pump configuration.

In their contribution titled "DNS of the Early Phase of Oblique Droplet Impact on Thin Films with FS3D" the authors from the Institute of Aerospace Thermodynamics, University of Stuttgart, investigate the impact of droplets at different angles onto walls covered with a thin film of the liquid—a process which can be seen as an elementary building block of droplet-wall interactions. In this work, the authors present a DNS study on 2.1 billion degrees of freedom of the early crown formation mechanisms present at an oblique droplet impact, as well as a grid study showing the resolution required to resolve the impact's details. Such investigations at high detail are intractable by experimental approaches. With the help of the simulations, two different mechanisms of crown formation were observed and could be explained. Besides these observations, the authors also present significant improvement of their parallelization concept. All computations are conducted with the well-established finite volume solver FS3D for the incompressible flow equations with volume-of-fluid extension for multispecies and interface treatment. The bottleneck with respect to parallelization is the solution of an elliptic pressure equation. Here, the authors propose improvements in the multigrid cycle strategies and present the newly implemented F-cycle as a more efficient alternative to either V- or W-cycles. This choice and an additional hybrid parallelization increase the achieved strong scaling considerably.

The final paper in this selection focuses on droplets as well, but in compressible flows. It is titled "A High-Order Framework for Compressible Droplet Dynamics" and is presented by the authors from the Institute of Aerodynamics and Gas Dynamics, University of Stuttgart. It is devoted to the numerical simulation of compressible two-phase flow systems using bot, a sharp and a diffuse interface approach. Both methods are implemented based on the high-order dG framework FLEXI. A dynamic load balancing procedure was developed to cope with the highly varying computational load per cell stemming from polynomial degree adaptivity, finite volume shock-capturing and complex Riemann solvers at phase boundaries. Up to 8192 cores have been used efficiently for this highly complex flow problem.

The selection of projects presented reflects the continuing progress of high-performance computing in the field of CFD using the resources continuously provided by the HLRS. High-resolution, scale-resolving numerical simulations on extremely fine computational grids including complex physical processes are crucial to gaining insight into flow physics. The availability of these large computing resources fosters both fundamental research and industrial design processes, where high-fidelity numerical methods on modern supercomputers help to enhance the reliability and the efficiency of the simulations, reduce development time, risks and costs and, thus, increase industrial competitiveness. They also support the generation of reduced order models that can help reduce computational costs in the future, in particular in the field of parameter studies or optimization problems. Many the presented projects and codes can take advantage of all available compute nodes at HLRS, that is of the "full HAWK", and run and handle computation on billions of degrees of freedom. While this is of course a positive development, it also highlights the challenges that all developers of CFD codes for large-scale computations are currently facing with increasing urgency: Exascale computing hardware, like the new generations of Hunter and Herder at HLRS, will be accelerator-based systems. This brings with it a re-thinking and often re-writing of large pieces of research software. While this step is absolutely necessary and worth it alone for ecological reasons, it will continue to be quite challenging for both commercial and research software. Here, as always, thanks are due to the staff of the HLRS computing center for their individual and custom-tailored continued support. Without their dedication, it would not have been possible and will not be in the future to maintain the high scientific quality we see in the projects.

Andrea Beck
Institut für Aerodynamik und Gasdynamik
Universität Stuttgart
Stuttgart, Germany
e-mail: andrea.beck@iag.uni-stuttgart.de
Peter Bastian
Interdisziplinäres Zentrum für Wissenschaftliches Rechnen
Universität Heidelberg
Heidelberg, Germany
e-mail: peter.bastian@iwr.uni-heidelberg.de

Towards Exascale CFD Simulations Using the Discontinuous Galerkin Solver FLEXI

Marcel Blind, Min Gao, Daniel Kempf, Patrick Kopper, Marius Kurz, Anna Schwarz, and Andrea Beck

Abstract Modern high-order discretizations bear considerable potential for the exascale era due to their high fidelity and the local computational load that allows for computational efficiency in massively parallel simulations. To this end, the discontinuous Galerkin (DG) framework FLEXI was selected to demonstrate exascale readiness within the Center of Excellence for Exascale CFD (CEEC) by simulating shock buffet on a three-dimensional wing segment at transsonic flight conditions. This paper summarizes the recent progress made to enable the simulation of this challenging exascale problem. For this, it is first demonstrated that FLEXI scales excellently to over 500 000 CPU cores on HAWK at the HLRS. To tackle the considerable resolution requirements near the wall, a novel wall model is proposed that takes compressibility effects into account and yields decent results for the simulation of a NACA 64A-110 airfoil. To address the shocks in the domain, a finite-volume-based shock capturing method was implemented in FLEXI, which is validated here using

---

M. Blind (✉) · M. Gao · D. Kempf · P. Kopper · M. Kurz · A. Schwarz · A. Beck
Institute of Aerodynamics and Gas Dynamics, University of Stuttgart, Stuttgart, Germany
e-mail: blind@iag.uni-stuttgart.de

M. Gao
e-mail: mg@iag.uni-stuttgart.de

D. Kempf
e-mail: kempf@iag.uni-stuttgart.de

P. Kopper
e-mail: kopper@iag.uni-stuttgart.de

M. Kurz
e-mail: m.kurz@iag.uni-stuttgart.de

A. Schwarz
e-mail: schwarz@iag.uni-stuttgart.de

A. Beck
e-mail: beck@iag.uni-stuttgart.de

T. Ludwig et al. (eds.), *High Performance Computing in Science and Engineering '23*,
https://doi.org/10.1007/978-3-031-91312-9_15

the simulation of a linear compressor cascade at supersonic flow conditions, where the method is demonstrated to yield efficient, robust and accurate results. Lastly, we present the TensorFlow-Fortran-Binding (TFFB) as an easy-to-use library to deploy trained machine learning models in Fortran solvers such as FLEXI.

1 Introduction

Over the last decades, the steadily increasing amount of computational resources has allowed computational sciences to tackle increasingly demanding and complex research questions by enabling larger and more impactful simulations. The most recent milestone in this development was reached in 2022 when Frontier became operational as the world's first exascale system. This upcoming generation of exascale systems offer tremendous computing power that promises to enable and accelerate scientific research and industrial innovation. The NASA CFD Vision 2030 Study [16] highlights the need to develop simulation software and algorithms to also harness these emergent computing resources efficiently in the field of Computational Fluid Dynamics (CFD). Recognizing this, a Center of Excellence for Exascale CFD (CEEC) has been launched under the European High Performance Computing Joint Undertaking (EuroHPC JU) umbrella. CEEC focuses on developing and adapting simulation software to enable European researchers and companies to perform simulations on an unprecedented scale to provide deeper insights, more accurate predictions, and to support engineering design optimization. The discontinuous Galerkin (DG) flow solver FLEXI[1] is one of the selected codes that should demonstrate exascale readiness by performing as a lighthouse case a Large Eddy Simulation (LES) of instationary shock buffet on a three-dimensional wing segment at realistic flight Reynolds numbers. While DG methods have matured considerably in the last decade, this application requires a range of features and properties that are crucial for complex applications in science and engineering, but are not yet readily available for modern high-order discretizations. For instance, the employed high-order DG method is suitable for scale-resolving simulations such as LES, however, requires special treatment to capture discontinuities such as shocks. Moreover, the resolution requirements necessary for such high Reynolds number flows are considerable, in particular near the walls, which renders the application of wall models a necessity. With the advent of scientific machine learning (ML) in recent years, it also remains an open research question how promising ML models can be efficiently integrated into high-performance computing (HPC) simulations at such a scale.

To address these difficulties, we present the following four building blocks that were developed, implemented and validated in order to facilitate the simulation of the defined lighthouse case. First, we report the scaling abilities of FLEXI in Sect. 2 and demonstrate that FLEXI can efficiently scale up to the maximum of 4096 compute nodes (over half a million CPU cores) provided by the flagship HAWK system at the

[1] https://github.com/flexi-framework/flexi.

High-Performance Computing Center Stuttgart (HLRS). The second building block addresses the requirement for wall modeling at flight Reynolds numbers to relax the enormous resolution demands in the near-wall region of the boundary layer. To this end, a compressible wall model is proposed and validated in Sect. 3 using the flow over a NACA 64A-110 airfoil. Next, the implemented shock capturing method based on the blending approach between a DG and a finite volume (FV) scheme by [6] is introduced in Sect. 4. This shock capturing approach is validated using the simulation of a transsonic, linear compressor cascade. Lastly, we present the TensorFlow-Fortran-Binding (TFFB) in Sect. 5, which provides an efficient interface between Fortran codes such as FLEXI and the ML framework TensorFlow [1] and is demonstrated to scale to tens of thousands of compute cores. Section 6 concludes the paper with a concise summary and an outlook on future developments.

2 Performance

In a first step, it is demonstrated that FLEXI scales excellently up to the full HAWK system. For this, a constant freestream is computed using Cartesian meshes with periodic boundary conditions ranging from 2^{10} to 2^{16} elements in order to simulate different problem sizes. The finer meshes are obtained from the coarsest one by repeatedly doubling the number of elements in a single dimension. The polynomial degree is chosen as $N = 7$, which corresponds to $8^3 = 512$ DOF per element and results in a maximum of over 34 billion DOF for the finest grid. The split-form DG formulation on Legendre–Gauss–Lobatto (LGL) nodes is employed, which is required to ensure stability in underresolved simulations like LES. The cases are computed on up to 4096 compute nodes, which corresponds to a maximum of over half a million compute cores. The number of nodes that can be applied for a given mesh size is limited by its number of elements, since FLEXI does not allow elements to be shared between processors. Conversely, the minimum number of nodes for a given problem size is determined by the required wall-time, since running the largest cases on a single compute node would require weeks to complete. The considered cases thus employ between 512 and 2 million DOF per core, which corresponds to the minimum and maximum load, respectively. Each simulation is advanced for 100 time steps by an explicit Runge–Kutta scheme and the performance is evaluated without initialization, analyzing routines or file I/O. The computation of each configuration is repeated 5 times to gather statistics, but only the mean results are reported here. For the test, FLEXI was compiled using the current software stack of `gcc/10.2.0`, `aocl/3.2.0` and `mpt/2.2.6`. The performance is measured using either the speedup or the performance index (PID). The speedup describes by which factor the walltime decreases if the number of employed nodes increases and is computed with respect to the lowest number of nodes employed for a given problem size. The PID can be defined as

$$PID = \frac{\text{wall-clock-time} \cdot \#\text{cores}}{\#\text{DOF} \cdot \#\text{time steps} \cdot \#\text{RK-stages}}, \quad (1)$$

and describes the time required to advance a single DOF for a single Runge-Kutta stage. Since the PID thus indicates the execution time, a smaller PID implies better performance. Hence, the PID allows to compare the performance for cases with different problem sizes and resources.

The obtained results are presented from two different perspectives in Fig. 1. First, the strong scaling (left) shows the speedup obtained by increasing the number of nodes employed for the simulation, while keeping the problem size, i.e. the mesh size, constant. Here, FLEXI demonstrates excellent scaling on up to 4096 compute nodes. The results also demonstrate that FLEXI can scale super-linearly in the strong case, since with an increasing number of nodes, the load per core decreases and an increasing portion of the relevant data fits into the fast CPU cache. As a consequence, the bottleneck of limited memory bandwidth becomes less important. The right panel in Fig. 1 illustrates the weak scaling properties of FLEXI, where the PID is shown over the specific load per core. For high loads, the memory bandwidth is a bottleneck and causes the performance to deteriorate, as discussed above. For low loads, however, the local work for each processor is not sufficient to hide the communication latency efficiently, which causes the cores to idle. As a consequence, FLEXI reaches its optimal performance between 10^3 and 10^4 DOF per core, where an optimum between caching efficiency and communication latency is reached.

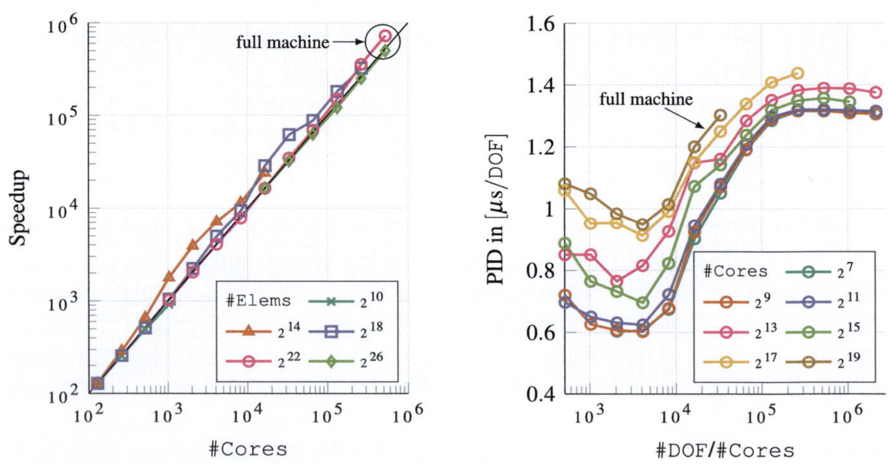

Fig. 1 Scaling of FLEXI on HAWK. Left: Strong scaling results as speedup for meshes comprising different number of elements. The ideal speedup is indicated in black. Right: PID over the specific load per core for different numbers of compute nodes. Only the mean is shown in both cases for the sake of clarity

3 Wall-Modeled Large Eddy Simulation

Wall-resolved LES (WRLES) is oftentimes infeasible for engineering applications due to the considerable resolution requirements close to walls. This stems from the necessity to accurately resolve the small-scale flow structures near the wall as well as the emergent velocity gradient in wall-normal direction. In wall-modeled LES (WMLES), the no-slip wall is not enforced explicitly, but the wall-shear stress in the LES domain is modified such that the no-slip condition is fulfilled by the underlying law of the wall. This means that the LES equations are solved in the entirety of the domain and the wall model is only imposed on the respective boundaries. For this, the wall model computes the properties of the boundary condition at the wall based on the instantaneous flow characteristics at a given interface point with distance h_{wm} from the wall. A wall model can thus be considered as a "black box" that yields a functional relationship between the local flow state and the output parameters which are required to impose the correct boundary condition at the wall.

3.1 *Wall Models for Compressible Flows*

Two of the most prominent wall models in the literature are the algebraic model based on an analytic law of the wall, e.g. Spalding [19], and a wall model that also considers compressibility effects by solving an ordinary differential equation (ODE). Algebraic wall models were one of the earliest proposed models and were derived by empirically fitting curves to experimental data. A prominent example is Spalding's law of the wall [19], which reads

$$y^+ = u^+ + 0.1108 \left[e^{0.4u^+} - 1 - 0.4u^+ - \frac{\left(0.4u^+\right)^2}{2!} - \frac{\left(0.4u^+\right)^3}{3!} - \frac{\left(0.4u^+\right)^4}{4!} \right], \tag{2}$$

and yields a single expression capable of capturing the viscous sublayer, the buffer layer and the turbulent core of an incompressible boundary layer. Here, y^+ and u^+ denote the wall-normal distance and the streamwise velocity in viscous units, respectively. Since this model was only formulated for incompressible flows, we present in the following an extension to compressible flows by modifying the input velocity at the interface exchange location h_{wm}.

Later, Van Driest proposed a transformation that allows to map compressible to incompressible boundary layers [20, 21]. Assuming the flow over a flat plate with perfect gas and a Prandtl number of Pr $= 1$, this transformation follows as

$$u_{\text{eq}} = \frac{u_e}{a} \arcsin\left(a\frac{u}{u_e}\right), \quad \text{with} \quad a = \sqrt{1 - \frac{T_e}{T_{aw}}}. \tag{3}$$

Here, u_{eq} denotes the equivalent incompressible velocity for a given compressible velocity u and u_e denotes the velocity at the boundary layer edge. Moreover, T_{aw} and T_e denote the temperature at the wall and the boundary layer edge, respectively. For a perfect gas with a heat capacity ratio of γ, the ratio of the temperatures at the wall and the edge of the boundary layer follows as

$$\frac{T_{aw}}{T_e} = \left(1 + \tfrac{\gamma-1}{\gamma}\sqrt[3]{\mathrm{Pr}}\right)\mathrm{Ma}_e^2 \tag{4}$$

with Ma_e as the Mach number at the boundary layer edge. Assuming that the velocity and temperature at the boundary layer edge match the values in the freestream yields an expression for $u_{\mathrm{eq}}(u)$ as

$$u_{\mathrm{eq}} = \frac{u_\infty}{b}\arcsin\left(b\frac{u}{u_\infty}\right), \quad \text{with} \quad b = \frac{\sqrt{\tfrac{\gamma-1}{2}\mathrm{Ma}_\infty^2\sqrt[3]{\mathrm{Pr}}}}{\sqrt{1+\tfrac{\gamma-1}{2}\mathrm{Ma}_\infty^2\sqrt[3]{\mathrm{Pr}}}}, \tag{5}$$

where u_∞ and Ma_∞ are the velocity and Mach number in the freestream, respectively.

In contrast to the algebraic Spalding wall model, the ODE model takes compressibility into account by design and, as a consequence, does not require the additional van Driest velocity transformation presented in Eq. (5). Moreover, the ODE model follows the conservation of heat and thus the conservation of energy. This renders the ODE wall model highly suitable for WMLES at high supersonic flow conditions. However, the ODE model requires solving an ODE with a suitable discretization on a separate one-dimensional grid at each solution point, which results in significantly higher computational cost than iteratively solving Eq. (2) for the wall shear stresses. To this end, the performance of both models is evaluated in the following based on the transsonic flow around a NACA 64A-110 airfoil and compared to a wall-resolved simulation.

3.2 Evaluation of the Wall Models

In a next step, the proposed wall model is compared with the ODE model using the flow around a NACA 64A-110 airfoil, which has already been investigated excessively in the literature [3, 4] and is shown in Fig. 2. The flow is transsonic with $\mathrm{Ma}_\infty = 0.72$ and a Reynolds number of $Re_c = 930\,000$ with respect to the chord length c. A geometric trip is positioned at $x = 0.05c$ to trigger the turbulent transition of the boundary layer. Four different configurations were investigated. First, a WRLES was computed, which serves as the underlying ground truth to assess the performance of the WMLES. Based on this, three WMLES were performed using the ODE model, the original Spalding model and the newly proposed extended

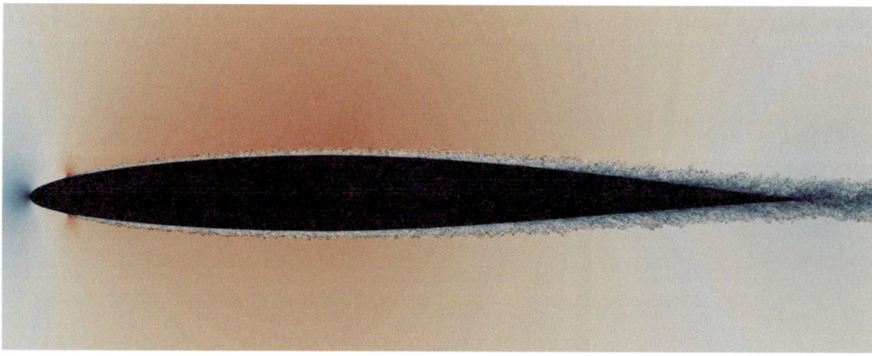

Fig. 2 Instantaneous flow field of the flow around the NACA 64A-110 profile with iso-surfaces of the Q-criterion and colored by the velocity magnitude. The regions of high velocity near the tip are caused by the geometric trip in the boundary layer

Table 1 Computational setup of the different configurations. The numbers referring to the Spalding-type model apply for both Spalding-type models investigated, i.e. the original model and the extended variant for compressible flows

Model	#Elems	#DOF	#Nodes	#Cores	PID (μs)	CPUh
WRLES	723 132	$370 \cdot 10^6$	512	65 536	1.0	5 000 000
WMLES-Spal.	368 800	$189 \cdot 10^6$	128	16 384	1.3	350 000
WMLES-ODE	368 800	$189 \cdot 10^6$	128	16 384	3.1	750 000

Spalding model for compressible flows, respectively. The computational setups and the obtained performance for all simulations are summarized in Table 1. It is important to note that all WMLES share the same computational mesh.

To assess the performance of the different wall models, Fig. 3 illustrates the velocity boundary layer profile on the pressure side at half the chord length, i.e. $x = 0.5c$ for all three wall models and the WRLES as ground truth. Moreover, the right plot of Fig. 3 shows a zoom of the log layer to highlight the differences between the models. Overall the differences between the considered wall models are small. However, the ODE model and the Spalding model with van Driest transformation clearly match the WRLES better in the log layer than the original Spalding model. This is to be expected and demonstrates the importance of compressibility effects in the boundary layer for transsonic flows. The van Driest transformation seems to shift the boundary layer profile of the WMLES closer towards the WRLES in comparison to the original Spalding model and thus improves its overall agreement with the reference. In contrast, the ODE model shows a different behavior, but tends to be generally closer to the WRLES. This is also expected since the ODE model includes compressibility effects by design.

While the ODE model thus yields the most accurate results of all tested wall models, it is also far more expensive than the Spalding-type models. For the inves-

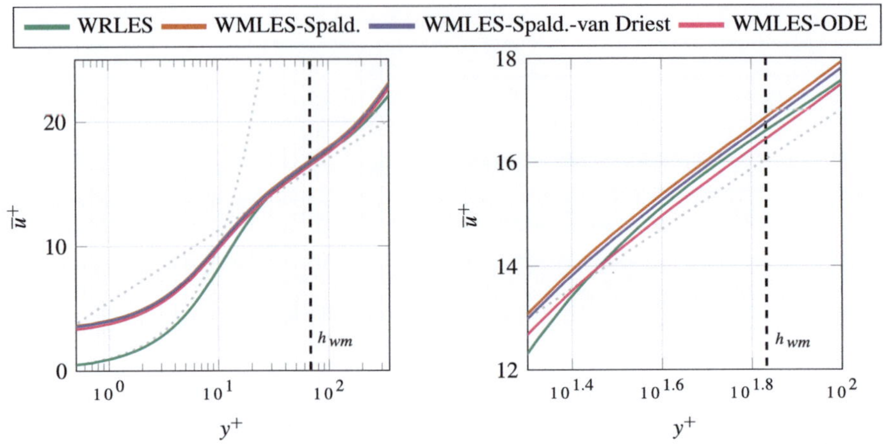

Fig. 3 Comparison of the mean velocity profile of a NACA 64A-110 airfoil on the pressure side at $x = 0.5c$ with a detailed view of the log layer on the right. The linear behavior in the viscous sub-layer and the log-law are depicted by dotted gray lines and the wall model exchange location h_{wm} is indicated by a dashed line

tigated case, the ODE model was more than twice as expensive as the algebraic Spalding model. In contrast, the van Driest transformation of the input velocity for the wall model comes at a negligible additional cost. Thus, the proposed van-Driest-transformed Spalding's model offers great accuracy with virtually no additionally computational cost. However, we note that this argument only holds for transsonic flows, since for the supersonic flow regime aspects such as heat conductivity are more relevant, which the ODE wall model can address by design. In summary, it has been demonstrated that the newly proposed van Driest transformation allows to extend the traditional Spalding model to transsonic flows and improves consistently on the incompressible formulation. Moreover, the new model has demonstrated to yield results which are comparable to the ODE model, while being considerably cheaper. With this building block in place, the following section introduces and validates the hybrid DG/FV subcell blending approach that is used to capture shocks within simulations.

4 Shock Capturing

The Navier–Stokes–Fourier equations admit strong gradients in the solution, which represent themselves as shocks in transsonic and supersonic flow regimes. Since high-order methods are inherently subject to oscillations in the vicinity of discontinuities or strong gradients, an adequate shock capturing approach is invaluable. This includes the detection of troubled-cells, i.e., cells in which the fluid solution is discontinuous, and the addition of local artificial viscosity to these cells. The shock detection can be addressed by analytical indicators, e.g., [15] or guided by machine learning [2]. The second step of the shock capturing is achieved via a locally h-refined low-order scheme [18] or a convex blending of a high-order with a lower-order solution [6], only to mention some possibilities. Overall, this procedure ensures a local application of the shock capturing approach such that only a minor computational overhead is introduced.

4.1 Hybrid DGSE/FV Subcell Scheme

A hard switching to a low-order scheme, as proposed by [18], induces numerical instabilities due to the different dispersion and dissipation properties of the high-order DG and the low-order FV scheme [5]. To reduce these numerical artifacts, a smooth transition between the low-order and the high-order discretization is required, which is achieved by the blending approach of [6]. This approach imposes locally additional numerical dissipation by performing an element-wise convex blending of a low-order FV subcell scheme, see e.g. [18], with the high-order DGSEM. For this, both operators are written in a similar manner, i.e., in the flux differencing form. Since the FV subcell scheme is inherently written in this form, only the DGSEM has to be modified. This is achieved using the split form DGSEM on Legendre–Gauss–Lobatto (LGL) points since the summation-by-parts property of the LGL nodes allows to write the volume operator in a flux-differencing form. Finally, the resulting hybrid scheme is written as

$$\mathbf{U}_t = \alpha^{\mathrm{FV}}\mathbf{U}_t^{\mathrm{FV}} + (1-\alpha^{\mathrm{FV}})\mathbf{U}_t^{\mathrm{DG}} \tag{6}$$

with the blending coefficient α^{FV} and $\mathbf{U}_t^{FV/DG}$ as the solution's time derivative as computed by the FV and DG operator, respectively. If a first-order FV operator is utilized, only the volume operator has to be blended, which enables a highly efficient scheme. However, this applies not to its second-order counterpart, where additionally the surface integral has to be blended. To avoid this constraint and any additional MPI communication, a first-order FV scheme is utilized in the outer subcells.

The blending coefficient α^{FV} is determined using a modified version of the modal indicator proposed by [15], which reads as

$$\alpha^{\mathrm{FV}} = \frac{1}{1+\exp(\frac{-s}{\mathbb{T}}(\mathbb{E}-\mathbb{T}))}, \quad \text{with } \mathbb{T} = 0.5\cdot 10^{-1.8(N+1)^{0.25}}, \tag{7}$$

where $\mathbb{E} \in [0, 1]$ denotes the highest mode energy indicator according to [15]. Although this indicator is equipped with empirical parameters, no additional tuning is required, i.e. it is applicable to arbitrary polynomial degrees and grid resolutions including unstructured meshes. To further increase the robustness of the hybrid scheme and to reduce the numerical artifacts induced through the coupling of both schemes, the blending coefficient is extended to the neighboring elements by a factor of 1/2. Altogether, this enables a robust scheme which is efficient and guarantees high-order accuracy in smooth regions.

4.2 Application

To demonstrate the applicability of the hybrid blending scheme, a linear compressor cascade equipped with NASA Rotor 37 profiles [13] at 50% span and supersonic flow conditions is employed. The rotor operates at design speed based on the reading 4192 in [13] with an inflow Mach number of $Ma = 1.4$ and a Reynolds number of $Re = 1\,472\,525$ based on the chord length $c = 0.0557$m. The computational domain is tessellated into 1 202 760 elements with $N = 4$, resulting in a total of 1.504×10^8 DOF. A wall-resolved LES is carried out on 128 nodes with 16 384 cores for around 4 000 000 CPUh. The PID is around 1.6μs. The hybrid DGSE/FV subcell scheme on LGL nodes is employed with a blending coefficient of $\alpha^{\mathrm{FV}} \in [0.01, 0.7]$, while for $\alpha^{\mathrm{FV}} < 0.01$, a pure DG method is utilized.

To assess the predictive performance of the convex blending scheme, the instantaneous flow field colored by the Mach number and the corresponding values of the blending coefficient are illustrated on the left and right of Fig. 4, respectively. A detached bow shock is visible at the leading edge of the rotor. This is followed by Prandtl-Meyer expansion waves and a passage shock wave system. Laminar boundary layers develop on the suction and pressure side which eventually transition to turbulent boundary layers at around 50% of the chord length. The results in Fig. 4 demonstrate that the blending scheme captures the shock waves well. However, also

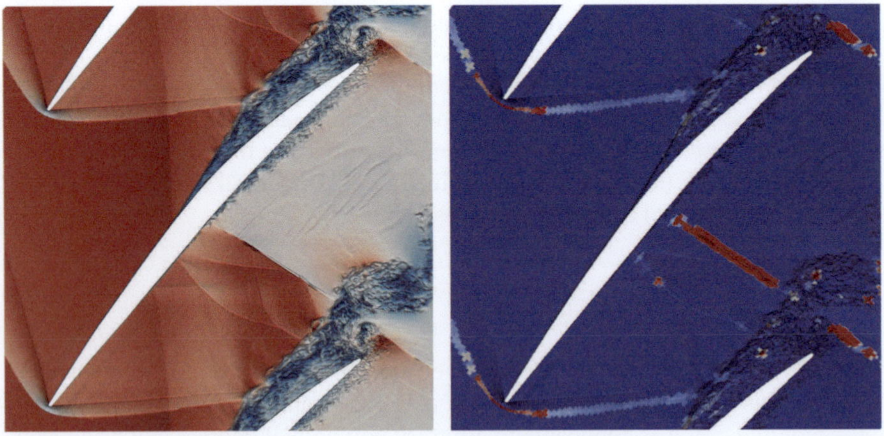

Fig. 4 Left: Instantaneous flow field around the NASA Rotor 37 blade colored by the Mach number. High Mach numbers, $Ma = 1.7$, are highlighted in red and low ones in blue. Right: Values of the blending coefficient. Low values, $\alpha^{\mathrm{FV}} = 0$, are colored in blue and high values, $\alpha^{\mathrm{FV}} = 0.7$, in red

turbulent structures are detected by the indicator. On the one hand, this is an inherent deficiency of shock indicators based on the modal representation or the jump terms of the nodal solution, since underresolved turbulent structures and discontinuities are similarly represented. On the other hand, this behavior is advantageous as the FV operator in the blending scheme introduces a form of dissipation and is able to reduce stability issues due to aliasing. Altogether, the hybrid discretization is able to capture all relevant physical phenomena, while maintaining the high-order accuracy in smooth regions and ensuring a robust scheme.

5 TensorFlow-Fortran-Binding

Deploying ML models within HPC simulations poses significant challenges, since both fields exhibit large differences in programming languages (interpreted vs. compiled) and computing hardware (GPU vs. CPU) even though the HPC community increasingly also adopts the potential of GPU hardware and interpreted languages like Julia and Python. This disparity is additionally reinforced by the fast pace of the ML community in both research and the development of ML frameworks like TensorFlow [1]. To this end, a variety of libraries have been proposed to bridge this chasm and to ease the use of ML models in established HPC codes at scale. The Fortran-Keras bridge [14] allows to evaluate Keras models by reimplementing the relevant functionalities directly in Fortran. Focussing more on reinforcement learning (RL), Relexi [9, 10] implements an RL training loop for HPC systems by coupling external flow solvers to TensorFlow using the SmartSim library. In this work, we present the open-source TensorFlow-Fortran-Binding (TFFB)[2] library as a novel approach that mainly focuses on the ease-of-use without sacrificing performance. The TFFB extends the interface approach proposed by Maulik et al. [12] by providing an additional Fortran interface that makes it easy to integrate into existing Fortran codes. The following sections, first, give implementation details and then demonstrate the good performance and scaling properties of the TFFB for a practical, ML-enhanced simulation.

5.1 Implementation

The TFFB is based on the approach by Maulik et al. [12], who used TensorFlow's C-API to call TensorFlow routines directly from OpenFOAM, which is written in C++. This approach comes with two major advantages over others proposed in the literature. First, all functionalities of TensorFlow are supported by design and, second, no additional overhead or dependencies are introduced, since the TensorFlow library is called directly using C/C++ code without requiring any calls to TensorFlow's Python

[2] https://github.com/flexi-framework/tffb.

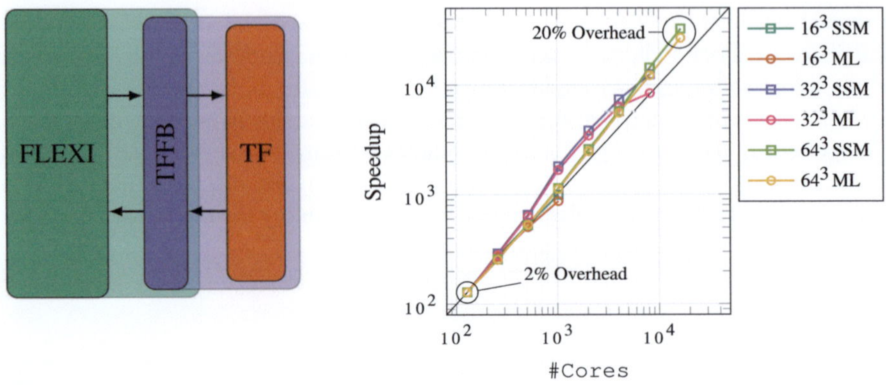

Fig. 5 Left: General sketch of dependencies when coupling FLEXI and TF using the TFFB. Right: Scaling results for the ML model using the TFFB (ML, circles) in comparison to the analytical turbulence model without the TFFB (SSM, squares). The three lines correspond to the meshes comprising either 16^3, 32^3 or 64^3 elements

API. The TFFB extends this approach to Fortran codes such as FLEXI as illustrated in Fig. 5. For this, the TFFB provides Fortran wrappers and an interface to the C++ code by Maulik et al. using the `ISO_C_Binding`, which defines a standardized interface between Fortran and C-type languages. Hence, the TFFB provides three different components. First, the code that calls the C-API using the C++ code from Maulik et al. and, second, an interface between the C++ code and Fortran using the `ISO_C_Binding`. The last component are Fortran wrappers that can be called directly from user code and take care of translating the input and output data such that it is suitable for the `ISO_C_Binding`. As a consequence, the user only has to call native Fortran routines from within the user code, while the translation through the interface and eventually to TensorFlow is handled by the TFFB. A major advantage of the TFFB is that does not have additional dependencies besides TensorFlow itself, which allows for easy integration into existing codebases either by linking against the standalone TFFB library or just incorporating and compiling the source files themselves into existing projects due to the permissive MIT license. The following section demonstrates that the proposed approach, while being lightweight and easy to integrate, is performant and scalable for HPC applications.

5.2 *Performance*

To demonstrate that the implemented interface is in fact performant, the TFFB is used to deploy a ML turbulence model developed in [8] in large-scale applications. The model was orginally trained using Relexi [9] and reinforcement learning (RL) to adapt the model coefficient of the analytical Smagorinsky turbulence model proposed in [17] dynamically in space and time for homogeneous isotropic turbulence.

In each time step, the model receives the local flow state within a single DG element as input and sets a scalar C_s parameter for the respective element, which is then used to compute the eddy viscosities at each solution point within the element. For this, the ML model comprises 5 layers of three-dimensional convolutions which results in a total of about 3300 model parameters. The reader is referred to [8] for details. It is important to stress that no performance optimization were performed in the sense that the model is employed as-is, i.e. without further optimizations like pruning or quantization, which could speed up the inference speed significantly. Moreover, a prebuilt version of the TensorFlow library is used, which does support only rudimentary optimizations in comparison to building it from source with all optimizations available for the given CPU architecture. The scaling tests performed are almost identical to the one discussed in Sect. 2. To specify different problem sizes, Cartesian meshes with either 16^3, 32^3 or 64^3 elements are employed. Using a polynomial degree of $N = 5$, this results in a total of 0.9, 7 and 56 million DOF, respectively. For each of the three problem sizes, the results for the strong scaling are obtained by increasing the number of compute nodes from 1 to 128, which corresponds to 16 384 compute cores at maximum. This procedure is performed once using the ML-informed model discussed above and once for the standard Smagorinsky model (SSM) with a constant C_s parameter for comparison.

For comparison, all lines are normalized based on the respective performance on a single node. The overhead introduced by the neural network is about 20% for the largest mesh running on the maximum of 128 compute nodes in comparison to only 2% when running the same case on a single node. These cases refer to the minimum and maximum load per core, respectively. The larger performance penalty in the low load case stems from the fact that FLEXI is typically bounded by the memory bandwidth. Since the evaluation of the ML model requires additional data to be loaded and held in the cache, the code does not run as cache-efficient as the case requiring less data for the simulation. In conclusion, the TFFB shows excellent scaling results for the given task on over 10 000 compute ranks, which renders it highly suitable for integrating ML models into HPC simulations. These results could be improved further by, first, optimizing the employed ML model using advanced techniques like pruning or quantization, see for instance [11], and, second, reducing the memory footprint of model evaluation to improve the cache efficiency in the case of very low loads.

6 Summary

This paper summarizes the recent developments in the open-source framework FLEXI in order to address the exascale lighthouse case defined within the context of the "Center of Excellence for Exascale CFD" (CEEC). This case is defined as simulating the shock buffet on a three-dimensional wing segment at transonic flight conditions at flight Reynolds numbers. To enable this challenging exascale simulation, FLEXI was first demonstrated to make efficient use of the computational resources provided by current HPC systems by scaling up to the maximum of 500 000 cores provided by the HAWK system at HLRS. To account for the prohibitive resolution demands near walls at such high Reynolds numbers, a novel wall model was proposed and validated using the simulation of a NACA 64A-110 airfoil. The model showed good accuracy in comparison to the wall-resolved simulation and the other wall models, while adding only negligible computational cost. In a next step, the newly implemented shock capturing approach of [6] was validated for a linear compressor cascade at supersonic flow conditions. This method was demonstrated to add only minor computational overhead, while yielding accurate and robust results. Lastly, the TensorFlow-Fortran Binding (TFFB) was presented as an convenient and HPC-suitable way to deploy trained ML models with large-scale simulations, and thus allows to take advantage of the recent success of data-driven models in scientific computing.

However, several future developments are still necessary to render the defined lighthouse case feasible. For instance, the novel TFFB library allows to incorporate ML models into HPC simulations from a technical perspective. However, the limited generalizability of ML models is a well-known and pressing issue in scientific ML. Future work should thus focus on improving the generalizability of ML models in order to apply them in realistic applications. Moreover, the upcoming generation of HPC systems increasingly focuses on GPU hardware, while FLEXI is optimized for CPUs. To this end, current developments focus on adapting the unstructured FLEXI code efficiently for GPUs to exploit the full potential of the upcoming successors of the HAWK system at HLRS.

Acknowledgements This work was funded by the European Union. This work has received funding from the European High Performance Computing Joint Undertaking (JU) and Sweden, Germany, Spain, Greece, and Denmark under grant agreement No 101093393 and from the Deutsche Forschungsgemeinschaft (DFG, German Research Foundation) with - EXC2075 – 390740016 under Germany's Excellence Strategy, the DFG project Rebound - 420603919 and in the framework of the research unit FOR 2895. We acknowledge the support by the Stuttgart Center for Simulation Science (SimTech) and the DFG International Research Training Group GRK 2160. We all truly appreciate the ongoing kind support by the HLRS in Stuttgart.

References

1. Abadi, M., Agarwal, A., Barham, P., Brevdo, E., Chen, Z., Citro, C., Corrado, G.S., Davis, A., Dean, J., Devin, M., Ghemawat, S., Goodfellow, I., Harp, A., Irving, G., Isard, M., Jia, Y., Jozefowicz, R., Kaiser, L., Kudlur, M., Levenberg, J., Mané, D., Monga, R., Moore, S., Murray, D., Olah, C., Schuster, M., Shlens, J., Steiner, B., Sutskever, I., Talwar, K., Tucker, P., Vanhoucke, V., Vasudevan, V., Viégas, F., Vinyals, O., Warden, P., Wattenberg, M., Wicke, M., Yu, Y., Zheng, X.: TensorFlow: large-scale machine learning on heterogeneous systems (2015). Software available from tensorflow.org
2. Beck, A.D., Zeifang, J., Schwarz, A., Flad, D.G.: A neural network based shock detection and localization approach for discontinuous Galerkin methods. J. Comput. Phys. **423**, 109824 (2020)

3. Blind, M., Schauerte, C., Schreyer, A.-M., Beck, A.: Numerical and experimental investigation of a NACA 64A-110 airfoil in transsonic flow regime. Submitt. AIAA J. (2023)
4. Blind, M.P., Kahraman, A.B., Larsson, J., Beck, A.: Residual estimation for grid modification in wall-modeled large eddy simulation using unstructured high-order methods. Comput. Fluids **254** (2023)
5. Gassner, G., Kopriva, D.A.: Comparison of the Gauss and Gauss-Lobatto discontinuous Galerkin spectral element method for wave propagation problems. SIAM J. Sci. Comput. **33**(5), 2560–2579 (2011)
6. Hennemann, S., Rueda-Ramírez, A.M., Hindenlang, F.J., Gassner, G.J.: A provably entropy stable subcell shock capturing approach for high order split form DG for the compressible Euler equations. J. Comput. Phys. **426**, 109935 (2021)
7. Krais, N., Beck, A., Bolemann, T., Frank, H., Flad, D., Gassner, G., Hindenlang, F., Hoffmann, M., Kuhn, T., Sonntag, M., Munz, C.-D.: FLEXI: A high order discontinuous Galerkin framework for hyperbolic-parabolic conservation laws. Comput. Math. Appl. **81**, 186–219 (2021)
8. Kurz, M., Offenhäuser, P., Beck, A.: Deep reinforcement learning for turbulence modeling in large eddy simulations. Int. J. Heat Fluid Flow **99**, 109094 (2023)
9. Kurz, M., Offenhäuser, P., Viola, D., Resch, M., Beck, A.: Relexi–a scalable open source reinforcement learning framework for high-performance computing. Softw. Impacts **14**, 100422 (2022)
10. Kurz, M., Offenhäuser, P., Viola, D., Shcherbakov, O., Resch, M., Beck, A.: Deep reinforcement learning for computational fluid dynamics on HPC systems. J. Comput. Sci. **65**, 101884 (2022)
11. Liang, T., Glossner, J., Wang, L., Shi, S., Zhang, X.: Pruning and quantization for deep neural network acceleration: a survey. Neurocomputing **461**, 370–403 (2021)
12. Maulik, R., Sharma, H., Patel, S., Lusch, B., Jennings, E.: Deploying deep learning in OpenFOAM with TensorFlow. In: AIAA Scitech 2021 Forum, p. 1485 (2021)
13. Moore, R.D., Reid, L.: Performance of single-stage axial-flow transonic compressor with rotor and stator aspect ratios of 1.63 and 1.78, respectively, and with design pressure ratio of 1.82. Technical report, NASA-TP-1338 (1978)
14. Ott, J., Pritchard, M., Best, N., Linstead, E., Curcic, M., Baldi, P.: A Fortran-Keras deep learning bridge for scientific computing. Sci. Program. **1–13**, 2020 (2020)
15. Persson, P.-O., Peraire, J.: Sub-cell shock capturing for discontinuous Galerkin methods. In: Proceedings of the 44th AIAA Aerospace Sciences Meeting and Exhibit (2006)
16. Slotnick, J., Khodadoust, A., Alonso, J., Darmofal, D., Gropp, W., Lurie, E., Mavriplis, D.: CFD vision 2030 study: a path to revolutionary computational aerosciences. NASA Langley Research Center; Hampton, VA, United States, Technical report (2014)
17. Smagorinsky, J.: General circulation experiments with the primitive equations: I. the basic experiment. Mon. Weather. Rev. **91**(3), 99–164 (1963)
18. Sonntag, M., Munz, C.-D.: Efficient parallelization of a shock capturing for discontinuous Galerkin methods using finite volume sub-cells. J. Sci. Comput. **70**(3), 1262–1289 (2017)
19. Spalding, D.B.: A single formula for the law of the wall. J. Appl. Mech. **28**(3), 455–458 (1961)
20. Van Driest, E.R.: Turbulent boundary layer in compressible fluids. J. Aero. Sci. **18**, 145–160 (1951)
21. White, F.M.: Viscous fluid flow. McGraw-Hill series in mechanical engineering. McGraw-Hill, Boston, Mass., 3rd ed. (2006)

DNS of the Early Phase of Oblique Droplet Impact on Thin Films with FS3D

Jonathan Lukas Stober, Johanna Potyka, Matthias Ibach, Bernhard Weigand, and Kathrin Schulte

Abstract Spray impacts occur in several environmental and technical applications. The impact of droplets at different angles onto walls covered with a thin film of the same liquid can be regarded as an elementary process here. Direct Numerical Simulations (DNS) provide an important contribution to the understanding and modelling of the impact outcome, which might be associated with the formation of a crown and ejection of secondary droplets. Thus, we gain detailed information about, e.g. the flow field and shape of the interface, which are not accessible in experiments. This chapter presents a DNS study of the early crown formation mechanisms present at an oblique droplet impact on a thin film, as well as a grid study showing the resolution required to resolve the impact's details. Highly resolved simulations in large domains require continuous development of the numerical solver's efficiency. The performance of different cycles of the multigrid (MG) solver for the solution of the pressure Poisson equation was compared, and a F-cycle was added. Furthermore, we implemented a hybrid MPI + OpenMP parallelisation, which both increases the scaling limit further. Additionally, studies on the strong and weak scaling are conducted. The choice of the F- and V- cycle in the MG-solver and the additional hybrid parallelisation increased the achieved computed cycles per hour (CPH) by a factor of 12.4 compared to the formerly employed setup. The nodes efficiently usable were increased by a factor of 16. Both, the close to linear scaling regime for the strong scaling and the almost constant performance regime for the weak scaling were increased by this factor.

J. L. Stober (✉) · J. Potyka · M. Ibach · B. Weigand · K. Schulte
Institute of Aerospace Thermodynamics (ITLR), University of Stuttgart, Stuttgart, Germany
e-mail: jonathan.stober@itlr.uni-stuttgart.de

K. Schulte
e-mail: kathrin.schulte@itlr.uni-stuttgart.de

T. Ludwig et al. (eds.), *High Performance Computing in Science and Engineering '23*,
https://doi.org/10.1007/978-3-031-91312-9_16

1 Introduction

Sprays occur in various environmental and technical applications, such as soil erosion, medical sprays and cleaning processes, with many droplets impacting at different angles and velocities onto solid walls covered with a film of the same liquid. The thickness of such films ranges from thin films to deep pools depending on the application. The impact of a single droplet might result in the formation of a crown and the ejection of secondary droplets. The comprehensive physical description of this elementary process is of great importance to optimise such applications. Many studies focus on the development of the crown shape as well as the differentiation between the impact regimes *deposition* and *splashing*. The latter results in the detachment of secondary droplets from the unstable rim of a crown. Most investigations focus on a normal droplet impact, i.e. with 90°, onto the wall film. Cossali et al. [4] conducted a large experimental study and derived an empirical correlation, which describes the deposition/splashing threshold as a function of the fluid properties, impact conditions and film thickness. The derivation of empirical thresholds for a large variation of impact conditions was also the aim of other studies [24], as well as the characterisation of the morphology after the impact [6, 21]. Yarin and Weiss [26] on the other hand, described the formation of the crown analytically with a kinematic discontinuity and derived a model predicting its expansion. Roisman and Tropea [22] extended this concept to oblique impacts by allowing arbitrary velocity fields in the wall film after impact. Their model predicts the velocity of the upward rising sheet, crown shape and height. Numerical studies on the crown height were conducted with different simulation approaches such as Lattice-Boltzmann method [3], the Coupled Level set and Volume of Fluid method (CLSVOF) [11], or the Volume of Fluid (VOF) method, which is applied in our code Free Surface 3D (FS3D) [9, 20].

In real applications, droplets impact not normally to the wall but at an angle α, see the sketch in Fig. 1A, which results in different crown shapes and splashing conditions [12]. Brambilla et al. [1] detected different asymmetric crown shapes in their numerical study with impact angles ranging from $10° < \alpha < 90°$ and a thin wall film, $h_f/D = 0.116$ with h_f as the film height and D the droplet diameter. Chen et al. [2] investigated the onset of splashing in their numerical study on thin films for different impact angles, $45° < \alpha < 75°$. Their results confirm an influence of the angle on the crown shape. However, no detailed investigation was carried out to compare and classify the mechanisms forming the crown in the early times after the impact, $\tau = tU/D < 0.5$, with t being the time and U the droplet's velocity. This early phase is of great importance, as it determines the further development of the impact morphology, but, on the other hand, it is difficult to investigate experimentally, because of its small temporal and spatial scales. Furthermore, information on the internal flow structure are required for a better understanding and prediction of the crown formation. Therefore, highly resolved Direct Numerical Simulations (DNS) are necessary, see an example in Fig. 1B, that allow us to identify two different crown formation mechanisms on both sides of the crown, explained in detail in Sect. 3.2. The simulation of the early crown formation is challenging due to small time scales,

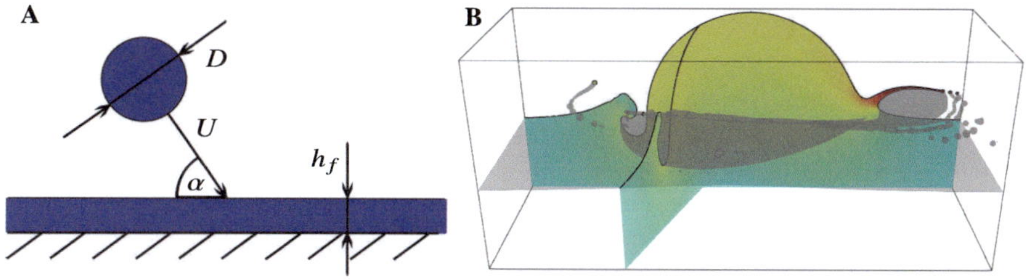

Fig. 1 **A** Sketch of an oblique droplet impact on a wall film, **B** 3D visualisation of the early phase ($\tau = tU/D = 0.5$) of impact

high velocities and the small size of the developing liquid structures. This requires high spatial and temporal resolutions, which can only be achieved with an efficient parallelisation of the code and the use of high performance clusters.

First, this work provides an overview of our in-house multi-phase simulation code Free Surface 3D (FS3D), see Sect. 2. After that, the computational setup for the simulation of an oblique droplet impact onto a wall film is presented, see Sect. 3.1. The simulation results are shown prior to the grid study to explain the different mechanisms of crown formation. The resolution required for reproducing relevant details of the impact is determined by comparing the shape of the interface. Recent performance enhancements are presented in the second part of this report, see Sect. 4, which enabled the study of such large and complex interactions by increasing the scaling limit. This includes a description of the different multi-grid cycles and the hybrid MPI + OpenMP parallelisation as well as the results of a strong and weak scaling.

2 Mathematical Description and Numerical Approach

The ITLR's in-house program package FS3D was developed since the 1990s for the direct numerical simulation of highly dynamic multiphase problems like droplet deformation [17], grouping [8], single and multicomponent droplet collisions [13], atomisation of liquid jets [16] as well as droplet impacts onto structured dry and wetted surfaces [9, 15]. To simulate such processes, a high spatial and temporal resolution as well as large computational domains are necessary. Therefore, the code is parallelised with MPI and OpenMP, and it is continuously optimised to increase its parallel efficiency and scaling on high performance clusters.

FS3D solves the incompressible Navier-Stokes equations for mass and momentum conservation

$$\partial_t \rho + \nabla \cdot (\rho \mathbf{u}) = 0, \tag{1}$$

$$\partial_t (\rho \mathbf{u}) + \nabla \cdot (\rho \mathbf{u} \otimes \mathbf{u}) = \nabla \cdot (\mathsf{S} - \mathsf{I} p) + \rho \mathbf{g} + \mathbf{f}_\gamma \tag{2}$$

on finite volumes. In Eqs. (1) and (2), $\mathbf{u}$ denotes the velocity vector, ρ the mass density, p the static pressure, $\mathbf{g}$ the gravitational acceleration, S the shear stress tensor and I the identity matrix. The surface tension forces are introduced as a body force, $\mathbf{f}_\gamma$, in cells close to the phase interface. Different surface tension models are implemented in FS3D, such as the continuous surface stress (CSS) model by Lafaurie et al. [10] applied in this study. An one-field formulation allows to solve one set of conservation equations for all phases with varying physical fluid properties across the interface. The VOF method by Hirt and Nichols [7] is applied to distinguish between the different phases by introducing the scalar indicator variable, f,

$$f(\mathbf{x}, t) = \begin{cases} 0 & \text{outside the liquid phase,} \\ (0, 1) & \text{at the interface,} \\ 1 & \text{inside the liquid phase} \end{cases} \tag{3}$$

that represents the volume fraction of the liquid in each cell. A transport equation for f needs to be solved in order to obtain the phases' distribution over time, according to

$$\partial_t f + \nabla \cdot (f\mathbf{u}) = 0. \tag{4}$$

To increase the accuracy of the corresponding f-fluxes, FS3D uses the Piecewise Linear Interface Calculation (PLIC) method by Rider and Kothe [18], which calculates the orientation and position of the interface in each cell. The local physical properties $\psi(\mathbf{x}, t)$ in each cell are defined by the f-variable and the fluid properties of the liquid and gaseous phase, ψ_l and ψ_g according to

$$\psi(\mathbf{x}, t) = \psi_l f(\mathbf{x}, t) + \psi_g (1 - f(\mathbf{x}, t)). \tag{5}$$

Further details on the numerical implementations and applications of FS3D are given in Eisenschmidt et al. [5]. A group of eleven researchers is currently developing and using this code. Within the report period from April 2022 until March 2023, approximately 45 million CPU core-hours were spent in total.

3 Simulation Results

First, this section presents the computational setup. After that, prior to the grid study, the simulation results are presented and the process of crown formation is explained. The subsequent comparison of the shape of the interface for different grid resolutions determines the required resolution necessary to reproduce relevant details of the impact.

3.1 Computational Setup

The previously described program package FS3D is employed to simulate the early phase of crown formation during an oblique droplet impact onto a thin film. Figure 2 shows a sketch of the impact of a droplet with a diameter $D = 2.4\,\text{mm}$ onto a wall film of the same liquid of height $h_f = 0.96\,\text{mm}$. The impact velocity is $U = 3.3\,\text{m/s}$ and the impact angle is $\alpha = 60°$. A cuboidal computational domain is used and it exploits the symmetry of the impact scenario using a symmetry boundary condition at $Z/D = 0$. For the bottom wall, $Y/D = 0$, a no-slip boundary was applied and all other boundaries use zero gradient conditions. The fluids of our choice are Hexadecane for the droplet and the film and air as ambient gas. The relevant material properties are given in Table 1. The impact conditions are summarised in the following dimensionless parameters, Weber number $We = \rho_l D U^2/\sigma = 750$, Ohnesorge number $Oh = \mu_l/\sqrt{\rho_l D \sigma} = 0.015$ and the non-dimensional film thickness $\delta = h_f/D = 0.4$, which represents the thin film regime. The centre of the droplet is initialised 1 D above the wall film. The initial contact with the wall film happens at $\tau = 0$.

The Cartesian grid consists of equidistant cubic cells within a refined domain marked by the red box in Fig. 2. Beyond the refinement, the cell size grows gradually to reduce computational cost. The total number of cells in x-direction is twice as large as in y- and z-direction. Fig. 2 shows the extension of the domain and of the grid refinement exemplarily for Grids I-IV. All parameters of the grids investigated in this study are summarised in Table 2. The number of cells increases for Grids

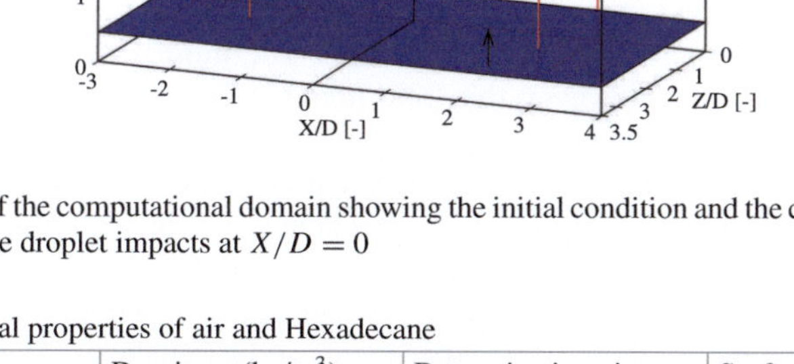

Fig. 2 Sketch of the computational domain showing the initial condition and the coordinate system for grid I-IV. The droplet impacts at $X/D = 0$

Table 1 Physical properties of air and Hexadecane

	Density ρ (kg/m^3)	Dynamic viscosity μ (Pa s)	Surface tension σ (N/m)
Air	1.225	$1.82 \cdot 10^{-5}$	$27.5 \cdot 10^{-3}$
Hexadecane	773.3	$3.40 \cdot 10^{-3}$	

Table 2 Parameters of the different computational grids investigated in the grid study

	Domain size $[1/D]$			Refinement domain $[1/D]$			Number of cells	Resolution $[\text{cells}/D]$
	X	Y	Z	X	Y	Z		
I	[–3, 4]	[0, 3.5]	[0, 3.5]	[–1.5, 2.5]	[0, 2.0]	[0, 2.0]	$256^3 \cdot 2$	87.5
II	[–3, 4]	[0, 3.5]	[0, 3.5]	[–1.5, 2.5]	[0, 2.0]	[0, 2.0]	$512^3 \cdot 2$	175
III	[–3, 4]	[0, 3.5]	[0, 3.5]	[–1.5, 2.5]	[0, 2.0]	[0, 2.0]	$768^3 \cdot 2$	262.5
IV	[–3, 4]	[0, 3.5]	[0, 3.5]	[–1.5, 2.5]	[0, 2.0]	[0, 2.0]	$1024^3 \cdot 2$	350
V	[–2, 2.6]	[0, 2.3]	[0, 2.3]	[–1, 1.7]	[0, 1.3]	[0, 1.3]	$1024^3 \cdot 2$	538.5
VI	[–1.5, 2]	[0, 2]	[0, 2]	[–0.75, 1.25]	[0, 1.1]	[0, 1.1]	$1024^3 \cdot 2$	700

I-IV at a constant domain size. A further increase of the resolution was realised by a reduction of the domain size for Grid V and VI. On the one hand, computational costs are reduced, on the other hand, the physical evaluation is limited to early time steps where the crown is still in the refined region. The resolutions range accordingly from 87.5 cells$/D$ up to 700 cells$/D$. A domain decomposition with 64^3 cells per process was used for all simulations. Grid IV and VI were computed with 8192 MPI processes on 128 nodes. Grid IV ran for approximately 33 h for a simulation time until $\tau = 2.0$, Grid VI 50 h until $\tau = 1.0$.

3.2 Results and Discussion

Figure 3 shows the evolution of the fluid interface at the symmetry plane in the early time ($\tau < 1$) after the droplet impact simulated with Grid IV. The differences in crown shape on the front, which is thin and bends outwards, compared to the thicker, inward bent crown at the back, are clearly visible. At the front, a small lamella forms right after impact and continues to grow outwards and forms the crown. On the back side, a different mechanism for crown formation is observed. Fig. 4 shows the interface and the velocity field in the fluid. The fluid is pushed outwards and forms an initial lamella. However, this fluid flow feeding the lamella is pinched-off at $\tau = 0.21$, because it is not able to follow the strong deflection into the lamella due to the horizontal impact velocity component. The fluid is then deflected sidewards and upwards with a larger radius, and ambient gas penetrates the wall film. At $\tau = 0.35$, the initial lamella bends inwards leading to the different crown shape. A validation with in-house experiments confirms these two mechanisms on both sides of the crown by comparing the morphology from experiments and simulations. The experimental recordings for $\tau = [0.19, 0.33, 0.55]$ confirmed the growth of an initial lamella and its receding as well as resulting in a similar crown shape on the back side.

For this highly dynamic and numerically challenging impact phase, a grid study is conducted to investigate the influence of the grid and to identify a sufficient resolution.

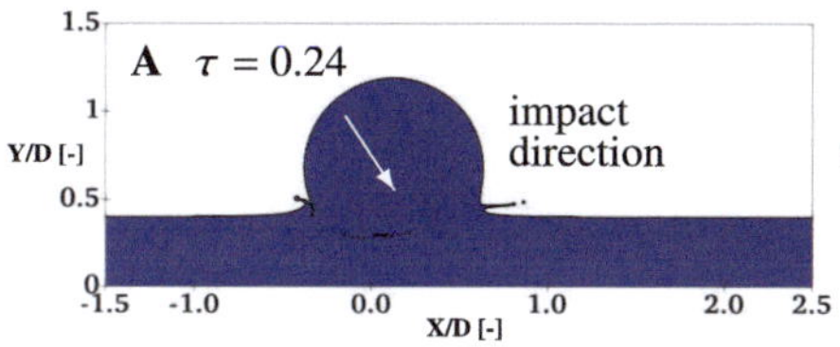

Fig. 3 Development of the interface at the symmetry plane for Grid D at two different dimensionless times $\tau = tU/D$

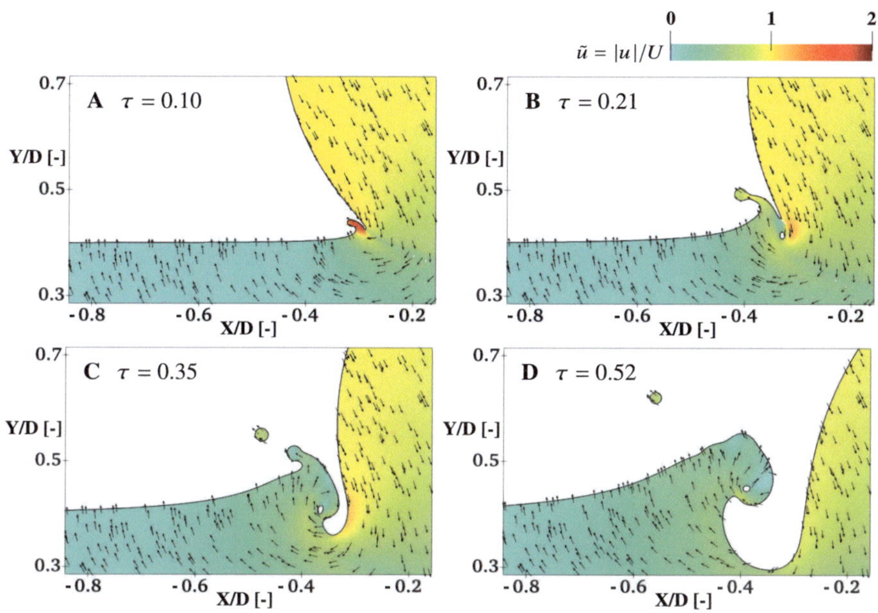

Fig. 4 Flow field at the symmetry plane with the dimensionless velocity magnitude $\tilde{u} = |u|/U$ at four moments in time

The development of the lamella and the formation of the crown was simulated with six different grid resolutions (Table 2). The results are compared by evaluating the fluid interfaces which are plotted in Fig. 5. Starting at an early time, $\tau = 0.13$, the lamella is very small and not captured by the lowest resolution (green), see Fig. 5A. All other simulations resolve the lamella, but its shape and length differ. At $\tau = 0.23$, the growing lamella collides with the bulk of the impacting droplet and leads to an entrapped tunnel of air just below the lamella, which is captured by Grid III (orange) and higher resolutions. This tunnel opens up again shortly after, at $\tau = 0.39$, for the simulations with Grid V (blue) and VI (cyan). For Grid III (orange), the tunnel opens only at $\tau = 0.53$ which leads to a complete different lamella shape at $\tau = 0.59$ as displayed in Fig. 5D. For Grid IV (magenta), the tunnel does not rupture in the investigated time, which leads to entrapped air. Comparing the resulting crown shape

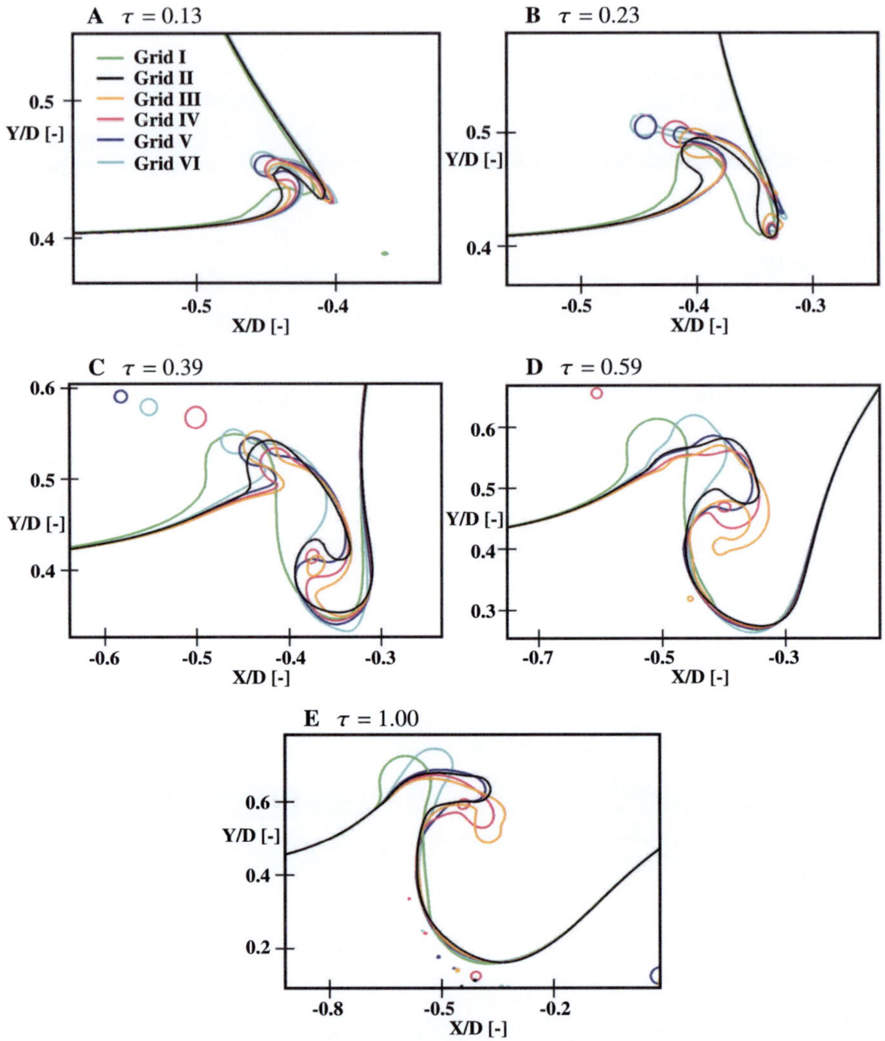

Fig. 5 Interface on the back side of the crown for six different numerical resolutions at five non-dimensional times

at $\tau = 0.59$ and $\tau = 1.00$, several finding become apparent. First, the simulation results are not yet grid independent. Second, the interface of Grid II (black) matches surprisingly well with the second finest resolution (Grid V, blue), while Grid III (orange) deviates the most from the other. This can be explained by the fact that the simulation with Grid III is the lowest resolution which captures the collision of the lamella with the droplet. However, the exact subsequent process of the tunnel of air and its rupture is not captured correctly. Therefore, the results for the later stages are inaccurate. The finer resolutions IV (magenta) and V (blue) converge towards the

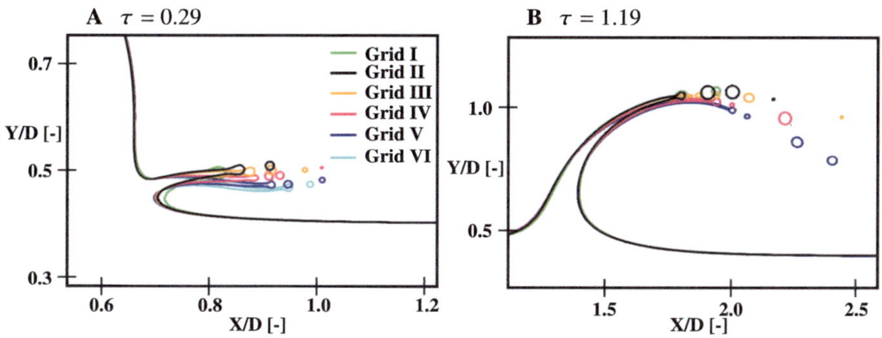

Fig. 6 Interface on the front side of the crown for six different numerical resolutions at $\tau = 0.29$ (left) and $\tau = 1.19$ (right)

finest VI (cyan). In the simulation with Grid II, the collision of the lamella with the droplet is not captured at all. Nevertheless, it results in a very similar shape of the crown for later times, $\tau \geq 0.59$.

Figure 6 shows the interface at the front side of the droplet impact for the different computational grids. At the early crown formation phase, see A, the thin lamella is shorter if the resolution is reduced with an unphysical rupture at the tip. The height of the lamella decreases for higher grid resolutions. The rupture is present also with the highest resolution in our simulations, but not observable in experiments. A complete reproduction of the dynamics of the early lamella, including the very thin and fast moving tip, is beyond the possible resolution.

In Fig. 6B, the interface is plotted for the developed crown at τ =1.19. The finest grid is not shown for this later time, because the crown has moved outside of the refined grid region. The shape of the crown is comparable for all grids. But, as in the early stage, a lower resolution leads to an earlier rupture and a shorter crown.

This grid study reveals that a detailed investigation of the early phase of crown formation after an oblique droplet impact requires high resolutions, and we could not achieve grid convergence for the resolution of all details. A grid resolution of $\geq$ 350 cells/D is necessary for the reproduction of the lamella at the back side and its collision with the droplet, which justifies the use of Grid IV for the previous investigation. The detachment at the tip of the lamella and the rim dynamic remain grid dependent also at later times ($\tau > 1.0$), which influences the size and number of secondary droplets, but the crown shape and the overall behaviour is comparable for different resolutions ($\geq$ 350 cells/D). Thus, these investigations lay the foundation for further studies about the influence of impact conditions, such as impact angle, impact velocity as well as film thickness and fluid properties, on the crown formation mechanism. The simulations allow a detailed view into the process of the droplet impact, which is not accessible from an outside view in the experiments. Furthermore, this study revealed the importance of a good scaling efficiency on supercomputers such as the HPE Apollo (Hawk), because a large number of cells is necessary to realise the required high resolutions of important details.

4 Computational Performance

Constant improvements of FS3D's performance are required to keep up with the supercomputing hardware development and increasingly larger simulation domains and resolutions while maintaining an efficient use of the resources. Multiple performance optimisations within FS3D dealt with the most time consuming routines [14] with focus on the multigrid (MG) solver [23]. The still most time-consuming MG-solver is worth further optimisation.

4.1 Scaling Bottleneck

In FS3D, the numerical solution of the incompressible Navier-Stokes equations necessitates the solution of a Poisson equation for the pressure. An MG-method, like the one implemented by Rieber [19] in FS3D, provides means to iteratively solve such huge systems of linear equations efficiently by solving the problem on consecutively coarser grids leading to faster convergence. This is due to the fact, that long wavelength errors, which converge slower on a fine grid, are solved on the coarser grids. The next coarser grid is obtained by dividing the number of cells in all three grid dimensions by two, i.e. eight cells are merged. This leads, in the extreme case, to a single remaining cell in which the problem can be solved directly. Within the coarsening, the solution is smoothed and short wave (high frequency) errors are damped. In our parallelised code, the restriction is performed on each MPI-process, until no further coarsening is possible anymore. After that, all values are collected on one single MPI-process and the restriction is executed again. The remaining matrix ($1 \times 1 \times 1$ for a cuboidal domain decomposition) is solved explicitly, and subsequently the refinement process is started. Prolongation steps with analogous smoothing and reduction of long wavelength errors follow until the grid reaches its original resolution. The coarsening and refinement steps in FS3D were, until now, performed by default with a robust W-cycle as illustrated in Fig. 7 (left). This description already reveals two major problems, leading to the bottleneck in the scaling of FS3D: In a strong scaling case, the domain on each process gets smaller. This leads to a smaller number of restriction steps that can be performed in parallel and the values have to be collected earlier on a single MPI-process. The number of MG-levels, which has to be solved on a single process thus scales with the number of MPI-processes. Additionally, also the communication effort scales with the number of MPI-processes. Furthermore, the number of cells that has to be solved on one MPI process is increased. Thus, an improved MG-solver should avoid steps on levels solved by one single MPI-process and increase the domain size per MPI-process.

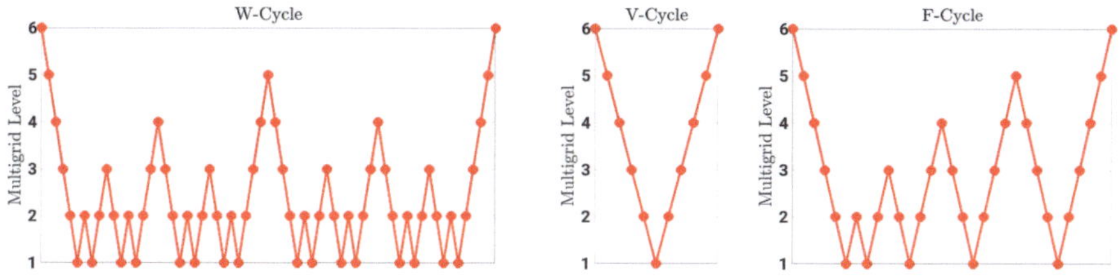

Fig. 7 Exemplary representation of the W-, V-, and F-cycle of the MG-solver for six MG-levels. The highest level corresponds to the finest, the lowest to the coarsest grid. The exemplary six levels would be present for 32^3 cells. The later presented strong scaling example with a domain size of 1024^3 cells globally has eleven levels in total, the number of levels where restriction and prolongation is executed in parallel decreases with an increasing number of MPI-processes

4.2 *Multigrid Cycle Methods*

The coarsening and refinement steps can be performed in different ways, as illustrated in Fig. 7, presenting the three main types of MG-cycles according to Wesseling [25] with varying trade-offs between the speed of solving one MG-iteration (entire cycle) and the rate of convergence of the solution, namely the V-, W-, and F-cycle.

The simplest one is the V-cycle in which the correction scheme is recursively applied only once at each successive MG-level, while for the W-cycle it is applied twice. The V-cycle obviously meets the requirement formulated above that less steps are to be performed on the single MPI-process levels. The number of steps in the W-cycle grows by a factor of two, if one MG-level is added, while for the V-cycle only two more steps, one additional restriction and one prolongation, are necessary. The drawback of the V-cycle, which is also implemented in FS3D, but was scarcely used until now, is that it converges slower, thus more iterations are necessary. Usually, this still leads to a large reduction in steps on the single MPI-process levels, but in cases with high pressure gradients, the V-cycle's convergence might fail due to a lower numerical robustness. The W-cycle is very robust, but as the discussion above shows, not well suited to reach a close to linear scaling to a large number of cores. Therefore, we now implemented a compromise between V- and W-cycle in FS3D: The F-cycle, which is also a compromise in terms of peak performance as shown later. In the cases used for the performance measurements, the averaged number of MG-iterations at peak performance for pure MPI (64 nodes) were ≈ 8.4 for the W-cycle, ≈ 9.1 for the F-cycle and ≈ 15.0 for the V-cycle, but, 15 iterations of V-cycle result in a shorter simulation time than the 8.4 iterations in the W-cycle. Additional to the reduction of steps, which cannot be solved in parallel, the all-to-one and one-to-all communication can be minimised.

The introduction of the F-cycle and newly found potential of the V-cycle reduce the bottleneck of serialised computations in the MG-solver for highly parallelised simulations.

4.3 Hybrid MPI + OpenMP Parallelisation

As discussed in Sect. 4.1, the number of remaining cells, which have to be solved by only one MPI-process, and the costly all-to-one communications govern the scaling limit. Reducing the number of MPI-processes increases the number of cells and thus the domain size per MPI-process. This is a sensible measure to reduce the number of MG-levels which finally have to be solved by a single processor. Implementing a hybrid parallelisation with MPI and additionally using OpenMP solves this problem by reducing the number of MPI-processes, while still employing all available cores. FS3D was parallelised with OpenMP before to run on vector computers in the past. Thus, we enhanced and revised the OpenMP parallelisation by implementing OpenMP on a loop level especially for the recently implemented efficient methods in the most time consuming routines of FS3D [14].

In addition to moving the levels in the MG-solver solved by one single MPI-process to coarser grids, the load balancing can be improved by a hybrid parallelisation, too. Especially on the coarsest levels, which were originally solved by only one process, are now solved by four OpenMP threads. Moreover, multiple costly MPI-communications are not required anymore and the shared memory regions reduce the amount of MPI-halos blocking memory in the caches. The halos are used for the boundary conditions of the local domain. As the memory is shared among OpenMP threads, no boundaries have to be exchanged within a shared memory sub-domain. However, one has to consider the limitations the hardware provides, as the performance can worsen significantly instead of the desired improvement: Four CPU-cores (CCX) share a common Level 3 (L3) cache on HPE Apollo (Hawk). Thus we tested two, four and eight OpenMP threads per MPI-process to investigate the impact of the shared cache. This revealed that four OpenMP threads per MPI-process improved the performance compared to using the same number of cores with only MPI-processes, while two or eight OpenMP threads worsened the performance compared to a pure MPI application solving the identical problem. Eight OpenMP threads do not share the Level 3 cache, leading to costly memory transfers as no "first-touch"-principle was implemented. Therefore, the following performance analysis employs four OpenMP threads per MPI-process.

4.4 Strong Scaling

The test case for the performance analysis is an oscillating water droplet within a cuboidal domain. A spatial resolution of 1024^3 grid cells was used to evaluate the strong scaling of FS3D. The strong scaling is shown in Fig. 8 (top). The figure presents the cycles per hour (CPH) achieved with this test case for the different numbers of nodes used for the parallelisation of the same case. The CPH is the number of time steps which can be computed in an hour, extrapolated from a 30 min run-time measurement. As each node on the employed infrastructure consists of

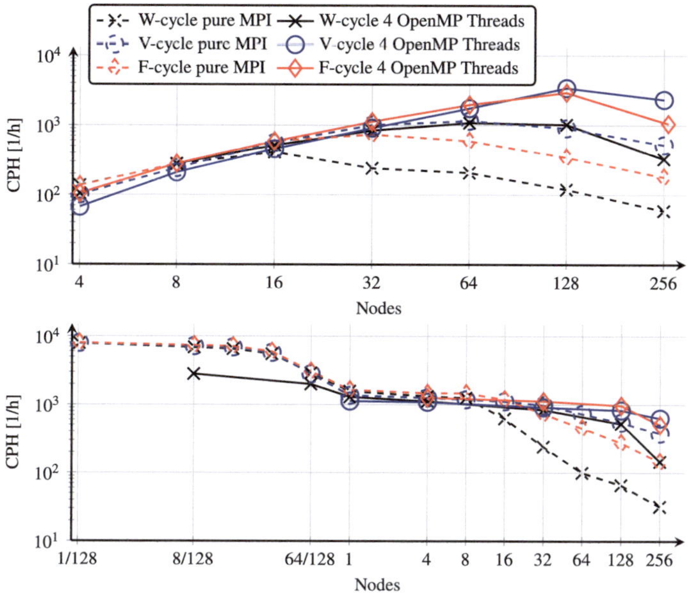

Fig. 8 Strong scaling (top): 1024^3 cells distributed on an increasing number of nodes. Weak scaling (bottom): 64^3 cells per core with an increasing number of nodes. Both with and without utilising OpenMP in addition to MPI showing estimated cycles per hour (CPH)

128 CPU-cores, the pure MPI cases were computed with one MPI-process per core, while the cases with a hybrid parallelisation with four OpenMP threads employed 32 MPI-processes per node with a stride of four OpenMP threads pinned to a CCX, i.e. four CPUs with a shared L3-cache. Therefore, we show a comparison of the usage of the same computational resources while solving the same problem size with an identical resolution with different parallelisation techniques and different MG-cycles. The number of nodes employed for the parallelisation was increased until the last combination tested showed a decrease in the performance, thus the largest number of nodes tested was 256 nodes, corresponding to 32^3 cells per core. This is thus a distribution of 32^3 cells per MPI-process for the pure MPI cases and $64 \times 64 \times 32$ cells per MPI-process with four OpenMP threads. Each case was repeated multiple times, to ensure a constant performance measurement.

Figure 8 (top) shows that there is not a single optimal choice for the parallelisation technique or MG-cycle method. Showing the strong scaling for less than four nodes was not possible with the used test case as the case did not fit into the memory of a single node. The W-cycle (pure MPI, dashed black line) serves as the reference case since it was formerly used in most of the conducted simulations shown in previous reports. In the following plots, different MG-cycles are denoted in different colours and symbols, pure MPI measurements are shown in dashed lines, hybrid calculations (MPI + OpenMP) in solid lines. For the cases where only MPI-parallelisation was used (dashed lines in Fig. 8 (top)), the W- and F-cycle outperform the V-cycle in terms of cycles per hour for parallelisation with up to eight nodes. The W-, F- and the V-cycle show an almost linear scaling up to 8, 16 and 32 nodes, respectively, before a non-ideal increase in CPH occurs followed by a decrease. Just by employing different MG-cycle schemes, a performance boost up to 3 times the CPH at 32 nodes for the F-cycle was achieved compared to the reference W-cycle case. The V-cycle even resulted in 5.5 times the cycles per hour at 64 nodes at peak performance.

The additional factor four in the scaling limit is reached by employing the hybrid MPI+OpenMP parallelisation, see the solid lines in Fig. 8 (top). The F-cycle in combination with the hybrid parallelisation performs best up to 64 nodes. Moreover, the new scaling limit is now 3420 CPH with the V-cycle, 128 nodes and a hybrid parallelisation, while the old close to linear scaling limit was at 276 CPH with the W-cycle, eight nodes and a pure MPI parallelisation.

4.5 Weak Scaling

The weak scaling employs the same test case, but with different domain decompositions. We kept the amount of grid cells per core constant at 64^3 and increased the number of used nodes. The pure MPI cases were run with 64^3 cells per MPI-process. With four OpenMP threads per MPI-process for the hybrid cases, the domain size per MPI-process was $128 \times 128 \times 64$. Thus, we again compare solving the same problem with the same resources.

We included the scaling on a single node in the left part of Fig. 8 (bottom). The weak scaling on a node level (fractions of a node on the Nodes axis) demonstrates that FS3D is memory bandwidth bound: With more than 32 cores (32/128 node) the performance drops. However, the speed-up comparing the performance of 128 instead of 32 or 64 cores is still positive. Instead of an ideal factor of four (compared to 32 cores) or two (compared to 64 cores) theoretically achievable, we obtain only a factor of 3.5 or 1.8 in the CPH. The performance plot also confirms the discussion in Sect. 4.3 that it does not make sense to employ a hybrid parallelisation on a low number of nodes. The problem of a large matrix to solve and many costly all-to-one communications is irrelevant at a low count of MPI-processes, but dominates at larger numbers. Beyond the single-node scaling, the performance stays roughly constant again with the pure MPI and F-cycle being the best choice for up to 16 nodes (2048 cores). Adding a hybrid parallelisation with four OpenMP threads at a larger number of nodes further increases the good scaling up to 128 nodes, again with the F-cycle. The V-cycle performs best among the test cases for 256 nodes, but the performance already drops below the (close to) constant scaling level observed for up to 128 nodes. Thus, the close to constant scaling limit increases by a factor of 16 in the usable nodes compared to the baseline with the W-cycle and pure MPI.

5 Conclusion

The early phase of crown formation during an oblique droplet impact onto a thin wall film was computed with the multiphase DNS code FS3D. The simulation provides additional information about the liquid gas interface and the internal flow field in the fluid, which can not be derived from experiments. With the help of the simulations, two different mechanisms of crown formation were observed and could be explained. The oblique mechanism on the back side of the impact is caused by a pinch-off of the initial lamella due to the horizontal impact velocity component, which prohibits the undisturbed growth of the lamella. A grid study for this highly dynamic and numerically challenging impact phase revealed that a detailed investigation requires a resolution $\geq$ 350 cells/D. At this resolution, the collision of the lamella with the droplet was adequately resolved. The detachment at the tip of the lamella remains to be grid dependent, however, the crown shape is resolved sufficiently accurate.

The choice of the F- and V- cycle in the MG-solver and an additional hybrid parallelisation increased the achieved computed cycles per hour (CPH) by a factor of $\approx$ 12.4 for the strong scaling compared to the formerly employed setup at the close to linear scaling limit. The strong- and weak scaling was increased to 16 times more nodes by employing the V- or F-cycle in the MG-solver instead of the W-cycle and introducing a hybrid MPI+OpenMP parallelisation with four OpenMP threads per MPI process.

Acknowledgements The authors kindly acknowledge the *High Performance Computing Center Stuttgart* (HLRS) for support and supply of computational resources on the HPE Apollo (Hawk) platform under the Grant No. FS3D/11142. Additionally, the authors gratefully acknowledge the financial support of the Deutsche Forschungsgemeinschaft (DFG, German Research Foundation) through the project DROPIT/GRK 2160/2 and under Germany's Excellence Strategy - EXC 2075 - 390740016.

References

1. Brambilla, P., Guardone, A.: Assessment of dynamic adaptive grids in volume-of-fluid simulations of oblique drop impacts onto liquid films. J. Comp. Appl. Math. **281**, 277–283 (2015). https://doi.org/10.1016/j.cam.2014.12.030
2. Chen, Z., Shu, C., Wang, Y., Yang, L.: Oblique drop impact on thin film: splashing dynamics at moderate impingement angles. Phys. Fluids **32**(3), 033303 (2020). https://doi.org/10.1063/5.0004142
3. Cheng, M., Lou, J.: Lattice Boltzmann simulation of a drop impact on a moving wall with a liquid film. Comp. Math. with Appl. **67**(2), 307–317 (2014). https://doi.org/10.1016/j.camwa.2013.07.003
4. Cossali, G.E., Coghe, A., Marengo, M.: The impact of a single drop on a wetted solid surface. Exp. Fluids **22**(6), 463–472 (1997)
5. Eisenschmidt, K., Ertl, M., Gomaa, H., Kieffer-Roth, C., Meister, C., Rauschenberger, P., Reitzle, M., Schlottke, K., Weigand, B.: Direct numerical simulations for multiphase flows: an overview of the multiphase code FS3D. J. Appl. Math. Comp. **272**(2), 508–517 (2016). https://doi.org/10.1016/j.amc.2015.05.095

6. Geppert, A., Terzis, A., Lamanna, G., Marengo, M., Weigand, B.: A benchmark study for the crown-type splashing dynamics of one- and two-component droplet wall-film interactions. Exp. Fluids **58**(12), 1–27 (2017). https://doi.org/10.1007/s00348-017-2447-2
7. Hirt, C.W., Nichols, B.D.: Volume of fluid (VOF) method for the dynamics of free boundaries. J. Comp. Phys. **39**(1), 201–225 (1981). https://doi.org/10.1016/0021-9991(81)90145-5
8. Ibach, M., Vaikuntanathan, V., Arad, A., Katoshevski, D., Greenberg, J.B., Weigand, B.: Investigation of droplet grouping in monodisperse streams by direct numerical simulations. Phys. Fluids **34**(8), 083314 (2022). https://doi.org/10.1063/5.0097551
9. Kaufmann, J., Geppert, A., Ertl, M., Bernard, R., Vaikuntanathan, V., Lamanna, G., Weigand, B.: Direct numerical simulations of one- and two-component droplet wall-film interactions within the crown-type splashing regime. ICLASS (2018)
10. Lafaurie, B., Nardone, C., Scardovelli, R., Zaleski, S., Zanetti, G.: Modelling merging and fragmentation in multiphase flows with SURFER. J. Comp. Phys. **113**(1), 134–147 (1994). https://doi.org/10.1006/jcph.1994.1123
11. Liang, G., Guo, Y., Shen, S.: Gas properties on crown behavior and drop coalescence. Num. Heat Transf., Part B: Fundam. **65**(6), 537–553 (2014)
12. Okawa, T., Shiraishi, T., Mori, T.: Effect of impingement angle on the outcome of single water drop impact onto a plane water surface. Exp. Fluids **44**(2), 331–339 (2008). https://doi.org/10.1007/s00348-007-0406-z
13. Potyka, J., Kromer, J., Liu, M., Schulte, K., Bothe, D.: Modelling and numerical simulation of binary droplet collisions under extreme conditions. In: Droplet Dynamics Under Extreme Ambient Conditions, pp. 127–147. Springer International Publishing Cham (2022)
14. Potyka, J., Stober, J., Wurst, J., Ibach, M., Steigerwald, J., Weigand, B., Schulte, K.: Towards DNS of droplet-jet collisions of immiscible liquids with FS3D. arXiv:2212.09727. (preprint)
15. Ren, W., Reutzsch, J., Weigand, B.: Direct numerical simulation of water droplets in turbulent flow. Fluids **5**(3), 158 (2020). https://doi.org/10.3390/fluids5030158
16. Reutzsch, J., Ertl, M., Baggio, M., Seck, A., Weigand, B.: Towards a direct numerical simulation of primary jet breakup with evaporation, Chap. 16, pp. 243–257. Springer, Berlin (2019)
17. Reutzsch, J., Kochanattu, G.V.R., Ibach, M., Kieffer-Roth, C., Tonini, S., Cossali, G., Weigand, B.: Direct numerical simulations of oscillating liquid droplets: a method to extract shape characteristics. In: ILASS-Europe (2019)
18. Rider, W.J., Kothe, D.B.: Reconstructing volume tracking. J. Comp. Phys. **141**(2), 112–152 (1998). https://doi.org/10.1006/jcph.1998.5906
19. Rieber, M.: Numerische Modellierung der Dynamik freier Grenzflächen in Zweiphasenströmungen. Dissertation, Universität Stuttgart (2004)
20. Rieber, M., Frohn, A.: A numerical study on the mechanism of splashing. Int. J. Heat Fluid Flow **20**(5), 455–461 (1999). https://doi.org/10.1016/S0142-727X(99)00033-8
21. Rioboo, R., Bauthier, C., Conti, J., Voue, M., De Coninck, J.: Experimental investigation of splash and crown formation during single drop impact on wetted surfaces. Exp. Fluids **35**(6), 648–652 (2003). https://doi.org/10.1007/s00348-003-0719-5
22. Roisman, I., Tropea, C.: Impact of a drop onto a wetted wall: description of crown formation and propagation. Fluid Mech. **472**, 373–397 (2002)
23. Steigerwald, J., Ibach, M., Reutzsch, J., Weigand, B.: Towards the numerical determination of the splashing threshold of two-component drop film interactions. Springer, Berlin (2022)
24. Vander Wal, R.L., Berger, G.M., Mozes, S.D.: Droplets splashing upon films of the same fluid of various depths. Exp. Fluids **40**(1), 33–52 (2005). https://doi.org/10.1007/s00348-005-0044-2
25. Wesseling, P.: An Introduction to Multigrid Methods. Wiley (1992)
26. Yarin, A.L., Weiss, D.A.: Impact of drops on solid surfaces: self-similar capillary waves, and splashing as a new type of kinematic discontinuity. J. Fluid Mech. **283**, 141–173 (1995)

Towards Large Scale Blunt Trailing Edge Noise Simulations

Kathrin Stahl, Jana Gericke, Harald Klimach, and Sabine Roller

Abstract Trailing edge noise is a significant contributor to the noise emissions of wind turbine blades. Thus, in this project we aim to study and understand this flow-induced noise emission with a detailed analysis of this sound source and possible mitigation strategies by employing the Lattice-Boltzmann method for a highly resolved simulation of the fluid flow and the resulting noise. We present the current status of the project including a preliminary study consisting of wind tunnel experiments and very large eddy simulations. Based on this, the objectives for the new study are defined. A new feature that shifts the communication of the global reductions in health checks into the background, was implemented recently and its impact is investigated in this report. We determine the setup as well as the operational parameters for production runs on *Hawk* to make best use of the machine.

1 Introduction

With the increasing proliferation of turbomachines such as propellers, fans, wind turbines, etc. the aerodynamically induced noise from airfoils became more and more important. Hence, during the last decades, the acoustics of lifting surfaces have been thoroughly studied. The aerodynamically induced noise from airfoils can be categorized into inflow noise and airfoil self-noise. Inflow noise is caused by the incoming turbulence impinging the leading edge. Mechanisms of self-noise are

K. Stahl (✉) · J. Gericke · H. Klimach · S. Roller
German Aerospace Center (DLR), Institute of Software Methods for Product Virtualization, Dresden, Germany
e-mail: Kathrin.stahl@uni-siegen.de

J. Gericke
e-mail: jana.gericke@dlr.de

H. Klimach
e-mail: harald.klimach@dlr.de

S. Roller
e-mail: sabine.roller@dlr.de

T. Ludwig et al. (eds.), *High Performance Computing in Science and Engineering '23*,
https://doi.org/10.1007/978-3-031-91312-9_17

stochastic velocity fluctuations in the turbulent boundary layers above the wetted surfaces, boundary layer separation, vortex shedding, and turbulent wake at the trailing edge (TE). Trailing edge noise (TEN) is one of the dominant sources in applications like wind turbines where the inflow noise is small or even negligible [10]. BROOKS ET AL. [2] classified four principally different TEN generation mechanisms relevant for two-dimensional airfoil sections. One of these mechanisms, blunt trailing edge noise (BTEN), will be investigated in detail within this study. BTEN caused by finite thickness of the TE produces a von Kármán-type quasi-periodic vortex shedding, which may lead to additional noise due to a tonal component in the acoustic spectra.

BTEN in general has been the subject of a large number of investigations. The physical mechanisms, underlying BTEN, have been investigated in numerous theoretical, numerical, and experimental studies—frequently dealing with a flat plate or a symmetric NACA0012 airfoil. Blades for turbomachinery, however, are cambered and in most cases asymmetrical. Their trailing edges have to be thickened for manufacturing reasons. Hence, BTEN and its contribution to the overall TEN is an issue that needs deeper investigation. Therefore, the present study is dedicated to the specific noise mechanism of BTEN using the Lattice-Boltzmann method (LBM). It is based on an existing study by STAHL [8], of which the methodology is explained in Sect. 2. Subsequently, partial results of it are shown and differences are addressed. The objective of the present study is to examine and identify possible causes for the differences found in the previous study by STAHL. Compared to the latter one, the degree of modeling is to be reduced in the present study. Thus, a higher resolution of the grid and an enlargement of the domain in spanwise direction are planned. Another objective of the investigation is the tripping band thickness and its reduction enabled by the smaller grid size due to the higher resolution.

For this the massively parallel in-house open-source solver *Musubi* [4] with large eddy simulation (LES) modelling of unresolved turbulent scales is employed. With regard to the *Musubi* flow solver, the physical objectives are to validate it on the basis of a large application-related case and to validate recently implemented wall functions. The economic goals of optimizing *Musubi* are also being pursued. This includes the consideration of load balancing, or as described in Sect. 3, a new feature to reduce synchronization points with non-blocking collectives.

2 LBM-VLES BTEN-Study

First, the joint setup of the existing study of STAHL and the present one are briefly described, before the progress and the next steps are outlined.

2.1 Airfoil and Parameters Varied

The complete analysis is based on the *DU93W210* airfoil. It has been developed by TIMMER AND VAN ROOIJ [9] at the TECHNICAL UNIVERSITY OF DELFT, originally for the mid-blade region of large wind turbine rotors.

As depicted in Fig. 1 the *DU93W210* is a cambered airfoil with a partly concave pressure side. The original airfoil has a trailing edge-thickness-to-chord ratio $t^* = t_{TE}/c = 0.005$ (blue/dark in Fig. 1)and is taken as the baseline geometry in this study. To investigate the effect of different t^*, the airfoil was truncated in three steps at the trailing edge. Truncation always implies that the effective chord length c becomes shorter than the baseline c_b. It is important to note that the t^* and the REYNOLDS number (*Re*) are always defined with the effective chord length c of the (truncated) airfoil.

In total, four different t^* were investigated under different loads and for different free-stream velocities u_∞. The different loads are realized by varying the angle of attack α_e. With $\alpha_e = 0°$, $1.5°$ and $3.5°$ investigated.

Two realistic values were chosen for the free-stream velocity u_∞. Which represent with $u_\infty = 30\,\mathrm{m/s}$ and $60\,\mathrm{m/s}$ a chord based moderate and high *Re*. The design of experiments with considered parameter combinations is presented in Table 1 with dimensioned data and dimensionless key numbers on the right. The MACH number (*Ma*) is given for classification purposes only and is not used further in the present study.

Most engineering applications have turbulent flows, like the applications for the airfoil utilized. Thus, the investigations are confined to a boundary layer being turbulent over the most part of the airfoil pressure and suction side. For both, the numerical simulation and the experiments, this requires careful tripping. Because a tripping band that is too high may result in a boundary layer too thick. In contrast to that a tripping band too low does not lead to a laminar to turbulent transition. Since α_e is varied during the study, the tripping positions have been chosen differently for suction and pressure side, but were kept the same for all configurations. The tripping

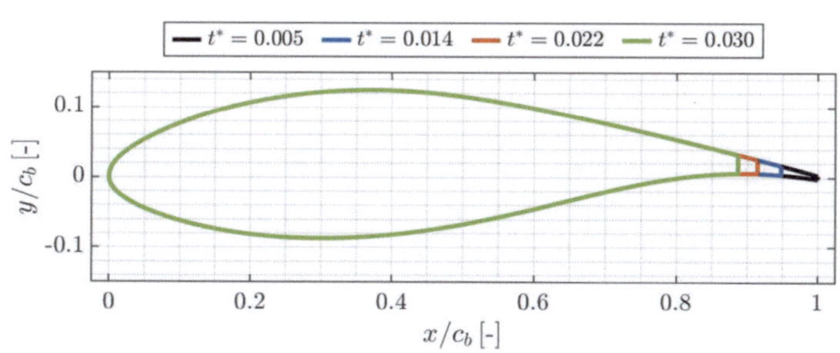

Fig. 1 DU93W210 airfoil (baseline with $t^* = 0.005$, black) and three blunt trailing edges ($t^* = 0.014$, blue, $t^* = 0.022$, red, $t^* = 0.030$, green); t^* is always defined with the effective chord length c of the truncated airfoil

Table 1 Parameters varied for experiment and LBM-VLES simulation for case I to XXIV. Cases marked with + have been simulated with LBM-VLES only. The § indicates the cases additionally simulated via the LBM-LES solver *Musubi*. The baseline case is highlighted in gray. (For a better overview ./. indicates that the parameter has not been changed with respect to the previous line)

Case	t_{TE} [mm]	c [mm]	α_{e} [°]	u_∞ [m/s]	t^* [–]	c/c_{b} [–]	$Re_{\mathrm{c}} \cdot 10^6$ [–]	$Re_{\mathrm{TE}} \cdot 10^6$ [–]	Ma [–]
I	1.5	300	0	30	0.005	1.000	0.6	3.1	0.087
II	./.	./.	1.5	./.	./.	./.	./.	./.	./.
III§	./.	./.	3.5	./.	./.	./.	./.	./.	./.
IV	./.	./.	0	60	./.	./.	1.2	6.2	0.175
V	./.	./.	1.5	./.	./.	./.	./.	./.	./.
VI	./.	./.	3.5	./.	./.	./.	./.	./.	./.
VII	4.0	284	0	30	0.014	0.947	0.6	8.2	0.087
VIII+	./.	./.	1.5	./.	./.	./.	./.	./.	./.
IX	./.	./.	3.5	./.	./.	./.	./.	./.	./.
X	./.	./.	0	60	./.	./.	1.2	16.4	0.175
XI+	./.	./.	1.5	./.	./.	./.	./.	./.	./.
XII	./.	./.	3.5	./.	./.	./.	./.	./.	./.
XIII+	6.0	274	0	30	0.022	0.913	0.6	12.3	0.087
XIV+	./.	./.	1.5	./.	./.	./.	./.	./.	./.
XV+	./.	./.	3.5	./.	./.	./.	./.	./.	./.
XVI+	./.	./.	0	60	./.	./.	1.1	24.6	0.175
XVII+	./.	./.	1.5	./.	./.	./.	./.	./.	./.
XVIII+	./.	./.	3.5	./.	./.	./.	./.	./.	./.
XIX	8.0	266	0	30	0.030	0.867	0.5	16.4	0.087
XX	./.	./.	1.5	./.	./.	./.	./.	./.	./.
XXI	./.	./.	3.5	./.	./.	./.	./.	./.	./.
XXII§	./.	./.	0	60	./.	./.	1.1	32.9	0.175
XXIII	./.	./.	1.5	./.	./.	./.	./.	./.	./.
XXIV	./.	./.	3.5	./.	./.	./.	./.	./.	./.

was applied using a zigzag tripping band on the suction side at $x/c_{\mathrm{b}} = 0.02$ and on the pressure side at $x/c_{\mathrm{b}} = 0.05$. The height of the tripping band for the baseline case is identical in the experiment and simulation. Depending on the configuration, a slight adjustment of the tripping band thickness in the simulation was necessary to force the laminar-turbulent transition. More detailed information on the selected tripping band thickness as well as its position are given in [8].

2.2 Experimental Setup and Measurement Techniques

The experimental investigations were carried out in the aeroacoustic wind tunnel of the UNIVERSITY OF SIEGEN. Figure 2 depicts the experimental setup for the measurement of the far-field acoustics in the wind tunnel. A precise description of the wind tunnel and its flow characteristics are provided in [1, 8].

The following data was collected in the wind tunnel experiments at the UNIVERSITY OF SIEGEN:

- Acoustic measurement of far-field noise using 12 microphones. By arranging the microphones in a circle around the trailing edge, it is possible to observe not only the far-field noise but also the directional characteristics.
- Measurement of the static pressure distribution around the airfoil via a total of 50 pressure measurement holes equipped with pressure transducers.
- Measurement of the transient and averaged flow data in the wake of the airfoil using hot-wire anemometry.
- Measurement of the transient alternating pressures near the trailing edge with a special probe microphone on the pressure and suction side.

2.3 Numerical Setup for LBM-VLES Simulations

The simulations are conducted for the *DU93W210* airfoil and the parameters are given in Sect. 2.1. Figure 3 represents the global simulation domain (not to scale). The domain is $100 \cdot c$ long in x- and y-direction and the trailing edge of the airfoil is placed

Fig. 2 Experimental setup in the wind tunnel at the UNIVERSITY OF SIEGEN: Microphone measurement of the acoustic far-field

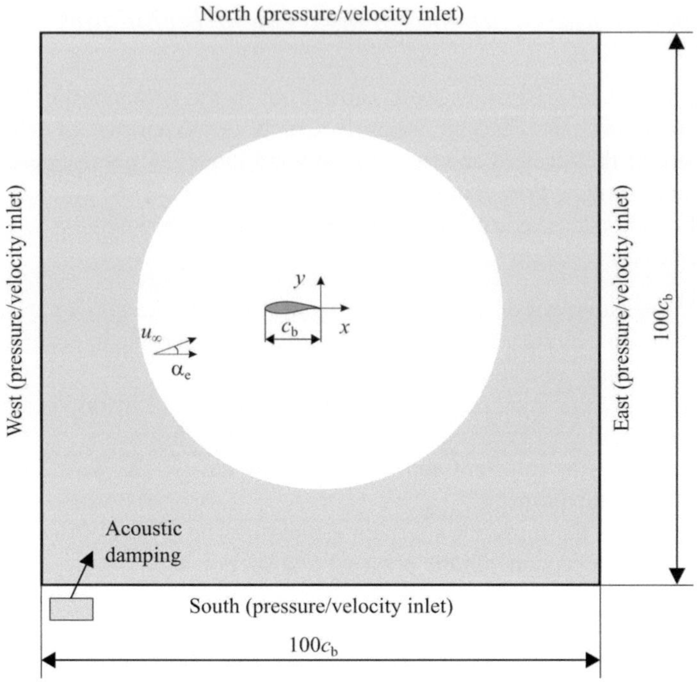

Fig. 3 Schematic representation of the LBM-VLES and LBM-LES setup with tenfold enlarged airfoil

at the center. In z-direction the airfoil has a span of 30 mm ($0.1 \cdot c$). Furthermore, the boundary conditions are set to periodic in this direction to limit computational costs.

To ensure a fully turbulent flow the airfoil is tripped close to its nose by a zigzag trip with a thickness of 0.7 mm on suction and pressure side, further details are given in Sect. 2.1. To avoid reflections, a sponge is deployed in the outer zone starting at $33 \cdot c$. The acoustic time, $T_{ac} = 50 \cdot c/c_s$, required by a wave traveling from the airfoil to the outer region of the domain is $T_{ac} = 0.04$ s. The convective time, $T_c = 100 \cdot c/u_\infty$, describes the time which a particle needs to travel through the domain from west to east. For the given setup the convective time is $T_c = 0.5\ s$ (for $u_\infty = 60$ m/s and $T_c = 1$ s for $u_\infty = 30$ m/s). Besides this, the time for the flow to pass the airfoil is called through-flow time, $T_f = c/u_\infty$, and lasts $T_f = 0.005$ s (0.010 s for $u_\infty = 30$ m/s). More information about the simulation domain, e.g. number of resolution zones, number of voxels and their sizes as well as the physical length of a time-step, are collected in Table 2.

The numerical flow solutions of the LBM-VLES are obtained by using the *SIMULIA PowerFLOW*™ solver, version 2020 R1. The simulation runs were conducted on the *OMNI* cluster at the UNIVERSITY OF SIEGEN and the *Niagara* supercomputer at the SCINET HPC Consortium in Canada. *PowerFLOW*™ is based on a cartesian mesh

Table 2 Parameters of the simulation domain

	Domain
Resolution zones	9
Finest voxel size	$5.86 \cdot 10^{-5}$ m
Coarsest voxel size	$1.5 \cdot 10^{-2}$ m
Time-step size	$9.941 \cdot 10^{-8}$ s
Number of voxels	$360 \cdot 10^6$ ($t^* = 0.005$)
	$350 \cdot 10^6$ ($t^* = 0.014$)
	$345 \cdot 10^6$ ($t^* = 0.022$)
	$340 \cdot 10^6$ ($t^* = 0.030$)
CPUh/T_f	$1.7 \cdot 10^3$
Physical simulated time	0.2 s

LBM with automatic mesh generation and without restrictions on the geometric complexity of the models that can be treated. The idea of LBM is the statistical tracking of the advection and collisions of fluid particles by an integer number of distribution functions aligned with predefined discrete directions given by the stencils. For each distribution function $f(\mathbf{x}, t, \mathbf{c})$ the Boltzmann equation is solved on the mesh. To model the sub-grid unresolved scales of turbulence the two equation $k - \epsilon$−model is applied by *PowerFLOW*™. However, the utilization of the *PowerFLOW*™ solver is limited by the licenses available to a total of 1, 024 processes.

2.4 Experimental and LBM-VLES Results

In the present study mainly the acoustic results from experiment and simulation are considered. Further results as well as a comparison between experimental and numerical results with regard to the pressure distribution, boundary layer characteristics, surface pressure fluctuations, flow characteristics in the airfoil wake etc., which show good agreement between experiment and simulation, are presented in [8]. The far-field noise for the different t^* also shows good qualitative agreement between experiment and simulation. They are depicted in Figs. 4 and 5 respectively. The trends and effects for variation of α_e and *Re* are identical for experiment and simulation. A detailed evaluation, interpretation and assessment of the results will be published in a separate study in the near future. However, especially in the cases where tonal components are observed in the far-field noise (at high t^* and low α_e), there are differences in frequency. A number of possibilities could be the cause:

- The boundary layer or the wake are not fully resolved
- Manufacturing inaccuracies of the test specimen
- Boundary layer tripping
- Inaccuracies when setting the angle of attack

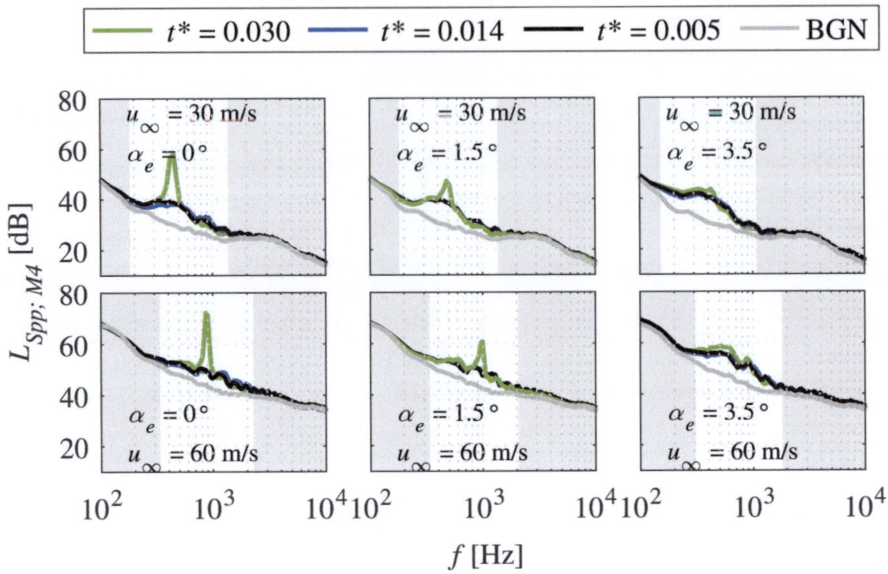

Fig. 4 Experimentally obtained power spectral density of far-field noise for different t^*, α_e and Re_c

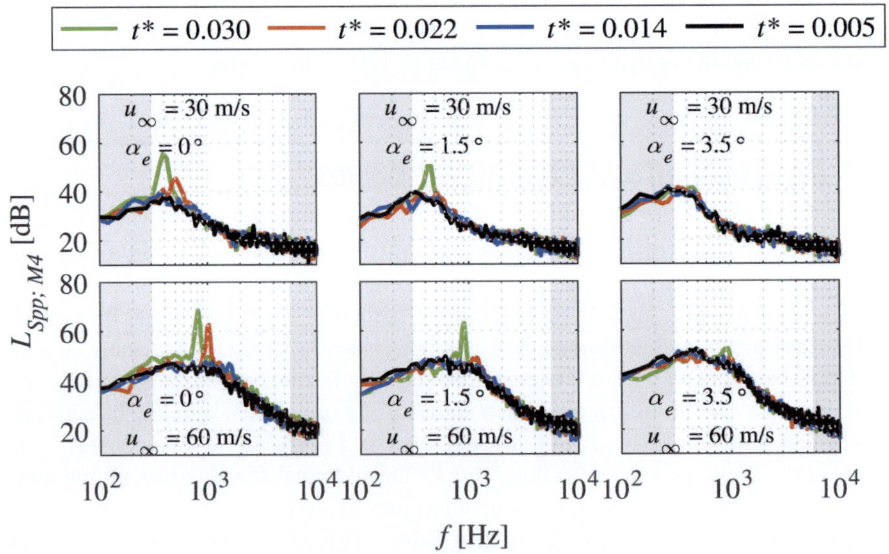

Fig. 5 Numerically obtained power spectral density of far-field noise for different t^*, α_e and Re_c

- Spanwise extension of the simulated airfoil is not sufficient
- Different inflow conditions (turbulence intensity, velocity distribution)
- Additional side plate—airfoil interaction noise in the experiment
- The turbulence model applied to model the unresolved scales of turbulence.

With regard to the resolution of the boundary layer as well as the airfoil wake and the sufficiency of the spanwise extension, it should be mentioned that a rudimentary mesh study is also presented in [8]. It was carried out in order to assess the influence whether and how (i) a grid refinement or (ii) an extension of the simulation domain in the spanwise direction would have an effect. Only the baseline case ($t^* = 0.005$, $\alpha_e = 0°$ and $Re_c = 1.2 \cdot 10^6$), which is highlighted in gray in Table 1, was used for the mesh study as reference. The grid refinement (i) is realized by adding a further multilevel resulting in a cell size of $2.93 \cdot 10^-5$ m for the finest cell. By that, the total number of elements is increased by a factor of ≈ 6 compared to the baseline case. In contrast to that, extending the simulation domain in the spanwise direction (ii) by doubling the span to $0.2 \cdot c$ results in a factor of 2. A comparison of the pressure distribution around the airfoil, the wall pressure spectrum close to the TE and the far-field noise of these two adjusted meshes (i) and (ii) with the baseline case revealed only minimal differences [8]. Furthermore, the results without additional spanwise extension or additional refinement (multilevel)—the baseline case—are satisfactorily good compared to the experiments. In addition, a trade-off between computational effort and accuracy should be made. Therefore, the baseline case—with coarser grid and the smaller spanwise extension compared to (i) and (ii)—was chosen for further investigations in the study by Stahl. At this point, the present study should extend the existing one to further narrow down the causes for the observed deviations between simulation and experiment. This will be done in several steps. Before changes are made to the setup, such as refining the mesh or extending the domain in spanwise direction especially for the configuration with a high degree of bluntness (e.g. case a of Table 1), a different turbulence model is applied to the baseline case. The realization is done with the help of the open-source solver *Musubi*.

3 LBM-LES BTEN-Study

For the LBM-LES simulation runs of the BTEN study we employ our open-source Lattice- Boltzmann solver *Musubi* on *Hawk*.

The supercomputer is a *HPE Apollo* system, where each node has 2 *AMD EPYC 7742* processors with 64 cores each, yielding 128 cores per node and operating at 2.25 GHz [5]. The *GCC* compiler 9.2 together with the *HPE MPI* librabry 2.23 is used to compile the solver.

At first, the setup and configuration as well as the simulation framework used to run the simulations are shortly introduced in Sect. 3.1. *Musubi* has been shown to scale to large numbers of *MPI* processes and without any license restrictions we would theoretically be able to make use of the whole machine for the large-scale computations. To use the available resources wisely and efficiently, we evaluate the performance of *Musubi* on *Hawk* and determine the operational parameters for production runs. The performance in terms of the problem size per node is assessed and the optimal configuration with respect to the distribution of parallelism to *OpenMP* threads and *MPI* processes is determined. To improve the scalability, a small and simple new feature was implemented and its effect on the performance is investigated in Sect. 3.4 along with the optimal configuration in terms of run time and number of nodes. The performance for LBM computations is expressed in terms of million lattice updates per second (MLUPs). As for each lattice update the same number of floating point operations is required, this metric also can be interpreted as a scaled measure for the floating point operations per second (FLOPs).

3.1 Lattice-Boltzmann Solver Musubi and Setup

The solver is part of the *APES* framework with dedicated pre- and post-processing tools like the mesh generator *Seeder* and the *TrElM* library as its centerpiece. *Musubi* is mainly written in *Fortran 2003* and hybrid parallel: it is primarily parallelized via *MPI*, and the loops of the compute kernels are also *OpenMP* parallel. Besides different collision schemes [7] and LES models, there are also various wall functions available. The key ingredients for wall-modeled LES were recently addressed in [6]. Based on those findings, we currently plan to use the Vreman LES model in combination with the Musker wall model and MRT collision scheme. With these settings we run the setup for *Musubi*, which is based on the one shown in Fig. 3 and described in Sect. 2.3 resulting in almost 588 million elements.

3.2 Performance Measurements

To evaluate the performance of the application on *Hawk* we use a simple, homogeneous setup with balanced partitions. The mesh of the cubic domain is internally generated by the solver during the simulation. The problem size varies between $8^3 = 512$ and $8^9 = 134{,}217{,}728$ elements in powers of 8. Where 8^9 is the largest power of 8 that fits onto a single node. Each simulation is executed for at least 5 minutes of running time on *Hawk*.

Figure 6 shows the node performance over the variation of the problem size with different thread-process relations, such that the product of threads and processes always equals the physical number of cores (128). We refer to this kind of plot as a performance map.

The performance for problem sizes smaller than 30,000 elements per node is low. For larger problems, there is a peak for around 1,000,000 elements per node. Up to this many elements the whole problem fits into the caches of the node (cache region). The peak is more pronounced, if there is no *OpenMP* parallelism and decreases with

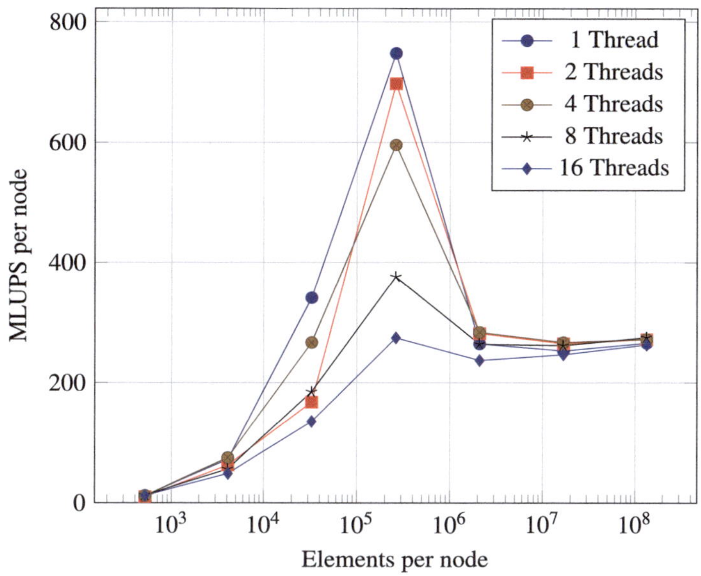

Fig. 6 Performance map on a single node for different number of threads utilizing all 128 cores

increasing thread counts. Problems involving more than 1,000,000 elements do not fit into the cache anymore and the performance is governed by frequent access to the main memory (memory region). Depending on the problem size (cache versus memory region), different strategies appear advisable. As most simulations operate in the memory region, we will concentrate on this area. There, the performance is relatively independent of the problem size for up to 8 *OpenMP* threads with around 250 MLUPs for these memory bound problem sizes. Thus, we utilize 8 *OpenMP* threads per process for the follow-on performance measurements on multiple nodes. To provide exactly the same problem sizes and shapes as in the single node evaluation, the number of nodes is increased by factors of 8.

The resulting performance map across multiple nodes is shown in Fig. 7 for up to 512 nodes.

The degradation of performance per node with increasing node numbers remains relatively low with a loss of less than 10% on 512 nodes and sufficiently large problems per node. However, a dependency on the problem size can be observed for this scaling. While for smaller problems per node a larger drop in performance per node is observed for 512 nodes, there is only little loss for the largest considered problem size per node.

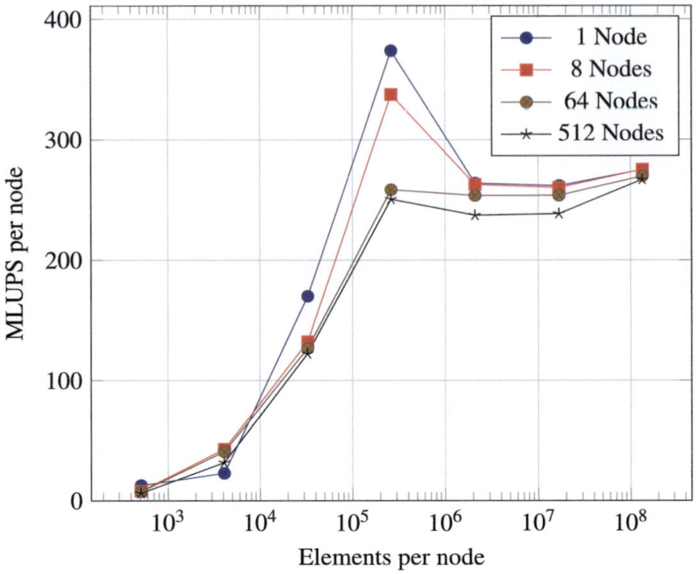

Fig. 7 Performance map across multiple nodes with 8 *OpenMP* threads per process, and 16 processes per node

3.3 *New Feature: Health Check with Non-blocking Reduction*

Musubi checks global conditions periodically to ensure a healthy simulation. These checks incur additional communication, not directly related to the computation itself. To minimize the impact of this additional communication on the scalability of the application, a new feature has been implement to allow for delayed checking of the global information [3]. Traditionally such checks are achieved by blocking collective reductions across all processes, leading to strict synchronization points. By using the non-blocking reduction from *MPI-3*, the synchronization requirements can be loosened and the interference with the productive computation minimized.

We realize the delayed evaluation of health checks with the non-blocking `MPI_Iallreduce` that directly replaces the previously used blocking operation `MPI_Allreduce`. Waiting on the completion of the non-blocking collective operation delayed until right before the next health check. Setting several status flags as logicals on each process, allows us to use a *logical or* as reduction operation in the *Allreduce* operation and provides the information if any process has encountered a specific status. The resulting communication pattern as seen in a tracing with Vampir in the blocking case is shown in Fig. 8a.

As elaborated, the non-blocking approach eases the synchronization requirements and allows for a greater variability in the execution of individual processes. This is illustrated in Fig. 8b.

(a) Trace of blocking algorithm

(b) Trace of non-blocking algorithm

Fig. 8 Comparison of traces for blocking and non-blocking (delayed) health check. Computations are green and blue, communication brown, *Allreduce* red

3.4 Comparison of Blocking and Non-blocking Reduction

A comparison of the delayed health check with the blocking reduction is done for a strong scaling measurement with 134,217,728 elements. Figure 9 illustrates the respective running times per iteration for the new non-blocking approach in orange and the blocking one in blue for up to 2,048 nodes. To maximize the number of involved *MPI* processes in this analysis, we use no *OpenMP* for the analysis and instead use as many *MPI* processes as there are physical cores in the system.

A nearly ideal scaling for up to 64 nodes or 8,192 processes can be observed in either case. Beyond that point the problem size per process gets sufficiently small to fit into the cache (as discussed earlier), and a superlinear speed-up is observed. An effect from the delayed health check is only observed beyond even that region, where the efficiency of the computation declines again with further increased core counts. For $1,024$ elements per process a benefit of around 40% is achieved, though further scaling diminishes the advantage again.

Finally, we investigate the effect of the interval length between health checks. The number of iterations between health checks is configurable in the solver, and

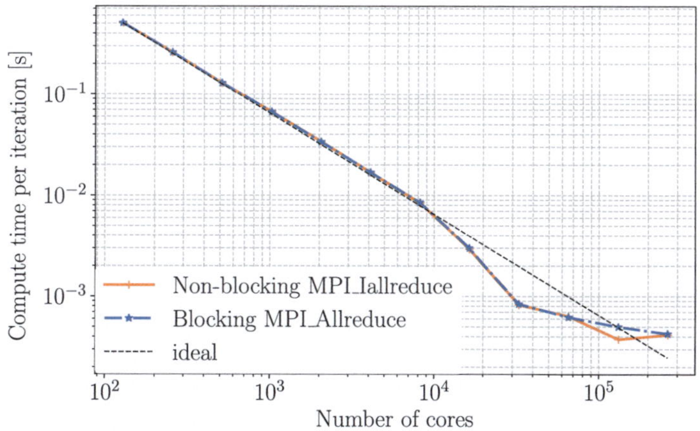

Fig. 9 Compute time per iteration of *Musubi* on *Hawk* using 1–2,048 nodes. Strong scaling measurement for 134,217,728 elements with non-blocking (orange) and blocking (blue) reduction. A *dashed black line* indicates ideal scaling

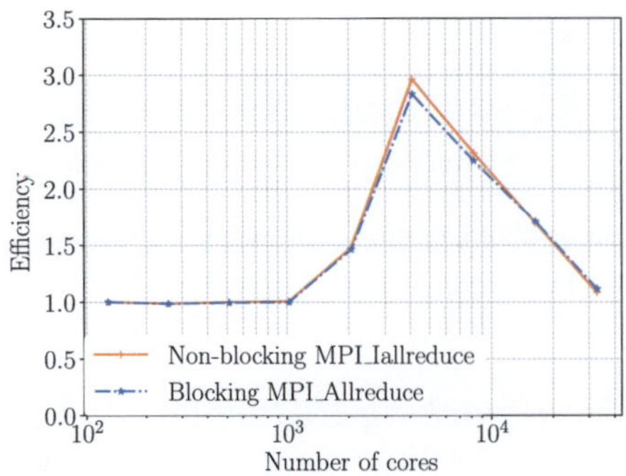

(a) Check interval = 1

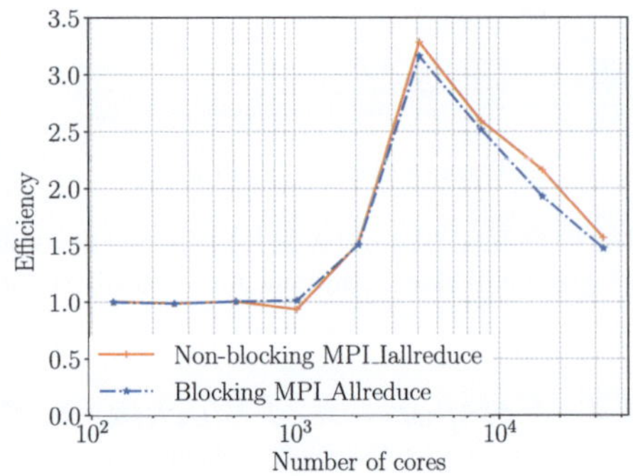

(b) Check interval = 5

Fig. 10 Efficiency of *Musubi* on *Hawk* using 1–256 nodes for different check intervals: 1 (left) and 5 (right). Strong scaling measurement for 16,777,216 elements with non-blocking (orange) and blocking health check (blue)

Fig. 10 shows the comparison of parallel efficiency between non-blocking and blocking reductions if there is just one iteration between checks (Fig. 10a) as in the previous analysis and when there are five iterations in between those checks (Fig. 10b).

With a slight increase in the interval length between the checks, more intermediate *MPI* calls are done. Thus, there is more time for the collective reduction to complete. This increases the benefit of the non-blocking approach. Especially for smaller partitions an improvement of the overall efficiency of up to 10% can be observed. This requirement of a sufficient amount of *MPI* calls in between to allow for progress to happen, has also been observed in a small dedicated benchmark that emulates the setup. More intermediate calls by a larger interval length can be used, but they do not improve the running times further. Testing the new feature with the target test case of the airfoil revealed that for a case with imbalances—due to boundary conditions, grid refinement etc.—its benefit is even higher.

Based on these findings we can specify the operational parameters for the baseline case with around 588 million elements. Keeping in mind, that we want to use more than 1 million elements per node, it is advisable to use at least 64 nodes resulting in 9.1 million elements per node. To decrease the time to solution, we plan to use 8 *OpenMP* threads per process on 256 nodes resulting in roughly 2.3 million elements per node. The newly implemented feature, the non-blocking reduction in health checks, should be used with an interval length of five or more.

Once the reference simulations, which are indicated via § in Table 1, with *Musubi* are done, the domain/mesh will be modified to investigate the aforementioned causes of refining the mesh or extending the domain in spanwise direction—with more elements on larger node counts, but with similar configurations per node.

4 Conclusion and Project Outlook

Experiments and LBM-VLES simulations to validate the setup were performed. A detailed evaluation of the data as well as a comparison between the experimental and numerical results with subsequent drawing of conclusions is planned as soon as all data have been produced. The setup for the baseline case with *Musubi* as well as the corresponding optimal configuration based on detailed performance measurements has been done. The setup for the further LBM-LES study using the solver *Musubi* has been designed and generated based on the BTEN LBM-VLES study and its findings. The next step is now to run the planned modified simulations, which will initially use a different turbulence model. Further on, a refinement of the mesh as well as an additional extension of the domain in spanwise direction is scheduled. We look forward to evaluating the results to identify further causes for the differences between experiment and the BTEN LBM-VLES study. Moreover, the objective is to validate the flow solver *Musubi* with an application-related case with relatively high Reynolds number. This also includes the validation of the wall model. At the same time, the economic efficiency of *Musubi* is being considered and optimized, including the analysis of load balancing and the implementation of new features that further improve the runtime of the solver.

Acknowledgements The authors gratefully acknowledge the Gauss Centre for Supercomputing e.V. (www.gauss-centre.eu) for funding this project by providing computing time on the GCS Supercomputer HAWK at Höchstleistungsrechenzentrum Stuttgart (www.hlrs.de).

References

1. Bald, A., Stahl, K., Foysi, H., Roller, S.: Attenuation of low-frequency pressure fluctuations within the test section of an aeroacoustic wind tunnel using Helmholtz resonators. In: Tagungsband: DAGA 2023 - 49. Jahrestagung für Akustik, Fortschritte der Akustik - DAGA 2023, pp. 1500–1503. Deutsche Gesellschaft für Akustik e.V. (2023)
2. Brooks, T.F., Pope, D.S., Marcolini, M.A.: Airfoil self-noise and prediction. Technical report, NASA Langley Research Center, Hampton (VA) (1989)
3. Gericke, J., Klimach, H., Ebrahimi Pour, N., Roller, S.: Using MPIs non-blocking allreduce for health checks in dynamic simulations. In: Park, J.S., Takizawa, H., Shen, H., Park, J.J. (eds.) Parallel and Distributed Computing, Applications and Technologies. Proceedings of PDCAT 2023, Lecture Notes in Electrical Engineering, vol. 1112, pp. 25–31. Springer Nature Singapore, Singapore (2023)
4. Hasert, M., Masilamani, K., Zimny, S., Klimach, H., Qi, J., Bernsdorf, J., Roller, S.: Complex fluid simulations with the parallel tree-based Lattice Boltzmann solver Musubi. J. Comput. Sci. **5**(5), 784–794 (2014)
5. HLRS: HPE Apollo (Hawk). https://www.hlrs.de/solutions/systems/hpe-apollo-hawk. Last accessed 10 July 2025 15:19:59
6. Spinelli, G.G., Gericke, J., Masilamani, K., Klimach, H.: Key ingredients for wall-modeled LES with the Lattice Boltzmann method: systematic comparison of collision schemes, SGS models, and wall functions on simulation accuracy and efficiency for turbulent channel flow. (in press)

7. Spinelli, G.G., Horstmann, T., Masilamani, K., Soni, M.M., Klimach, H., Stück, A., Roller, S.: HPC performance study of different collision models using the Lattice Boltzmann solver musubi. Comput. Fluids **255**, 105,833 (2023)
8. Stahl, K.: Investigation of blunt trailing edge noise from an asymmetric airfoil under different aerodynamic loads. Berichte aus der Strömungstechnik. Ph.D. Thesis. Universität Siegen. Shaker Verlag, Düren (2023). https://doi.org/10.2370/9783844092851
9. Timmer, W., Van Rooij, R.P.J.O.M.: Summary of the Delft university wind turbine dedicated airfoils. J. Sol. Energy Eng. **125**(4), 488–496 (2003)
10. Wagner, S., Bareiß, R., Guidati, G.: Wind Turbine Noise. Springer, Berlin Heidelberg (2012)

Computational Aspects for DNS of Compressible Turbulent Boundary Layers on up to Full Machine Size

Tobias Gibis, Jason Appelbaum, and Christoph Wenzel

Abstract In this study, a representative benchmark case for the IAG in-house DNS code NS3D is presented, which has been developed specifically with the aim of evaluating code performance on Hawk and future machines on up to full machine size. The case is representative for simulations of compressible turbulent flat plate boundary layers, which are intended to provide both reference data and a better understanding of the physical processes within realistic boundary layers. Weak and strong scaling behaviour is shown up to 4096 nodes, demonstrating good scaling performance. Additionally, NS3D's newly developed internal performance monitoring tools are presented and the statistics of the computational performance of several important routines are analysed. The performance monitoring gives indications of potential problems during the simulation runtime, such as nodes that are too slow, and can be configured to abort the simulation in case of unacceptable performance. Finally, for the weak scaling cases, the scaling performance of the recently implemented parallel HDF5 I/O is shown.

1 Introduction

Over the past few decades, advances in computing have taken direct numerical simulations (DNS) of turbulent boundary layers from barely possible to sufficient scale. Larger scale DNS has in turn contributed greatly to a better understanding of the physical processes in realistic boundary layers in recent years, thus improving understanding and modelling for practical applications. Compared to their incompressible counterparts, progress in DNS of compressible turbulent boundary layers lags behind

---

T. Gibis · J. Appelbaum · C. Wenzel (✉)
Institute of Aerodynamics and Gas Dynamics, Stuttgart, Germany
e-mail: wenzel@iag.uni-stuttgart.de

T. Gibis
e-mail: gibis@iag.uni-stuttgart.de

J. Appelbaum
e-mail: appelbaum@iag.uni-stuttgart.de

T. Ludwig et al. (eds.), *High Performance Computing in Science and Engineering '23*,
https://doi.org/10.1007/978-3-031-91312-9_18

due to their increased complexity. Because of the additional coupling of the energy equation, these boundary layers have a larger parameter space, including effects due to wall temperature and effects of compressibility, leading to more difficult design of experiment. The ability to accurately extract flow quantities and to clearly define boundary conditions has improved the reference data dramatically—especially in the compressible regime—which has led to improvements in the fundamentals of the field.

1.1 Goals of This Study

Within the LAMTUR project, several studies on compressible turbulent boundary layers have been carried out, among others. In the past, these studies mainly have been focusing on data quality and scaling laws for adiabatic zero pressure gradient cases [31], adiabatic sub- and supersonic turbulent boundary cases with pressure gradients [8, 28] and additional heat transfer [29, 30]. Currently, the simulation work (on compressible turbulent boundary layers) mainly includes large-scale simulations to further investigate pressure gradient and heat transfer effects, and very large-scale simulations of a $M = 2.0$ turbulent boundary layer up to very high Reynolds numbers. For the largest cases, thousands of nodes are used simultaneously, which represents a large portion of the Hawk supercomputer. Parallel scaling to this extent is a major challenge for the simulations, not only because of the necessary ability of the code to scale to this level, but also due simply to performance differences between nodes, which can have a significant impact on the overall speed as well as stability of simulations at such scales. In order to overcome these issues, additional monitoring tools have been developed and implemented in NS3D to obtain real-time data on node performance and possible hardware failures in a non-intrusive way. In addition, online monitoring also allows jobs that are significantly under-performing to be killed.

This work is dedicated to describing in detail a realistic, representative case setup for a compressible turbulent boundary layer DNS that has been used—with some modifications—for all of the above mentioned work in the last few years. This representative case setup was recently developed in collaboration with HLRS as a very production-near benchmark case, providing a platform for optimising the solver code and investigating other aspects of performance. The results presented are intended to characterise NS3D as well as the Hawk system and thus are considered to be of interest to many users. In the following section, a brief introduction to turbulent compressible boundary layers is given to the interested reader to quickly outline the research topic under consideration. Afterwards, the representative test case is introduced, the scaling behaviour on Hawk up to 4096 nodes is shown, the monitoring results are presented, and some remarks on I/O are made.

1.2 Physical Introduction

Fluid flow is generally described by three coupled, partial differential equations for mass, momentum and energy transport. For flow conditions in which it can be reasonably assumed that mass density is constant, such as low Mach number flows, the equations can be significantly simplified, being referred to as 'incompressible'. In their full form, the set of equations is referred to as 'compressible'. Compressible turbulent flows have more complex behaviour compared to incompressible cases, where the momentum equation is usually decoupled from the energy equation. Most fundamentals of turbulence theory have been developed using the incompressible equations, and thus the basic strategy for understanding compressible cases is often rooted in relating them to their incompressible counterparts. For laminar boundary layers, several analytical transformations have been derived (under certain assumptions) that transform a compressible boundary layer solution into an incompressible one [32]. However, the multi-scale nature of turbulence prevents a similar direct transformation for turbulent flows. Nevertheless, selected semi-empirical relationships have been found which form the basis of our understanding of compressible turbulent boundary layers, although they are often assumed to be only strictly valid for cases without pressure gradients or at adiabatic wall conditions.

The first of these building blocks is the Strong Reynolds analogy [18]. This analogy is based on the similarity of the momentum and energy equations for a parallel flow (with $Pr = 1$ and zero pressure gradient) and relates fluctuating temperature and velocity fields to each other. For the temporal mean values, a relationship similar to the Walz relation is obtained. The second building block is Morkovin's hypothesis [18], which states that compressibility effects are essentially passive. This statement is based on the idea that the total shear stress within the boundary layer is identical between two comparable boundary layers with different Mach numbers. This results in a transformation for the turbulent stresses between different Mach numbers. The third building block is the van Driest transformation [5], which transforms the mean streamwise velocity and the friction coefficient c_f, see for example [32]. The transformation uses a mixing length model and a Walz relation to obtain a van Driest velocity relating compressible to incompressible flows.

While Morkovin's hypothesis has been found to be valid for a wide range of parameters, the Strong Reynolds analogy and the Van Driest transformation are known to be inaccurate under certain conditions. To name one example, the Strong Reynolds analogy gives the incorrect prediction of zero total enthalpy fluctuations for adiabatic walls [11]. Therefore, considerable research effort has been put into improving all mentioned building blocks, resulting in considerable improvements of e.g. the Strong Reynolds Analogy [7, 13, 33], and improved compressibility transformations [3, 10, 25, 27].

Recent work included within the LAMTUR project is primarily based on the above mentioned research questions. Thus, it is aimed at bridging the gap between 'simple' adiabatic turbulent boundary layers and more complex cases, such as those under the influence of pressure gradients and with wall heat transfer.

2 Test Case and Numerical Method

All cases presently considered are run using the in-house DNS code NS3D. The code solves the compressible Navier-Stokes equation for a calorically perfect gas with a constant $\gamma = c_p/c_v = 1.4$ and a constant Prandtl number of $Pr = 0.71$ on a block structured curvilinear grid. Sutherland's law is used to model the temperature dependence of viscosity, with a linear law below Sutherland's temperature. An explicit classical 4th order Runge-Kutta scheme is used for the time discretisation. For the spatial discretisation, a compact tridiagnonal subdomain 6th order scheme is used [14, 16]. To decouple the linear systems of the subdomains, the dependence of the compact ghost derivatives (derivative of the first ghost point) on the next derivative is replaced by an explicit 8th order scheme [14]. This allows all derivatives within each MPI subdomain to be computed by a compact scheme and sufficiently reduces the influence of the explicit scheme [14]. Other schemes such as explicit finite differences are also available within NS3D, but are typically used not used in the supersonic regime. For the convective derivatives, alternating forward and backward biased finite differences are used in each Runge-Kutta sub timestep, see [2, 16]. The viscous derivatives are treated in a Laplace formulation using a direct second derivative stencil when the derivative is not mixed. In addition, a compact 10th order filter is used to stabilise the simulation [26].

2.1 Case Setup and Boundary Conditions

Figure 1 shows a schematic of a representative case setup. The simulated case can be divided into an inflow region, a usable region (domain of interest) and an outflow region. Such a case setup is used for supersonic flow simulations with Mach number $M_\infty = 1.2$ to $M_\infty = 2.5$. In the inflow region, the artificially introduced turbulent

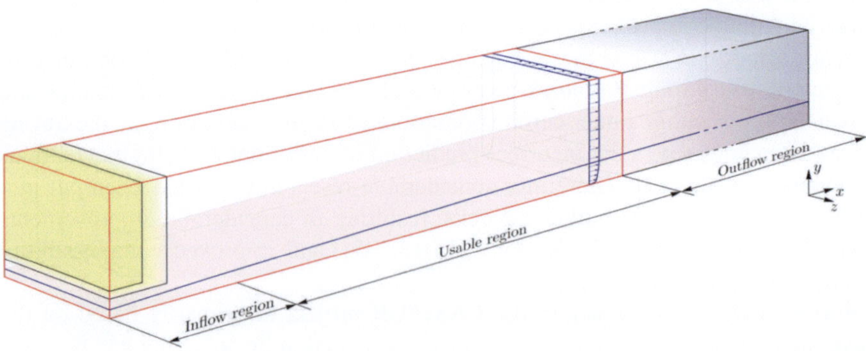

Fig. 1 Schematic of the case setup. The yellow region marks the location of the inflow sponge. The blue line shows the evolution of the boundary layer thickness within the domain

field at the inflow plane relaxes towards a physically realistic, nearly self-similar turbulent boundary layer. This region usually extends more than 100 inflow boundary layer thicknesses (usually 150 inflow boundary layer thicknesses are used) if very high data quality is to be achieved ($\lesssim$1% deviation from established references) [4]. For pressure gradient cases, this region is usually longer as the pressure gradients must be applied carefully to obtain a self-similar velocity profile. To reduce artificial noise and—for supersonic cases—to remove the shock caused by physical inaccuracies of the inflow condition, a numerical sponge is used to force the solution towards a prescribed baseflow, see [1]. Furthermore, the numerical grid is stretched slightly in the streamwise direction in the first few boundary layer thicknesses, which helps to reduce noise stemming from the digital filtering process and has no negative effect on transitional behaviour. For the usable region, the numerical grid must be well-resolved and large enough to allow for the physical evolution of the boundary layer. Usually, the domain width and grid resolution follow the guidelines established in [21]. Finally, the outflow region is responsible for reducing the potentially detrimental influence of the outflow boundary condition on the region of interest. To do so, a weak numerical sponge is applied and the grid is smoothly stretched in the streamwise direction throughout the outflow region.

The most important boundary conditions are briefly described below.

Inflow—Turbulent boundary condition: The turbulent inflow used in the present studies is generated by a digital filtering method. Details of the method can be found in [15, 23, 24, 31]. The boundary conditions use a Mersenne twister to generate pseudo-random numbers for the three velocity components (either the NS3D version, or generators from the AOCL or MKL math libraries), which are generated as normal distributions using the Box-Muller theorem. For the present test case, AOCL is used to generate the pseudo-random numbers. To generate the appropriate length scales, filters are used to obtain a turbulent structure. The tensor product property of convolution is used to reduce the computational cost. For the streamwise direction, a temporal correlation is used. The resulting fluctuation field is transformed using the Cholesky transformation to obtain the correct Reynolds stress distributions [17]. The strong Reynolds analogy is used to obtain the turbulent temperature field. In the case of a cooled wall, the generalised Reynolds analogy [33] is utilised, where the required dT/du is calculated using the improved Walz relation [33]. This helps to improve the stability of the boundary condition, as temperature and velocity fluctuations are not strictly anti-correlated as predicted by the original Strong Reynolds Analogy.

Alternative inlet approaches such as the rescaling recycling technique [17] are implemented but require a different setup than the one presently described.

Freestream—Characteristic boundary condition: The upper boundary condition is important for prescribing the pressure distribution in pressure gradient cases. A characteristic subsonic boundary condition is commonly used for supersonic cases [9, 22] whereas a fixed pressure gradient condition is typically used for subsonic cases [28]. For supersonic cases, a boundary condition based on spatial characteristics [12] is also implemented. Fixing pressure gradient here works well for favorable pressure gradients [20].

Wall—Adiabatic/isothermal wall boundary condition: The wall is treated as a no slip wall with all velocity components $u, v, w = 0$. The pressure is usually set by setting $dp/dn|_w = 0$, which is handled by an optimised 5th order stencil that sets the wall value assuming a zero derivative [16]. Note that for some setups involving transpiration, the option to use a 4th order extrapolation for the pressure is recommended as $dp/dn|_w \neq 0$. Wall temperature is either handled by an adiabatic condition $dT/dn|_w = 0$, realised analogously to the zero-gradient pressure condition, or simply by fixing the wall temperature.

Outflow condition—Constant d/dt and parabolic outflow: The outflow boundary is usually defined through one of two available conventions. The first is 'constant d/dt', which sets the time derivative at the outflow plane to the time derivative one point before, or 'parabolic outflow', which assumes a null second derivative at the boundary. For supersonic cases, the outflow is a common source of instability along with the inflow condition. To stabilise the outflow, a weak forcing term realised by the numerical sponge directly upstream of the outflow plane attenuates potential back-propagating disturbances.

3 Computational Aspects

Similar to [19, 20], multi-node scaling tests were performed using the realistic, representative test case with and without turbulent inflow boundary conditions to quantify the scaling behaviour of production cases. Hereafter, the case without the turbulent inflow is referred to as the 'simplified case' and the case with the turbulent inflow is referred to as the 'production case'. Boundary conditions are not computed on all MPI ranks and are thus a source of imbalance in the code. Accordingly, effort has been invested in recent years to minimise their impact on overall performance. In the following paragraphs, the strong and weak scaling characteristics of the test case will be evaluated (Tables 1 and 2).

The scaling tests start at a reference size of 64 nodes. The strong scaling tests are run up to 2048 nodes and the weak scaling tests are run up to 4096 nodes, the current maximum realisable job size on Hawk. This represents the actual range for current NS3D production cases; smaller cases are unusual. All cases considered were run using 32 MPI processes per node and 4 OpenMP threads per MPI process. Such allows for good OpenMP parallel efficiency and also minimises the impact of domain boundaries [19]. To evaluate the performance of the runs, the NS3D monitoring tools were used to obtain rank runtimes for each MPI rank, allowing the distribution of individual rank and node performance to be evaluated, see Sect. 3.1.

Table 1 Summary of numerical parameters for the strong scaling cases

Nodes	Grid points	Grid points per rank	MPI ranks
64	[9472 × 544 × 672]	[296 × 68 × 84]	[32 × 8 × 8]
128	[9472 × 544 × 672]	[296 × 34 × 84]	[32 × 16 × 8]
256	[9472 × 544 × 672]	[296 × 34 × 42]	[32 × 16 × 16]
512	[9472 × 544 × 672]	[148 × 34 × 42]	[64 × 16 × 16]
1024	[9472 × 544 × 672]	[74 × 34 × 42]	[128 × 16 × 16]
2048	[9472 × 544 × 672]	[37 × 34 × 42]	[256 × 16 × 16]

Table 2 Summary of numerical parameters for the weak scaling cases

Nodes	Grid points	Grid points per rank	MPI ranks
64	[2464 × 376 × 408]	[77 × 47 × 51]	[32 × 8 × 8]
128	[4928 × 376 × 408]	[77 × 47 × 51]	[32 × 8 × 8]
256	[4928 × 376 × 816]	[77 × 47 × 51]	[64 × 8 × 16]
512	[4928 × 376 × 1632]	[77 × 47 × 51]	[64 × 16 × 32]
1024	[9856 × 752 × 816]	[77 × 47 × 51]	[128 × 16 × 16]
2048	[9856 × 752 × 1632]	[77 × 47 × 51]	[128 × 16 × 32]
4096	[19712 × 752 × 1632]	[77 × 47 × 51]	[256 × 16 × 32]

Figure 2 shows the scaling characteristic of NS3D. In (a) the speedup for strong scaling is shown. The strong scaling shows a continuous superlinear speedup from 64 nodes to 2048 nodes, both with and without turbulent boundary conditions. The scaling efficiency for strong scaling is thus always well above 1. This is due to caching effects, as NS3D's performance is highly dependent on memory bandwidth. The order of magnitude of the scaling efficiency is similar to previous results in [19]. However, due to improved system stability in comparison to [19], fewer underperforming nodes were encountered, resulting in better results for large node counts. For example, during acquisition of the test benchmark data for 2048 and 4096 nodes, slow nodes were detected through a quick NS3D test run, after which they were offlined before running a full NS3D benchmark test.

Weak scaling was performed with 5906208 points per node, which is quite close to most production setups. Weak scaling is shown with and without turbulent boundary conditions, which does not affect the scaling. Since the cases were only run for about 800 s or 2700 iterations rather than a full 24 h, only the main loop timer is considered, illustrated by the bottom two red curves. To get a first impression of the I/O performance, the main loop timer without I/O is also plotted, illustrated by the top curves. It can be seen that for weak scaling, the performance drop is 8% for 4096 nodes (with I/O) and 2–3% without I/O, indicating that I/O is the main factor for the performance drop in this case. Note that the 4096 node performance was only determined by one measurement, but the rank specific timers suggest a nominal run

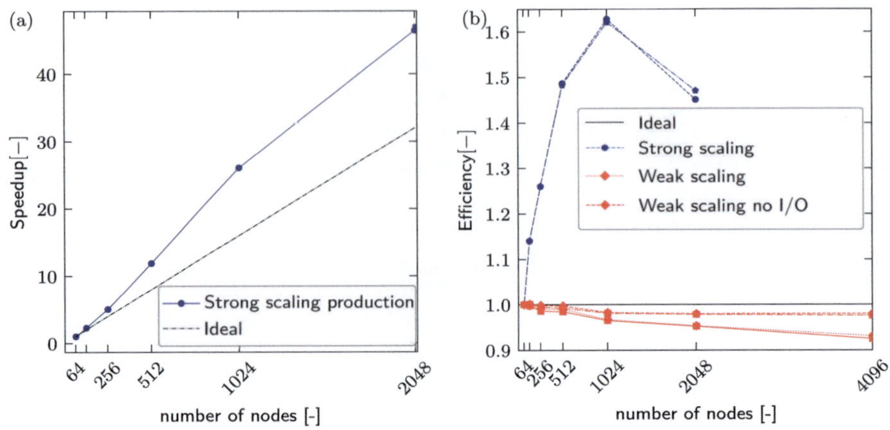

Fig. 2 Multinode performance **a** Strong scaling speedup for the production case (with turbulent boundary condition) and the simplified case (without turbulent inflow condition) **b** Weak and strong scaling efficiency for both the production case and the simplified case. For weak scaling, the efficiency is also given without I/O time

free of issues. The weak scaling makes it easy to evaluate the iteration performance, where it can be seen that the relative performance deficit for the given size is only 3–4%.

In summary, the NS3D has strong parallel performance as seen in previous reports and is fit to handle production cases spanning the full cluster size. The main issue in practice is the reliability of the network and rank performance on cluster-sized jobs with $O(10^6)$ ranks and the respective amount of cores.

3.1 Rank Statistics

The clusters that NS3D runs on are getting larger, and it is therefore crucial that the code has detailed timers to allow for the debugging of potential performance issues that may occur for large scale production jobs. For example, for the largest job size within the present study, the code runs on 4096 nodes with 131072 ranks, each running in a core complex with shared L3 cache. Hence, the large size of the job makes encountering underperforming hardware more probable.

NS3D takes timings of parts of the code using `MPI_Wtime`. In addition to global statistics, it is possible to write out the timings for each rank. This allows a quick analysis of the performance of the code on very large production jobs, although it cannot provide the in-depth analysis of specialised profiling tools. In order to have no measurable overhead, only parts with long runtimes and infrequent or one-off calls are timed, and the rank timings are only collected at certain intervals. For the present tests, the collection period was 2700 timesteps. In the following section, the timings of the computational phases with the exception of boundary conditions are analysed, as these are mostly identical on all ranks and do not contain any communication. The timers analysed include the computation of the spatial derivatives and the assembly of the time derivative, which is the routine with the highest runtime, as confirmed by multiple profiling tools.

As an example, the weakly scaled 1024 node case is analysed in more detail. Six runs were performed, and the basic stochastic information is given in Table 1. Of the runs performed, numbers 3, 5 and 6 could be considered to have normal performance, judging both by the total main loop time as well as the timer distribution shown in Fig. 3. Runs 1 and 2 are abnormal, while run 4 has a slight abnormality. Comparing the simple stochastic information, one sees somewhat counterintuitively that the normal runs have a higher mean and a higher standard deviation than the abnormal runs. Also, both parameters seem to be fairly consistent for the normal runs (Table 3).

In order to explain this somewhat counterintuitive behaviour, the histogram of the rankings is shown as a probability density function in Fig. 3. The histogram bins were chosen between the minimum and maximum values according to the Freedman-Diaconis rule [6]. A fitted normal distribution is also shown for reference. Again we can see three similar runs (which are the normal runs 3, 5 and 6), two runs (1 and 4) with similar distributions and one run (2) with a characteristically different distribution.

Starting with the normal runs, the distribution has a very broad peak but essentially no tail, resulting in a high standard deviation. It is obvious that the distribution is not well approximated by a normal distribution. Such makes finding problematic nodes difficult, as most simple outlier tests assume a normal distribution. The reason for this behaviour is that in the current setup, 4 ranks share a memory lane and thus interfere with each other. This slows down the timed routines when the 4 ranks often arrive at the same time. However, it also allows latecomers to catch up as they often do not have to share the memory bandwidth.

Runs 1 and 4 most likely had a problem with memory. Using 32 MPI ranks per Hawk node yields 4 ranks per RAM channel. Therefore, a single underperforming RAM channel (or L3 cache) presents itself through equally poor performance on 4 MPI ranks. Such is the case for both runs. For the rest of the ranks, a double peak structure follows, with a main peak at a much faster speed than in the other runs. The exact origin is difficult to identify with certainty, but a likely explanation is that the a slow rank forces its neighbours to wait for communication. While a receiving neighbour rank waits, the ranks that share resources with it have less competition for memory access and thus statistically overperform.

Lastly, run 2 has a distribution with very wide tails, even though the standard deviation and the maximum and minimum computation times are significantly lower than the other runs. This run most likely suffered from network problems, as the subsequent test (without the turbulent boundary condition) within the run experienced an MPI crash. From experience, many simulations that suffer from this problem exhibit a degradation in performance prior to the actual crash.

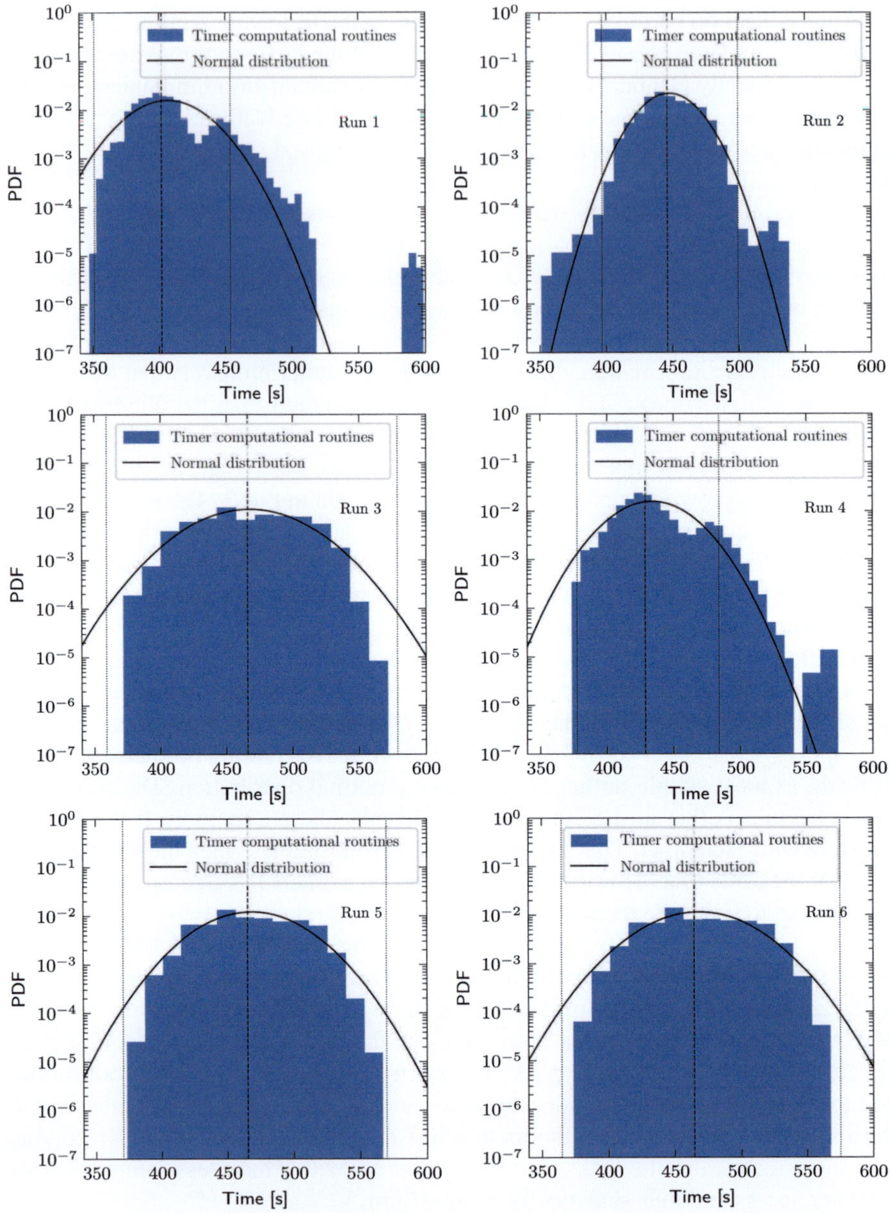

Fig. 3 The timer of the main computational routines (which do not include communication time) for the 1024 node runs, plotted as a probability density function (PDF). The normal distribution fitted to the data is shown for reference. The dashed line marks the mean. The dotted lines mark 1.5 times the interquartile range (IQR)

Table 3 Stochastic properties of the weak scaling case with 1024 nodes. Shown are the total times for the main loop, the mean time of the main computational (communication free) routines (spatial derivatives, assembly of the time derivative), the standard deviation of the computational routine times, and the minimum and maximum times.

Run	Main loop (s)	Mean (s)	Std deviation (s)	Min (s)	Max (s)
1	780.44	406.79	24.87	352.08	594.14
2	825.80	447.44	18.00	358.11	533.74
3	754.73	467.43	35.50	377.90	563.44
4	767.29	433.56	25.24	374.57	572.34
5	755.21	468.27	32.37	382.04	558.98
6	754.10	468.13	34.26	374.71	560.86

3.2 Parallel HDF5 I/O

In the last report [20], the new HDF5 based file I/O was introduced. This allows the number of files to be drastically reduced for large jobs. For a 64 node case, the previous report gave recommendations for ws10. In the present report, I/O rates have been gathered from the weak scaling tests which have been performed on the new ws11.

In general, the recommendations from the last report, namely to use ROMIO with `ROMIO_FSTYPE_FORCE="ufs:"`, a Lustre stripe size of 16MB and a stripe count of 16 are still recommended. For the larger cases (>512 nodes), the stripe count should be increased to 32, which was used for large cases within the present study. The use of the default file system driver in ROMIO is not recommended on Hawk, as it was also found to yield only a fraction of the I/O bandwidth on the workspace 'ws11', as well as on 'ws10'.

In the last year, minor optimisations have been implemented mostly focused on scaling the write operations for large scale jobs. The I/O bandwidth scaling for the weak scaling tests is shown in Fig. 4a. Note that this bandwidth includes everything from the moment the I/O routines are called; the actual write speed the file system sees is higher (see the red curve). During runtime, two files are written: a restart file and a single precision output file.

For the present study, the I/O rates start at an average of 5.4GB/s. The main reason for this is the small overall I/O size, resulting in only ~2s of I/O time. This is generally in line with the results obtained in [20]. For the larger cases, the I/O bandwidth increases. However, for cases larger than 1024 nodes, the I/O scaling reaches a plateau. A major reason for this is the long time it takes to create and open the files (see the 'Create + open' part of the total write time in blue). Nevertheless, even on 4096 nodes, the I/O time for the restart and output files is not particularly critical for production cases. In absolute numbers, the restart with about 1.1TB is written in 30s, which would typically result in ~5–10min of I/O for 24 h runs, a comparatively small fraction of overall wall time.

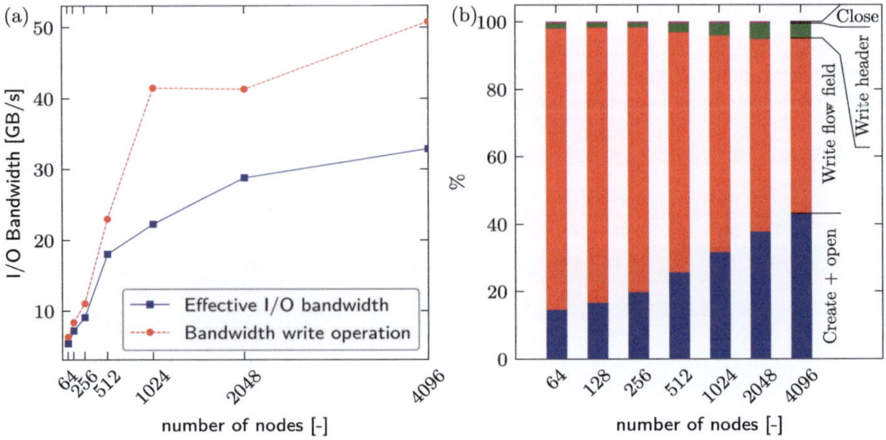

Fig. 4 **a** I/O bandwidth for the weak scaling cases. The effective bandwidth is the bandwidth from the code perspective (including opening and internal data preparation) and the write bandwidth is closer to what the file system actually sees. **b** The I/O time split into its components (create & open, write flow field, write header, close)

In summary, the use of `ROMIO_FSTYPE_FORCE="ufs:"` on both workspaces is highly recommended on Hawk. For very large cases, the I/O performance without `ROMIO_FSTYPE_FORCE="ufs:"` will result in unacceptably poor I/O performance. Secondly, the NS3D I/O scheme works very well for large case files. There is still room for improvement for small files output from large cases, for example writing only small subsets of the flow field. Some optimisations could be made here in the future, as the settings chosen currently favour larger data size per rank. Possible improvements include using features of newer HDF5 versions, such as asynchronous I/O, writing multiple datasets at once, or taking a closer look at the MPI I/O settings, as the data size per node is quite small in NS3D production cases on Hawk.

4 Summary and Conclusion

A large part of the computing resources allocated to LAMTUR have been used for fundamental research on compressible turbulent boundary layers. Greater resource availability and continual code improvements have made larger production cases run with the in-house code NS3D practically possible, allowing the pursuit of further state-of-the-art research topics. For example, the number of grid points used in the case setup from [31] was 400 million, whereas production cases are being run as of today with 24 billion grid points, a factor of 60 increase. The case setups are instruments of different research goals, nevertheless the comparison illustrates the successful application of NS3D to new, more computationally demanding problems with a constant focus on performance and reliability at scale.

From an HPC perspective, this dramatic increase in numerical grid points used in a single simulation has led to different challenges over the years. A few years ago, the biggest challenges were pure code efficiency, leading to code refactoring and repeated optimisation of the then-new turbulent inflow conditions. With HPC clusters continually growing in capacity, and with it the possible case size, the primary challenge is currently to ensure reliable and fast simulation execution even for large scale simulations where minor node or network problems are inevitable purely due to the large number of HPC hardware components involved.

The setup utilised in the present study is representative of the what is currently being used to conduct research on compressible turbulent boundary layers. Therefore, the coarse dimensions (leading to a certain network topology), boundary conditions and computational settings are comparable to real production cases, but designed in such a way that weak and strong scaling analysis is easily possible. The current setup thus makes it possible to quickly generate realistic cases for all sizes on Hawk and potentially also on Exascale computers, which has motivated its creation.

In addition to the scaling analysis for Hawk, the study includes the results of NS3D's internal performance monitoring. This gives insight into individual node performance and highlights potential problems with the system used. Finally, the scaling of I/O performance is evaluated, showing good behaviour up to 1024 nodes and stagnating thereafter. Favourable strong scaling behaviour makes it advantageous to use a large number of nodes, reducing data size per rank. Since collective I/O usually performs more efficiently when the component MPI ranks write large data sizes, the I/O bandwidth does not achieve the theoretical bandwidth of the Lustre file system, about 200GB/s. Nevertheless, for most realistic jobs, total I/O time as a fraction of computational time is small. In summary, a meaningful benchmark case has been formulated and discussed for NS3D, which can serve as a basis for future collaboration between HLRS and IAG for future machines.

Acknowledgements The simulations were performed on the national supercomputer HPE Apollo '*Hawk*' at the High Performance Computing Center Stuttgart (HLRS) under grant GCS-Lamt.

References

1. Appelbaum, J., Ohno, D., Rist, U., Wenzel, C.: DNS of a turbulent boundary layer using inflow conditions derived from 4D-PTV data. Exp. Fluids **62**, 1–21 (2021). https://doi.org/10.1007/s00348-021-03287-4
2. Babucke, A.: Direct Numerical Simulation of Noise-Generation Mechanisms in the Mixing Layer of a Jet. Verlag Dr. Hut (2009)
3. Bai, T., Griffin, K.P., Fu, L.: Compressible velocity transformations for various noncanonical wall-bounded turbulent flows. AIAA J. **60**(7), 4325–4337 (2022). https://doi.org/10.2514/1.J061554
4. Ceci, A., Palumbo, A., Larsson, J., Pirozzoli, S.: Numerical tripping of high-speed turbulent boundary layers. Theoret. Comput. Fluid Dyn. **36**(6), 865–886 (2022). https://doi.org/10.1007/s00162-022-00623-0

5. Driest, E.V.: Turbulent boundary layer in compressible fluids. J. Aeronaut. Sci. **18**(3), 145–160 (1951)
6. Freedman, D., Diaconis, P.: On the histogram as a density estimator:L2 theory. Zeitschrift für Wahrscheinlichkeitstheorie und Verwandte Gebiete **57**(4), 453–476 (1981). https://doi.org/10.1007/BF01025868
7. Gaviglio, J.: Reynolds analogies and experimental study of heat transfer in the supersonic boundary layer. Int. J. Heat Mass Transf. **30**(5), 911–926 (1987). https://doi.org/10.1016/0017-9310(87)90010-X
8. Gibis, T., Wenzel, C., Kloker, M., Rist, U.: Self-similar compressible turbulent boundary layers with pressure gradients. Part 2. Self-similarity analysis of the outer layer. J. Fluid Mech. **880**, 284–325 (2019). https://doi.org/10.1017/jfm.2019.672
9. Giles, M.B.: Nonreflecting boundary conditions for Euler equation calculations. AIAA J. **28**(12), 2050–2058 (1990). https://doi.org/10.2514/3.10521
10. Griffin, K.P., Fu, L., Moin, P.: Velocity transformation for compressible wall-bounded turbulent flows with and without heat transfer. Proc. Natl. Acad. Sci. **118**(34), e2111144118 (2021). https://doi.org/10.1073/pnas.2111144118
11. Guarini, S.E., Moser, R.D., Shariff, K., Wray, A.: Direct numerical simulation of a supersonic turbulent boundary layer at Mach 2.5. J. Fluid Mech. **414**, 1–33 (2000)
12. Harris, P., Fasel, H.: Numerical investigation of unsteady plane wakes at supersonic speeds. In: 34th Aerospace Sciences Meeting and Exhibit, p. 686 (1996)
13. Huang, P.G., Coleman, G.N., Bradshaw, P.: Compressible turbulent channel flows: DNS results and modelling. J. Fluid Mech. **305**, 185–218 (1995). https://doi.org/10.1017/S0022112095004599
14. Keller, M.A., Kloker, M.J.: Direct numerical simulation of foreign-gas film cooling in supersonic boundary-layer flow. AIAA J. **55**(1), 99–111 (2016)
15. Klein, M., Sadiki, A., Janicka, J.: A digital filter based generation of inflow data for spatially developing direct numerical or large eddy simulations. J. Comput. Phys. **186**(2), 652–665 (2003). https://doi.org/10.1016/S0021-9991(03)00090-1
16. Kloker, M.J.: A robust high-resolution split-type compact FD scheme for spatial direct numerical simulation of boundary-layer transition. Appl. Sci. Res. **59**(4), 25 (1997). https://doi.org/10.1023/A:1001122829539
17. Lund, T.S., Wu, X., Squires, K.D.: Generation of turbulent inflow data for spatially-developing boundary layer simulations. J. Comput. Phys. **140**(2), 233–258 (1998). https://doi.org/10.1006/jcph.1998.5882
18. Morkovin, M.V.: Effects of compressibility on turbulent flows. Mécanique de la Turbul. **367**, 380 (1962)
19. Ohno, D., Selent, B., Kloker, M.J., Rist, U.: Direct numerical simulation of bypass transition under free-stream turbulence for compressible flows. In: Nagel, W.E., Kröner, D.H., Resch, M.M. (eds.) High Performance Computing in Science and Engineering '21, pp. 223–238. Springer International Publishing, Cham (2023). https://doi.org/10.1007/978-3-031-17937-2_13
20. Peter, J., Gibis, T., Kloker, M.J.: Favorable-pressure-gradient influence on supersonic film cooling with turbulent main flow. In: Nagel, W.E., Kröner, D.B., Resch, M.M. (eds.) High Performance Computing in Science and Engineering '22 (to Be Published). Springer Cham (2023)
21. Poggie, J., Bisek, N.J., Gosse, R.: Resolution effects in compressible, turbulent boundary layer simulations. Comput. Fluids **120**, 57–69 (2015). https://doi.org/10.1016/j.compfluid.2015.07.015
22. Poinsot, T., Lele, S.: Boundary conditions for direct simulations of compressible viscous flows. J. Comput. Phys. **101**(1), 104–129 (1992). https://doi.org/10.1016/0021-9991(92)90046-2
23. Touber, E.: Unsteadiness in shock-wave/boundary layer interactions. Ph.D. thesis, University of Southampton (2010)
24. Touber, E., Sandham, N.D.: Large-eddy simulation of low-frequency unsteadiness in a turbulent shock-induced separation bubble. Theoret. Comput. Fluid Dyn. **23**(2), 79–107 (2009). https://doi.org/10.1007/s00162-009-0103-z

25. Trettel, A., Larsson, J.: Mean velocity scaling for compressible wall turbulence with heat transfer. Phys. Fluids **28**(2), 026102 (2016)
26. Visbal, M.R., Gaitonde, D.V.: On the use of higher-order finite-difference schemes on curvilinear and deforming meshes. J. Comput. Phys. **181**(1), 155–185 (2002)
27. Volpiani, P.S., Iyer, P.S., Pirozzoli, S., Larsson, J.: Data-driven compressibility transformation for turbulent wall layers. Phys. Rev. Fluids **5**(5), 052602 (2020). https://doi.org/10.1103/PhysRevFluids.5.052602
28. Wenzel, C.: DNS of compressible turbulent boundary layers: pressure-gradient influence and self-similarity. University of Stuttgart, Disseration (2019)
29. Wenzel, C., Gibis, T., Kloker, M.: About the influences of compressibility, heat transfer and pressure gradients in compressible turbulent boundary layers. J. Fluid Mech. **930**, A1 (2022). https://doi.org/10.1017/jfm.2021.888
30. Wenzel, C., Gibis, T., Kloker, M., Rist, U.: Reynolds analogy factor in self-similar compressible turbulent boundary layers with pressure gradients. J. Fluid Mech. **907** (2021). https://doi.org/10.1017/jfm.2020.876
31. Wenzel, C., Selent, B., Kloker, M., Rist, U.: DNS of compressible turbulent boundary layers and assessment of data/scaling-law quality. J. Fluid Mech. **842**, 428–468 (2018)
32. White, F.M.: Viscous Fluid Flow. McGraw-Hill Education (2005)
33. Zhang, Y.S., Bi, W.T., Hussain, F., She, Z.S.: A generalized Reynolds analogy for compressible wall-bounded turbulent flows. J. Fluid Mech. **739**, 392–420 (2014). https://doi.org/10.1017/jfm.2013.620

HPC-Simulations of Unsteady Turbomachinery Flows

Simon Hummel, Nikola Kovachev, Christopher Fuhrer, and Damian M. Vogt

Abstract Two turbomachinery test cases are investigated numerically within the scope of this paper. In the first test case, the effect of different operating points on the aerodynamic excitation of the wheel is investigated in three similar vaneless turbocharger turbines. Therefore, many URANS simulations at different operating conditions covering the entire operating map have been performed on high-fidelity models. The analysis of the evaluated excitation maps suggests similar forcing mechanisms. In the second test case, the complex flow field at partload conditions of an axial gas turbine diffuser with two different struts are investigated. The flow phenomena such as smooth wall separation and recirculation for example are predicted performing URANS and SBES simulations. By applying SBES as a scale resolving model, the agreement of the diffuser performance and flow characteristics with experimental data are improved depending on the operating point and strut geometry compared to an URANS approach.

1 Introduction

Turbomachines are a key technology for the transition to a greener power generation and transportation systems. Due to their wide application, a reduction of overall engine weight and size, reduction of fuel consumption, increasing efficiency, increased reliability, reduction of manufacturing cost and inspection intervals can

S. Hummel · N. Kovachev · C. Fuhrer (✉) · D. M. Vogt
ITSM, Stuttgart, Germany
e-mail: christopher.fuhrer@itsm.uni-stuttgart.de

S. Hummel
e-mail: simon.hummel@itsm.uni-stuttgart.de

N. Kovachev
e-mail: nikola.kovachev@itsm.uni-stuttgart.de

D. M. Vogt
e-mail: damian.vogt@itsm.uni-stuttgart.de

T. Ludwig et al. (eds.), *High Performance Computing in Science and Engineering '23*,
https://doi.org/10.1007/978-3-031-91312-9_19

make a large impact on reaching the climate goals faster. In order to meet these requirements, a great scientific effort on improving the numerical prediction has been deployed, both in academia and in industry, such that production and development time and costs can be further reduced. Especially, stationary machines for power generation have to deal with increased part-load operation, since the share of fluctuating renewable energy is ever increasing.

The investigations in this paper are part of the research effort at the Institute of Thermal Turbomachinery and Machine Laboratory (ITSM) and aim to contribute to two highly relevant areas in the field of turbomachinery—aeromechanics and diffuser flow modeling. The following chapters present the motivation for each study, the numerical approach and the currently achieved results. Subsequently, the computational aspects regarding the usage of HPC ressources are discussed.

2 Aeromechanics

High cycle fatigue failure of blades is a major issue when designing turbomachines. A root cause can be found in forced response blade vibrations induced by the inherently unsteady flow field. In case of a resonance, the vibration amplitudes can reach excessive values if not sufficiently damped and lead to a stand-still or even to a total machine failure. Therefore, a large effort is put already in the design phase to deal with the phenomena. Forced response is usually numerically evaluated at a small number of operating conditions for new designs due to the large computational costs required. However, in the case of machines with variable speed critical operating conditions might not be considered, which can result in severe machine failures.

In the current study, the research intent is to evaluate the aerodynamic excitation inducing the forced response vibrations in the entire operating map of a turbocharger radial turbine for different resonance crossings. Based on the results, a thorough analysis of the local and the global excitation is conducted in order to find approaches for simplified prediction of the operating point influence. The findings are going to be validated with performance and vibration test data and with investigations on two further turbocharger turbines of a similar scale. In the following, the numerical model and the investigation methodology are presented. Moreover, results of the predicted excitation maps of the three test objects as well as of the analysis of the local excitation are presented.

2.1 Investigation Methodology

Vaneless turbocharger turbines for heavy duty truck applications of similar sizes are used as test cases. A large number of unsteady 3D CFD simulations at different operating conditions spanning over the entire operating map are conducted and evaluated regarding the aerodynamic excitation. This is quantified by means of the generalized force criterion, Eq. (1), yielding the scalar product of the harmonic pressure and blade mode shape corresponding to the investigated resonance crossing.

$$F_{gen} = \int p_{harm} \cdot \Phi d A \tag{1}$$

Multiple resonance crossings of the first structural mode (M1) with the low Engine Orders EO5-EO7 are considered in order to assess the different impact of the operating point on each of them. A similar approach was utilized in an investigation of a vaned radial turbine for an excitation by the blade-passing frequency by Kovachev et al. [1]. A thorough analysis on the harmonic flow field at each predicted operating point is conducted in order to find correlations with parameters, accessible through low-fidelity approaches. Hereby, the aerodynamic excitation is evaluated in a statistical manner to account for blade-to-blade and temporal fluctuations of the generalized force. The harmonic pressure amplitudes evaluated for all blades from the last revolution are aggregated to reference blade coordinates and its statistical measures like a mean value are calculated at each nodal location. The findings are transferred to the other two test objects and validated with a high-fidelity prediction in order to verify their general validity.

2.2 Numerical Model

The numerical model of each turbine consists of the turbine's volute, rotor and diffuser as well as the outlet piping up to the measurement location on the test bench. The volute is meshed by tetrahedral elements for the main flow field and prism layers at the boundaries. All other components feature hexahedral elements and are meshed in a block-structured manner. Besides, the rotor mesh features a high level of detailing by comprising the non-constant tip gap, the blade fillet, the rotor bulb and the back-disc cavity. A final mesh containing 18 Mio. cells and a time step size corresponding to 540 time steps per revolution are selected based on the mesh and time step sensitivity studies, presented in [5]. Furthermore, an RMS criterion of 10^{-5} and a limit for the number of inner loops of 10 has proven to deliver accurate results and was therefore adopted in the model.

The numerical calculation is performed with the commercial solver Ansys CFX R19.5. The URANS model closure is achieved by using the Shear-Stress Turbulence (SST) model. The numerical model is divided into two domains at a relative motion—the volute and the rotor—connected by a transient sliding mesh interface. Each unsteady simulation is initialized from a preceding RANS simulation. The boundary conditions at the inlet feature total temperature and total pressure according to the prescribed operating point as well as assumed turbulence intensity of 5% and a turbulent viscosity ratio of 10. An averaged static pressure is defined at the outlet. The walls are assumed to be adiabatic and smooth with an exception of the volute walls, which are defined with a roughness according to data from the manufacturer. The fluid is modeled as air with ideal gas properties.

A total of 6 revolutions is calculated at each operating point based on the results of the preliminary studies. During the simulation, the surface pressure of each blade is exported in csv-format every 10th time step. These is then used in an offline code to assess the aerodynamic excitation and the convergence of the simulation. In this case, a single simulation takes about 150 h on 3 nodes with 128 cores on the HPE Apollo Hawk and produces approximately 20 GB of data.

2.3 Results

A total number of 207 CFD simulations have been conducted and evaluated on the three test objects. Hereby, the operating map of the reference turbine is covered by 87 operating points and of the other two turbines—by 60 operating points with identical parameters.

The excitation maps for three turbines at the resonance crossing of M1EO5 are depicted in Fig. 1. The generalized force is plotted as a function of the operating parameter in terms of Expansion Ratio (ER) and reduced speed. The sample data and its standard deviation from a statistical evaluation of the excitation are included. A smoothing surface is fitted on to the data to approximate the continuous distribution of the generalized force in the operating map.

The significant influence of the operating conditions on the generalized force for the turbine 1 is evident. The ER seems to have a dominant role and higher values for this parameter lead in nearly all cases to an increase of the excitation. On the other hand, the reduced speed changes the generalized force in a non-linear manner. The same relations are observed for the other two turbines as well, indicating similar forcing mechanisms. The locations of maximum excitation changes though. The shape of the excitation maps is further substantially different. This has to be to a large extent an effect of the local excitation, which is currently under investigation.

The local excitation at diverse operating points is analyzed. So far, it has been found out that similarities among certain operating conditions exists. This is illustrated in Fig. 2 on the example of the blade loading at EO5 at mid-span. The pressure coefficient, yielding the harmonic pressure amplitude normalized by the inlet dynamic pressure, and the pressure phase for three operating points are depicted. Both parameters are obviously similar to a large extent. Differences are observed at the trailing edge (TE) on the suction side (SS) of the blade, where a shock is present. Based on this observation, the excitation at the three operating points is expected to scale nearly linear with inlet pressure with a deviation dependent on the impact of the transsonic effect. This is resembled in the excitation maps shown above, in which a linear increase of the generalized force is found for some regions. Further analysis of the local excitation is expected to contribute to a better understanding of the relations between the global excitation at different operating conditions. The

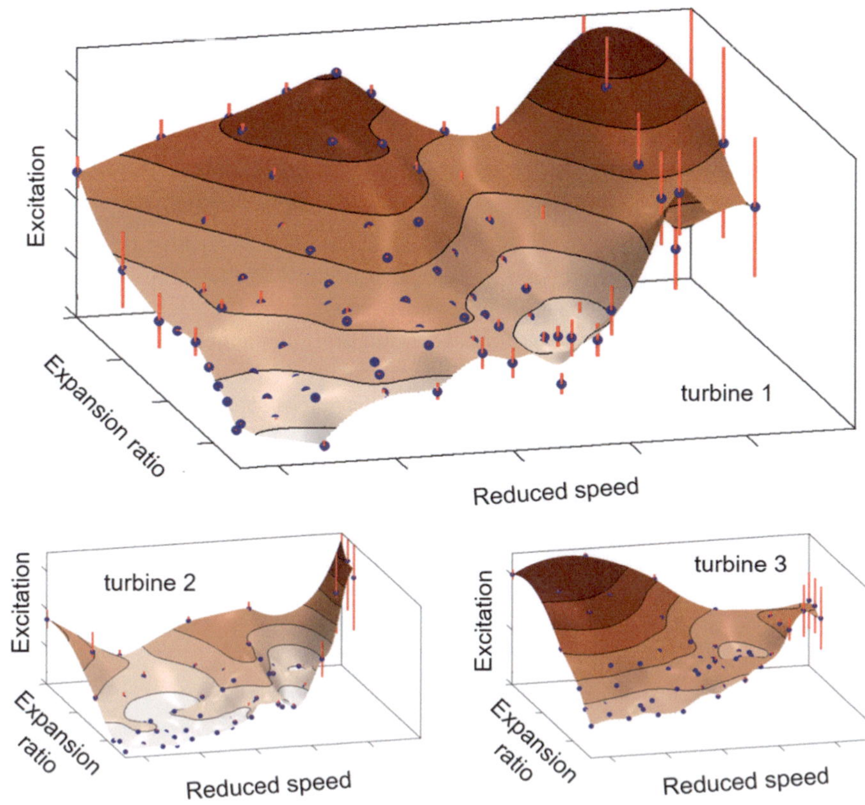

Fig. 1 Excitation maps of the M1EO5 resonance crossing

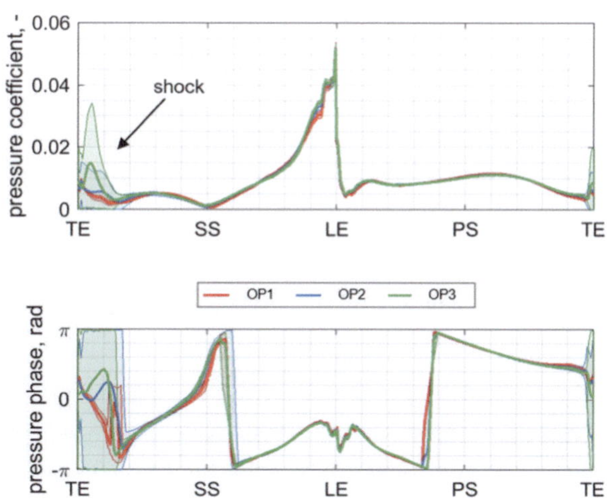

Fig. 2 Local excitation at similar operating points at mid-span for the reference turbine and EO5

validation with the predicted data for the other two turbines is intended to confirm an universality of the findings for different geometries.

3 Diffuser Flow

Nowadays, the investigation of off-design conditions of power-generating gas turbines is of particular importance due to the transformation of the energy sector. The operation of the investigated axial diffuser at off-design conditions leads to complex flow phenomena such as flow separation, recirculation, reattachment and free shear layer mixing. Preliminary studies [2, 3] based on the RANS approach in steady-state and unsteady mode demonstrate that the prediction of separation and mixing of the flow is very challenging with the used turbulence model and show that methods resolving turbulence, as presented in [5], are needed. In this work, the Stress-Blended Eddy Simulation (SBES) model is used as a hybrid RANS-LES model. Two representative partload operating points for an heavy duty gas turbine and two diffuser supporting strut geometries are investigated. The impact using a scale-resolving model compared to the unsteady RANS (URANS) approach is shown and numerical results are compared with experimental data obtained with the ITSM in-house diffuser test rig.

3.1 Numerical Method

Numerical simulations of the axial diffuser flow are performed using the commercial solver Ansys CFX R19.5. For the spatial discretization, the bounded second order central differencing scheme [4] in the SBES simulations and the high resolution scheme in the URANS simulations are used. The second order backward Euler scheme is applied for the temporal discretization of both models. For the SBES and URANS model, the spatial discretization of the convection terms of the turbulence model is carried out with the high resolution scheme.

The SBES model is a hybrid RANS-LES model developed by Menter. The concept of the model is based on the application of a blending or shielding function f_{SBES} to automatically switch between a RANS and LES model. For eddy viscosity based models, the blending of the SBES model eddy viscosity is defined as

$$\nu_{\mathrm{t}}^{\mathrm{SBES}} = f_{\mathrm{SBES}} \nu_{\mathrm{t}}^{\mathrm{RANS}} + (1 - f_{\mathrm{SBES}})\, \nu_{\mathrm{SGS}}^{\mathrm{LES}}. \tag{2}$$

In this study, the Shear Stress Transport (SST) turbulence model is used for the RANS part $\nu_{\mathrm{t}}^{\mathrm{RANS}}$ and for the LES subgrid-scale eddy viscosity $\nu_{\mathrm{SGS}}^{\mathrm{LES}}$ the Wall-Adapting Local Eddy-viscosity (WALE) model is applied. The turbulence in the URANS simulations is completely modelled with the SST model.

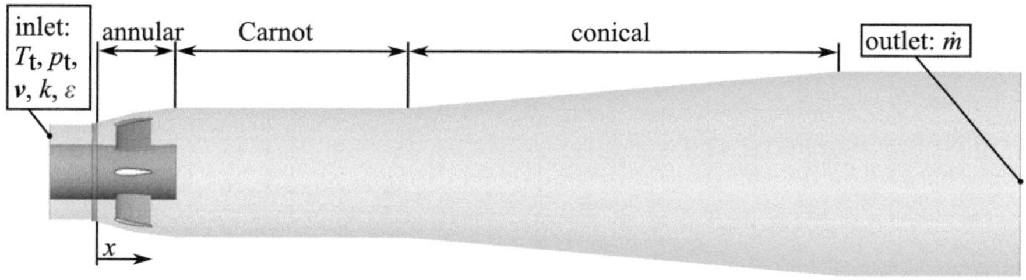

Fig. 3 Computational model of the diffuser

3.2 *Computational Model*

The computational model of the investigated axial diffuser is shown in Fig. 3. The diffuser geometry can be divided in three sections: an annular section including six supporting struts, a Carnot and conical section. Additionally, a tip leakage flow at $x = 0$ is modelled. At the model inlet, total temperature T_t, total pressure p_t, velocity direction $\mathbf{v}$ and the turbulence quantities k and ϵ are defined, which result from a precursor simulation of a swirler domain with experimental inlet boundary condition derived from representative turbine outflow conditions. As an outlet boundary condition of the diffuser, the measured massflow $\dot{m}$ is used. The mesh used for the URANS simulations consists of 36 millions hexahedral elements and is increased to 75 millions elements for the SBES simulation. To resolve the boundary layer, the first cells have an averaged dimensionless wall distance of $\bar{y}^+ < 1$. However, the mesh resolution in streamwise and spanwise direction is approximately one order above to fulfill the requirements of a wall-resolved LES. In comparison to the URANS model, the time step of the SBES model is reduced to $\Delta t_{\text{SBES}} = 1/40 \Delta t_{\text{URANS}} = 2.5 \cdot 10^{-6}$s ensuring the CFL number is less than unity in the regions of interest. At least two flow-through times are run before time-averaging is started to ensure the solution has reached a statistically steady state. Several flow-through times are needed to obtain statistically meaningful mean and turbulence properties, which are monitored during the simulation.

3.3 *Results*

The two investigated off-design operating points are representative for the operation in partload of a heavy duty gas turbine. Figure 4 shows the computed inlet conditions defined with the inlet section of the diffuser test rig at ITSM. The Mach number is reduced from $Ma = 0.5$ to $Ma = 0.3$ at midspan between the first partload operating point (PL1) and the second (PL2), whereas the swirl angle is approximately doubled from 30° to 60°. The total pressure increases towards the casing at both partload conditions. Flow separations at the supporting struts or at the hub in the annular dif-

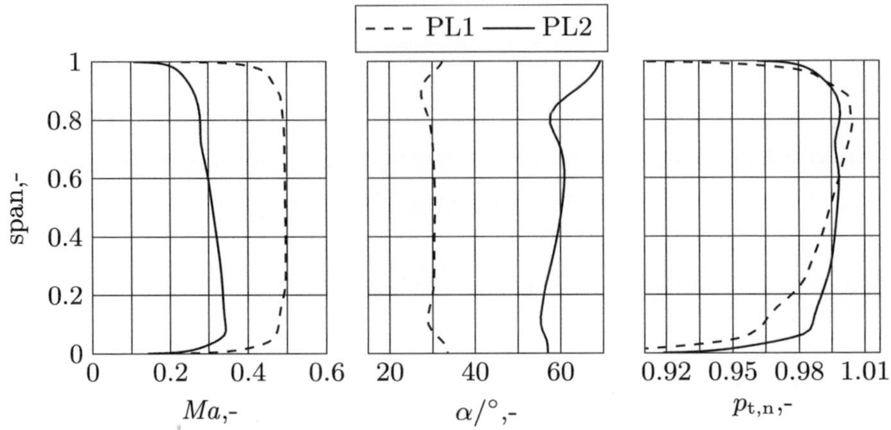

Fig. 4 Radial distribution of circumferentially averaged Mach number *Ma*, swirl angle α and total pressure $p_{t,n}$ at the diffuser inlet

fuser are mainly caused by a high inlet swirl. The resulting complex flow phenomena have a significant influence on the diffuser performance characteristics. The pressure recovery coefficient along the diffuser length evaluated with the casing wall pressure at both partload conditions for strut1 is illustrated in Fig. 5. The experimental data were obtained with the ITSM in-house diffuser test rig. The influence of the inlet swirl on the diffuser performance can be clearly identified comparing PL1 and PL2. With increased inlet swirl at PL2, the flow in the near wall region at the hub is weakened and separates before entering in the diffuser at $l_n = 0$, which results in a pressure drop. In the annular diffuser, hub separation is superimposed by strut separation and the pressure recovery further decreases. A large scale strut separation also occurs at PL1 causing a reduced pressure in the strut region. Downstream of the blunt hub end, the hub wake differs between PL1 and PL2. The hub wake at PL2 extends over almost the entire length and closes shortly before the diffuser outlet compared to a hub wake closure in the Carnot section at PL1. This leads to the trend of C_p downstream of the hub. The maximum pressure recovery at PL2 is already achieved at the end of the Carnot section, whereas at PL1, recovering of the residual kinetic energy also exists in the conical diffuser section. Due to the resolution of turbulent structures of high kinetic energy in the SBES model, the pressure distribution along the casing at both operating points is in very good agreement with the experiment, especially downstream of the hub end. From a global point of view, it can be stated that for PL2, the pressure recovery at the diffuser outlet is already very well predicted with a URANS simulation, whereas for PL1 the pressure rise is underestimated. At the diffuser inlet, it is challenging for both models to predict the pressure increase when the main flow and the tip flow are mixed and the cross section is widened. However, the advantage of the SBES in computing the hub separation at PL2 is recognizable.

The resulting distribution of the pressure recovery by changing the strut geometry is given in Fig. 6. Leading and trailing edge of strut2 are inclined in flow direction

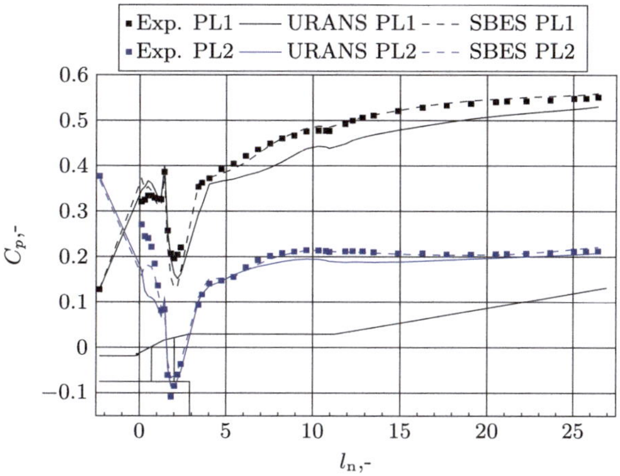

Fig. 5 Pressure recovery coefficient along the casing wall for experimental data and numerical results of URANS and SBES for strut1

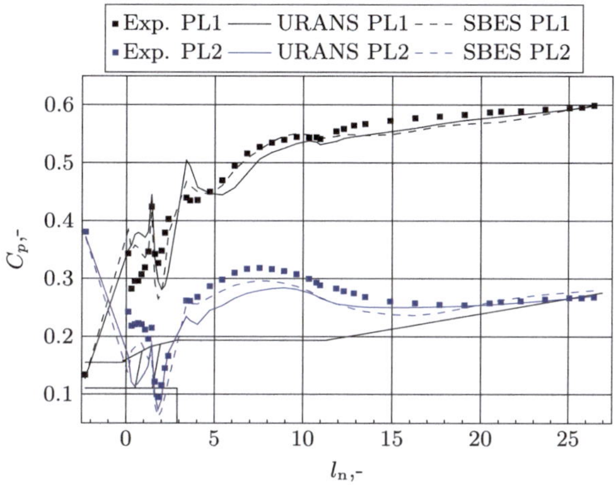

Fig. 6 Pressure recovery coefficient along the casing wall for experimental data and numerical results of URANS and SBES for strut2

with the same angle. Compared to strut1, the length of strut2 is shortened and the maximum thickness of the NACA profile is increased to keep the cross sectional area constant. The pressure recovery is increased at both operating points applying strut2 (see Figs. 5 and 6). The strut separation is reduced by the new strut design resulting in a lower pressure drop in the strut section. At the diffuser inlet, the redistribution of mass flow towards the casing by sweeping the strut is better predicted with the SBES at PL1 and PL2, but there are still differences compared to the experimental data. The axial position of the maximum pressure recovery downstream of the hub end at PL2 agrees well between experiment and SBES. Further downstream, there are slight deviations between experiment, URANS and SBES. At PL1, the pressure enhancement due to the closing hub wake is more accurately predicted by the SBES model. In the conical diffuser, the trend is similar for URANS and SBES.

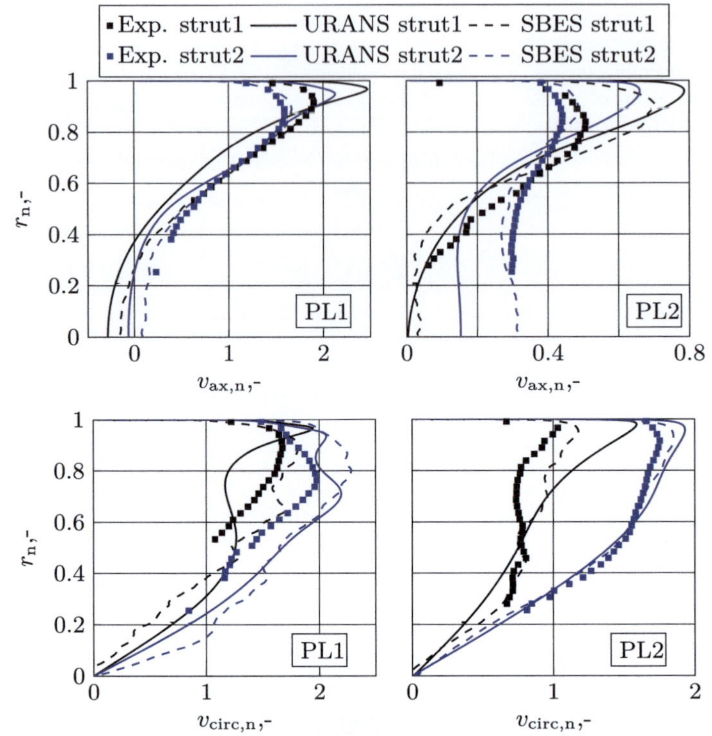

Fig. 7 Experimental and numerical radial distribution (URANS and SBES) of normalized axial (top) and circumferential velocity (bottom) in the Carnot diffuser section at $l_n = 6.7$ for strut1 and strut2 diffuser. PL1 on the left and PL2 on the right hand side

Figure 7 depicts radial profiles of normalized axial and circumferential velocity in the Carnot diffuser section at $l_n = 6.7$ for experimental data and numerical results at both operating points and both struts. It should be noted that a lack of measured traverse data near the diffuser centerline indicates very low velocities outside the probe calibration range or flow recirculation zones. The change in velocity profiles varying the strut is less pronounced in PL1. The same trend of a more uniform distribution of axial velocity and increased circumferential velocity at strut2 occurs in the experiment, URANS and SBES model. In general, a better agreement with the experimental data is achieved with the SBES model. Here, the increased axial and circumferential velocity near the casing is reduced.

4 Computational Resources

All simulations were performed on HPE Apollo Hawk system at the HLRS Stuttgart using ANSYS CFX in combination with the "Open MPIHAWK" parallelization method. For the diffuser test case, at least two flow-through times are needed to get rid of the initial oscillations followed by several flow-through times for time-averaging. This leads to a simulation time of up to 20 d for the SBES model with 75M elements on 1536 cores. A single simulation in the aeromechanic test case took 150 h with 18M elements on 384 cores. In the time frame 04.2022–03.2023, a total of 18 million CPUh were used.

The scaling ability for the aeromechanic and diffuser test case is given in Fig. 8 by comparing the realized speedup factor with the ideal one for different levels of parallelization. The absolute CPU-time per time step is used for the evaluation. The results for both test cases show a similar degree of parallelization of the utilized solver. The speedup factor is reduced with increasing number of CPUs for a simulation. Especially with a high number of computational nodes, the scaling ability of the code is not ideal.

The domain partitioning for the diffuser test case is shown in Fig. 9 in terms of a histogram and a box plot, representing the 25 and 75% percentiles. The median values for the different investigated levels of parallelization in (b) depicted in red are found to match well with an ideal uniform partitioning of the domain, which is represented by a normalized value of 1. In general, the 25 and 75% percentiles as well as the histograms indicate a small scatter range in the partition size, being a decisive prerequisite for the scaling ability (see Fig. 8). For the diffuser test case, the scatter range in partition size is increased using more cores. Furthermore, it is visible, that a few partitions with a lower number of normalized elements per core occur. These outliers are plotted as red symbols. Supposedly, these outliers lead to a limited scaling capability of the employed solver.

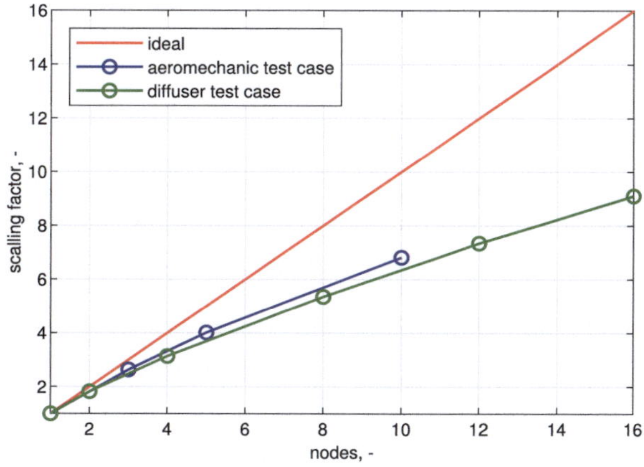

Fig. 8 Scaling ability for the aeromechanic and diffuser test case

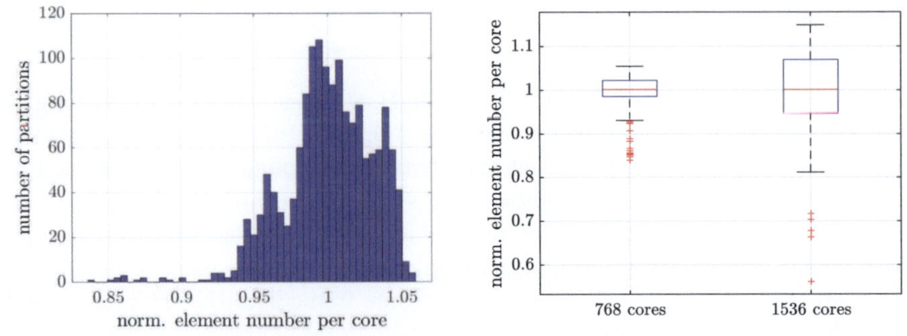

a) Partitioning histogram for 1536 cores

b) Partitioning box plot for 768 and 1536 cores

Fig. 9 Domain partitioning by means of **a** histogram and **b** boxplot for the diffuser test case and different parallelization levels

5 Conclusions

For the aeromechanics study, high-fidelity CFD models for the prediction of the aerodynamic excitation has been obtained for three vaneless radial turbines. These are used to evaluate a large number of operating points by means of URANS simulations in order to assess the excitation maps at different resonance crossings. Both operating parameters in terms of expansion ratio and reduced speed are found to have substantial impact on the excitation. Furthermore, significant differences between test objects are observed. However, the governing change mechanisms are presumed similar. An analysis of the underlying local forcing field enables a better understanding of the nature of the differences. The usage of a HPC system makes it for the first time possible to have such a detailed insight in the forcing behavior over a wide range of operating conditions.

For the diffuser test case, flow simulations at off-design operation were performed using the commercial solver ANSYS CFX. The URANS and SBES approach of two partload conditions and two strut geometries were investigated and compared to experimental data. The scale-resolving simulations showed a more accurate prediction of the diffuser flow field in terms of pressure recovery and velocities compared to experimental data. The simulations applying scale resolving models, like the SBES, require large computational resources due to the computation of several flow-through times to get time-averaged results of the flow quantities and the need of a high number of grid elements to resolve large eddies of high turbulent energy.

Investigations on the scaling ability of the utilized solver on the HPE Apollo Hawk system has shown comparable results for both test cases. The observed speedup factors, however, are limited and stagnate already at low node counts resulting in longer total computational times.

References

1. Kovachev, N., Vogt, D.M.: Numerical investigation of the effect of operating point on the aerodynamic excitation in a radial turbine stage. In: Proceedings of the ASME Turbo Expo 2020: Turbomachinery Technical Conference and Exposition, vol. 10A: Structures and Dynamics. Virtual, Online. Sept 21–25. ASME, GT2020-16045 (2020)
2. Bauer, M., Hummel, S., Schatz, M., Kegalj, M., Vogt, D.M.: Investigation of the flow field and the pressure recovery in a gas turbine exhaust diffuser at design, part-Load, and over-Load conditions. J. Turbomach. (2022). https://doi.org/10.1115/1.4053836
3. Hummel, S., Bauer, M., Vogt, D.M.: Study of gas turbine exhaust diffuser flow and performance characteristics at off-design conditions. In: Proceedings of 14th European Conference on Turbomachinery Fluid Dynamics & Thermodynamics. ETC14, ETC2021-590 (2021)
4. Menter, F.R.: Best practice: scale-resolving simulations in ANSYS CFD, Version 2.0 (2015)
5. Hummel, S., Kovachev, N., Fuhrer, Ch.:HPC-simulations of unsteady turbomachinery flows (2022). Annual report on project ACID44166

CFD-Driven Surrogate-Based Optimization

Fabian Hübenthal, Xiao Shao, Matthias Meinke, and Wolfgang Schröder

Abstract The potential of spanwise traveling transversal sinusoidal surface waves as an active drag reduction strategy for turbulent boundary layer flows is investigated using wall-resolved large-eddy simulations (LES). LES predictions are performed for Mach numbers $M = 0.2$ and 0.7 for a fixed momentum-thickness-based Reynolds number of $Re_\theta = 1000$ to cover various flow conditions and to extend a previous data set beyond Mach numbers of $M = 0.1$. These initial LES data points are planned according to a near-random space-filling sampling method, i.e., Latin hypercube sampling (LHS), within the space of promising actuation parameters. The main goal of surrogate-based optimization is to choose the actuation parameters such that the objective of drag reduction for given flow conditions is maximized by exploiting and exploring the surrogate model, which is fitted to the data and enriched with prior knowledge. Two surrogate-based optimization (SO) strategies using support vector regression (SVR) and Gaussian process regression (GPR) are investigated. To compare both strategies, the simulation data for $M = 0.1$ are used to model the dependence of the objective drag reduction on the actuation parameters. A previous purely exploitative approach of SVR-based ridgeline optimization (RO) for $M = 0.1$ is extended to GPR-based Bayesian optimization (BO) to further automate and efficiently guide the CFD-based optimization of the actuation parameters.

F. Hübenthal (✉) · X. Shao · M. Meinke · W. Schröder
Chair of Fluid Mechanics and Institute of Aerodynamics, Aachen, Germany
e-mail: f.huebenthal@aia.rwth-aachen.de

X. Shao
e-mail: x.shao@aia.rwth-aachen.de

M. Meinke
e-mail: m.meinke@aia.rwth-aachen.de

W. Schröder
e-mail: office@aia.rwth-aachen.de

T. Ludwig et al. (eds.), *High Performance Computing in Science and Engineering '23*,
https://doi.org/10.1007/978-3-031-91312-9_20

1 Introduction

A significant portion of the energy demand and associated greenhouse gas emissions originate from the transportation sector. As fuel-burning jet engines will remain the primary propulsion source for large civil aircraft for decades to come and rising energy costs pose an economic challenge for high-speed trains, reducing energy consumption and associated costs requires aerodynamic optimization. For that purpose, various types of drag reduction techniques are being developed. In contrast to passive drag reduction devices, active methods such as in-plane oscillation [11], blowing and suction [12], plasma actuator [13], and wall-deformation traveling waves [1–3, 14] are tuneable to practical application scenarios, but are less well studied.

In particular, spanwise traveling transversal surface waves were proven to manipulate the near-wall turbulent boundary layer to reduce viscous drag and achieve positive net power savings. The observed maximum drag reduction and maximum net power saving are 13.4 and 12.2%, respectively for turbulent channel flows [15], while in the turbulent boundary layer flow, 26% maximum drag reduction and 10% maximum net power saving are found in [2]. The design space is spanned by the actuation parameters wavelength, amplitude, period, and the given flow conditions in terms of Reynolds and Mach numbers. This design space is too vast to be investigated by grid search methods motivating the usage of surrogate-based optimization techniques, such as ridgeline and Bayesian optimization.

In a previous study, LES results for a flat plate turbulent boundary layer flow for the Mach number $M = 0.1$ were smoothly interpolated using support vector regression (SVR), focusing on the dependence of the drag reduction on the actuation parameters. The resulting SVR model is used to identify the a-priori assumed ridgeline, i.e., the curve connecting the wavelength-parameterized optima locations, by applying sparse nonlinear symbolic regression. Based on the prior knowledge of self-similarity and the Tomiyama and Fukagata scaling [2, 6] for the local skin friction coefficient, the ridgeline is extrapolated 66% beyond the initial wavelength range, revealing an unexplored but subsequently numerically validated optimization potential [6].

This follow-on study focuses on extending the previous CFD-based data set and the optimization pipeline using an SVR-based ridgeline surrogate to an automated optimization loop. First, wall-resolved LESs are performed and analyzed to provide initial data sets for Mach numbers $M = 0.2$ and $M = 0.7$, which are beyond the previously studied Mach number of $M = 0.1$ and closer to real-world applications. Second, Bayesian Optimization (BO) with the Gaussian process (GP) prior as a surrogate is chosen to address the exploration-exploitation trade-off to close the tuning loop. The prior knowledge utilized for and obtained by the SVR-based ridgeline optimization is incorporated into the GP prior to efficiently guiding the optimization of the actuation parameters.

This paper is structured as follows: First, the CFD-based and surrogate-based optimization methodologies employed in this study are described. Then, numerical results are presented and discussed from the CFD and optimization perspective. Finally, conclusions of the findings are given.

2 CFD-Based Methodologies

In this section, the numerical method of the finite volume (FV) structured solver of the in-house solver framework m-AIA (multiphysics Aerodynamics Institute Aachen) is introduced. The simulation setups are then described in terms of domain size, grid resolution, and the entire actuation parameter space for different Mach numbers. Finally, scaling results on HAWK are presented.

2.1 Numerical Method

All actuated and reference cases are simulated by LES based on the body-fitted deformable structured grid-based FV solver of the in-house solver framework m-AIA. An arbitrary Lagrangian-Eulerian formulation of the conservation equations is used to simulate the flow field over moving boundaries. A second-order accurate FV approximation of the governing equations is used in which the inviscid fluxes are computed by the advection upstream splitting method (AUSM) using a monotonic upstream-centered scheme for conservation laws (MUSCL) to reconstruct the cell-center values. The viscous fluxes are discretized by a modified cell-vertex scheme at second-order accuracy. The time integration is performed via a five-stage Runge-Kutta scheme with second-order accuracy. The volume fluxes are formulated such that the geometry conservation law (GCL) is satisfied. Further details on the numerical method can be found in Albers et al. [2].

2.2 Computational Setup

The simulation data are based on a validated zero-pressure gradient (ZPG) flat plate approximation of the active drag reduction technique using spanwise traveling transversal surface waves as depicted in Fig. 1.

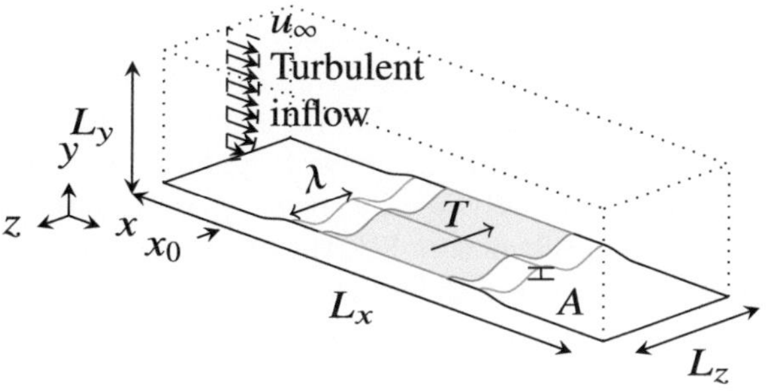

Fig. 1 Sketch of the LES domain of the actuated turbulent flat plate flow [2]

The LES domain has a rectangular shape with lengths in Cartesian dimensions denoted as L_x, L_y, L_z. The actuation parameters of the spanwise traveling waves are the wavelength λ, the period T and the amplitude A. At the inflow of the domain the reformulated synthetic turbulence generation (RSTG) method is used to initiate a turbulent boundary layer (TBL) flow. The onset of the surface actuation is located at x_0, where a fully developed turbulent boundary layer is established. The surface area A_{surf} for the integration of the wall-shear stress τ_{w} is shaded in gray in Fig. 1. Periodic boundary conditions are used in the spanwise direction z, characteristic outflow conditions are applied on the downstream and upper boundaries and the no-slip condition is imposed on the wall [6].

The actuation is characterized by the non-dimensional actuation parameters wavelength $\lambda^+ = \lambda u_\tau/\nu$, amplitude $A^+ = Au_\tau/\nu$ and time period $T^+ = Tu_\tau^2/\nu$, such that the wall-normal coordinate of the surface wave is described by Equation (1). For the three-dimensional turbulent flow, the wave parameters are non-dimensionalized and given in inner units $(\bullet)^+$ using the friction velocity u_τ and the kinematic viscosity ν both at x_0 averaged in the spanwise direction. The piecewise-defined function $g(x^+)$ ensures a smooth spatial transition in the streamwise direction from the non-actuated to the actuated surface area and vise versa.

$$y^+_{\text{wall}}\left(x^+, z^+, t^+|\lambda^+, T^+, A^+\right) = g(x^+)\, A^+ \cos\left(\frac{2\pi}{\lambda^+}z^+ + \frac{2\pi}{T^+}t^+\right) \tag{1}$$

The simulations are conducted at a momentum thickness-based Reynolds number $\text{Re}_\theta = \frac{u_\infty \theta}{\nu} = 1000$ at x_0 and various Mach numbers $\text{M} = \frac{u_\infty}{a} = 0.1, 0.2$, and 0.7 using the free stream velocity u_∞, the momentum-based boundary layer thickness θ, the kinematic viscosity ν and the isothermal ideal gas speed of sound. Based on a grid convergence study by Albers et al. [2], the mesh resolution for all Mach numbers is $\Delta x^+ = 12$ resp. $\Delta x^+ = 10$ in the streamwise, $\Delta y^+_{\text{wall}} = 1$ in the wall-normal direction gradually coarsening off the wall up to $\Delta y^+_{\text{edge}} = 16$ at the boundary layer edge, and $\Delta z^+ = 4$ in the spanwise direction. This yields a DNS-like resolution near the wall rendering these simulations wall-resolved LES. For $\text{M} = 0.1$, 80 LESs, i.e., one reference/non-actuated and 79 actuated cases, were performed for grid-like distributed actuation parameter combinations within the bounds $[200, 20, 10]^{\text{T}} \leq [\lambda^+, T^+, A^+]^{\text{T}} \leq [3000, 120, 78]^{\text{T}}$. Details of the simulation parameters, i.e., grid sizes and resolutions can be found in Table 1. For $\text{M} = 0.2$, and 0.7, 44 and 36 cases are performed consuming 7,464,652 and 3,114,298 core-hours separately for the parameters space generated by a near-random space-filling sampling method, i.e., Latin hypercube sampling (LHS). As shown in Table 2, a wider range of parameters is used for the higher Mach number except for the wavelength, since the drag reduction effect is significantly inferior for $\lambda^+ \leq 1000$ compared to those greater than 1000, cf. Table 2 in Albers et al. [2]. The parameter spaces for the different Mach numbers are visualized in Fig. 2. Further details on the numerical method, the computational setup, the validation of the LES model and a table of the simulation data points for $\text{M} = 0.1$ can be found in Albers et al. [2].

Table 1 Numerical discretization details for the present LES. L_x, L_y, and L_z are the Cartesian dimensions of the domain in terms of the momentum-based thickness θ

M	$L_x \times L_y \times L_z$	Δx^+	Δy^+_{wall}	Δz^+	$N_{\text{cell},L_{z,\max}}$
0.1	$190\theta \times 105\theta \times$ $[21.65\theta, 25.98\theta, 34.64\theta, 38.97\theta, 64.95\theta]$	12.0	1.0	4.0	72×10^6
0.2	$361\theta \times 101\theta \times$ $[21.65\theta, 32.47\theta, 43.30\theta, 54.19\theta, 64.94\theta]$	10.0	1.0	4.0	168×10^6
0.7	$361\theta \times 101\theta \times$ $[21.65\theta, 32.47\theta, 43.30\theta, 54.19\theta, 64.94\theta]$	10.0	1.0	4.0	168×10^6

Table 2 Actuation parameters sets for the wavelength and bounds for the period and amplitude of the turbulent boundary layer simulations for three flow conditions

Re_θ	M	λ^+	T^+	A^+
1000	0.1	{200, 500, 600, 900, 1000, 1600, 1800, 3000}	[20, 120]	[10, 78]
1000	0.2	{1000, 1500, 2000, 2500, 3000}	[14, 158]	[10, 79]
1000	0.7	{1000, 1500, 2000, 2500, 3000}	[14, 158]	[10, 79]

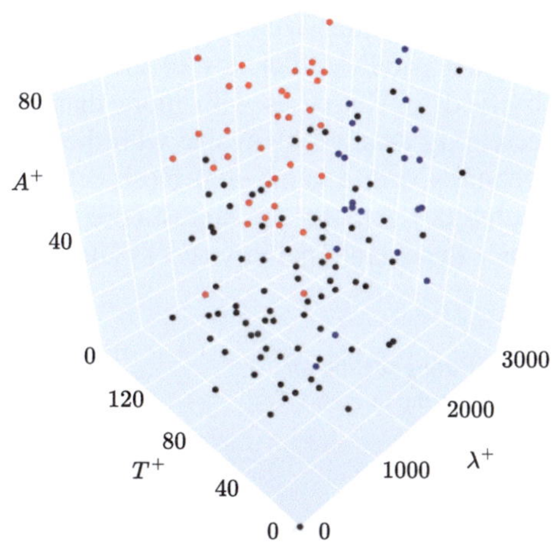

Fig. 2 LES data points in the actuation parameter space with black points for M = 0.1, blue for M = 0.2 and red for M = 0.7

To obtain quasi-steady flow fields above the actuated surface, the simulation is conducted in two phases. Initially, the non-actuated reference cases are run until a quasi-steady state is reached for which more than 4000 convective time units t_{conv} are needed for each wavelength. Subsequently, the flow field is averaged for $t_{\text{conv}} = tu_\infty/\theta > 3000$ convective time units. The actuated simulations are initialized with an instantaneous solution of the corresponding reference cases. The wave motion is smoothly introduced by multiplying the wave amplitude with the function $1 - \cos(ta/\theta)$. Once a quasi-steady state for the viscous drag is observed, the flow field is averaged.

The relative skin friction reduction Δc_f and the relative drag coefficient reduction Δc_d in the streamwise direction are defined as shown in Eqs. (2) and (3) averaging the wall-shear stresses $\tau_{\mathrm{w},x}$ over the surface area A_surf in Fig. 1 when a quasi-steady flow state is achieved. Here, the indices ref and act denote the non-actuated reference and actuated cases, respectively [2, 6].

$$\Delta c_f = \frac{\tau_{\mathrm{w},x,\mathrm{ref}} - \tau_{\mathrm{w},x,\mathrm{act}}}{\tau_{\mathrm{w},x,\mathrm{ref}}} \tag{2}$$

$$\Delta c_\mathrm{d} = \frac{c_{\mathrm{d,ref}} - c_{\mathrm{d,act}}}{c_{\mathrm{d,ref}}} \tag{3}$$

2.3 Scaling Tests

A strong scaling test using the m-AIA FV solver for benchmark problems is shown in Fig. 3, and Table 3 on the HAWK system, ranging from 32 nodes up to the maximum possible allocation size of 4096 nodes, i.e., 524288 compute cores. The results show excellent parallel efficiency of the m-AIA code, where a superlinear speedup is observed for the cases with more than 1024 nodes. This is probably due to the decreasing local problem size with the increasing core number. The local problem size starts with a number of mesh cells on the order of $\mathcal{O}(100.000)$ which drops to $\mathcal{O}(1000)$ cells for the largest core number used, with higher data localities that reduce the overall memory access latencies.

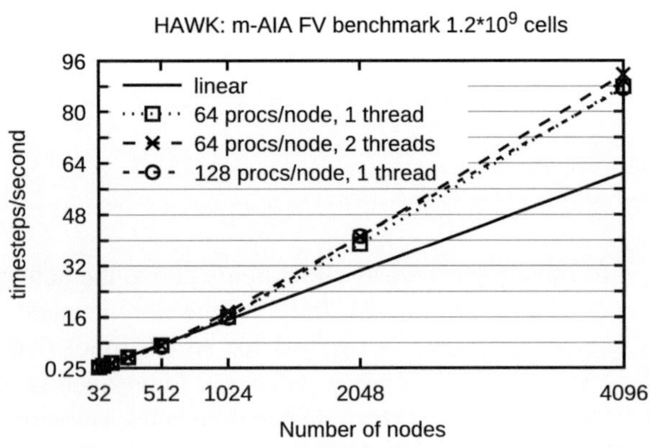

Fig. 3 Strong scaling on HAWK for the m-AIA FV CFD solver

Table 3 Numerical data of the strong scaling on HAWK for the m-AIA FV CFD solver

128 procs/node		Speedup: procs/node, #threads			
#Cores	DOF/core	Linear	128/1	64/1	64/2
4096	299520	1	1.00	1.00	1.00
8192	149760	2	1.93	1.96	1.96
16384	74880	4	3.76	3.84	3.83
32768	37440	8	7.47	7.51	7.57
65536	18720	16	14.68	15.26	15.42
131072	9360	32	34.15	33.87	37.56
262144	4680	64	89.54	81.53	87.70
524288	2340	128	188.94	184.55	195.64

3 Surrogate-Based Optimization Methodologies

Given the flow conditions specified for example for the TBL flow over the flat plate by the Reynolds number Re_θ and the Mach number M collected in the parameter vector $\boldsymbol{p} = [\mathrm{Re}_\theta, \mathrm{M}]^\mathrm{T}$, the goal is to choose the actuation parameters in the optimization vector $\boldsymbol{x} = [\lambda^+, T^+, A^+]^\mathrm{T}$ with dimension $d_{\boldsymbol{x}} = 3$ such that the relative drag reduction coefficient objective $f(\boldsymbol{x}|\boldsymbol{p}) = \Delta c_\mathrm{d}$ is maximized, cf. Eqs. (5) and (3). Inequality and equality constraints, i.e. $\boldsymbol{g}$ and $\boldsymbol{h}$, are placed on the optimization variables in $\boldsymbol{x}$ forming the feasible set $\boldsymbol{\mathcal{X}} = \{\boldsymbol{x} \in \mathbb{R}^{d_x} \,|\, \boldsymbol{g}(\boldsymbol{x}|\boldsymbol{p}) \leq \boldsymbol{0}, \boldsymbol{h}(\boldsymbol{x}|\boldsymbol{p}) = \boldsymbol{0}\}$. These constraints may be based on material deformation limits, technical actuation bounds, or further aerodynamic requirements depending on the application, like the lift-to-drag ratio L/D for an airfoil use case as analyzed in [3]. In general, a non-convex, nonlinearly constrained optimization problem is obtained, i.e., a nonlinear program (NLP), which is stated in the Eq. (4) [2, 6].

$$\boldsymbol{x}^* = \arg\max_{\boldsymbol{x}\in\boldsymbol{\mathcal{X}}} f(\boldsymbol{x}|\boldsymbol{p}) \tag{4}$$

$$\text{with } f(\boldsymbol{x}|\boldsymbol{p}) = \Delta c_\mathrm{d} \tag{5}$$

The main goal of surrogate-based optimization (SO) in the present context is to perform as few expensive sequential LES at HPC scale as possible, while providing sufficiently many insightful data for the characterization of the problem and guiding the choice of the actuation parameters by leveraging a surrogate model, i.e., to apply efficient simulation planning. Prior knowledge Π_n and data $D_n = \{(\boldsymbol{x}_i, y_i)|1 \leq i \leq n\}$ after having performed n LES, with noisy output $y_i = f(\boldsymbol{x}_i|\boldsymbol{p}) + \varepsilon_i$ and assumed Gaussian noise $\varepsilon_i \sim \mathcal{N}(0, \sigma_\mathrm{n}^2)$, are incorporated into the surrogate model $\hat{f}(\boldsymbol{x}|\boldsymbol{p})$ to approximate the expensive-to-evaluate LES-based objective function $f(\boldsymbol{x}|\boldsymbol{p})$, i.e., $\hat{f}(\boldsymbol{x}|\boldsymbol{p}) \approx f(\boldsymbol{x}|\boldsymbol{p})$. From the perspective of sequential optimization, the idealized goal is to asymptotically reach the true optimum $\boldsymbol{x}^*$ for an increasing number n of

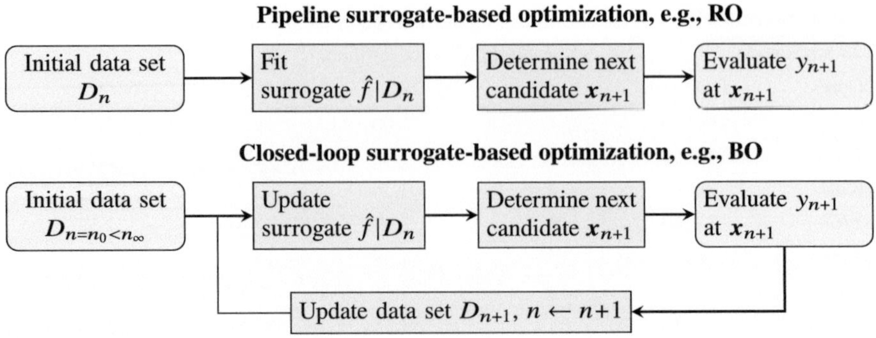

Fig. 4 This figure compares pipeline to closed-loop surrogate-based optimization

original objective function evaluations, i.e., $\lim_{n \to n_\infty < \infty} \boldsymbol{x}_n \approx \boldsymbol{x}^*$, with n_∞ indicating the final evaluation resp. simulation.

In the following subsections, two examples of sequential surrogate-based optimization strategies, i.e., SVR-based ridgeline optimization (RO) and GPR-based Bayesian optimization (BO) are introduced as illustrated in Fig. 4.

3.1 SVR-Based Ridgeline Optimization

In a previous study, the data for $\mathrm{M} = 0.1$ were smoothly interpolated using support vector regression (SVR) via scikit-learn [7]. The SVR model $\Delta\hat{c}_{\mathrm{d,SVR}}$ unveiled a ridgeline curve parameterized by the wavelength λ^+. That means that each λ^+-plane features a single optimum, i.e., $T_{\mathrm{r}}^+, A_{\mathrm{r}}^+ = \arg\max_{T^+, A^+} \Delta\hat{c}_{\mathrm{d,SVR}}(T^+, A^+|\lambda^+)$, such that the ridgeline $T_{\mathrm{r}}^+(\lambda^+)$, $A_{\mathrm{r}}^+(\lambda^+)$ is the curve in the parameter space connecting all these optima. Combining the prior knowledge of an existing ridgeline with the Tomiyama and Fukagata scaling and self-similarity for wavelengths $\lambda^+ \geq 1000$ [2, 6], sparse symbolic nonlinear regression using PySINDy [8] was applied to successfully extrapolate the data up to wavelengths of $\lambda^+ = 5000$, which is 66% beyond the initial data range. The resulting symbolic ridgeline surrogate model $\Delta\hat{c}_{\mathrm{d,RM}}$ was exploited and validated with one on- and one off-ridgeline LES. This approach is called ridgeline optimization (RO) and follows a pipeline-like procedure of generating an initial data set, fitting the ridgeline model and exploiting the model to determine the next actuation parameters to be simulated as illustrated in Fig. 4 [2, 6].

3.2 GPR-Based Bayesian Optimization

In contrast to the exploitative strategy of RO, explorative-exploitative approaches like Bayesian optimization (BO) address the exploration-exploitation trade-off by not directly optimizing the surrogate model prediction. Instead, a stochastic surrogate model, classically a Gaussian process regression (GPR) model, is utilized, which not only provides a prediction as the posterior mean $m|D_n(\boldsymbol{x}) = \hat{f}(\boldsymbol{x}|\boldsymbol{p})|D_n$, but also the corresponding uncertainty in the prediction as the posterior standard deviation $s|D_n(\boldsymbol{x})$. Both the prediction as well as the prediction uncertainty are incorporated into a so-called acquisition function α, which trades off exploration, i.e., performing LESs in regions of high uncertainty (global search), and exploitation, i.e., querying LESs in regions of already promising surrogate model predictions (local search). An example for the acquisition function is given in Eq. (13). The difference of exploitative pipeline, cf. Eq. (6), and explorative-exploitative loop, cf. Eq. (7), surrogate-based optimization strategies to determine the next queried LES is highlighted in the following equations, which are reformulations of the original problem in Eq. (4) to sequential optimization problems.

$$\boldsymbol{x}_{n+1} = \arg\max_{\boldsymbol{x}\in\mathcal{X}} \hat{f}|D_n(\boldsymbol{x}|\boldsymbol{p}) \tag{6}$$

$$\boldsymbol{x}_{n+1} = \arg\max_{\boldsymbol{x}\in\mathcal{X}} \alpha[\hat{f}|D_n(\boldsymbol{x}|\boldsymbol{p}), s|D_n(\boldsymbol{x}|\boldsymbol{p})\,|\boldsymbol{\theta}_\alpha] \tag{7}$$

When focusing on GPR-based BO, the objective function $f(\boldsymbol{x})$ is modeled as a Gaussian process $\mathcal{GP}$ with prior mean function $m(\boldsymbol{x}|\boldsymbol{\theta}_m)$ and covariance resp. kernel function $k(\boldsymbol{x}, \boldsymbol{x}'|\boldsymbol{\theta}_k)$ assuming that the original objective $f(\boldsymbol{x})$ is distributed as a Gaussian process, i.e., $f(\boldsymbol{x}) \sim \mathcal{GP}\big(m(\boldsymbol{x}), k(\boldsymbol{x}, \boldsymbol{x}')\big)$.

The simulation data D_n is arranged as a matrix $\boldsymbol{X} = [-\boldsymbol{x}_i-] \in \mathbb{R}^{n\times d_x}$ for the inputs $\boldsymbol{x}_i$ and a vector $\boldsymbol{y} = [y_i] \in \mathbb{R}^{n\times 1}$ for the noisy outputs y_i, such that $D_n = \{\boldsymbol{X}, \boldsymbol{y}\}$. Using the Gaussian process prior model on the data, $\boldsymbol{y}$ is a joint multivariate normal distribution $\mathcal{N}$ with the data prior mean vector $\boldsymbol{m}(\boldsymbol{X}) = [m(\boldsymbol{x}_i)]$ and data prior covariance matrix $\boldsymbol{K}(\boldsymbol{X}, \boldsymbol{X}') = [k(\boldsymbol{x}_i, \boldsymbol{x}_j)]$ evaluated at each $\boldsymbol{x}_i, \boldsymbol{x}_j$ of $\boldsymbol{X}$, i.e., $\boldsymbol{y} \sim \mathcal{N}(\boldsymbol{m}(\boldsymbol{X}), \boldsymbol{K}(\boldsymbol{X}, \boldsymbol{X}'))$. The prior belief expressed as the prior mean function $m(\boldsymbol{x})$ and the prior kernel $k(\boldsymbol{x}, \boldsymbol{x}')$ is conditioned on the data D_n using Bayes' theorem to obtain the Gaussian process posterior with the posterior mean function $m|D_n(\boldsymbol{x})$, the posterior kernel $k|D_n(\boldsymbol{x}, \boldsymbol{x}')$ and the posterior standard deviation $s|D_n$ given in Eqs. (8)–(10). Based on the posterior, a single prediction at $\boldsymbol{x}$ given data D_n can be made as $y|\boldsymbol{x}, \boldsymbol{X}, \boldsymbol{y} = y|\boldsymbol{x}, D_n \sim \mathcal{GP}(m|D_n(\boldsymbol{x}), k|D_n(\boldsymbol{x}, \boldsymbol{x}'))$.

$$\hat{f}(\boldsymbol{x}|\boldsymbol{p}) = m|D_n(\boldsymbol{x}) = m(\boldsymbol{x}) + \boldsymbol{k}(\boldsymbol{X}, \boldsymbol{x})^T \boldsymbol{K}(\boldsymbol{X}, \boldsymbol{X}')^{-1}(\boldsymbol{y} - \boldsymbol{m}(\boldsymbol{X})) \tag{8}$$

$$k|D_n(\boldsymbol{x}, \boldsymbol{x}') = k(\boldsymbol{x}, \boldsymbol{x}') - \boldsymbol{k}(\boldsymbol{X}, \boldsymbol{x})^T \boldsymbol{K}(\boldsymbol{X}, \boldsymbol{X}')^{-1}\boldsymbol{k}(\boldsymbol{X}, \boldsymbol{x}') \tag{9}$$

$$s|D_n(\boldsymbol{x}) = \sqrt{k|D_n(\boldsymbol{x}, \boldsymbol{x}' = \boldsymbol{x})} \tag{10}$$

By choosing the prior mean and the prior kernel functions, prior knowledge can be incorporated into the stochastic model of the objective function. Here, the prior mean function $m(\boldsymbol{x})$ is either set to the data mean mean$(\boldsymbol{y})$ or to the ridgeline model $\Delta\hat{c}_{\mathrm{d,RM}}(\boldsymbol{x})$, cf. Equation (11). The squared exponential kernel with automatic relevance determination (ARD) is chosen like in [6], cf. Eq. (12), assuming a smooth, i.e., infinitely differentiable objective. In the case of using the data mean or the ridgeline model, there are no prior mean hyperparameters to tune, i.e., $\boldsymbol{\theta}_m = \varnothing$. However, the squared exponential kernel with the scale matrix $\boldsymbol{M} = \mathrm{diag}(\boldsymbol{l})^{-2}$ possesses hyperparameters $\boldsymbol{\theta}_k = \{\sigma_f, \boldsymbol{l}, \sigma_n\}$ including the signal variance σ_f, length-scales $\boldsymbol{l} = [l_\lambda, l_A, l_T]^{\mathrm{T}}$ for each actuation parameter in $\boldsymbol{x}$ and the output noise hyperparameter σ_{n} using the indicator function $\delta(\boldsymbol{x} - \boldsymbol{x}') = \{0 \text{ for } \boldsymbol{x} \neq \boldsymbol{x}', 1 \text{ for } \boldsymbol{x} = \boldsymbol{x}'$. The hyperparameters $\boldsymbol{\theta} = \{\boldsymbol{\theta}_m, \boldsymbol{\theta}_k\}$ are tuned by maximum likelihood estimation (MLE) with multi-restart to best fit the data with respect to the Gaussian likelihood function $p(\boldsymbol{y}|\boldsymbol{X}, \boldsymbol{\theta})$, i.e., $\boldsymbol{\theta}^* = \arg\max_{\boldsymbol{\theta}\in\boldsymbol{\Theta}} \ln p(\boldsymbol{y}|\boldsymbol{X}, \boldsymbol{\theta})$, using scikit-learn [7]. A detailed explanation of the background of GPR can be found in Rasmussen et al. [4].

$$m(\boldsymbol{x}) = m(\boldsymbol{x}|\boldsymbol{\theta}_m) = \mathrm{mean}(\boldsymbol{y}) = \mathrm{const} \quad \text{or} \quad \Delta\hat{c}_{\mathrm{d,RM}}(\boldsymbol{x}) \tag{11}$$

$$k(\boldsymbol{x}, \boldsymbol{x}') = k(\boldsymbol{x}, \boldsymbol{x}'|\boldsymbol{\theta}_k) = \sigma_f^2 \exp\left(-\frac{1}{2}(\boldsymbol{x} - \boldsymbol{x}')^T \boldsymbol{M} (\boldsymbol{x} - \boldsymbol{x}')\right) + \sigma_{\mathrm{n}}^2 \delta(\boldsymbol{x} - \boldsymbol{x}') \tag{12}$$

The upper confidence bound (UCB) with hyperparameter $\boldsymbol{\theta}_\alpha = \{\beta\}$ is chosen as the acquisition function for its simplicity in addressing the exploration-exploitation trade-off when testing BO, cf. Eq. (13). The hyperparameter β decides on how exploratory or exploitative the surrogate-based optimization is performed, with $\beta = 0$ being purely exploitative and $\beta \to \infty$ being entirely explorative. An overview of acquisition functions is provided in Shahriari et al. [5] and state-of-the-art implementations can be found in BoTorch [10] using GPyTorch [9] for GPR.

$$\alpha(\boldsymbol{x}|\boldsymbol{\theta}_\alpha) = \alpha_{\mathrm{UCB}}(\boldsymbol{x}|\beta) = m|D_n(\boldsymbol{x}) + \sqrt{\beta}\, s|D_n(\boldsymbol{x}) \tag{13}$$

The next query resp. simulation with the actuation parameter settings $\boldsymbol{x}_{n+1}$ is determined by maximizing the acquisition function according to Eq. (7). After evaluating the compute-intensive LES-based objective to obtain y_{n+1}, the data set is updated to D_{n+1}. The optimization loop is closed by updating the surrogate model to D_{n+1}, including the hyperparameter tuning step. The closed-loop is compared to the pipeline surrogate-based optimization procedure in Fig. 4.

4 Results

The initial phase of the study demonstrates the control effectiveness of spanwise transversal traveling wave techniques under various flow scenarios and highlights the influence of Mach numbers on the control outcomes. Furthermore, a comparison between the reference case and the actuation case reveals distinctive features in their instantaneous flow fields, confirming the efficacy of the applied control method. The performed simulations provide an insightful and diverse data set for the development of optimization strategies targeting the drag reduction. In this work, the purely exploitative SVR-based ridgeline optimization is extended and compared to the explorative-exploitative GPR-based Bayesian optimization incorporating the previously obtained prior knowledge for M = 0.1.

4.1 Skin Friction Reduction

Figure 5 shows a scaling for the skin friction coefficient Δc_f against $A^+\sqrt{2\pi/T}$ combining the amplitude of the actuation velocity and the thickness of the Stokes layer [15] for all three Mach numbers. For the lower Mach number M = 0.1, the skin friction drag reduction exhibits scattered distribution for $\lambda^+ \leq 1000$ when plotted against the scaling parameter $A^+\sqrt{2\pi/T}$. For larger wavelengths $\lambda^+ > 1000$, the skin friction reduction scales almost linearly over the entire range. Beyond a certain threshold exceeding 20, the previously observed linear trend in reducing skin friction begins to deteriorate. The hypothesis is that this degradation is attributed to the excessive injection of momentum into the boundary layer caused by the fast propagation of large surface wave amplitudes. Consequently, this results in an increased spanwise velocity component, leading to an intensified turbulent exchange [2]. Based on the results obtained from M = 0.1, it can be observed that longer wavelengths have a greater potential for generating larger shear stress reduction. Therefore, the wavelengths were always chosen larger than 1000 for the Mach number variation. Comparing Fig. 5b and c, it can still be seen that M = 0.2 has a similar wavelength-independent linear trend as for M = 0.1, despite the fact that M = 0.2 contains more cases. The scaling factor in the initialization parameter space for M = 0.7 remains within a range smaller than 25, which is different from the case in M = 0.2 where it is kept below 30. This observation precisely indicates that the spanwise traveling waves have different effects on the flow conditions between M = 0.2 and M = 0.7.

4.2 Flow Statistics

The information from the instantaneous flow field is crucial for a deeper understanding of the drag reduction mechanisms. This section presents the figures obtained by applying the λ_2^+-criterion, which was developed by Jeong and Hussain in 1995 [16] and is a widely used criterion in fluid mechanics to characterize and analyze flow structures and turbulence. Figure 6 shows the contours of the λ_2^+-criterion. It is evident that the total number of vortex structures in the near wall region is significantly reduced for the actuated flow. Extended regions with little or no structures appear in Fig. 6b. Closer inspection reveals that, unlike the structures of the non-actuated

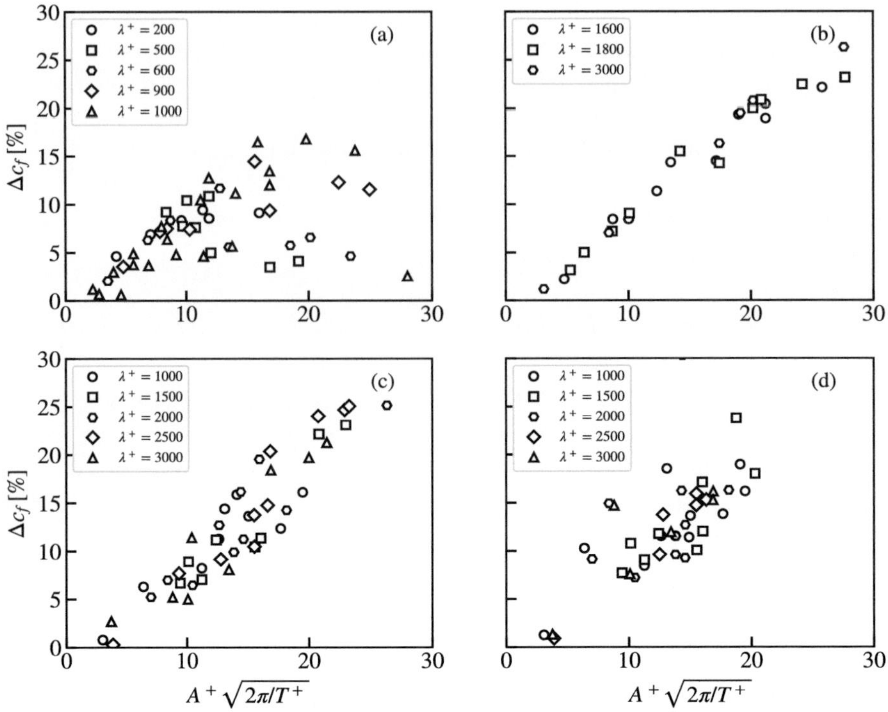

Fig. 5 Relative skin friction reduction of all cases as a function of the scaling factor for **a** $\lambda^+ \leq 1000$ for $M = 0.1$ [2], **b** $\lambda^+ > 1000$ for $M = 0.1$ [2], **c** $\lambda^+ > 1000$ for $M = 0.2$, and **d** $\lambda^+ > 1000$ for $M = 0.7$

reference flow in Fig. 6a, the structures of the actuated flow are tilted to the left and right depending on the phase angle of the traveling wave. Moreover, the distribution is less homogeneous, and the preferential angle of inclination of the structures depending on the wave phase persists [2].

4.3 GPR-Based Bayesian Optimization

Figure 7a–d provides an overview of the predictions in the parameter space of the SVR model trained on the entire data set, i.e., $n = 80$, the symbolically regressed ridgeline model according to Fernex et al. [6], the GPR posterior model using the data mean prior and the GPR posterior model with the ridgeline prior. Both GPR models have been conditioned resp. trained on the data set reduced to LES data points with wavelengths $\lambda^+ \geq 1000$, i.e., $n = 47$, which are more promising to achieve higher drag reductions. This choice is made for comparison reasons of the different GPR models, since the ridgeline model is only valid for wavelengths $\lambda^+ \geq 1000$ [2, 6].

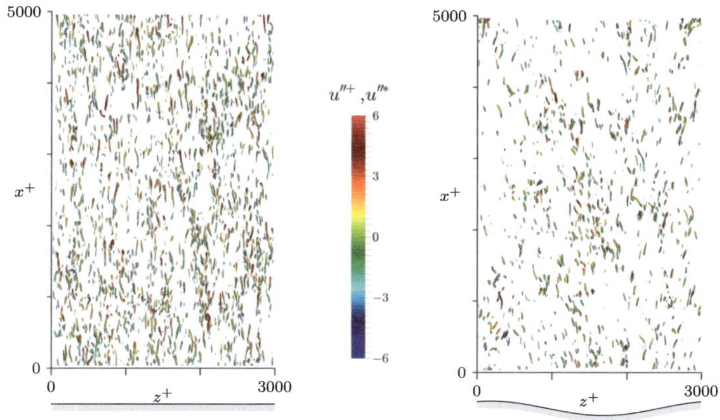

(a) (b)

Fig. 6 Contours of the λ_2^+-criterion colored by the random velocity fluctuations u''^+ in the near-wall region $y^+ < 20$ of **a** non-actuated reference case, and **b** actuated case with the highest drag reduction for $M = 0.1$, i.e., $\lambda^+ = 3000$, $T^+ = 50$, and $A^+ = 78$, normalized by the friction velocity of the reference case $u_{\tau,\text{ref}}$ [2]

Therefore, the isosurfaces of the drag reduction $\Delta\hat{c}_\text{d}(\boldsymbol{x})$ are plotted for these larger wavelengths. Each LES is represented by a single point in the actuation parameter space spanned by the wavelength λ^+, the period T^+ and the amplitude A^+. The non-actuated resp. reference LES is marked as a red point. The on- and off-ridgeline validation LESs used in [2, 6] are not part of the training data set, but are also included in the subplots at $\lambda^+ = 5000$. To facilitate the comparison between the models, the ridgeline proposed by Fernex et al. [6] is highlighted by a solid blue line in each subplot. The ridgeline of the uninformed GPR model, i.e., using the data mean prior, and of the ridgeline-informed GPR model are indicated as solid black lines in the corresponding subplots (c) and (d) in Fig. 7.

The SVR model and the uninformed GPR model provide a similar smooth interpolation of the data points near the maximum drag reduction. However, distant from the data points the uninformed GPR posterior tends to fall back to the data mean prior in the amplitude and period direction faster than the SVR model. Except for the period $T_\text{r}^+(\lambda^+)$, the amplitude $A_\text{r}^+(\lambda^+)$ coordinate of the ridgeline differs in the curvature and overall trend. The next query point $\boldsymbol{x}_{n+1}$ using UCB with $\beta = 1$ is determined close to the maximum GPR prediction, thus more exploiting the model rather than exploring the parameter space.

In the case of the ridgeline model and the ridgeline-informed GPR model, both models show a high similarity of the ridgeline and isosurface shapes. The ridgeline pathways start to diverge at wavelengths $\lambda^+ \geq 2000$ and stabilize on an almost parallel course with a period difference of $\Delta T_r^+ = 6$ for wavelengths $\lambda^+ \geq 3000$. Moreover, the shape of the isosurfaces for smaller wavelengths around $\lambda^+ = 1000$ start to differ more due to the reduced validity of the ridgeline model, which is corrected by the GPR model using data D_n. The ridgeline-informed GPR demonstrates

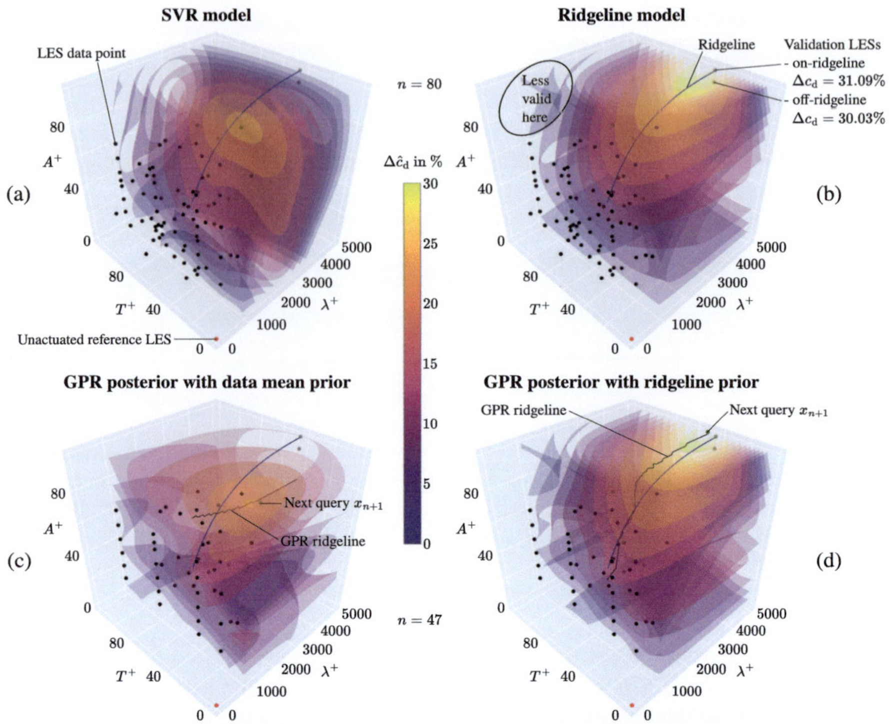

Fig. 7 Isosurface plots of the relative drag reduction predictions $\Delta \hat{c}_d$ of **a** the SVR model, **b** the ridgeline model [6], **c** the GPR model with data prior mean and **d** the GPR model with ridgeline prior in the actuation parameter space for $\mathrm{M} = 0.1$

enhanced extrapolation capabilities by relying on the ridgeline prior mean when making predictions far away from data points, as for the validation LESs, i.e., on-ridgeline $\Delta c_d = 31.09\%$ and off-ridgeline $\Delta c_d = 30.03\%$. For the ridgeline-informed GPR model the next query $\boldsymbol{x}_{n+1}$ is located directly at the upper optimization bound of the wavelength, i.e., $\lambda^+ = 5000$, unveiling the importance of and the reliance on the choice of optimization bounds when closing the loop using Bayesian optimization with a prior mean-informed GPR model.

Based on the GPR lengthscale hyperparameters for the standardized input data, the inverse relevance resp. extrapolation capabilities in the direction of the respective actuation parameter can be quantified, i.e., $\boldsymbol{l}^*_{\varnothing} = [l^*_\lambda \approx 3.62, l^*_A \approx 0.812, l^*_T \approx 0.885]^\mathrm{T}$ for the GPR model with data mean prior and $\boldsymbol{l}^*_{\mathrm{RM}} = [l^*_\lambda \approx 7.013, l^*_A \approx 0.406, l^*_T \approx 0.469]^\mathrm{T}$ for the ridgeline-informed GPR model. While the λ-lengthscale is almost doubled, the T- and A-lengthscales are approximately halved for the ridgeline-informed model compared to the data mean prior GPR model. In general, the wavelength actuation parameter appears to be the more reliable direction to extrapolate for both models. This is especially true for the ridgeline-informed

GPR model, where the ridgeline prior is modified by the GPR far beyond the initial data range, cf. Fig. 7d. However, the halved T- and A-lengthscales indicate that extrapolation off the ridgeline must be performed with caution, hence the uncertainty quantified by the GPR model increases progressively.

To further automate this ridgeline-informed GPR-based BO approach, the prior ridgeline, more generally, the prior mean identification using symbolic regression needs to be automated as well to successively update the informed GPR model for closed-loop optimization. Alongside, the optimization bounds need to be adapted according to the validity of the prior mean model, here the ridgeline model.

5 Conclusions

The existing database of spanwise-traveling-wave-induced drag reduction at $\mathrm{M} = 0.1$ has been extended to include a variation of the Mach number for $\mathrm{M} = 0.2$ and 0.7 by means of LESs. Three different sets of initialized parameter spaces yielded significant reductions in skin friction, reaching the maximum of 26, 25, and 23.7%, respectively. Utilizing the λ_2^+-criterion, the visual representation of instantaneous vortex structures reveals the effective suppression of near-wall streamwise vortices.

Based on the data for $\mathrm{M} = 0.1$, the previous purely exploitative SVR-based ridgeline optimization is extended to an explorative-exploitative approach using GPR-based Bayesian optimization to close the CFD-based optimization loop for achieving a higher degree of automation than the exploitative pipeline technique. Results of one BO query step show that incorporating the prior knowledge of the ridgeline model as the prior mean, i.e., ridgeline-informed GPR model, significantly improves the extrapolation capabilities, especially in the wavelength direction, compared to the vanilla uniformed GPR model. At the same time, the extrapolation quality in the period and amplitude directions is reduced, so that the informed GPR model relies on the ridgeline prior to make the predictions in these directions and on the predefined optimization bounds for the next simulation queries. In conclusion, ridgeline-informed GPR-based Bayesian optimization is a promising approach to efficiently optimize the actuation parameters when a prior mean model, such as the ridgeline prior, and the feasible set, i.e., the optimization bounds, are either given or carefully modeled and successively updated.

Acknowledgements The research was funded by the Deutsche Forschungsgemeinschaft (DFG) in the framework of the research projects SCHR 309/52 and SCHR 309/68. The authors are grateful to the Gauss Center for Supercomputing e.V. (www.gauss-centre.eu) for funding this project by providing computing time on the GCS supercomputer HAWK at the Höchstleistungsrechenzentrum Stuttgart (www.hlrs.de).
The authors gratefully acknowledge the CoE RAISE project for supporting this research in the context of use case 3.1, which receives funding from the European Union's Horizon 2020 –Research and Innovation Framework Programme 2020-INFRAEDI-2019-1–under grant agreement no. 951733 (https://www.coe-raise.eu/).

References

1. Albers, M., Meysonnat, P.S., Schröder, W.: Actively reduced airfoil drag by transversal surface waves. Flow, Turbul. Combust. (2019). https://doi.org/10.1007/s10494-018-9998-z
2. Albers, M., Meysonnat, P.S., Fernex, D., Semaan, R., Noack, B.R., Schröder, W.: Drag reduction and energy saving by spanwise traveling transversal surface waves for flat plate flow. Flow, Turbul. Combust. (2020). https://doi.org/10.1007/s10494-020-00110-8
3. Albers, M., Schröder, W.: Lower drag and higher lift for turbulent airfoil flow by moving surfaces. Int. J. Heat Fluid Flow (2021). https://doi.org/10.1016/j.ijheatfluidflow.2020.108770
4. Rasmussen, C.E.: Gaussian Processes for Machine Learning. In: Rasmussen, C.E., Williams, C.K.I. The MIT Press, Cambridge, Massachusetts, USA, ISBN 0-262-18253-X (2006)
5. Shahriari, B., Swersky, K., Wang, Z., Adams, R.P., De Freitas, N.: Taking the human out of the loop: a review of Bayesian optimization. Proc. IEEE (2015). https://doi.org/10.1109/JPROC.2015.2494218
6. Fernex, D., Semaan, R., Albers, M., Meysonnat, P.S., Schröder, W., Noack, B.R.: Actuation response model from sparse data for wall turbulence drag reduction. Phys. Rev. Fluids (2020). https://doi.org/10.1103/PhysRevFluids.5.073901
7. Pedregosa, F., Varoquaux, G., Gramfort, A., Michel, V., Thirion, B., Grisel, O., Duchesnay, É. Scikit-learn: machine learning in Python. J. Mach. Learn. Res. (2011). https://doi.org/10.5555/1953048.2078195
8. de Silva, B.M., Champion, K., Quade, M., Loiseau, J.C., Kutz, J.N., Brunton, S.L.: Pysindy: a python package for the sparse identification of nonlinear dynamics from data. J. Open Source Softw. (2020). https://doi.org/10.21105/joss.02104
9. Gardner, J., Pleiss, G., Weinberger, K.Q., Bindel, D., Wilson, A.G.: Gpytorch: blackbox matrix-matrix gaussian process inference with GPU acceleration. Adv. Neural. Inf. Process. Syst. (2018). https://doi.org/10.5555/3327757.3327857
10. Balandat, M., Karrer, B., Jiang, D., Daulton, S., Letham, B., Wilson, A.G., Bakshy, E.: BoTorch: a framework for efficient Monte-Carlo Bayesian optimization. Adv. Neural. Inf. Process. Syst. (2020). https://doi.org/10.5555/3495724.3497531
11. Jung, W.J., Mangiavacchi, N., Akhavan, R.: Suppression of turbulence in wall-bounded flows by high-frequency spanwise oscillations. Phys. Fluids A (1992). https://doi.org/10.1063/1.858381
12. Park, J., Choi, H.: Effects of uniform blowing or suction from a spanwise slot on a turbulent boundary layer flow. Phys. Fluids (1999). https://doi.org/10.1063/1.870167
13. Jukes, T., Choi, K.S., Johnson, G., Scott, S.: Turbulent drag reduction by surface plasma through spanwise flow oscillation. In: 3rd AIAA Flow Control Conference (2006). https://doi.org/10.2514/6.2006-3693
14. Mäteling, E., Albers, M., Schröder, W.: How spanwise travelling transversal surface waves change the near-wall flow. J. Fluid Mech. (2023). https://doi.org/10.1017/jfm.2023.54
15. Tomiyama, N., Fukagata, K.: Direct numerical simulation of drag reduction in a turbulent channel flow using spanwise traveling wave-like wall deformation. Phys. Fluids (2013). https://doi.org/10.1063/1.4826887
16. Jeong, J., Hussain, F.: On the identification of a vortex. J. Fluid Mech. (1995). https://doi.org/10.1017/S0022112095000462

Soot Oxidation in a Turbulent Temporally Evolving Non-premixed Air/ethylene Jet

Gandolfo Scialabba, Lukas Berger, and Heinz Pitsch

Abstract A direct numerical simulation (DNS) of a turbulent temporally evolving non-premixed air/ethylene jet, specifically designed for soot oxidation, is presented in this work. The performed DNS is currently the largest available data set on soot formation and consumption in a turbulent environment. In addition, it is the first large-scale simulation to cover the later stages of soot evolution. A detailed description of the soot evolution, with a particular focus on soot oxidation, is provided. Contributions from both OH radicals and molecular oxygen to the soot particles' depletion and their distribution in mixture fraction space are analyzed. In addition, an assessment of existing subfilter models to account for soot-gas phase interaction, particularly relevant for soot oxidation, is performed, and possible improvements are inferred.

1 Introduction

The formation and evolution of carbonaceous nanoparticles such as soot adversely affect the environment and human health. The detrimental effects of soot emissions are linked not only to its mass but also to chemical composition and particle number distribution [1] leading to increasingly stringent regulations on soot emissions. Achieving these ultra-low emission targets requires high-fidelity numerical models that reliably predict soot formation, transport, and consumption.

Due to the wide timescale range of soot formation and consumption, from a very slow formation process arising from polycyclic aromatic hydrocarbons (PAH) to significantly faster particles' depletion through surface reactions, modeling of soot particles' evolution is a challenging task. Additionally, in turbulent environments typical of combustion devices, highly non-linear interactions between the flow field and soot particles need to be taken into account.

G. Scialabba · L. Berger · H. Pitsch (✉)
Institute for Combustion Technology, RWTH Aachen University, Aachen, Germany
e-mail: h.pitsch@itv.rwth-aachen.de

T. Ludwig et al. (eds.), *High Performance Computing in Science and Engineering '23*,
https://doi.org/10.1007/978-3-031-91312-9_21

Obtaining experimental data for soot in turbulent conditions is a challenging task and the collected data is sometimes limited to narrow regions of the flow domain [2, 3]. For this reason, numerical simulations serve as a valuable complement to experimental measurements in reducing the costs for developing cleaner combustion devices, for example using Large-Eddy Simulations (LES) [4], where the subfilter interactions must be modeled. To make large-scale simulations of turbulent combustion feasible, it is necessary to reduce the computational time for the estimation of chemistry. A typical example of reduced order combustion models are the flamelet progress variable (FPV) models [5], where only a reduced set of tracking scalars are solved and used as mapping variables for all the chemical species present. However, these models do not account for the disparate timescales characteristic of soot evolution and, although some improvements in the context of soot prediction have been suggested [6, 7], an accurate prediction of the soot-turbulence interaction is still an open question, hence the need of further model development. For this purpose, fully resolved simulations (Direct Numerical Simulations, DNS) of turbulent combustion, due to their extensive spatially and temporally resolved data, are a fundamental benchmark tool.

DNS data sets for soot formation and evolution under turbulent conditions including detailed chemistry and soot models are extremely sparse, some examples being the works of Lignell et al. [8, 9] and Attili et al. [10, 11]. These data sets are an invaluable tool for soot subfilter model development, in particular for soot particles' formation [12–14]. However, the main focus of these works was mostly on soot formation processes with little data on particles' consumption due to oxidation, which interacts with the gas phase in a substantially different way due to time scale differences. LES simulations performed by Wick et al. [15] showed that oxidation was severely overpredicted when compared with experimental measurements. Although some approaches to include the interaction between gas phase and soot in subfilter models have been later suggested [13], modeling of soot oxidation in an LES context is still an open question, and specific data for soot oxidation modeling in a turbulent environment is required. In this work, a DNS of a turbulent temporally evolving non-premixed air/ethylene jet specifically designed to obtain sufficient soot formation and almost complete consumption of soot due to strong surface oxidation is performed. This DNS is the first one to extensively cover the later stages of soot evolution in a turbulent environment and, with over 1.5 billion data points, it is currently one the largest available data set of soot evolution in a turbulent flame. A detailed description of the flame evolution, with a particular focus on soot particles' evolution and consumption is given. Finally, an assessment of the currently available soot subfilter models is provided.

2 Code Description and Numerical Methods

The DNS has been performed with a high-fidelity in-house code (CIAO), which has been successfully used in several reactive DNS studies of similar configurations in the past, e.g. [10, 16]. It is based on the numerical algorithms developed by Desjardins et al. [17], which solve the reacting Navier-Stokes equations in the low-Mach limit [18]. CIAO is a massive-parallel, higher-order semi-implicit finite difference code that employs Crank-Nicolson time advancement and an iterative predictor corrector scheme. Spatial and temporal staggering is used to increase the accuracy of stencils. The Poisson equation for the pressure is solved by the multi-grid HYPRE solver. Species and temperature equations are discretized with a WENO scheme. The temperature and species equations are advanced by utilizing an operator splitting according to Strang. The chemistry operator uses a time-implicit backward difference method implemented in the stiff ODE solver CVODE. The message-passing interface (MPI) standard is used for parallelization.

Chemical reactions are modeled with a similar reduced mechanism adopted in the DNS from Attili et al. [10], with 41 species and 269 reactions. The mechanism is based on the detailed model from Blanquart et al. [19], which was specifically designed for high-temperature combustion at engine-relevant conditions including PAH pathways. In this reduced model, the main pathways for naphthalene formation have been retained to model soot inception and PAH adsorption. A unity Lewis number (ratio of thermal and mass diffusivities) has been selected for all gas phase species. The soot particle size distribution is described by the Number Density Function (NDF) and its evolution is governed by the Population Balance Equation (PBE). Its direct solution is usually too expensive, thus, different approaches to simplify the PBE are employed, with different levels of detail and cost. Two examples are sectional methods [20], for which the NDF is discretized in multiple sections, and moment methods [21], where instead statistical moments of the NDF are solved for. In this work, a bivariate moment method, where soot particles and aggregates are described by their volume and surface, is employed to describe soot particle evolution [22]. The bivariate moments are defined as:

$$M_{x,y} = \sum_{j=0}^{\infty} V_j^x S_j^y N_j \, ,$$

where V, S, and N are the volume, surface, and particle number density of soot aggregates of class j. Soot particles are characterized by a high Schmidt number [23], therefore diffusive soot transport is neglected. The resulting advection-reaction equation for a generic bivariate soot statistical moment is:

$$\frac{\partial M_{x,y}}{\partial t} + \nabla \cdot (\mathbf{u} M_{x,y}) = \dot{M}_{x,y}. \tag{1}$$

where $\dot{M}_{x,y}$ is a source term describing the various particulate internal processes: particle inception, coagulation, surface growth due to the HACA growth mechanism [24] and PAH adsorption, and particle consumption through surface oxidation reactions. The rate of change of the individual statistical moments depends on other additional moments, therefore a closure approach is necessary. The hybrid method of moments (HMOM) [25] is used with a set of four statistical moments, $M_{0,0}$, $M_{1,0}$, $M_{0,1}$, and N_0, which describe soot particles number density ($M_{0,0}$), soot volume fraction ($M_{1,0}$), soot surface density ($M_{0,1}$), and the number of smallest particles at inception size (N_0). An in-house stand-alone C++ library is used for the single-point computation of the soot moment source terms and their respective closure. The library is also integrated into FlameMaster [26] to compute 1D Flamelet solutions used to initialize the DNS chemical composition. Due to the lack of a diffusive term and the stiffness of the soot moment equations, a Lagrangian particle method [27] is employed for their solution, solving the following equations for each Lagrangian point particle:

$$\frac{d\mathbf{x}^p}{dt} = \mathbf{u}^p, \quad \frac{d\mu^p_{x,y}}{dt} = \frac{\dot{M}^p_{x,y}}{\rho^p}, \quad p = 1, \ldots, N \tag{2}$$

where $\mathbf{x}^p$,$\mathbf{u}^p$ and $\mu^p_{x,y}$ are the particle location, the Eulerian fluid velocity, and the set of mass-based soot moments at particle position, respectively. This method has been successfully applied in the DNS from Attili et al. [10, 11] and a more detailed description of its implementation and the projection of the lagrangian solution on the Eulerian grid for a two-way coupling can be found in [10, 28].

2.1 Code Performance

The high-fidelity, in-house code CIAO is fully parallelized using MPI, and its good scaling behavior has been demonstrated on several high-performance computers:

- SuperMUC/LRZ: Scaling has been shown up to 131,072 cores on SuperMUC Phase 1 during the Extreme Scaling Workshops 2013 and 2014. This study was part of the Gauss project PR45DI.
- JUQUEEN/JSC: IBM Blue Gene/Q system, which was utilized up to the full machine with 458,752 cores with almost perfect scaling. CIAO is a member of High-Q-Club since 2015.
- Hazel Hen/HLRS: Cray XC40 system, which was utilized up to 20,000 cores as part of the Gauss project GCS-MRES.

CIAO was evaluated for strong scalability in the PRACE Extreme Scaling Workshop 2014 SuperMUC Phase 1. The code demonstrated an excellent scaling up to half of the machine size and a reasonably well scaling on the whole machine. Finally, the good scaling capabilities of CIAO for the presented work performed using HAWK are shown in Fig. 1. The most critical operations in terms of computational cost in the

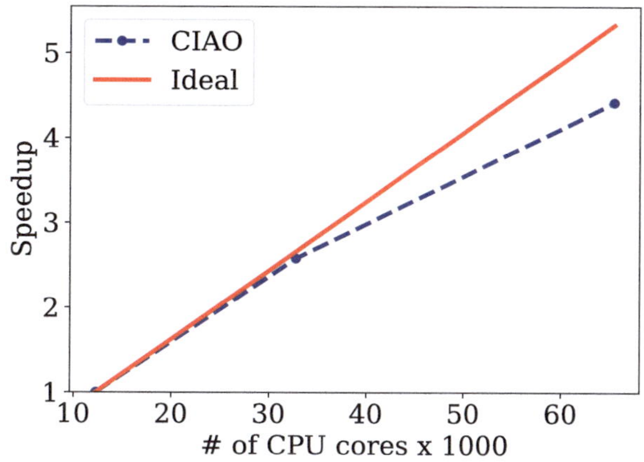

Fig. 1 Strong scaling of CIAO on HAWK for the presented configuration

simulations are the solution of the elliptic pressure equation, the transport of scalars, in this case including the solution of Lagrangian particle trajectories for soot, and finally the chemistry/combustion solver. For this case in particular, chemistry and soot account for up to 60% of the cost. The solution of both chemical and soot moment source terms does not require any information exchange between processes, therefore good scalability is observed. The performed DNS significantly benefits from this aspect since a large detailed chemical mechanism is employed, increasing the chemistry solver's relative cost. Parallel I/O operations are implemented with HDF5, allowing for efficient use of the file system and a limited number of temporary and restart files. Details on the resources used for this work are provided in the next section.

3 Results

3.1 DNS Configuration and Resources Usage

For this work, the original configuration of Attili et al. [10] has been used as a reference for the design of a non-premixed ethylene/air temporally evolving turbulent jet specific for soot oxidation. A schematic of the DNS configuration, in particular of the fuel distribution, is shown in Fig. 2. The central fuel jet of Attili et al. is replaced with a mixture fraction distribution of two narrow fuel stripes placed on top of the generated mixing layer. Compared to the original setup from Attili et al., this configuration allows for complete and rapid fuel consumption, hence almost complete oxidation of the produced soot within the simulation time. For Z=1, the fuel composition is nitrogen-diluted ethylene (67% N_2 and 33% C_2H_4) at 400K, and air at 800K is used as oxidizer (Z $=$ 0) to minimize the possibility of flame extinction due to turbulence. This translates to a stoichiometric mixture fraction of $Z_{st} = 0.17$. Initial conditions

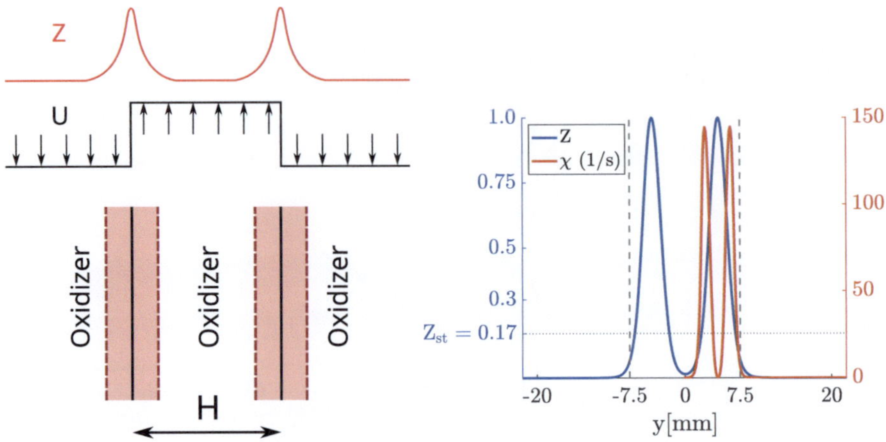

Fig. 2 Schematic of DNS configuration. Left: a sketch of the mixture fraction and velocity arrangement. Right: actual mixture fraction profile for fuel stripes and corresponding scalar dissipation rate χ (shown only for one fuel stripe for better visualization). Vertical dashed lines indicate the initial mixing layers position ($U = 0$, $y_{ML} = |7.5|$ mm)

for the chemical species are taken from a 1D flamelet simulation with the same mixture and boundary conditions and calculated with the same kinetic mechanism. The species distribution is mapped onto the imposed mixture fraction profile and the scalar dissipation rate of the initial flamelet solution is $\chi_{st} \approx 0.11 \cdot \chi_{st,q} = 74s^{-1}$, where $\chi_{st,q} = 680s^{-1}$ is the extinction stoichiometric scalar dissipation rate for the reference flamelet manifold. The smooth initial profile of Fig. 2 is taken such that the stoichiometric scalar dissipation χ_{st} at Z_{st} matches the one from the flamelet used for initialization.

The simulation domain consists on a box with dimensions $L_x \times L_y \times L_z = 186.36 \times 308.34 \times 62.16$ mm. Periodic boundary conditions in the streamwise (x) and spanwise direction (z) are employed, while an outlet boundary condition is imposed on the crosswise (y) direction. The initial jet width is H = 15 mm. A uniform spacing of $\Delta = 120\mu$m is used across the x and z directions and for the region closer to the jet in the y direction (|y| < 90 mm). Closer to the outlets (90 mm < |y| < 154.17 mm), the grid is stretched to ensure that the flame remains far from the domain boundaries, avoiding possible numerical issues, while minimizing computational time waste. This results in a total of approximately 1.5 billion grid points (1554 × 1860 × 519). Throughout the whole simulation, the flame remains within the unstretched mesh region (|y| < 90 mm). The mixing layers of the temporal jet are located at |y| = 7.5 mm. An optimal fuel stripes' center position of |y| = 4.5 mm was selected from preliminary investigation to avoid an excessive lateral spreading of the jet.

The initial velocity field is generated similarly to the one from Attili et al. [10], where two realizations of a turbulent channel flow at $Re_\tau = 240$ were taken and rearranged such that the same temporally evolving jet configuration is obtained.

The mean centerline jet velocity in the x direction is $U_c = 60$ m/s, while in the outer region ($|y| > H$) a constant velocity of $-U_c$ is imposed. This yields a total $\Delta U = 120$ m/s, which, using the centerline viscosity value, results in a jet Reynolds number of 16000. The minimum Kolmogorov scale is 131 μm, which ensures an η/Δ ratio higher than one, and a minimum of 8 grid points across the OH layer is ensured. The overall physical simulation time is around 10 ms, with a maximum $\Delta t = 1.1\mu$s. Finally, for the solution of the soot moment equations, the domain was uniformly initialized with two particles per Eulerian cell to properly resolve the soot fields and allow the projection of the Lagrangian solution onto the Eulerian grid, for a total of 3 billion point particles.

The simulation was performed at the High-Performance Computing Center in Stuttgart (HLRS), specifically on HPE Apollo (HAWK). The entire granted resources for this first phase, 43.7 million core hours, were used to perform the simulation, including preliminary test runs at lower resolutions and the generation of the two turbulent channel flows necessary for the velocity profile initialization. Up to 65536 cores were used for the simulation (512 nodes with two AMD EPYC 7742 CPUs each, 128 cores per node). However, to obtain an optimal number of I/O operations per run and a more efficient usage of the resources, the simulation runs were often limited to 32678 cores (256 nodes), as the required wall time for I/O operations did not show a significant improvement between 256 and 512 nodes. An average of 2–3 I/O operations per run were performed, with each HDF5 output file requiring 1.3 TB of space, plus an additional binary file of 300 GB containing the Lagrangian particles data. A total of 40 time steps, one every 0.25 ms, were saved. Although the simulation is characterized by a large number of chemical species and by the additional Lagrangian particles for soot moment equations, the full number of cores per node has been used for the 256 and 512 node runs, meeting the necessary memory per core requirements. Test runs with fewer computational nodes (e.g., 128 nodes) were performed, however, due to memory limitations, the full number of cores per node could not be employed in this case. Among these different configurations, a very good scaling of the code was observed, with the only exception being the I/O performance between 256 and 512 nodes which often depended on the overall load on the file system. Including initialization and I/O operation, an effective average wall time per step of 364 s was observed when using 32768 cores (256 nodes). For reference, the wall time per step when using 128, 256 and 512 nodes are included in Table 1. For the 128 nodes case, only 96 cores per node could be used due to memory limitations.

Table 1 Wall time per step for different configurations

Number of nodes	Number of cores	Wall time per step [s]
128	12288 (96 per node)	955
256	32768 (128 per node)	364
512	65536 (128 per node)	216

3.2 Gas Phase and Soot Evolution

The two fuel stripes are initialized with a burning flamelet solution; therefore, four flame fronts are present in the early stages of the simulation, two in between the fuel stripes and two external fronts. This results in a fast spreading of the mixture fraction, as can be seen in Fig. 3. In this figure, cuts for different time instances across the simulation are shown, from a very early stage (0.25 ms) to the end of the simulation (10 ms). The internal flame fronts are quickly suppressed after a few milliseconds both due to the rapid consumption of the central portion of oxidizer and the fast mixing of the fuel stripes induced by turbulence. Average mass fractions of O_2 and OH (not shown here) in the crosswise direction clearly show this trend, as OH radical concentration decreases quickly in the jet core. Turbulent mixing is caused by the Kelvin-Helmoltz instabilities at $|y| = 7.5\,\text{mm}$, where the mixing layer between the positive and negative velocity streams is generated. Overall, due to the fast mixing, the mixture fraction value mostly drops below Z = 0.4 already at 2 ms (second cut of Fig. 3). This value indicates approximately the mixture fraction location of peak naphthalene production in the one-dimensional flamelet used for initialization. Even though the fuel consumption is rapid and PAH formation requires longer residence times, enough time for sufficient naphthalene production is provided in the simulation to allow for soot formation. In later stages, the mixture fraction peak value continues dropping, with most of the mixture being below stoichiometric value ($Z_{st} = 0.17$), indicating an almost complete fuel consumption. PAH is mainly formed between two and four milliseconds, with only a few pockets left afterward and completely depleted before the end of the simulation.

The formation of PAH in the first half of the simulation translates into a significant soot particle formation, which can be observed in Fig. 4 (top), where cuts of soot number density are shown across the simulation time. Soot number density peaks are reached at around 3-4 ms, similarly to PAH, and later the total number of particles drops significantly. This happens mainly for three reasons: first, due to the lack of

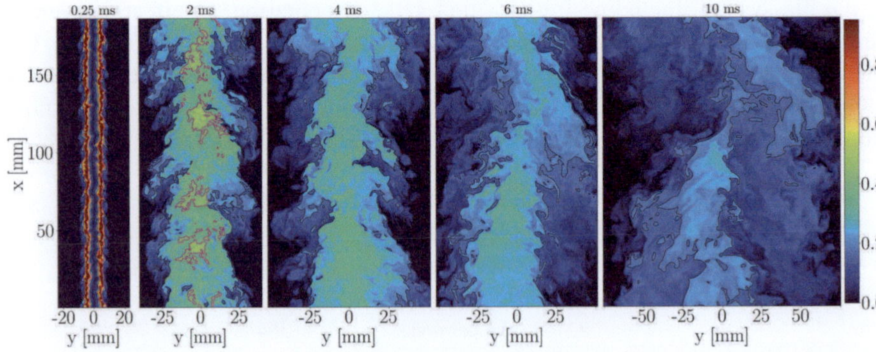

Fig. 3 Two-dimensional cuts of the mixture fraction distribution at 0.25, 2, 4, 6, and 10 ms. The black line indicates the stoichiometric mixture fraction (Z = 0.17)

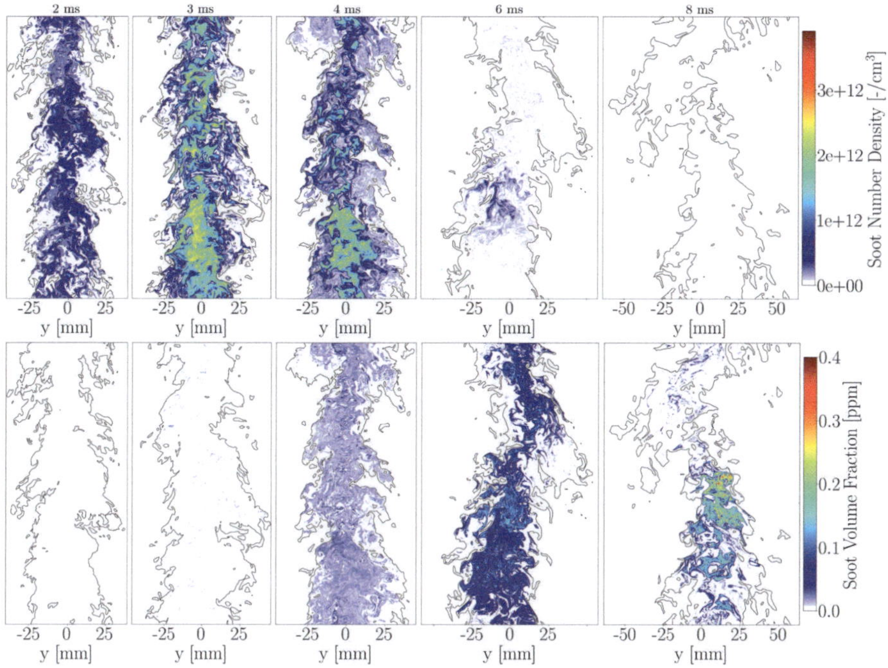

Fig. 4 Two-dimensional cuts of soot number density (top), and soot volume fraction (bottom) at 2, 3, 4, 6, and 8 ms

PAH in the second half of the simulation, no new incipient particles are formed. Second, after formation and initial growth, soot particles collide with each other and merge into larger particles (coagulation), leading to a further drop in the overall number density. Finally, as soot particles are transported towards leaner mixtures, and towards the flame fronts, they are consumed by surface oxidation, both from OH radicals and molecular oxygen.

Conversely, the overall soot volume fraction follows a different trend, as shown in Fig. 4 (bottom). In the first half of the simulation, a large number of particles is formed; however, most of the particles are at incipient size (which in the soot inception model used are assumed to be spherical particles with a diameter of approximately 1 nm), therefore not contributing significantly to the overall soot volume growth. Later, after 4 ms, surface growth starts to become predominant and peak values of 0.4 ppm are reached in the final stages of the simulation. Simultaneously, the region of high concentration of soot becomes more and more narrow (e.g., cut at 8 ms) at higher mixture fractions above stoichiometry. This is due to most of the domain being characterized by a leaner mixture with strong oxidation, thus particle shrinking and depletion. The overall time evolution of soot number density and soot mass is shown in the left plot of Fig. 5, where it can be seen that the two quantities' peaks are observed in different instants and that, by the end of the simulation, both quantities are small, indicating almost complete oxidation.

Generally, the main mechanisms responsible for soot surface growth are PAH adsorption, which happens through the collision of PAH dimers on the surface of existing soot particles, and the HACA growth mechanism. As PAH is mostly consumed before any significant soot mass is formed, the HACA growth mechanism remains the sole contributor to the formation of larger soot agglomerates. This is inherently different from all other existing DNS data sets of turbulent sooting flames where growth is mainly driven by PAH adsorption due to the continuous formation of PAH and incipient soot particles which drift towards richer mixtures. This makes this presented data set unique also for future analysis of HACA growth in turbulent environments.

The individual contribution of the different soot volume fraction source terms and their evolution is shown in the right plot of Fig. 5, together with the total net rate. Although limited, contributions from inception and PAH adsorption are predominant in the first phase, where the richer mixture favors these two processes. Later, soot is found at lower mixture fractions, and oxidation and HACA are predominant. Contributions from both source terms become increasingly important, but the overall soot oxidation overcomes the growth phase after 6 ms, leading to a total negative net rate until the end of the simulation. A more in-depth analysis of soot oxidation is provided in the next subsection. Finally, the joint probability density function (jPDF) of soot volume fraction and mixture fraction is shown in Fig. 6 for four different instants. In the first plot (4 ms) only smaller soot volume fractions, below 0.05 ppm, are most likely to be found and appear over a wider range of mixture fractions (from Z = 0.2 up to Z = 0.4) as the different physical phenomena contribute together to soot formation and growth. Later, at both 6 and 8 ms, most of the soot volume fraction is located between Z = 0.25 and Z = 0.3, where the HACA species such as C_2H_2 are still largely present, promoting surface growth. In addition, as the fuel is fully consumed, the peak mixture fraction becomes smaller and soot is gradually

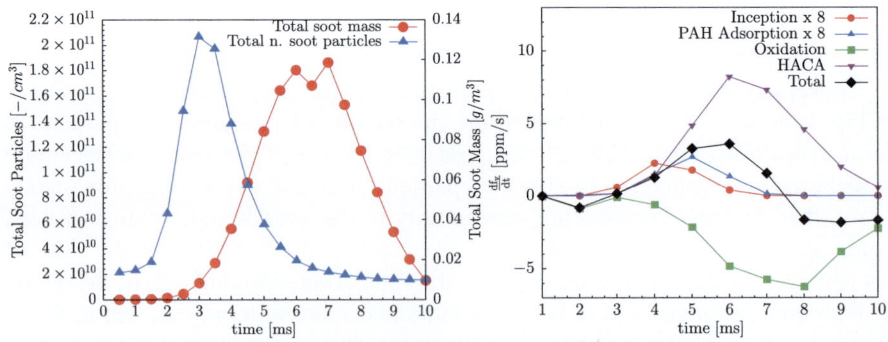

Fig. 5 Left plot: time evolution of the total number of soot particles (red) and total soot mass (blue). Right plot: cumulative contribution of the individual soot volume fraction source terms. The black line indicates the net contribution

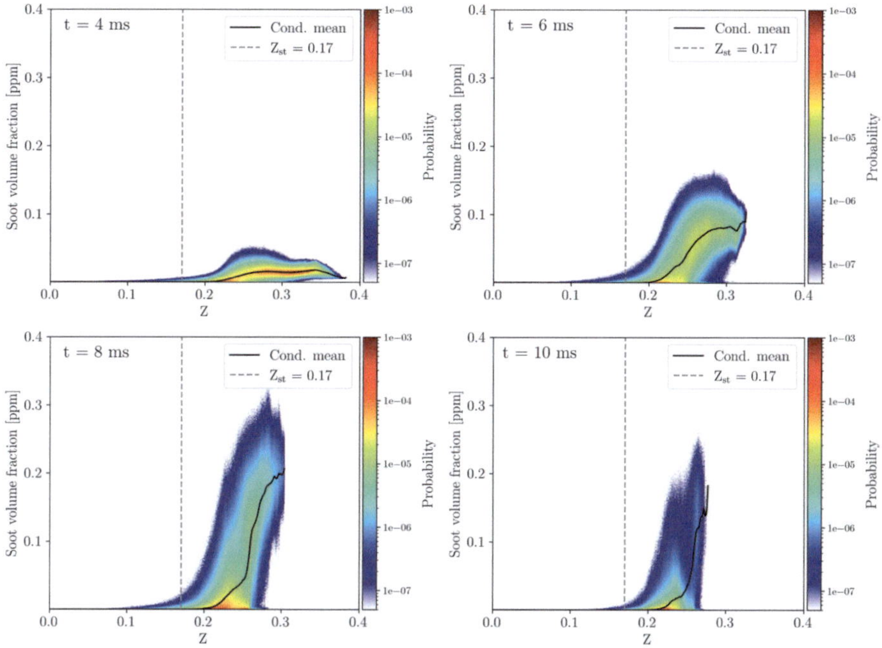

Fig. 6 Joint PDF of soot volume fraction and mixture fraction at different times. The black dashed line indicates the stoichiometric mixture fraction. The black continuous line indicates the conditional mean of the field with respect to mixture fraction

transported towards lower mixtures ($Z < 0.2$) where oxidation becomes predominant. At 10 ms, the higher soot volume fraction values are less likely to be found as only a few pockets of soot are left, while the rest has been oxidized.

It is worth noting that only a little amount (roughly 1% of the total soot mass) of soot survives the oxidation phase and drifts toward and below the stoichiometric mixture fraction ($Z_{st} = 0.17$). The DNS was intentionally designed such that a strong oxidation and complete particle depletion could be obtained, however in typical combustion devices, higher turbulence is usually present and, as also shown in the experiments of Kruse et al. [3] and discussed by Attili et al. [11] in their DNS runs with different Damköhler number, pockets of larger soot agglomerates can survive through the flame front and be transported towards leaner mixtures, for example in case of local extinction phenomena. A more in-depth study on soot breakthrough phenomena is the main focus of this project's follow-up application, where soot particles' leakage into leaner mixtures will be enforced with a parameter variation, namely the Damköhler number.

3.3 Soot Oxidation

The soot oxidation term for a generic soot moment can be written as follows:

$$\dot{M}^{\mathrm{Ox}}_{x,y} = f\left(M_{x,y}\right) \cdot \left(f(T, Y_{\mathrm{OH}}) + f(T, Y_{\mathrm{O_2}})\right) \tag{3}$$

where the first term includes all the terms dependent on the statistical moments, while the second includes the contribution from the gas phase quantities. In particular, the second term is split into two different contributions, one soot oxidation rate coefficient due to OH radicals ($k_{\mathrm{Ox,OH}}$) and the second due to molecular oxygen ($k_{\mathrm{Ox,O_2}}$). For the soot moment equation that describes soot volume fraction ($M_{1,0}$), the first term simplifies to the overall soot surface density ($M_{0,1}$). For further analysis on soot oxidation in the following sections, the DNS data at 8 ms has been selected, as it is characterized by significant oxidation levels, both for larger aggregates and smaller particles. Soot oxidation is usually localized in narrow layers with sharp peaks, and in case of strong oxidation, smaller pockets of soot hardly survive the oxidation process towards leaner mixtures. Nonetheless, in the joint PDF of soot oxidation rate and mixture fraction (Fig. 7), although higher oxidation values are localized slightly above stoichiometric conditions (Z = 0.2), moderate oxidation phenomena are observed in a broader mixture fraction range, with a non-negligible portion of oxidation happening below stoichiometric conditions. Joint statistics of the individual contributions from OH and O_2 to the soot oxidation rate coefficient (Fig. 8) with respect to mixture fraction, clearly show that both $k_{\mathrm{Ox,OH}}$ and $k_{\mathrm{Ox,O_2}}$ have the same order of magnitude. These coefficients depend only on the gas phase, with OH oxidation acting on a broader mixture fraction range and, more importantly, slightly above stoichiometric conditions. Molecular oxygen, instead, only contributes below stoichiometric conditions. Therefore, when soot is convected towards leaner mixtures from the growth region, it is mostly oxidized by means of OH radicals

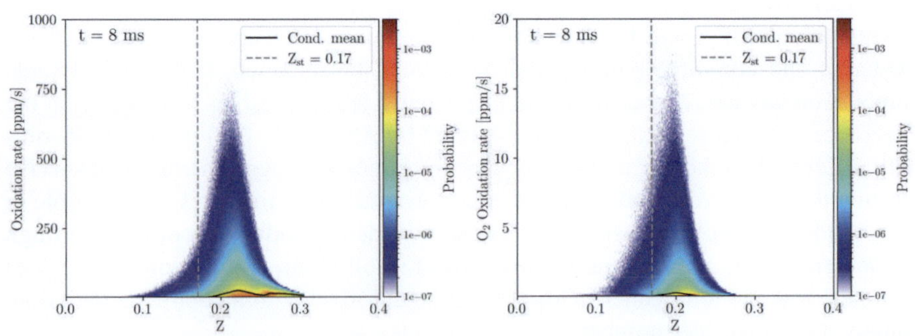

Fig. 7 Joint PDF of the soot oxidation rates (OH and O_2 contribution on the left, O_2 only on the right) and mixture fraction at 8 ms

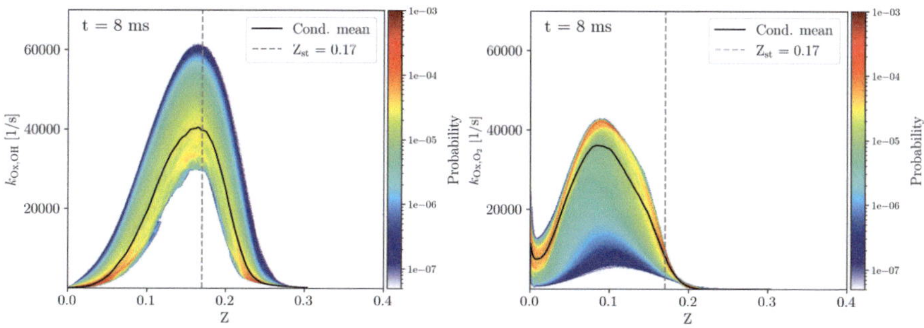

Fig. 8 Joint PDF of soot oxidation rate coefficient due to OH (left) and molecular oxygen (right) at 8 ms

acting on the particle's surface, as suggested by the overall oxidation joint statistics. Nonetheless, these results suggest that, in case of breakthrough events, soot oxidation due to molecular oxygen might become more important.

3.4 Soot Subfilter Models Analysis

An assessment of existing soot subfilter models is given in this section. For the computation of the filtered soot moment source term, the joint PDF of the soot moments $M_{x,y}$ and of the thermochemical properties ξ is necessary. Using Bayes' theorem, this term can be rewritten as follows $P(M, \xi) = P(M|\xi)P(\xi)$, where the first term is the PDF of the soot moments conditioned on the thermochemical properties. A first model was proposed by Mueller and Pitsch [12] which assumed that soot formation time scales are much larger than chemical time scales. It follows that the two variables are independent, hence the conditional PDF $P(M|\xi)$ can be rewritten as $P(M|\xi) = P(M)$, where this last term is a presumed subfilter PDF that takes soot intermittency into account with a 2-δ formulation. An improved formulation for $P(M)$ has been suggested by Berger et al. [14], but it was shown that this formulation does not affect soot oxidation prediction. In fact, in case of fast soot evolution process, such as oxidation, the different time scales assumption is no longer true and the soot-gas phase subfilter interactions must be considered. Wick et al. [15] showed that the 2-δ model led to a significant overprediction of soot oxidation comparing LES with experimental results. An improved model for $P(M|\xi)$ has been introduced by Yang et al. [13], where the correlation with the gas phase is accounted for by means of a cut-off mixture fraction (Z_{soot}), for which it is assumed that soot is no longer oxidized below this value. A recommended value for it, which can be chosen arbitrarily, is the mixture fraction value at which surface growth and oxidation have the same magnitude. Figure 9 shows small DNS cuts (33x33 points) of soot oxidation quantities, where the stoichiometric mixture fraction (Z_{st}), and the cut-off mixture value defined as $Z_{soot} = Z_{avg} = (Z_{sg,max} + Z_{ox,max})/2$ are indicated with

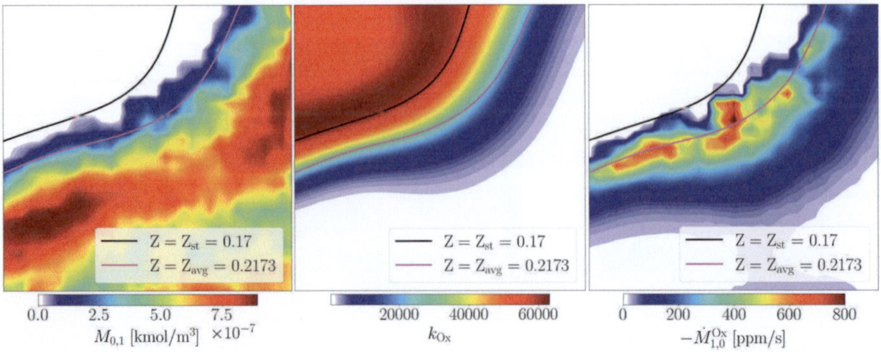

Fig. 9 33 × 33 cuts of DNS data for characteristic oxidation quantities: soot surface (left), soot oxidation rate coefficient (center), and soot oxidation rate (right). The cuts are centered at mixture fraction $Z = 0.22$. The two isolines indicate the stoichiometric mixture fraction (white) and the cut-off value $Z_{Soot} = Z_{avg}$ (red)

black and magenta isolines, respectively. It can be observed that, depending on the cut-off value, only part of the soot oxidation region is accounted for.

The conditional mean with respect to the Favre-averaged mixture fraction of the filtered DNS data at a filter size of $\Delta = 32$ is compared with the different models in Fig. 10. For all cases, the oxidation rate coefficient k_{Ox} has been replaced with its mean conditioned on mixture fraction, to remove any additional error arising from its misprediction. Reference filtered DNS data is indicated in blue, while the green line indicates the 2-δ model, here referred to as the "Marginal PDF" model. Several cut-off mixture fractions have been tested for the cut-off model (here referred to as the "Conditional PDF" model), but only two values have been reported here: a lower cut-off value ($Z_{st} = 0.17$), in orange, and an optimal cut-off value ($Z_{soot} = 0.216$)

Fig. 10 Conditional average on Favre averaged mixture fraction of filtered DNS data and two different models (Marginal PDF and Conditional PDF models) at filter sizes of $\Delta = 32$. The blue line indicates the reference solution (filtered DNS)

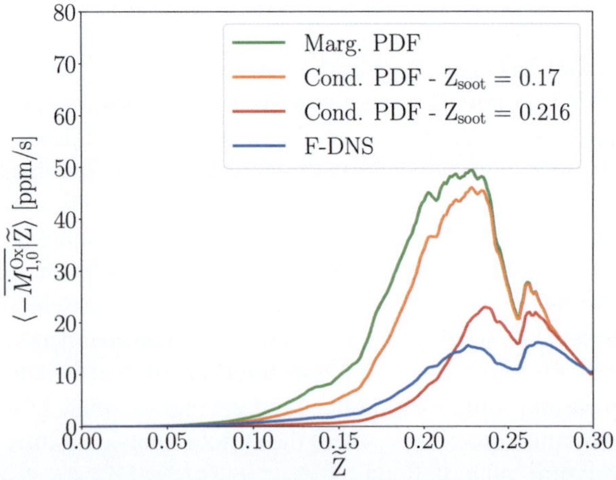

which recovers the average oxidation level of the filtered DNS, in red. The 2-δ model shows a strong overprediction compared to the reference solution, as the turbulence-soot interactions are not accounted for by this model. The conditional PDF model with a lower cut-off mixture fraction value (Z_{st}) does not significantly improve the predictions and almost recovers the overpredicted 2-δ results. The conditional PDF model with the optimal cut-off mixture fraction $Z_{soot} = 0.216$, instead, is closer in predicting the conditional average of the filtered DNS. The optimal cut-off value is similar to the average cut-off level (Z_{avg}) shown in Fig. 9, which can be easily estimated from 1D counterflow simulations. Although the average oxidation level can be recovered with an appropriate cut-off value, reducing significantly the spurious oxidation from the 2-δ model, oxidation values are overpredicted above the cut-off value and underpredicted below the cut-off value (leaner region) by the conditional PDF model.

Although the conditional PDF model improves the oxidation prediction, it is worth noting that from the DNS data, as seen in the joint PDF in Fig. 7 or from the small cuts in Fig. 9, the oxidation is not necessarily located just above the cut-off value, but actually, a significant amount of oxidation happens around stoichiometric conditions, therefore the similar level of oxidation between the conditional PDF model and the filtered DNS might result from some error compensation across the cut-off mixture fraction. In the case of soot breakthrough events in leaner mixtures, the assumption of a cut-off value for oxidation might lead to a severe misprediction of soot oxidation and of the general soot emissions in highly turbulent environments characteristic of combustion devices. The influence of soot breakthrough events and an improved model to account for these effects will be investigated in depth in a follow-up project.

4 Summary and Conclusions

A new DNS focused on soot oxidation in a turbulent environment was performed and analyzed. The evolution of soot particles across the mixture fraction space was discussed and implications for further soot subfilter models were considered. The presented data set is one of the largest available on soot formation and oxidation in a turbulent environment and the first large-scale simulation to cover the later stages of soot evolution. HACA growth was shown to be the main growth mechanism, compared to other data sets where PAH adsorption is predominant due to the richer conditions. An almost complete soot particle depletion is reached and soot oxidation was found to be mostly driven by surface reaction with OH radicals, as most of the consumption happens above stoichiometric conditions and very few particles penetrate the flame front, where O_2 oxidation would become important. Existing soot subfilter models to account for subfilter interaction between soot and gas phase were tested a-priori on the DNS data. The conditional PDF model showed the best results when compared with filtered DNS data. However, this might result from some error compensation, as oxidation phenomena are observed also below the cut-off value in the DNS data and the cut-off value is chosen in an arbitrary way. Moreover, this model does not account for the possibility of soot breakthrough events which may occur in case of strong turbulence, typical of combustion devices. More research on soot subfilter models is necessary, in particular concerning soot oxidation and soot breakthrough.

Acknowledgements Generous support of the Deutsche Forschungsgemeinschaft (DFG) under grant number PI 368/25-1 and the Research Association for Combustion Engines (FVV) under grant number 6013970 is gratefully acknowledged. The research leading to these results has also received funding from the European Union's Horizon 2020 research and innovation program under the Center of Excellence in Combustion (CoEC) project, grant agreement no. *952181*. The authors gratefully acknowledge the Gauss Centre for Supercomputing e.V. (www.gauss-centre.eu) for funding this project by providing computing time on HAWK at Höchstleistungsrechenzentrum Stuttgart (www.hlrs.de).

References

1. Lighty, J.S., Veranth, J.M., Sarofim, A.F.: Combustion aerosols: factors governing their size and composition and implications to human health. J. Air Waste Manag. Assoc. **50**(9), 1565–1618 (2000)
2. Park, O., Burns, R.A., Buxton, O.R., Clemens, N.T.: Mixture fraction, soot volume fraction, and velocity imaging in the soot-inception region of a turbulent non-premixed jet flame. Proc. Combust. Inst. **36**, 899–907 (2017)
3. Kruse, S., Ye, J., Sun, Z., Attili, A., Dally, B., Medwell, P., Pitsch, H.: Experimental investigation of soot evolution in a turbulent non-premixed prevaporized toluene flame. Proc. Combust. Inst. **37**, 849–857 (2019)
4. Pitsch, H.: Large-eddy simulation of turbulent combustion. Annu. Rev. Fluid Mech. **38**, 453–482 (2006)
5. Pierce, C.D., Moin, P.: Progress-variable approach for large-eddy simulation of non-premixed turbulent combustion. J. Fluid. Mech. **504**, 73–97 (2004)
6. Pitsch, H., Riesmeier, E., Peters, N.: Unsteady flamelet modeling of soot formation in turbulent diffusion flames. Combust. Sci. Technol. **158**, 389–406 (2000)
7. Mueller, M.E., Pitsch, H.: Large eddy simulation subfilter modeling of soot-turbulence interactions. Phys. Fluids **23**, 115104 (2011)
8. Lignell, D.O., Chen, J.H., Smith, P.J., Lu, T., Law, C.K.: The effect of flame structure on soot formation and transport in turbulent nonpremixed flames using direct numerical simulation. Combust. Flame **151**, 2–28 (2007)
9. Lignell, D.O., Chen, J.H., Smith, P.J.: Three-dimensional direct numerical simulation of soot formation and transport in a temporally evolving nonpremixed ethylene jet flame. Combust. Flame **155**, 316–333 (2008)
10. Attili, A., Bisetti, F., Mueller, M.E., Pitsch, H.: Formation, growth, and transport of soot in a three-dimensional turbulent non-premixed jet flame. Combust. Flame **161**, 1849–1865 (2014)
11. Attili, A., Bisetti, F., Mueller, M.E., Pitsch, H.: Damköhler number effects on soot formation and growth in turbulent nonpremixed flames. Proc. Combust. Inst. **35**, 1215–1223 (2015)
12. Mueller, M.E., Pitsch, H.: LES model for sooting turbulent nonpremixed flames. Combust. Flame **159**, 2166–2180 (2012)
13. Yang, S., Lew, J.K., Mueller, M.E.: Large eddy simulation of soot evolution in turbulent reacting flows: Presumed subfilter PDF model for soot-turbulence-chemistry interactions. Combust. Flame **209**, 200–213 (2019)

14. Berger, L., Wick, A., Attili, A., Mueller, M.E., Pitsch, H.: Modeling subfilter soot-turbulence interactions in large eddy simulation: An a priori study. Proc. Combust. Inst. **38**, 2783–2790 (2021)
15. Wick, A., Priesack, F., Pitsch, H.: Large-eddy simulation and detailed modeling of soot evolution in a model aero engine combustor. In: Turbo Expo: Power for Land, Sea, and Air, vol. 50848, p. V04AT04A020. American Society of Mechanical Engineers (2017)
16. Berger, L., Attili, A., Pitsch, H.: Synergistic interactions of thermodiffusive instabilities and turbulence in lean hydrogen flames. Combust. Flame **244**, 112254 (2022)
17. Desjardins, O., Blanquart, G., Balarac, G., Pitsch, H.: High order conservative finite difference scheme for variable density low mach number turbulent flows. J. Comput. Phys. **227**, 7125–7159 (2008)
18. Tomboulides, A., Lee, J., Orszag, S.: Numerical simulation of low Mach number reactive flows. J. Sci. Comput. **12**, 139–167 (1997)
19. Blanquart, G., Pepiot-Desjardins, P., Pitsch, H.: Chemical mechanism for high temperature combustion of engine relevant fuels with emphasis on soot precursors. Combust. Flame **156**, 588–607 (2009)
20. Saggese, C., Ferrario, S., Camacho, J., Cuoci, A., Frassoldati, A., Ranzi, E., Wang, H., Faravelli, T.: Kinetic modeling of particle size distribution of soot in a premixed burner-stabilized stagnation ethylene flame. Combust. Flame **162**, 3356–3369 (2015)
21. Frenklach, M.: Method of moments with interpolative closure. Chem. Eng. Sci. **57**, 2229–2239 (2002)
22. Mueller, M.E., Blanquart, G., Pitsch, H.: A joint volume-surface model of soot aggregation with the method of moments. Proc. Combust. Inst. **32**, 785–792 (2009)
23. Scialabba, G., Langer, R., Attili, A., Berger, L., Pitsch, H.: The relevance of soot diffusion in counterflow laminar flame configurations. In: 10th European Combustion Meeting (2021)
24. Frenklach, M., Wang, H.: Detailed mechanism and modeling of soot particle formation. In: Bockhorn, H. (ed.) Soot Formation in Combustion, pp. 165–192. Springer (1994)
25. Mueller, M., Blanquart, G., Pitsch, H.: Hybrid method of moments for modeling soot formation and growth. Combust. Flame **156**, 1143–1155 (2009)
26. Pitsch, H.F.: A C++ computer program for 0D combustion and 1D laminar flame calculations, v4.2.1
27. Koumoutsakos, P.: Multiscale flow simulations using particles. Annu. Rev. Fluid Mech. **37**, 457–487 (2005)
28. Attili, A., Bisetti, F.: Application of a robust and efficient lagrangian particle scheme to soot transport in turbulent flames. Comput. Fluids **84**, 164–175 (2013)

Modeling of Coupled Turbulent Channel-Porous Media Flow Through a Deep Autoencoder-Echo State Network Framework

Xu Chu, Sandeep Pandey, Yanchao Liu, and Bernhard Weigand

Abstract In this study, we propose a novel approach, namely the combined Convolutional Deep AutoencoderâŁ"Echo State Network (CDAE-ESN) model, for the analysis and forecasting of dynamics and low-order statistics in coupled turbulent channel-porous media flows. Such systems find wide applications in industrial settings, including transpiration cooling and smart interface engineering. However, the complex geometry of coupled flow systems presents additional challenges for purely data-driven models. Our results demonstrate that the integration of deep autoencoder and echo state network techniques enables effective modeling and prediction of dominant flow behaviors, particularly within the porous domain exhibiting laminar regimes. To enhance the model's applicability across a broader range of data domains, we further employ fine-tuning on a dataset encompassing varying porosities. The achieved average statistics exhibit a reasonable agreement, underscoring the efficacy of our proposed approach.

1 Introduction

Turbulent flow over permeable interfaces is ubiquitous in nature and engineering applications, from sediment transport to the transpiration cooling in gas turbines. Porous media are generally featured with complex topology in space [7]. A wide range of properties has been identified to have a significant effect on the mass and momentum transport across the porous medium-free flow interface. However, understanding and optimizing macro/micro-scale characteristics of porous media remains

X. Chu (✉)
Cluster of Excellence SimTech, University of Stuttgart, Stuttgart, Germany
e-mail: xu.chu@simtech.uni-stuttgart.de

S. Pandey
Institute of Thermodynamics and Fluid Mechanics, Technische Universität Ilmenau, Ilmenau, Germany

Y. Liu · B. Weigand
Institute of Aerospace Thermodynamics (ITLR), Stuttgart, Germany

T. Ludwig et al. (eds.), *High Performance Computing in Science and Engineering '23*,
https://doi.org/10.1007/978-3-031-91312-9_22

a challenge. This is due to both the difficulty of experimental measurements at the pore scale [29] as well as the high costs of numerical simulations to resolve the full spectrum of scales.

Early studies of turbulent flows over permeable beds have addressed the effect of varying permeability on surface flows. The turbulent surface flow is similar to that of a canonical boundary layer when the interfacial permeability is low, as the near-wall structures are less disrupted. However, as the permeability increases, large-scale vortical structures emerge in the surface flow. This is attributed to Kelvin-Helmholtz (KH) type instabilities from inflection points of the mean velocity profile. More recently, the application of porous media for drag reduction inspired a series of in-depth studies on anisotropic permeability.

Direct Numerical Simulation (DNS) exhibits an edge of observing and analyzing turbulent physics in a confined small space, not only for the canonical cases as channel flows and pipe flows [4, 21]. Suga et al. [28] investigated the influence of the anisotropic permeability tensor of porous media at a higher permeable regime. It was found that streamwise and spanwise permeabilities enhance turbulence whilst vertical permeability itself does not. In particular, the enhancement of turbulence is remarkable over porous walls with streamwise permeability, as it allows the development of streamwise large-scale perturbations induced by the KelvinâŁ"Helmholtz instability. However, DNS is still an expensive solution when one extend it for the parametric study often desired in the industrial context. It is still common to use analytical models or advanced turbulence models for this purpose which provide a reasonable accuracy in established applications. However, these models can diverge and model calibration is a time-consuming process.

Machine learning (ML) is on surge where it has successfully tackled many complex tasks ranging from self-driving cars, drug discovery, weather prediction, and more recently state-of-the-art large language models such as GPT for natural language processing. Machine learning has been democratized heavily in the last decade and some of the reasons include vast availability of open-source libraries, active community, huge volume of incoming data from physical as well as numerical experiments and an ease in access to high-performance computing resources via cloud. In recent years, ML has shown potential in the modelling of various fluid flow use-cases [2, 11, 23, 24]. Typically, big data ML problems rely upon some kind of feature extraction process e.g. conventional feature selection process, long-standing proper orthogonal decomposition (POD), relatively new dynamic mode decomposition (DMD) [15] or ML based methods [27]. Feature extractor transform the input raw data into information preserving feature vector with a primary goal of reducing the dimensionality of the data. Several attempts have been made in the past few years to combine POD or DMD with machine learning [22]. However, these methods have certain limitation especially when the flow advances in the turbulent flow regimes. Therefore, an end-to-end machine learning system is being suggested where POD and DMD is replaced by a non-linear DNN [16, 22]. A common architecture is deep convolution autoencoder (DAE) to extract the features for further processing and decision making. Turbulent flow problems often involve tempo-spatial data, therefore, extracted features could be used to train a downstream network to build a data-driven dynamic model. A common choice is a recurrent neural network (RNN) which preserves the sequential relationship in a time series. A combination of such approaches has shown exemplary results in terms of data compression and modeling.

In this work, we make an attempt to model complex flow behaviors in a porous medium with the help of a data-driven approach. The flow is in a turbulent regime and due to the presence of porosity, geometry is resolved to a finer scale. This setup makes the end-to-end ML task challenging and equally interesting. The goal is to combine the deep learning based feature extraction step followed by a recurrent neural network to model temporal evolution. Echo state network was chosen as the RNN due to its proven efficiency in modeling multi-dimensional tempo-spatial data. The trained models are restricted to the domain of training and work poorly beyond the trained data domain which could be a specific parameter such as Reynolds number. A commonly used strategy called transfer learning is used, where the objective is to transfer the gained knowledge from the training of source task to related target task with a limited dataset [35]. Fine tuning, which can be considered as a part of transfer learning, is one methodology where an already trained and optimized model tuned with another similar set of data for a similar task. This could decrease the training time significantly in addition to neural architecture search and training data requirements. Fine-tuning can be performed on the entire network level where we allowed to retrain all the layers while initializing the parameters with the pretrained model. Alternatively, several layers can be freeze and retraining is allowed on only few layers. The former method is suitable when network size is very large and data is limited. While first approach could lead to the overfitting [34]. Therefore, we also present the result from fine-tuning where trained model from a given data domain was transferred to a different data domain.

2 Synthetic Data Generation

In our DNS, the three-dimensional incompressible Navier–Stokes equations are solved in non-dimensional form,

$$\frac{\partial u_j}{\partial x_j} = 0 \tag{1}$$

$$\frac{\partial u_i}{\partial t} + \frac{\partial u_i u_j}{\partial x_j} = -\frac{\partial p}{\partial x_i} + \frac{1}{Re_D}\frac{\partial^2 u_j}{\partial x_i \partial x_j} + \Pi\delta_{i1} \tag{2}$$

where Π is a constant pressure gradient in the mean-flow direction. The governing equations are normalized using the half-width of the whole simulation domain H (Fig. 1) and the averaged bulk velocity U_b of the channel region ($y/H = [0, 1]$). Hereafter, the velocity components in the streamwise x, wall-normal y, and spanwise z directions are denoted as u, v, and w, respectively. The domain size ($L_x/H \times L_y/H \times L_z/H$) is $10 \times 2 \times 0.8\pi$ in all cases. The lower half ($y/H =$

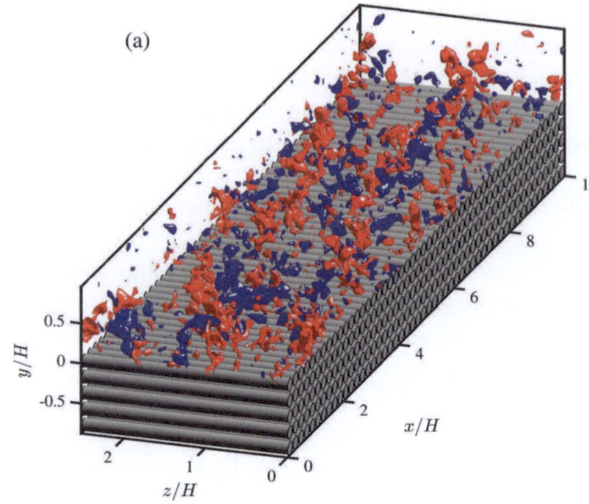

Fig. 1 Configuration of the computational domain (C05), the permeability in three directions can be found in Table 1. The blue and red isosurfaces show the wall-normal fluctuation v'/u_τ^p at level -0.4 and 0.4, respectively

$[-1, 0]$) contains the porous media, and the upper half ($y/H = [0, 1]$) is the free channel flow. The porous layer consists of 50 cylinder elements along the streamwise direction and 5 rows in the wall-normal direction, as illustrated in Fig. 1. The distance D between two nearby cylinders is fixed at $D/H = 0.2$. No-slip boundary conditions are applied to the cylinders, the upper wall, and the lower wall. Periodic boundary conditions are used in both streamwise and spanwise directions.

The spectral/*hp* element solver Nektar++ is used to solve equations (1), (2) [1, 5, 6, 8, 9, 20, 31, 32]. The geometry in the x-y plane is discretized into quadrilateral elements with local refinement near the cylinders (see Fig. 1). Local element expansions are applied based on the modified Legendre basis [13]. We used flexible polynomial orders across the wall normal range in a continuous Galerkin projection. The polynomial order in the free flow region $y/H = [0.2, 1]$ is $P = 6$-7. The near wall region and the top two layers of cylinders $y/H = [-0.4, 0.2]$ are enhanced with a higher order of $P = 8$-9. A lower order of $P = 5$ is selected in the deeper positions of the cylinder array ($y/H = [-1, -0.4]$). The spanwise direction is extended with a Fourier spectral method. The 2/3 rule is used to avoid aliasing errors. The time-stepping is performed with a second-order mixed implicit-explicit (IMEX) scheme. The time step is fixed at $\Delta T/(H/U_b) = 5 \times 10^{-4}$.

Four DNS cases are performed with varying porosity $\varphi = 0.5$ and 0.6, which is defined as the ratio of the void volume to the total volume of the porous structure. The parameters of the simulated cases are listed in Table 1, where the cases are named after their respective porosity. The superscripts $(\cdot)^p$ and $(\cdot)^s$ represent permeable wall and smooth wall side variables, respectively. Variables with superscript $^+$ are scaled by friction velocities u_τ of their respective side and viscosity ν.

For an arbitrary variable ϕ, we denote the time averaged value as $\bar{\cdot}$, i.e., $\overline{\phi} = 1/T \int_0^T \phi \mathrm{d}t$, and spatial averaged value in x-z plane as $\langle\cdot\rangle$, i.e., $\langle\phi\rangle = 1/A_f \int_{A_f} \phi \mathrm{d}A$, A_f being the fluid area. The instantaneous turbulent fluctuation is $\phi' = \phi - \overline{\phi}$, and

Table 1 Simulation parameters. The porosity of the porous medium region is φ. The friction Reynolds numbers are Re_τ^p and Re_τ^s for the porous and impermeable top walls respectively. $\sqrt{K_{\alpha\alpha}}^{p+}$ and $C_{\alpha\alpha}^{p+}$ are the diagonal components of the permeability tensor and Forchheimer coefficient, respectively, in the direction of α ($\alpha \in \{x, y, z\}$), which are normalized by wall units. r_c^{p+} is the radius of the cylinders

Case	φ	Re_τ^p	Re_τ^s	$\sqrt{K_{xx}}^{p+}$,	$\sqrt{K_{zz}}^{p+}$ $\sqrt{K_{yy}}^{p+}$	C_{xx}^{p+}, C_{yy}^{p+}	r_c^{p+}	C_f^p	C_f^s
C05	0.5	336	180	4.55	8.86	0.37	42	0.0112	0.0084
C06	0.6	464	190	9.34	15.23	0.58	48	0.0149	0.0085

form-induced fluctuation is $\tilde{\overline{\phi}} = \overline{\phi} - \langle\overline{\phi}\rangle$. For the permeable wall side, the total shear stress at the interface can be derived as:

$$\tau_w^p = \left(\mu \frac{\partial \langle\overline{u}\rangle}{\partial y} - \rho\langle\overline{u'v'}\rangle - \rho\langle\overline{u}\rangle\langle\overline{v}\rangle - \rho\langle\tilde{\overline{u}}\tilde{\overline{v}}\rangle\right)_{y=0}. \tag{3}$$

For the smooth wall side,

$$\tau_w^s = \left(\mu \frac{\partial \langle\overline{u}\rangle}{\partial y}\right)_{y=H}, \tag{4}$$

The friction velocity can then be calculated from total shear stress at both walls $u_\tau = \sqrt{\tau_w/\rho}$. The friction coefficient is defined as $C_f = \tau_w/(\frac{1}{2}\rho U_b^2)$, which is listed in Table 1.

Note that the distance between cylinders is fixed, and the porosity is changed by varying the radius of the cylinders. The normalized cylinder radius is in the range $r_c^{p+} = r_c u_\tau^p/\nu =$ 42-48 for all the cases tested (see Table 1), such that the effect of surface roughness is assumed to be at a similar level. For all the cases, the Reynolds number of the top wall boundary layer is set to be $Re_\tau^s = \delta^s u_\tau^s/\nu \approx 180$ (δ is the distance between the position of maximum streamwise velocity and the wall). In this manner, changes in the top wall boundary layer are minimized. On the top smooth wall side, the streamwise cell size ranges from $4.1 \leq \Delta x^{s+} \leq 6.3$ and the spanwise cell size is below $\Delta z^{s+} = 5.4$. On the porous media side, Δz^{p+} is below 8.4, whereas Δx^{p+} and Δy^{p+} are enhanced by polynomial refinement of local mesh. The total number of grid points ranges from 88×10^6 (C05) to 595×10^6 (C08). Each cylinder in the porous domain is resolved with 80 to 120 grids along the perimeter. The spatial resolution of current work is close to those of previous DNS researches. For example, [25] conducted DNS of a turbulent flow over sediment beds using the immersed boundary method, where the diameter is discretized into 36-50 points. The mesh of [14] has 26 grid points along the diameter of each grain.

By performing parameter tests on porous media and fitting the Darcy-Forchheimer equation, the permeability tensor K and Forchheimer coefficient C may be obtained, and the details of the computation can be referred to [30]. The intensity profiles of u', v', and w' are shown in Fig. 1b–d, respectively. The turbulent intensity grows with porosity for both the channel and the porous medium region. In the current study, we focus on the up and down-welling motions at the upper surface of the porous medium, which is represented by the fluctuation v'.

3 Data-Driven Modelling

Data driven modelling (DDM) is an approach to harness the available data about a system and establish a connection between the system state variables (input, internal and output variables) without explicit knowledge of the physical behaviour [26]. In this work, we have the synthetic data derived from the DNS while the end model should be based upon this data and should able to provide an model to predict temporal evoliution of a state variable. Figure 2 depicts our end-to-end 2-step ML framework. To have an extendable framework, we chose to first extract the features with the help of a deep encoder and feed the extracted features to a second level recurrent neural network (RNN). This architecture is inspired from our earlier work [24].

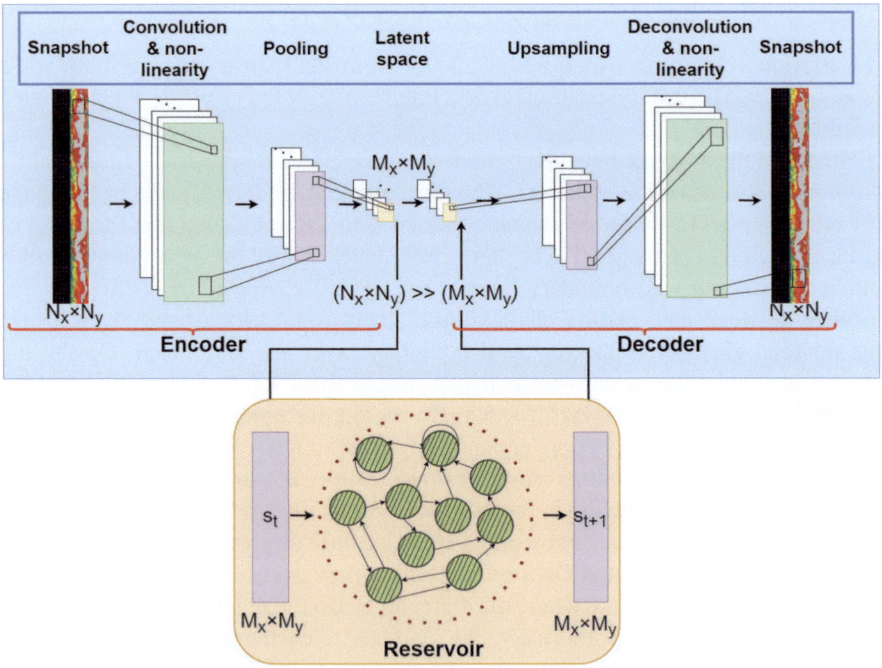

Fig. 2 Two-step machine learning architecture for end-to-end ROM model for porous media. In the first step, DAE is trained as feature extractor cum decoder while the second step trains the extracted feature in autoregressive manner

3.1 Feature Extraction with Autoencoder

The deep convolution autoencoder (DAE) is an implementation of an unsupervised machine learning paradigm, where the training of the model does not rely upon the labeled data. The objective of learning is to extract efficient representations or features in a low-dimension hyper-plane that can reproduce the data with minimal reconstruction error. That is to say, an autoencoder encodes input data $\boldsymbol{I}$ into $\boldsymbol{c} = \text{encode}(\boldsymbol{I})$ with $\boldsymbol{c} \in \mathbb{R}^l$. And it decodes $\boldsymbol{c}$ back into $\hat{\boldsymbol{I}} = \text{decode}(\boldsymbol{c})$ subsequently. The reconstruction error is calculated with $\hat{\boldsymbol{I}} - \boldsymbol{I}$. Here, 2-dimensional data were obtained from the DNS as described in Sect. 2. Thus autoencoder consists of several convolution layers resulting in a deep architecture. Each convolution layer in the network acts as a local feature extractor and shares the weights. The convolution layer is followed by a pooling layer, max pooling in this case, which reduces the dimension and focuses on the non-repeating key features. It gives a unique property of translation invariance to the network. At the end of the encoder network, there is a bottleneck layer that has the latent space or extracted features. These features are then passed to a decoder section which will attempt to recreate the original DNS data from the features. The training starts with random initialization of all the weights in the network and learns them as per the loss function via backpropagation.

3.2 Echo State Network (ESN)

ESN is closely related to reservoir computing (RC) and it shines out with sequential data where temporal relationship is present. In a typical ESN setup, input data is fed to a input layer which is connected to a big sparse reservoir which is finally connected to an output layer. ESN differs from typical RNN in terms of not requiring a back-propagation, therefore, it is inherently fast in training. More details can be found in our earlier work [24]. In this work, ESN acts as a downstream network which learns to predict the temporal evolution of the features obtained by the encoder section of DAE in an auto-regressive manner. The predicted values can be fed back to the decoder section of DAE to reproduce the feature of interest. Therefore, the during the deployment, we are only interested in the trained ESN and decoder while encoder can be thrown away.

4 Results and Discussion

4.1 Data Driver Flow Model

As mentioned earlier, we first trained a DAE network to extract the feature in a lower-dimensions. For the same, input and output layer has the same dataset i.e.

Table 2 DAE architecture used at first level for feature extraction. The table summarizes the encoder and decoder architectures (Conv = convolution, Max Pool = max pooling, Upsamp = upsampling). Symbol E2 denotes for example encoder hidden layer No. 2

Encoder		Decoder	
Layer	Output size	Layer	Output size
Encoder input	1501 × 109 × 1	Decoder input	24 × 2 × 34
2D Conv-E1	1501 × 109 × 256	2D Conv-D1	24 × 2 × 34
Max Pool-E1	751 × 55 × 256	2D Upsamp-D1	48 × 4 × 34
2D Conv-E2	751 × 55 × 256	2D Conv-D2	48 × 4 × 32
Max Pool-E2	376 × 28 × 256	2D Upsamp-D2	96 × 8 × 32
2D Conv-E3	376 × 28 × 128	2D Conv-D3	96 × 8 × 64
Max Pool-E3	188 × 14 × 128	2D Upsamp-D3	192 × 16 × 64
2D Conv-E4	188 × 14 × 64	2D Conv-D4	192 × 16 × 128
Max Pool-E4	94 × 7 × 64	2D Upsamp-D4	384 × 32 × 128
2D Conv-E5	94 × 7 × 32	2D Conv-D5	384 × 32 × 256
Max Pool-E5	47 × 4 × 32	2D Upsamp-D5	768 × 64 × 256
2D Conv-E6	47 × 4 × 34	2D Conv-D6	768 × 64 × 256
Max Pool-E6	24 × 2 × 34	2D Upsamp-D6	1536 × 128 × 256
		Output with 2D Conv	1536 × 128 × 1
		Output with Cropping	1501 × 109 × 1

the snapshots from the DNS consist of u-velocity component with a dimension of $1501 \times 109 \times 1$. The neural network has multiple convolution layers combined with a max pooling layer while the decoder section has a convolution layer and a up-sampling layer, giving a converging-diverging shape. The features or latent modes are obtained at the end of the encoder section and it has a shape of $24 \times 2 \times 34$, thereby giving a dimensionality reduction of 99%. This could increase further on the expense of reconstruction loss. Table 2 illustrates the various layers and their shape. After fixing the DAE architecture, various hyperparameters were obtained by using Bayesian optimization. This particular approach is part of the AutoML paradigm [12], and it assists in achieving an optimized set of parameters in a relatively less number of iterations when compared to random or grid search. It is worth to mention that the data were scaled between 0-1 before the training.

Once DAE is trained, we extract the features for the entire dataset using its encoder part. These encoded data which has 99% reduced dimension were further used to train the second level of RNN i.e. ESN. Unlike DAE, training an ESN is relatively straightforward and computationally cheaper because of straightforward design choices. The optimized network of ESN has 3726 reservoirs, 0.93 as spectral radius, 0.94 as leaking rate and 0.16 as reservoir density. For the best results, the look back history period is 4 timesteps i.e., model requires 4 timesteps data as an initialization then the system can predict auto-regressively. As a first step of validation, we visualize the low-dimensional features from the DAE along with the generated data from ESN

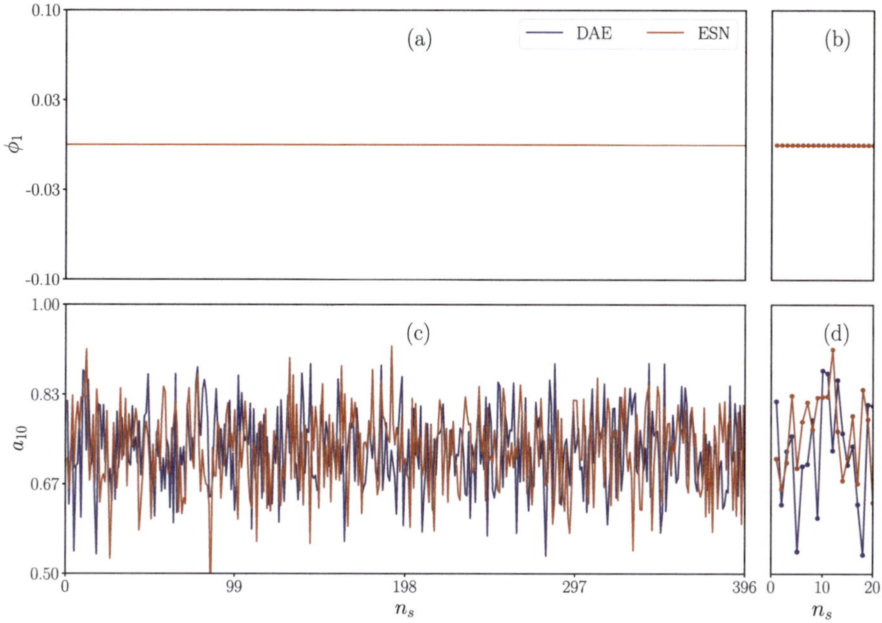

Fig. 3 A comparison of extracted features from DAE which acts as ground truth for the training, and prediction from ESN for 2 of the modes

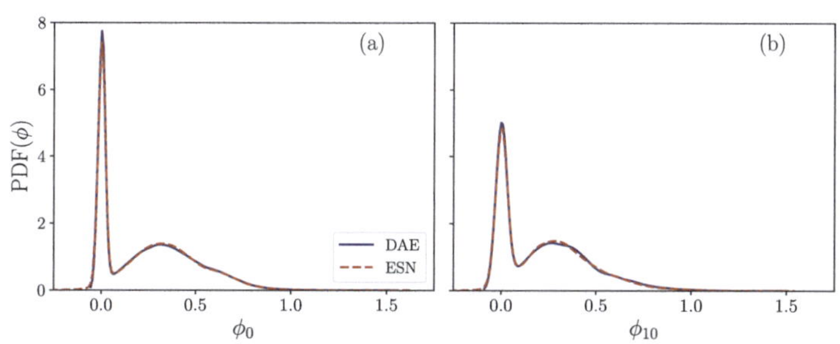

Fig. 4 Comparison between DAE and ESN modes in terms of pdf for 2 of the modes

on the blind validation set. We call the method generative because the model works auto-repressively once we provide the initialize data. Figure 3 shows such time-series for 2 of the modes out of 1632. It is impressive that model maintain a value of 0 without any minor fluctuations for ϕ_1 while series fluctuates with the right phase and amplitude for ϕ_{10}. Figure 4 further illustrates the probability density function (PDF) of these 2 modes for the DAE and ESN. It reaffirms the generating capability of ESN where it is able to match the overall distribution with the ground trouth.

As a next step of validation is the comparison in high-dimensional space where flow physics need to be captured. Figure 5 displays contour plots of the original DNS data in panels (c) and (f), reconstructed data from the features space of the DAE in

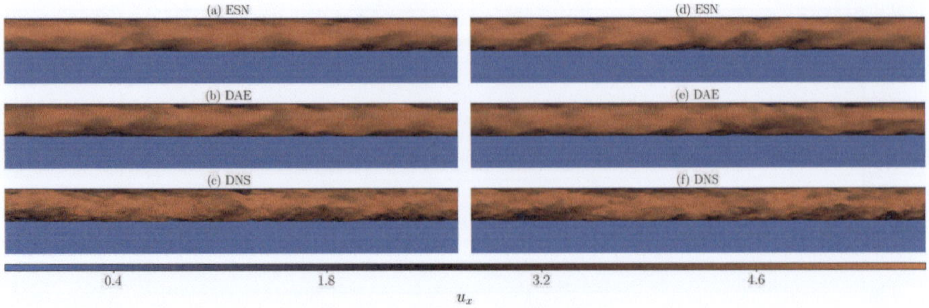

Fig. 5 Contours of u_x by **a**, **d** ESN (final end to end model), **b**, **e** DAE (feature extractor) and **c**, **f** DNS (GT) at 2 different timestamps

panels (b) and (e), and the decoded value from the prediction of ESN is shown in panels (a) and (d). One can observe a good qualitative agreement in terms of capturing the flow fields. Both regions i.e. main channel flow and porous flow can be predicted. Both DAE and ESN were trained from the DNS, and are not able to fully capture the strong turbulent field in the channel. While the flow in the porous media section is much better because it falls within the laminar region mostly. The obvious reason is the limited capability of DAE to capture the entire flow field because it discards 99% of input data and usage only 1% of the features, similar to conventional reduced-order modelling approach such as POD. The derived model with ESN shows an excellent result while matching field compared to its GT i.e. DAE. It suggests that the entire model can further be improved if a state-of-the-art feature extractor such as transformer model can be used instead of DAE. It is worth to mention DAE surpasses conventional as well as some advance feature extractors e.g., variational autoencoder for this use-case, which was observed in our experimentation. Often, average and fluctuations are critical values for the analysis in industrial applications. Therefore, we further validate the results with mean and fluctuations profile over the cross-section of the channel. Figure 6 shows these profiles and a mean velocity profile overlap with the DNS for both DAE and ESN. The fluctuation of velocity is slightly deviated in the main channel while they remain zero in the porous section due to dominating laminar flow. The distinction made by the model is also great where it can predict the interface where flow is transiting from laminar to turbulent.

4.2 Fine Tuning

The trained model has shown a good agreement in terms of quantitative as well as qualitative characteristics of the flow. However, the trained model is limited to a given data domain that was used for the training which means that the model will not predict accurate results as we use different conditions. This is one of the biggest downsides of such a data-driven model. There are several approaches that one can

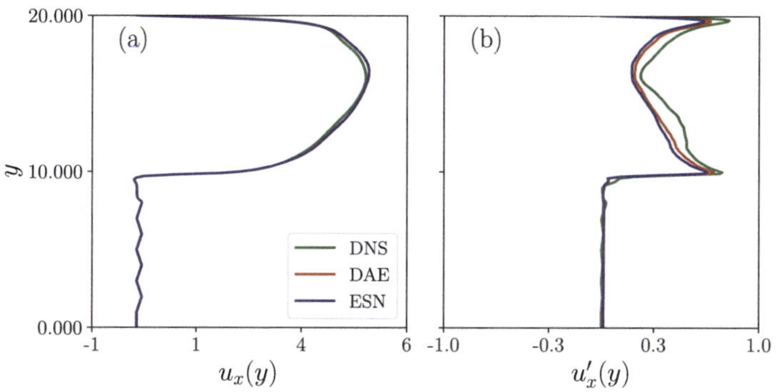

Fig. 6 Comparison among ground truth (DNS), reduced order model (DAE) and temporal model for DAE modes using ESN. **a** Average fields **b** Fluctuations

Table 3 Various cases considered for fine tuning

Case name	Description	Training time	Mean Square Error
DNS	Simulation results which acts as GT for case C06	–	–
DAE90	Model trained from scratch similar to case C05	$O(10^3)$	0.025
DAE91	Direct prediction using model from case C05 without any training	0	0.081
DAE92	Model from case C05 retrained with all data points for case C06	$O(10^2)$	0.030
DAE93	Model from case C05 retrained with 10% data points for case C06	$O(10^1)$	0.049

use to increase the applicability of such a model to a wider data domain. In this work, we assess the fine-tuning which is widely used in the machine-learning community. Fine-tuning makes the use of existing trained model and tune it with the newer limited data. As mentioned earlier, DAE is a computationally expensive component of this workflow, therefore, we further assess DAE with the fine-tuning. We tested the DAE model with a different set of data. Here, we used case C06 which has a higher level of porosity compared therefore a higher Reynolds number as mentioned in Table 1.

Table 3 shows the cases used for the fine tuning study and corresponding results are shown in Figs. 7 and 8. For the averaged streamwises velocity profile and its fluctuation, the reduced-order model shows comparable results as in its original validation, which proves the validity of the methodology of the fine-tuning.

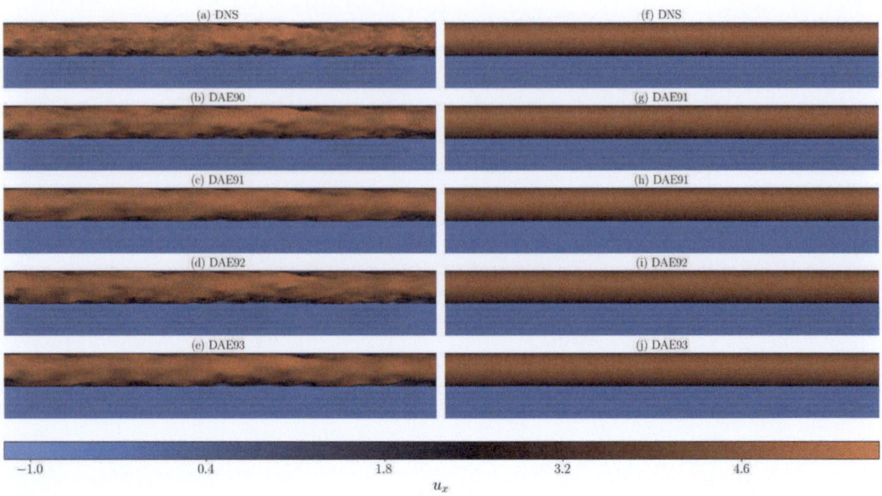

Fig. 7 Contours of (**a**, **b**, **c**, **d**): Instantaneous velocity, **f**, **g**, **h**, **i** Temporal average velocity. Labels are as per Table 3

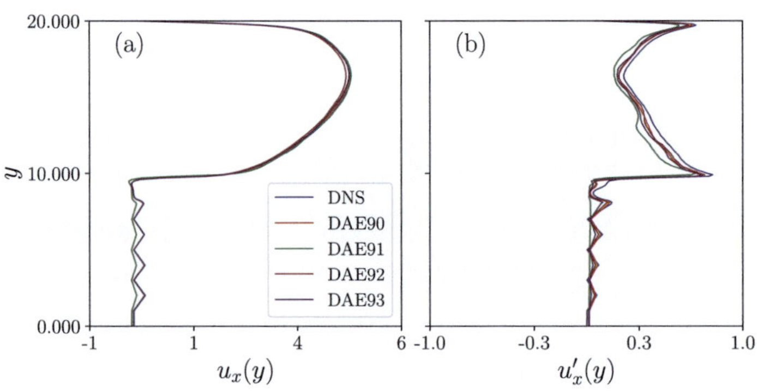

Fig. 8 Comparison among ground truth (DNS), reduced order model (DAE) and temporal model for DAE modes using ESN. **a** Average fields **b** Fluctuations

5 Computational Performance

The supercomputing systems used for the DNS was *HAWK* located at the High Performance Computer Center Stuttgart (HLRS). The new flagship machine *HAWK*, based on the Hewlett Packard Enterprise platform running AMD EPYC 7742 processor code named *Rome*, has a theoretical peak performance of 26 petaFLOPs, and consists of a 5,632-node cluster. One AMD EPYC 7742 CPU consists of 64 cores, which leads to 128 cores on a single node sharing 256 GB memory on board. This means that a total of 720,896 cores are available on the HAWK system (Fig. 9).

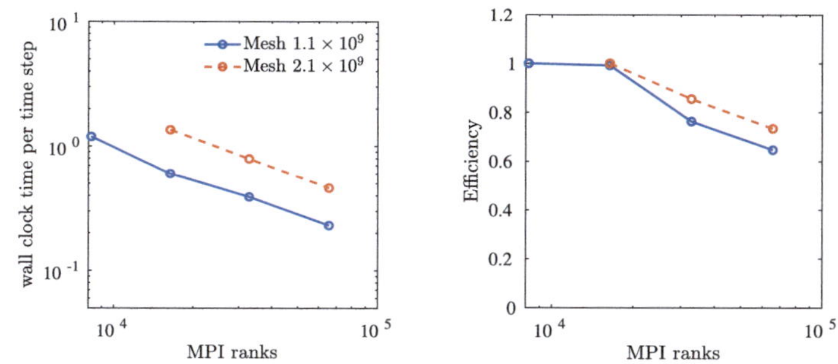

Fig. 9 Strong scaling behavior of the solver on 'HAWK', wall clock time per time step on the left side. On the right side the efficiency for different numbers of MPI ranks

The scalability of our direct numerical simulations, employing a high-order solver, was thoroughly evaluated on the advanced HPC system, *HAWK*, utilizing CPU cores ranging from 8,192 to 65,536. By leveraging ample computational resources, we achieved remarkable performance, with a wall clock time per time step ranging from 0.2 to 0.4 s. This optimal computational efficiency enables efficient direct numerical simulations of turbulent flows. Notably, when utilizing the maximal number of MPI ranks, our simulations demonstrated a minimum efficiency of 70%. This achievement highlights the exceptional scalability of our high-order solver on the state-of-the-art HPC cluster, *HAWK*, setting it apart from previous scalability tests conducted on the low-order Finite Volume Method (FVM) code OpenFOAM [3, 4, 10, 17–19, 33] using earlier HPC machines at HLRS. Our findings underscore the outstanding scalability exhibited by our current high-order solver on the advanced HPC system, offering enhanced capabilities for conducting direct numerical simulations. This scalability is crucial for effectively studying turbulent flows and holds significant promise for advancing research in the field.

6 Conclusions

Our study introduces a combined Convolutional Deep Autoencoder-Echo State Network (CDAE-ESN) model as a powerful tool for directly analyzing and forecasting the dynamics and low-order statistics of coupled turbulent channel-porous media flows. The significance of such a flow model lies in its applicability to various industrial systems, including transpiration cooling and smart interface engineering. We used fully-resolved direct numerical simulation to generate the data. By integrating deep autoencoder and echo state network techniques, we overcome the challenges posed by the complex geometry of coupled flow systems, enabling effective modeling and prediction of dominant flow behaviors, particularly within the laminar regime of the porous domain. The obtained results demonstrate the capability of our proposed approach to capture and represent the intricate flow dynamics. Furthermore, to broaden the model's applicability, we conducted fine-tuning on a dataset with varying porosity. This step aimed to enhance the generalizability of the CDAE-ESN model across a wider range of data domains. The reasonable agreement observed in the average statistics further validates the robustness and versatility of our approach. In summary, the combined CDAE-ESN model presents a promising avenue for analyzing and predicting the behavior of coupled turbulent channel-porous media flows. Its potential applications in diverse industrial systems underscore its significance and highlight the future prospects for incorporating such data-driven models in fluid mechanics research.

Acknowledgements The study has been financially supported by SimTech (EXC 2075/1âŁ" 390740016) and SFB-1313 (Project No.327154368) from Deutsche Forschungsgemeinschaft (DFG). The financial support by the Deutsche Forschungsgemeinschaft with Grant No. SCHU 1410/30-1 and in parts by the project "DeepTurbâŁ"–Deep Learning in and of Turbulence" by the Carl Zeiss Foundation is deeply appreciated. The authors gratefully appreciate the access to the high performance computing facility *Hawk* at HLRS, Stuttgart of Germany. We would like to extend our sincere appreciation to Prof. Jörg Schumacher for his fruitful discussions.

References

1. Cantwell, C.D., Sherwin, S.J., Kirby, R.M., Kelly, P.H.: From h to p efficiently: strategy selection for operator evaluation on hexahedral and tetrahedral elements. Comput. Fluids **43**(1), 23–28 (2011)
2. Chu, X., Chang, W., Pandey, S., Luo, J., Weigand, B., Laurien, E.: A computationally light data-driven approach for heat transfer and hydraulic characteristics modeling of supercritical fluids: From dns to dnn. Int. J. Heat Mass Trans. **123**, 629–636 (2018)
3. Chu, X., Laurien, E., McEligot, D.M.: Direct numerical simulation of strongly heated air flow in a vertical pipe. Int. J. Heat Mass Trans. **101**, 1163–1176 (2016)
4. Chu, X., Laurien, E., Pandey, S.: Direct numerical simulation of heated pipe flow with strong property variation. In: High Performance Computing in Science and Engineering' 16, pp. 473–486. Springer (2016)
5. Chu, X., Müller, J., Weigand, B.: Interface-resolved direct numerical simulation of turbulent flow over porous media. In: High Performance Computing in Science and Engineering '19, pp. 343–354. Springer International Publishing (2021)
6. Chu, X., Wang, W., Yang, G., Terzis, A., Helmig, R., Weigand, B.: Transport of turbulence across permeable interface in a turbulent channel flow: interface-resolved direct numerical simulation. Trans. Porous Media **136**(1), 165–189 (2021)
7. Chu, X., Weigand, B., Vaikuntanathan, V.: Flow turbulence topology in regular porous media: from macroscopic to microscopic scale with direct numerical simulation. Phys. Fluids **30**(6), 065102 (2018)
8. Chu, X., Wu, Y., Rist, U., Weigand, B.: Instability and transition in an elementary porous medium. Phys. Rev. Fluids **5**(4), 044304 (2020)
9. Chu, X., Yang, G., Pandey, S., Weigand, B.: Direct numerical simulation of convective heat transfer in porous media. Int. J. Heat Mass Trans. **133**, 11–20 (2019)
10. Evrim, C., Chu, X., Laurien, E.: Analysis of thermal mixing characteristics in different t-junction configurations. Int. J. Heat Mass Trans. **158**, 120019 (2020)
11. Fang, R., Sondak, D., Protopapas, P., Succi, S.: Neural network models for the anisotropic Reynolds stress tensor in turbulent channel flow. J. Turb. 1–19 (2019)
12. He, X., Zhao, K., Chu, X.: Automl: a survey of the state-of-the-art. Knowl.-Based Syst. **212**, 106622 (2021)

13. Karniadakis, G.E., Karniadakis, G., Sherwin, S.: Spectral/hp element methods for computational fluid dynamics. Oxford University Press on Demand (2005)
14. Karra, S.K., Apte, S.V., He, X., Scheibe, T.D.: Pore-resolved simulations of turbulent boundary layer flow over permeable and impermeable sediment beds (2022). arXiv:2204.13875
15. Kutz, J.N., Brunton, S.L., Brunton, B.W., Proctor, J.L.: Dynamic mode decomposition: data-driven modeling of complex systems. SIAM (2016)
16. Nakamura, T., Fukami, K., Hasegawa, K., Nabae, Y., Fukagata, K.: Convolutional neural network and long short-term memory based reduced order surrogate for minimal turbulent channel flow. Phys. Fluids **33**(2), 025116 (2021)
17. Pandey, S., Chu, X., Laurien, E.: Investigation of in-tube cooling of carbon dioxide at supercritical pressure by means of direct numerical simulation. Int. J. Heat Mass Trans. **114**, 944–957 (2017)
18. Pandey, S., Chu, X., Laurien, E.: Numerical analysis of heat transfer during cooling of supercritical fluid by means of direct numerical simulation. In: High Performance Computing in Science and Engineering'17, pp. 241–254 (2018)
19. Pandey, S., Chu, X., Laurien, E., Weigand, B.: Buoyancy induced turbulence modulation in pipe flow at supercritical pressure under cooling conditions. Phys. Fluids **30**(6), 065105 (2018)
20. Pandey, S., Chu, X., Weigand, B., Laurien, E., Schumacher, J.: Relaminarized and recovered turbulence under nonuniform body forces. Phys. Rev. Fluids **5**(10), 104604 (2020)
21. Pandey, S., Laurien, E., Chu, X.: A modified convective heat transfer model for heated pipe flow of supercritical carbon dioxide. Int. J. Thermal Sci. **117**, 227–238 (2017)
22. Pandey, S., Schumacher, J.: Reservoir computing model of two-dimensional turbulent convection. Phys. Rev. Fluids **5**(11), 113506 (2020)
23. Pandey, S., Schumacher, J., Sreenivasan, K.R.: A perspective on machine learning in turbulent flows. J. Turbu. **21**(9–10), 567–584 (2020)
24. Pandey, S., Teutsch, P., Mäder, P., Schumacher, J.: Direct data-driven forecast of local turbulent heat flux in rayleigh-bénard convection. Phys. Fluids **34**(4), 045106 (2022)
25. Shen, G., Yuan, J., Phanikumar, M.S.: Direct numerical simulations of turbulence and hyporheic mixing near sediment water interfaces. J. Fluid Mech. **892**, A20 (2020)
26. Solomatine, D., See, L.M., Abrahart, R.: Data-driven modelling: concepts, approaches and experiences. Practical hydroinformatics: Computational intelligence and technological developments in water applications, pp. 17–30 (2008)
27. Storcheus, D., Rostamizadeh, A., Kumar, S.: A survey of modern questions and challenges in feature extraction. In: Feature Extraction: Modern Questions and Challenges, pp. 1–18. PMLR (2015)
28. Suga, K., Okazaki, Y., Kuwata, Y.: Characteristics of turbulent square duct flows over porous media. J. Fluid Mech. **884** (2020)
29. Terzis, A., Zarikos, I., Weishaupt, K., Yang, G., Chu, X., Helmig, R., Weigand, B.: Microscopic velocity field measurements inside a regular porous medium adjacent to a low reynolds number channel flow. Phys. Fluids **31**(4), 042001 (2019)
30. Wang, W., Chu, X., Lozano-Durán, A., Helmig, R., Weigand, B.: Information transfer between turbulent boundary layer and porous media. J. Fluid Mech. **920**, A21 (2021)
31. Wang, W., Lozano-Durán, A., Helmig, R., Chu, X.: Spatial and spectral characteristics of information flux between turbulent boundary layers and porous media. J. Fluid Mech. **949**, A16 (2022)
32. Wang, W., Yang, G., Evrim, C., Terzis, A., Helmig, R., Chu, X.: An assessment of turbulence transportation near regular and random permeable interfaces. Phys. Fluids **33**(11), 115103 (2021)
33. Yang, G., Chu, X., Vaikuntanathan, V., Wang, S., Wu, J., Weigand, B., Terzis, A.: Droplet mobilization at the walls of a microfluidic channel. Phys. Fluids **32**(1), 012004 (2020)
34. Yosinski, J., Clune, J., Bengio, Y., Lipson, H.: How transferable are features in deep neural networks? Adv. Neural Inf. Proc. Syst. **27** (2014)
35. Zhuang, F., Qi, Z., Duan, K., Xi, D., Zhu, Y., Zhu, H., Xiong, H., He, Q.: A comprehensive survey on transfer learning. Proc. IEEE **109**(1), 43–76 (2020)

Optimization of a Non-standard Pump from Galvanic Industry with the Island Model

Ines Waldsteiner, Alexander Tismer, and Stefan Riedelbauch

Abstract A fully automated evolutionary optimization of a non-standard pump from galvanic industry is presented applying the island model for a high degree of parallelization on large supercomputers. The aim of this optimization is to improve the efficiency and the head of the pump at a specified operating point. The investigated pump is parameterized by nine degrees of freedom. The results of three optimization runs are compared with the baseline. In three optimization runs, the initialization, the migration strategy as well as the number of fully connected islands and the population size per island are varied. In addition, different definitions of fitness are used. The results of the optimization runs yield increased efficiencies and wider operating ranges than the baseline. The definition used for calculating fitness depends on the focus of the desired performance improvement.

1 Introduction

The electroplating industry utilizes submersible pumps with geometries that substantially deviate from classic centrifugal pumps. The share of energy costs of a plant in the galvanic industry is typically in the range of 7–20% of all costs, which is significantly higher than the industry average [1]. The investigated pump is robustly built, but exhibits a relatively small efficiency. This results in significant potential savings by achieving better efficiency or extending the operating range of the pump. The aim of this work is to increase the efficiency and the head of this pump by applying a fully automated, evolutionary, parallel optimization. The pump under investigation is installed in dipping baths where the electrolytes have to be continuously circulated, filtered and refilled to ensure optimal coating. That application results in special requirements for the design of the pump. To identify the pump performance characteristics, a flow field simulation analysis for the pump within a large reservoir is

I. Waldsteiner (✉) · A. Tismer · S. Riedelbauch
Institute of Fluid Mechanics and Hydraulic Machinery, Stuttgart, Germany
e-mail: ines.kuhnle@ihs.uni-stuttgart.de

T. Ludwig et al. (eds.), *High Performance Computing in Science and Engineering '23*,
https://doi.org/10.1007/978-3-031-91312-9_23

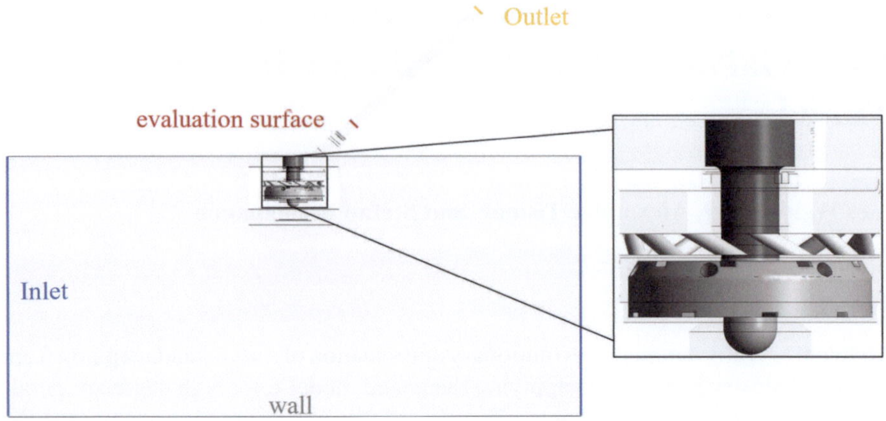

Fig. 1 View of the whole simulation area with the shaft in dark grey, the runner in medium grey, the junction pipes in lighter grey and the remaining simulation area in light grey

performed. That particular simulation setup ensures well-posed inflow and outflow boundary conditions for the pump itself, Fig. 1.

There is only limited knowledge about the design possibilities for this kind of electroplating pumps. For this reason, it is advisable to generate an automatic design process. The creation of an automated design process in combination with optimization algorithms reduces the individual effort for a manual design. To reduce the required design time, the optimization process is parallelized for a high number of computational nodes. The parallelization is performed with the island model for evolution algorithms [4].

2 Workflow

In order to perform an optimization run, the parameters of the optimization and the flow simulations have to be specified. The problem to be studied and the aim of the optimization have to be precisely defined as well in order to obtain a successful result.

2.1 *Optimization Process*

For the optimization of the electroplating pump the in-house design tool *Python Design Tool Object-Oriented* (pyDtOO) is used [6]. Only the migration, *dtPagmo2Binding* and the database from this design tool are used. The optimization process is fully automated. Therefore, different scripts and programs as well as program languages to connect the optimization with the flow field simulations are needed. For the optimization process a python script with different classes is applied. One class describes the problem and calculates the fitness function. In this class different functions are created, like functions for generating the geometry and meshes of the runner and the junction pipe or specifying the boundary conditions and simulation parameters for the flow field simulation. There are also functions to create the scripts for generating a definition file for the CFD Simulation with Ansys CFX or to start the steady state simulations of the investigated operating point with Ansys CFX as well as a function for evaluating the simulation results and calculating the fitness value. One additional function within this class is created to execute all the described functions in the required sequence and return the calculated fitness value to the optimization. All functions, which are called in this function are started as subprocesses with the python module subprocess. The ICEM and Ansys CFX Pre- and Postprocessing processes are always started as subprocesses, too. The generation of the geometries and meshes cannot be parallelized, because one task in ICEM is restricted to serial execution. Another class of this python script is the *loadbalancer*, which handles the number of requested cores as working nodes, distributes the flow field simulation on the working nodes and creates a queue if there are more working nodes needed than available. A last class is used to setup the optimization. This includes creating the desired number of islands and initializing the *loadbalancer* as well as the islands. The python script, the generation of geometries and meshes, the Pre- and Postprocessing of the simulations and the calculation of the fitness are all done on the master node. Only the time and computational intense simulations of the pump are done on the worker nodes. In this work three different optimization runs are presented. All optimizations exploit a differential evolution algorithm [5]. To connect the in-house design tool to the optimization algorithm the python library pygmo is used [2]. The optimization is parallelized using the island model [3]. The first optimization run (run 1) is generated with twelve islands containing ten individuals per island. The islands are all connected to each other and initialized very homogeneously. This means that the parameters are only covering a small range of each possible parameter range. For run 1 the migration algorithm implemented in pygmo is used, which, in this case, migrates the top five individuals of an island to another island after each generation. When a restart occurs, the last individuals of the islands are pushed back onto the islands. For the second (run 2) and the third (run 3) optimization run the number of islands is increased to 36 islands and the number of individuals on one island is decreased to five. The initialization of these two runs is more diverse than it is in run 1. The islands are also all connected to each other. In the last two optimization runs the *dtCreateMultiCulti* migration algorithm from [6] is used. This migration algorithm favors a divers parameter range for each parameter and exchanges the population of an island by picking from the whole pool of individuals of the run. As with the migration algorithm, individuals are pushed onto the islands after a restart. In run 1 and run 2 the fitness values are defined the same way. The only difference between run 2 and run 3 is the way the fitness values are calculated.

2.2 Numerical Methods

For the CFD simulations the commercial code Ansys CFX version 20.2 is used. All flow simulations are incompressible, single-phase, steady-state simulations with no consideration of gravity. For the flow simulation of the pump, the SST turbulence model with a high resolution scheme and near wall treatment is chosen. The rotating domain is coupled to the non-rotating domains with frozen rotor interfaces. At the inlet a constant mass flow rate is applied to set the operating point. The opening boundary type with an applied constant pressure is used at the outlet. Each flow simulation lasts for 300 time steps with a rotation angle per time step of 40°. The results are averaged over the last 100 time steps to compensate deviations and oscillation. The used advection scheme for the flow simulations is high order and all flow simulations are performed isothermal at a temperature of 20 °C. Each flow simulation uses 64 cores and is started with the *Open MPIHAWK Distributed Parallel* start method to parallelize the time-consuming flow simulation. All meshes for the flow simulations consist of a prism layer in boundary areas, with at least ten cells over the height. The unstructured meshes for the geometry of each individual are generated automatically. Meshes which only contain geometries that remain unchanged in the optimization process, such as the inlet pipe, are not changed during the optimization. Some of those constant meshes are block structured to generate better results.

2.3 Problem Description

The geometry of the baseline design for this optimization consists of an inlet pipe, two chambers and an outlet pipe. The runner is placed in the first chamber and is composed of a cylinder with six inclined boreholes and nine grooves at the top and the bottom of the cylinder. The grooves guide the fluid back into the first chamber to achieve less mass flow around the shaft and less backflow into the inlet pipe. The first chamber also consists of a step to minimize the backflow into the inlet pipe. The second chamber of the pump functions as a pressure chamber. The two chambers are connected with eleven inclined junction pipes and guide the fluid to the outlet pipe. The components of the pump and a view of the flow, that also includes the relative flow rates within, are shown in Fig. 2. The motor turns the impeller and sucks the liquid into the pump. The impeller increases the total pressure of the fluid. The fluid then has to flow through the junction pipes and the second chamber towards the outlet pipe. Between the outlet of the impeller and the outlet of the outlet pipe the flow is redirected by the junction pipes, the second chamber and the outlet pipe, which contribute to total pressure losses. Additionally, the flow in the gaps result in mass flow losses within the pump. In the current optimization only the geometry of the impeller and the junction pipes are changed. This results in nine degrees of freedom for this pump. There are four degrees of freedom in the impeller geometry

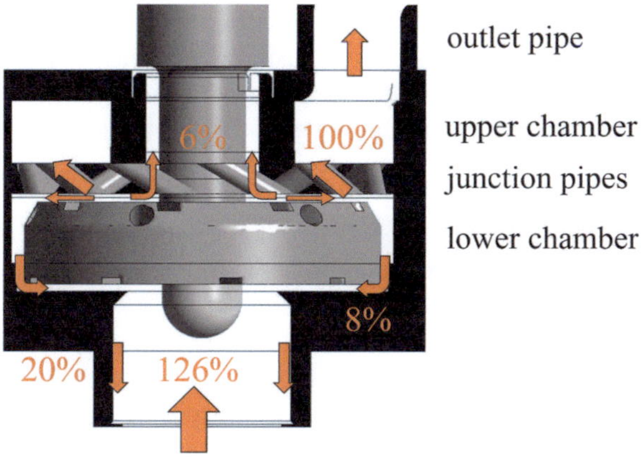

Fig. 2 Mass flow through the baseline referenced to the mass flow at the outlet of the pump

and five degrees of freedom in the junction pipe geometry. The parameterization of impeller and junction pipe geometry of the investigated pump is presented in Fig. 3.

The parameters in the runner are the radius of the borehole at the entrance of the runner r_1 and the radius of the borehole at the end of the runner r_2 as well as the inclination angle relative to the horizontal α and the inclination angle between the entrance and the end of the boreholes in circumferential direction β. The number of boreholes in the runner and the number of grooves at the top and the bottom of the runner is constant over the whole optimization process. The parameters in the junction pipes are the inclination angle of the junction pipes relative to the horizontal γ, the radius of the junction at the outlet of the first chamber x_1 and the radius of the junction pipe at the inlet to the second chamber x_2. Also, the position of the junction pipes in radial direction in the two chambers R_1 and R_2 are parameters in the junction pipe geometry. The number of junction pipes is constant over the whole optimization process. With an independent variation of the radii of the junction pipes and the radial positions of the junction pipes in the chambers the created geometries can fail due to junction pipes sticking out of the chambers. For this reason, the radii of the junction pipes are varied as a percentage after run 1 and are denoted by ξ_1

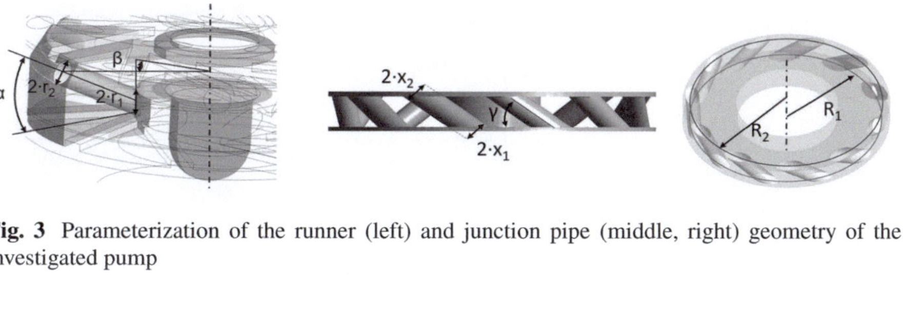

Fig. 3 Parameterization of the runner (left) and junction pipe (middle, right) geometry of the investigated pump

and ξ_2 in the following. The radii of the junction pipes are varied between 99 % and 1 % of the minimal distance from the position of the junction pipes to the inner and outer edges of the chambers. Inside the optimization process the radii of the junction pipes are calculated to design the geometry of the junction pipes. The aim of this optimization is to find a pump design, which is as robust as the baseline, with a better efficiency and a higher head at the same operating point. The fitness is calculated in the optimization process and defines the quality of the design. The fitness is minimized in the optimization process. The fitness contains the efficiency and the head at the operating point. It consists of an efficiency term and a head term. Because of the minimization of the fitness the efficiency is deducted from 1. For the head term the head at the operation point is divided by the best head of the baseline over the whole operating range and also deducted from 1. For the first two optimization runs the efficiency and the head term are multiplied to calculate the fitness (fitness·). In the last optimization run the fitness is calculated by adding the two terms (fitness+). Both fitness functions depend on the efficiency and the head of the pump at the investigated operation point. In Fig. 4 different constant fitness values for fitness+ and fitness· are plotted versus efficiency and head of the pump.

Individuals with the same fitness value are considered equal in the optimization process. The constant fitness values of the two fitness functions behave differently. In case of fitness·, the fitness value changes only slightly at constant head even with a strong change in efficiency. Only at high efficiencies the fitness value decreases stronger due to small changes in the efficiency. This means that the influence on the fitness value of the head is greater at low efficiencies and the influence of the efficiency increases at higher efficiencies. With fitness+, the gradient of the constant fitness values is constant over the entire range. Therefore, the influence of the head and the efficiency remains the same over the entire range. In case of fitness+, there is the possibility that individuals with high head and low efficiencies have the same fitness value as individuals with small head and high efficiencies. In contrast, fitness· requires a large head in order to achieve small fitness values.

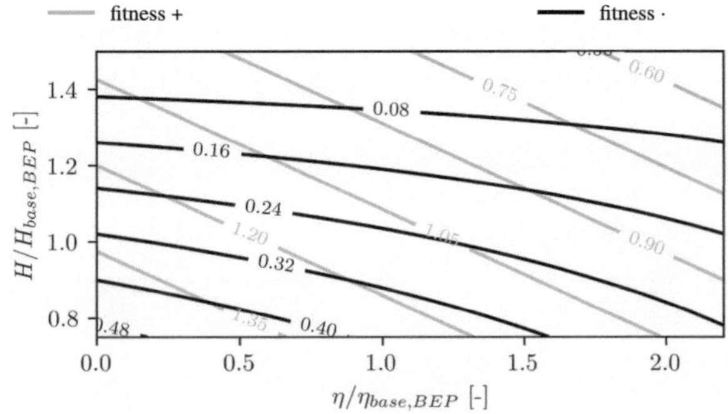

Fig. 4 Trend of constant fitness values in the efficiency-head diagram for the fitness· function in black and the fitness+ function in lightgrey

3 Results

For the optimization runs it is important to see the coverage of the parameter range of all parameters to identify if the optimization tries all possible variations. Normalized parameter values for each individual of run 1 and run 2 versus each individual geometry parameter are displayed in Fig. 5.

Every individual of the optimization run is represented by a line. For run 1 the range of the parameters is not completely covered. Especially for the position of the junction pipes in the first R_1 and second chamber R_2 small values are not covered and for the radius of the junction pipes at the start x_1 and the end x_2 of the junction pipes big values are not completely covered. Because of this, a different more diverse initialization and a different migration strategy is implemented in run 2 and run 3. For run 2 and run 3 the parameter range of the degrees of freedom are mostly covered. In Fig. 6 the parameters for all successful individuals of run 1 and 2 are shown.

The coloring of the line represents the fitness value of the individual. For each optimization run the parameter values of the parameters lie in similar ranges. In all three optimization runs, almost the entire parameter range is covered for the runner bore inclination angle β, even for individuals with small fitness values. This indicates that the influence of this parameter is less relevant for the fitness value of the individuals. For all optimization runs, the values of the inclination angle γ of the junction pipes for individuals with small fitness values are small. This indicates that

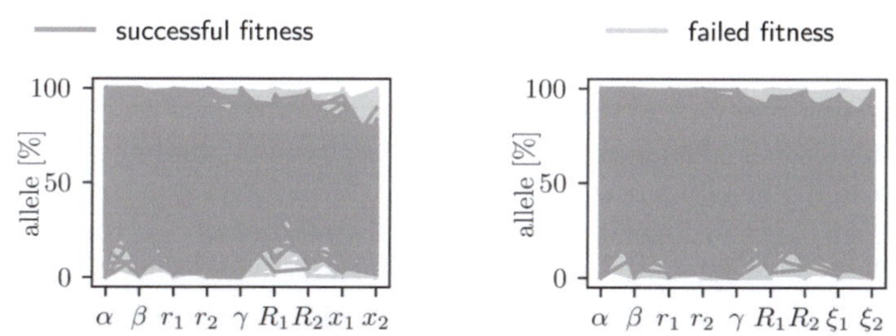

Fig. 5 Standardized parameter values of all individuals of run 1 (left) and run 2 (right)

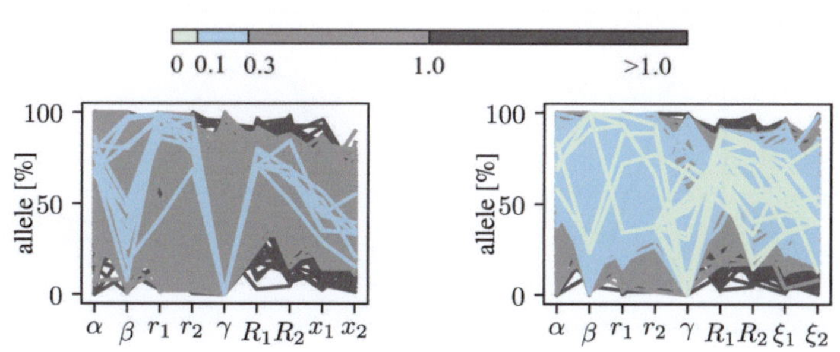

Fig. 6 Standardized parameter values of all successful individuals of run 1 (left) and run 2 (right) coloured by fitness

this parameter has a dominating influence on the fitness and that small values for this parameter result in small fitness values. This result agrees with a previous sensitivity analysis. In the following, only the best individuals of the last two optimization runs are analysed, because of the smaller achieved fitness values. For run 2, fitness is calculated by multiplication, the individual with the best efficiency is considered ($\eta\cdot$). Based on results of run 3, the individual with the best head ($H+$) is selected. In run 3 the fitness is calculated by addition. In Fig. 7 the normalized parameters for $\eta\cdot$, $H+$ and the baseline are shown.

Note that radius values for the junction pipes x_1 and x_2 are used in this figure instead of the percentages that are used in the optimization runs. It is noticeable that for all shown individuals the inclination angle γ is set to the smallest possible value. This is consistent with the previous results that this parameter has a dominant influence on the fitness value and that the fitness value improves the smaller this parameter value is set. The parameter values of the junction pipe geometry for all shown individuals lie in a similar area. The optimized individuals have greater boreholes in the runner as the baseline (see r_1 and r_2). The inclination angle α of the boreholes in the runner lie at larger values for all individuals shown, but differ among the individuals. The inclination angle β differs even more. This is also consistent with the previous results. The efficiency and the head of the best individuals of run 2 and 3 across the whole operating range indicate well visible differences in peak efficiency and width of the operating range, Fig. 8.

The efficiency and the head are normalized to the value of the baseline at the investigated operating point. The operating range is referred to the mass flow at the outlet of the pump. The mass flow is normalized to the maximum value of the baseline. The way the fitness values are defined has an influence on the outcome of the optimization run. Due to the larger influence of efficiency on the fitness value for large efficiencies when using the fitness definition with multiplication (fitness$\cdot$), individuals with high efficiencies are favored in run 2, when a good head is achieved. In fitness+, efficiency and head are weighted differently, which results in a preference for individuals with high head even with lower efficiency. This is also reflected

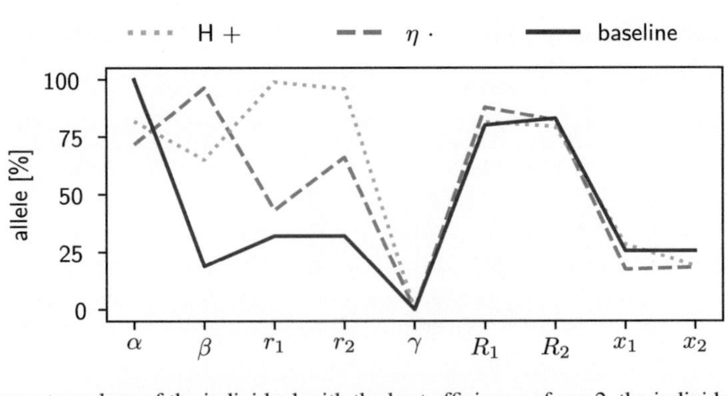

Fig. 7 Parameter values of the individual with the best efficiency of run 2, the individual with the best head of run 3 and the baseline

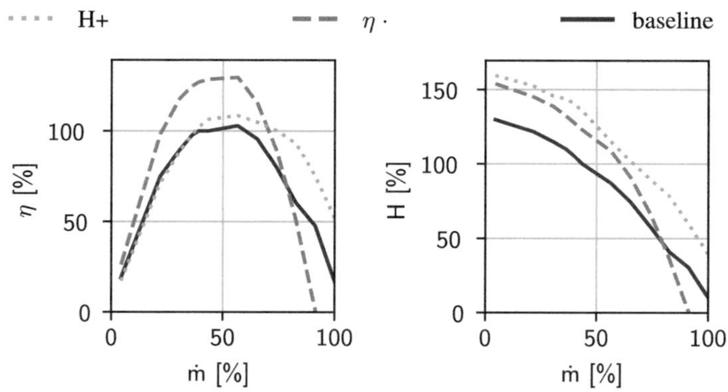

Fig. 8 Efficiency (left) and head (right) of the best individuals of the optimization runs and the baseline across the whole operating range

in the trend of the head and the efficiency across the entire operating range. The individuals of the optimization run with a calculated fitness value by multiplication provide a much better efficiency but have a slightly smaller operating range than the baseline. The maximum efficiency improvement for $\eta\cdot$ is 28,4%. The individuals of the optimization run with a calculated fitness value by addition have a wider operating range than the baseline but the improvement of the efficiency at the operating point is smaller. The size of the operating range is related to the head at the operating point under consideration. This results in a wider working range for individuals with greater head. The maximum improvement of the head compared to the baseline for $H+$ is 37,3%. Only in the case of a large mass flow, the head of $\eta\cdot$ is smaller than the baseline. For the most part of the operating range, a larger mass flow is pumped by all optimized individuals at the same head. By optimizing the nine pump parameters studied at a given operating point, efficiency and head are improved across the entire operating range. The difference of total pressure of the optimized individual with the best head of run 3 and the baseline at a plane through the pump at the investigated operating point show an increase in pressure through the pump, Fig. 9. The variable total pressure is calculated as the difference of the total pressure at the investigated location and the area averaged total pressure at the inlet. The long outlet pipe is shown shifted and reduced by 50%.

$H+$ increases the total pressure in both chambers and in the outlet compared with the baseline. The increase in pressure through the runner with larger boreholes is higher. In the investigated operating point, the grooves above the runner additionally lead to an increase in total pressure. The total pressure differences due to the vortices in the outlet are seen in all individuals, even in the baseline. However, the total pressure in the vortices increases and the loss through the outlet pipe becomes smaller. The flow through the pump also changes when comparing an optimized individual with the baseline. Figure 10 shows the flow in one borehole of the runner for the

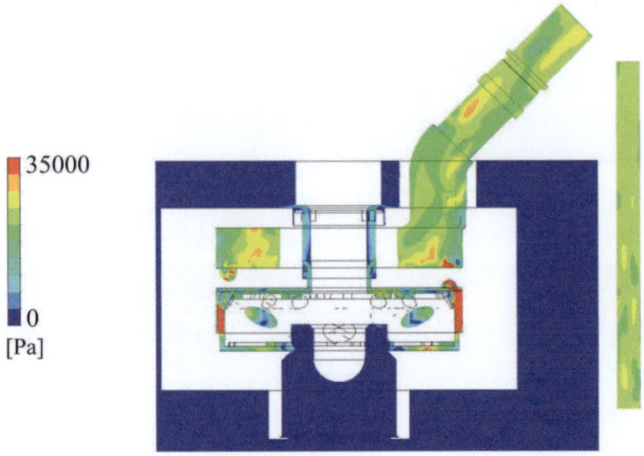

Fig. 9 Difference of the total pressure of the optimized pump and the baseline in the investigated operating point

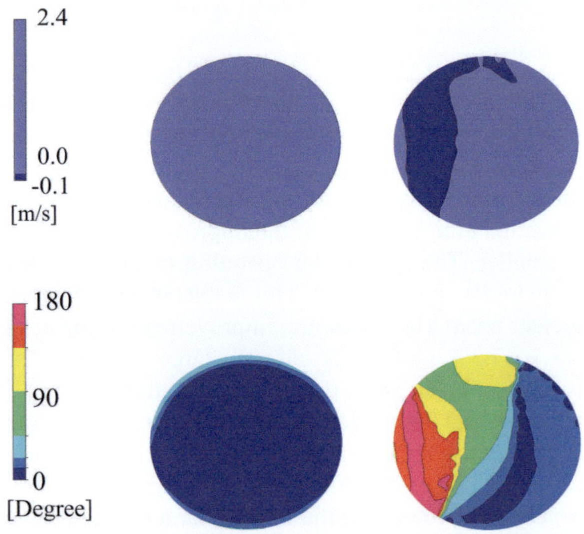

Fig. 10 Velocity (top) and the angle in relation to the normal (bottom) of the surface on a plane in the middle of one borehole in the baseline (left) and in $H+$ (right)

baseline and $H+$. The upper part shows the velocity of the flow in the borehole, with a dark blue colour representing backflow. The lower part of Fig. 10 shows the difference between the flow angle and the bore inclination.

Due to the increase of the borehole radii the flow in those boreholes changes. In some of the geometries this leads to turbulence and backflow in those boreholes like for $H+$ at the studied operating point. On the plane in the middle of one borehole the flow in the baseline usually moves in the same direction as the normal of the plane with a positive velocity. In $H+$ the stream is flowing in different directions, even backflow is observed. Due to the geometry optimizations for some optimized individuals the backflow in the inlet pipe is smaller than in the baseline. This reduces the needed mass flow in the inlet pipe. Figure 11 shows the flow through the gap at

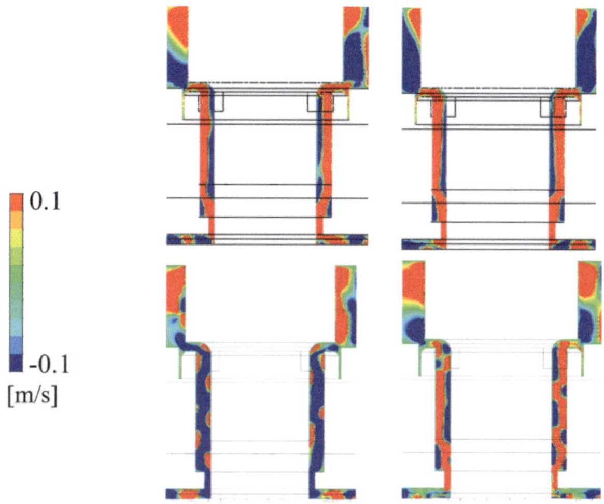

Fig. 11 Flow through the gap at the shaft for the baseline (left) and *H+* (right), for a small mass flow (top) and for a large mass flow (bottom)

the shaft for different operating points. Regions where fluid flows from the pump into the tank are marked in red, regions where fluid is sucked into the pump through the gap at the shaft are marked in blue.

The flow through the gap at the shaft changes for different operating points. For the investigated operating point and smaller flowrates fluid flows from the first chamber through the gap at the shaft back in the tank, for all shown individuals and the baseline. For large mass flows, the direction of flow through the gap on the shaft changes in the baseline but not for the best individuals of run 2 and run 3. This allows the optimized pumps to operate at lower immersion depths, as no medium enters the pump through the gap at the shaft, even at high mass flows.

4 Computational Resources

Each optimization run uses seven 128-core nodes. The optimization process uses 64 cores, the master cores, during the optimization run. The rest of the cores are used for the simulations, the simulation cores. All serial processes such as the meshing of each individual and the optimization script run on the master cores. On the simulation cores only the time consuming simulations of the individuals run in Ansys CFX. For evaluating the computational resources, the speedup of the numerical flow simulation in Ansys CFX, the occupancy of the simulation cores and the speedup of the optimization process with usage of the island model is investigated. An approximately 16.8 M cell mesh is used for a typical simulated individual. In Table 1 the speedup values of the simulation are presented with different numbers of used cores.

In the first column the used cores are given, the second column shows the needed time for the simulation, and the other columns show the speedup values with the

Table 1 Wall-clock time for a representative candidate's simulation using different number of cores; speedup s calculated for different references; measurement performed on a single node

# cores [–]	Time [s]	s_{16} [–]	s_{32} [–]	s_{64} [–]	s_{128} [–]
16	17640.0	b2/[−1,0]			
32	9120.3	b2/[−1,0]	b3/[−2,0]		
64	4380.8	b2/[−1,0]	b3/[−2,0]	b4/[−3,0]	
128	2220.7	b2/[−1,0]	b3/[−2,0]	b4/[−3,0]	b5/[−4,0]
256	1320.4	b2/[−1,0]	b3/[−2,0]	b4/[−3,0]	b5/[−4,0]

Table 2 Occupancy of the cores used for the simulations for each optimization run

Optimization run [–]	Maximum cumulated time [s]	Used time [%]
Run 1	19129474	29.2
Run 2	16430388	94.6
Run 3	11249875	93.2

referenced number of cores as index. The time shown for the simulation does not include the time needed to create or mesh the geometry - only the simulation time. The number of used cores doubles for each row, the ideal speedup for each row is accordingly to be doubled. The most speedups lie below the ideal speedup values. The speedup s_{32} has for the most cores the best speedup values and shows for 64 cores an over ideal speedup. In the optimization runs each flow field simulation uses 64 cores that the needed time for the simulation matches the time, which is needed to create the geometries and meshes of one individual. From the cores requested for each optimization run, minus the master cores, 13×64 cores remain for simulations. To evaluate simulation core occupancy the cumulative time of all simulation cores in use is compared to the maximum possible usage time of all simulation cores together. The occupancy of the thirteen times 64 cores for the simulations for the three different optimization runs is tabulated in Table 2.

Run 1 ran for around seventeen days but the simulation cores are only occupied around 29% of the time. To get a better occupancy of the same number of simulation cores the number of islands is increased in run 2 and run 3. The number of islands increase by a factor of three to 36 islands. By increasing the number of islands, a better utilization of up to 94.6 % of the simulation cores is achieved. For evaluating the whole optimization process the fitness speedup is used. The fitness speedup compares the minimum reached fitness values for the optimization runs at different numbers of individuals. In Fig. 12 the minimal fitness values of all optimization runs over the number of individuals are shown. Note that the fitness values are plotted logarithmic.

For the first thousand individuals the fitness values get smaller very quick for all optimization runs. The speedup at the beginning of the optimization runs are therefore large. Depending on the optimization run, a different number of individuals is needed

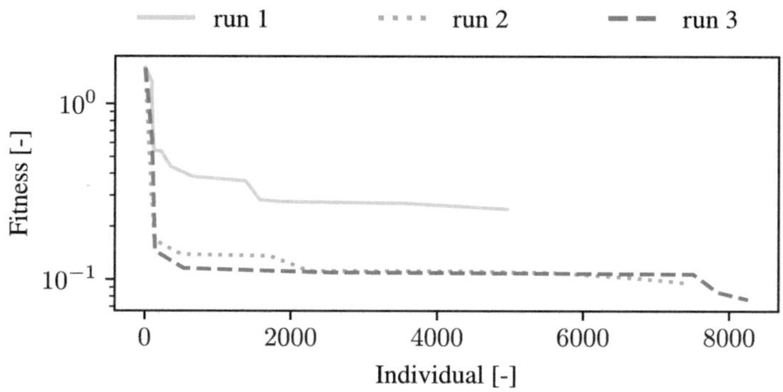

Fig. 12 Trend of the minimum fitness values over the number of individuals for all optimization runs

until an individual with lower fitness is found. The more individuals are generated in an optimization run, the smaller the fitness speedup becomes. Only in the last optimization run there is a bigger speedup at larger numbers of individuals. Runs 2 and 3 create more individuals while reducing the optimization runtime to fourteen (run 2) and ten days (run 3).

5 Conclusion and Outlook

A fully automated evolutionary optimization of a non-standard pump from galvanic industry is presented. Nine degrees of freedom parameterize the pump. The results of three optimization runs are compared. In the first two optimization runs the fitness values are calculated identical by multiplication of the efficiency and the head term. In run 3 the fitness values are calculated by adding those terms. The first optimization run consists of twelve fully connected islands with a population of ten on each island. The initialization of this optimization run is homogeneously. Due to a low occupancy of the simulation cores the number of islands for the last two optimization runs is increased to 36 and the population number is decreased to five. Also, the initialization is adapted to be more divers, since the parameter range of each parameter in run 1 is not completely covered. The influence of each parameter on the fitness value is not uniform. The inclination angle γ of the junction pipes has a big influence on the fitness value. The inclination angle β of the runner boreholes has a small influence on the fitness value. The results of all optimization runs show a change in the flow through the pump. The mass flow losses through the pump, as well as the flow through the boreholes of the runner and the gap at the shaft change as a result of the optimization. In all three optimization runs the efficiency and the operating range are improved. At the same time the way the fitness value is calculated has an effect on the results of the optimization run. By calculating the fitness value through multiplication of the fitness terms, a better efficiency is achieved, in case of addition a wider operating range. This is consistent with the fitness function characteristics resulting from the different definitions. Both achievements make a difference for the operation of this pump in the galvanic industry.

In this study only degrees of freedom of the runner and the junction pipes are taken into account. A next step could be to select additional or different degrees of freedom such as a different shape of the second chamber or bent boreholes in the runner or junction pipes.

The shown optimization runs in this work ran between ten and seventeen days. To shorten the duration of the optimization runs, the workflow of the optimization process could be further parallelized. This could either be done by using more nodes and more islands in the simulation process or by using more cores for each flow field simulation.

Acknowledgements The simulations were performed on the national supercomputer HPE Apollo Hawk at the High Performance Computing Center Stuttgart (HLRS) under the grant number 44196. Furthermore, the authors gratefully acknowledge financial support from the German Federal Ministry for Economic Affairs and Energy (BMWi) within the framework of the Central Innovation Program for SMEs based on a resolution of the German Bundestag and the cooperation with Bohncke GmbH.

References

1. Effiziente Energienutzung in der Galvanikindustrie. Guideline (2003)
2. Biscani, F., Izzo, D.: A parallel global multiobjective framework for optimization: pagmo. Journal of Open Source Software **5**(53), 2338 (2020). https://doi.org/10.21105/joss.02338
3. Cohoon, J.P., Hegde, S.U., Martin, W.N., Richards, D.: A parallel genetic algorithm. Proceedings of the 2nd International Conference on Genetic Algorithms and Their Application pp. 148–154 (1987)
4. Izzo, D., Ruciński, M., Biscani, F.: The Generalized Island Model, pp. 151–169. Springer, Berlin, Heidelberg (2012). https://doi.org/10.1007/978-3-642-28789-3_7
5. Storn, R., Price, K.: Differential evolution—a simple and efficient heuristic for global optimization over continuous spaces. J. Glob. Optim. **11**, 341–359 (1997). https://doi.org/10.1023/A:1008202821328
6. Tismer, A.: Entwicklung einer Softwareumgebung zur automatischen Auslegung von hydraulischen Maschinen mit dem Inselmodell. Ph.D. thesis (2020)

CFD of Transient Operations of a Pump Turbine

Marco Zorn, Joshua Tholen, and Stefan Riedelbauch

Abstract The workflow and the setup to simulate transient operations of a pump-turbine are investigated and improved within the scope of this study. A first simulation of a load increase with the guide vane angle opening from 6° to 15° is carried out to identify the validity of the workflow and numerical setup. The results indicate the necessity to correct the measured guide vane angles that are used within the simulation. Additionally, a small initial timeframe of a startup is used to further investigate the numerical setup. The convergence criteria for the mesh deformation, the correction of the measured guide vane angle, the frequency of the remeshing process as well as inlet and outlet boundary conditions are investigated and improved.

1 Introduction

The expansion of renewable but volatile energies like wind and solar power requires a stabilization of the electrical grid which can be achieved by a flexible operation of hydropower plants. Unfortunately, this flexibility requires the operation under severe off-design conditions including deep partial load operation or transient operations like start and stop procedures. Both come with higher pressure fluctuations that lead to an accumulation of damage within the runner and other components [1–3].

Studies on the flow field [4] and structural analysis [5] of steady state operation at deep part load conditions have already been conducted at the Institute of Fluid Mechanics and Hydraulic Machinery (IHS) and are currently extended to further investigate the occurring flow phenomena at these operating conditions. This paper, however, only describes the current work on the CFD-simulation of transient operations. The future objective here is to simulate a full transient operation and use the simulated pressure fields as input for a subsequent structural analyses to enable a comparison between the damage of a steady state off-design condition like deep part load and a transient operation. Prior to this simulation a workflow is developed

M. Zorn (✉) · J. Tholen · S. Riedelbauch
Institute of Fluid Mechanics and Hydraulic Machinery, Stuttgart, Germany
e-mail: marco.zorn@ihs.uni-stuttgart.de

T. Ludwig et al. (eds.), *High Performance Computing in Science and Engineering '23*,
https://doi.org/10.1007/978-3-031-91312-9_24

and multiple studies are carried out to improve the accuracy of the simulation and investigate possibilities to reduce the required numerical effort.

2 Workflow for CFD Simulations of Transients

The developed workflow to simulate transient operations of pump-turbines is adapted and partially extended based on the findings and descriptions of previous works within this topic [6–9]. Compared to the simulation of a steady operating point, as described in [4] three new topics need to be addressed: The motion of the guide vanes, the acceleration and deceleration of the runner rotational speed and the inlet and outlet boundary conditions that might be influenced by transient effects within the penstock (see Fig. 1).

2.1 Guide Vane Motion

During a transient operation of a pump-turbine the discharge and consequently the power output is controlled by adjusting the angle of the guide vanes. Depending on the guide vane angle two different approaches are used. The threshold guide vane angle that decides which approach is used depends on the smallest guide vane angle that can be meshed with reasonable quality.

2.1.1 Mesh Deformation and Remeshing

If the guide vane angle is above that value, the mesh deformation and remeshing approaches of ANSYS CFX are used. Within the mesh deformation the mesh boundary cells representing the guide vanes are displaced according to the prescribed guide vane motion. However, if only the boundary cells are moved this will decrease the mesh quality and will lead to inverted mesh elements. Consequently, the displacement of the boundary cells needs to be diffused into the inner mesh elements. To do this, a diffusion equation is solved that calculates the displacement of every mesh node. Using a diffusion coefficient within this equation that depends on the wall

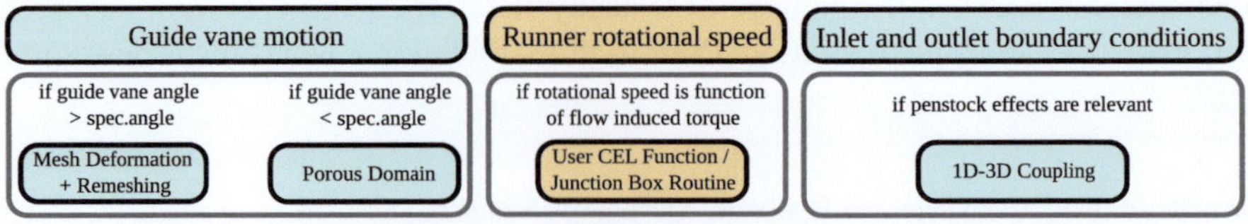

Fig. 1 Workflow for a CFD simulation of transient operations

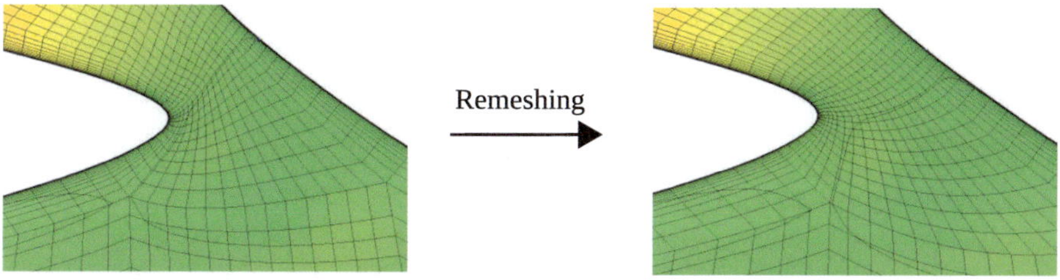

Fig. 2 Deformed and remeshed mesh between the guide vanes

distance and the cell size helps to control the displacement of the mesh nodes. It is usually preferred to have the same displacement for all nodes close to walls to not deform them and maintain the same boundary layer resolution. Most of the deformation of the mesh is usually preferred to happen away from the boundary layers in larger cells, where this deformation will affect the cell quality less. If the mesh displacement becomes larger the mesh quality will significantly decrease at some point. Then a new mesh (see Fig. 2), that is adjusted to the current guide vane angle, is required. To do this the remeshing approach of ANSYS CFX is used as soon as a certain criterion is reached. This criterion can be based on mesh quality parameters, fixed guide vane motion angles or other. Within the remeshing process an external script is started to create a new block-structured mesh for the guide vanes within ICEM CFD. Both the current guide vane angle and mass flow are used as parameters within this script to create a mesh with the correct angle and to adjust the y^+ values of the mesh.

2.1.2 Porous Domain

If the guide vane angle is below the previously mentioned threshold, the motion of the guide vane is considered by using a porous domain within the gap of the guide vanes (see Fig. 3).

Using only the quadratic term of the Forchheimer equation, the pressure gradient within the porous domain depends on the loss coefficient K_{loss}, the density ρ and the velocity u within the porous domain.

$$-\frac{\mathrm{d}p}{\mathrm{d}x} = K_{loss}\frac{\rho}{2}u_{porous}^2 \tag{1}$$

The idea is to adjust K_{loss} dependent on the real guide vane angle. As an first approach, a formula derived from the energy equation was developed within a master thesis [10] at the IHS to approximately get the correct mass flow. This formula only requires information on the real guide vane angle γ, the shortest real gap distance $w_{gap}\left(\gamma_{porous}\right)$ between the guide vanes dependent on γ and the width w_{porous} and

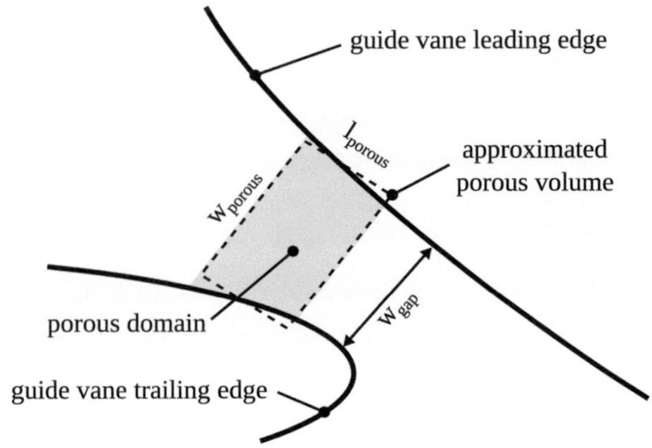

Fig. 3 Porous domain within the gap between two guide vanes

length l_{porous} of an approximated rectangular region that fits to the region of the porous domain (see Fig. 3).

$$K_{loss} = \frac{1}{l_{porous}} \left(\frac{w_{porous}}{w_{gap}\,(\gamma))} \right)^2 \left[1 - \left(\frac{w_{gap}\,(\gamma)}{w_{gap}\,(\gamma_{porous})} \right) \right]^2 \tag{2}$$

It must be mentioned that this formula is only delivering a good approximation of the mass flow. Local velocities and thus local kinetic energy within the gap between the guide vanes and shortly downstream of it is lower than in reality since the guide vane gap within the used mesh is larger than in reality.

2.2 Runner Rotational Speed

During, for example, a start or stop of a pump-turbine, its rotational speed changes. To calculate the change of the rotational speed and thus the current rotational speed during each timestep within the simulation, a so-called Junction Box Routine and User CEL functions within ANSYS CFX are used. The equation for this calculation requires the rotational speed ω_{t-1} and runner torque $T_{Ru,t-1}$ of the previous timestep as well as the timestep size dt and the moment of inertia I of the rotating parts.

$$\omega_t = \omega_{t-1} + \frac{T_{Ru,t-1}}{I} \cdot \mathrm{d}t \tag{3}$$

Using this approach multiple things need to be considered. First, the runner itself is simulated within a rotating frame, which means the Centrifugal, Coriolis and Euler force need to be considered. Whereas the Centrifugal and Coriolis force depend on the rotational speed, the Euler force is a function of angular acceleration which is only calculated within the User CEL functions but not available to the solver itself.

Consequently, the Euler force needs to be added as a source term. Moreover, to get the correct angular acceleration the torque needs to be of high accuracy. This requires a scaling of the simulated torque at model scale to the prototype scale of the measurements and further a correction to consider the influence of the runner gap, which is neglected within the mesh.

2.3 Inlet and Outlet Boundary Conditions

As described in [9] a fast transition between different operating points can cause water hammer waves that are travelling through the penstock. This influences the pressure at the inlet of the pump-turbines spiral casing, which means, that a boundary condition with a constant static or total pressure value at the extended spiral case inlet is no longer suitable to represent the reality. If no measurements of the static pressure are available these pressure waves need to be simulated. Whereas one option could be to treat the fluid as compressible instead of incompressible which would lead to an inacceptable increase in simulation effort [11]. Another option is the coupling between a 1D incompressible simulation of the penstock and the 3D CFD simulation of the pump-turbine [7]. This is not done yet in the present work but must be mentioned for the sake of completeness. Moreover, this issue can apply to the outlet boundary condition as well.

3 Numerical Setup

The basic setup of the following investigations is described within this section. All setup changes relevant to the different investigations are described in the respective sections. The pump-turbine is meshed with a block-structured approach in ICEM CFD and is simulated using the ANSYS-CFX software. Single phase incompressible simulations are carried out at model scale and include the spiral casing, the entire stay and guide vane area, the entire 360° runner without the gap, the draft tube and the test rig tailwater tank. As boundary conditions a constant total pressure at the extended inlet of the spiral casing and an opening pressure of 0 Pa at the tank outlet are specified. The interfaces between rotating and stationary domains are set as transient rotor stator interface. All simulations start from an initial solution of one steady operating point that shows convergence of the transient statistics. The Stress Blended Eddy Simulation (SBES) turbulence model [12] is used with a timestep equivalent to 0.2° of runner rotation and a mesh size of 27 million nodes. Within this 27 M mesh the average boundary layer resolution is below $y^+ = 1$. Within one timestep four inner loops are used. A bounded central difference scheme for spatial discretization and a second order backward Euler scheme for temporal discretization is used. Measurements of the guide vane angle are used to move the guide vanes and replicate the transient operation. Below a guide vane angle of 0.3° the porous domain approach is used. Above, the mesh deformation and remeshing approach is applied.

4 First Results of a Load Increase

A first investigation of the mesh deformation and remeshing approach is carried out by simulating the load increase from 6° to 15° of guide vane angle. To gain first knowledge faster, the timestep is increased to 3.6° of runner rotation, which needs to be considered evaluating the results. The remeshing is started every 0.5° of guide vane rotation. As visible in Fig. 4 the simulated torque shows a similar course as the measured runner torque, indicating that the mesh deformation and remeshing approach is suitable for this simulation. However, a significant shift in the torque values is visible within the entire timeframe but especially within the first 20 revolutions. Possible reasons for this deviation are the larger timestep, the neglect of the runner gap and especially the accuracy of the measured guide vane angle, which is known to be low at prototype scale.

All three influences are investigated within further simulations. First, the influence of the runner gap is determined by comparing simulations with and without the runner gap at a fixed guide vane angle of 6° to each other for both model and prototype scale. The mesh including the runner gap consists of around 40 million nodes. To simulate prototype scale the model scale meshes are directly scaled to prototype scale. Due to the different Reynolds number the boundary layer resolution changes to an average value of $y^+ = 12$ at the runner inside surfaces and $y^+ = 33$ within the runner gap surfaces. As the automatic wall treatment of ANSYS CFX is made to handle these values, the results are expected to enable an estimation of the difference between model and prototype scale. Still, it must be kept in mind, that the boundary layer resolution is not the same. This is not adjusted due to the increasing computational effort that would be accompanied by that. To avoid an influence of the timestep, the timestep is decreased back to 0.2° of runner rotation.

Fig. 4 Simulated and measured torque during a load increase from 6° to 15° of guide vane angle

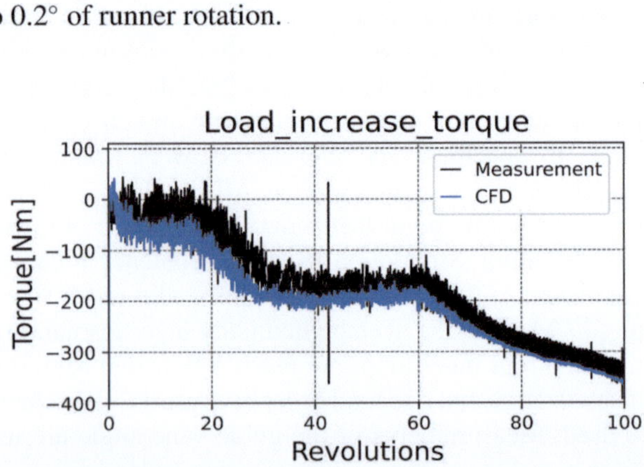

Table 1 Influence of runner gap on runner torque at model and prototype scale

Run gap	Model scale	no gap	Proto scale	no gap
Runner torque [%T_{opt}]	5.05	6.68	7.88	8.79
Diff. proto to model	–	–	2.83	2.1
Friction losses in gap [%T_{opt}]	1.4	–	0.76	–
Other losses due to gap [%T_{opt}]	0.23	–	0.14	–

In Table 1 the results of this investigation is presented. It is seen that the gap has a larger influence on the torque at model scale than in prototype scale, which is in line with theory, since the friction coefficient decreases with higher reynolds number. Moreover, most of the torque losses are due to friction within the gap, whereas other influence like mainly losses due to leakage flow have a smaller impact. The difference in runner torque between prototype and model is close to the analytical value of +1.93%. The simulated losses within the gap of in total −0.9% at prototype scale are considered in all following simulations as well as the efficiency correction of +1.93% from model to prototype. Regarding, the large torque deviation in Fig. 4 the neglect of the runner gap, however, does not explain this. Consequently, the influence of the guide vane angle is investigated.

Multiple simulations with different guide vane angles are carried out to find the guide vane angle correction, that leads to a better fit between simulated and measured torque. The main focus is to achieve the correct torque at a measured guide vane angle of 6.0° which presents the speed no load condition. This is important to get the correct rotational speed during a start of the pump-turbine. A correction of −0.75° is found to be a good assumption.

5 Investigation of Setup Parameters for a Startup

5.1 Study on Required Inner Loops Within the Mesh Deformation

At the start of each timestep the diffusion equation of the mesh deformation is solved iteratively prior to the solution of the flow-relevant quantities. The influence of the accuracy of the solution of the mesh deformation is investigated by comparing three simulations that use root mean square residual values of 1e-4, 1e-5 and 1e-6 as convergence criteria for the inner loops. A guide vane opening from 6.0° to 12.0° within 12 revolutions is simulated. Within Fig. 5 the simulated torque and the minimum mesh cell angle is presented. The simulated torque is similar for all setups with the exception, that the higher frequency fluctuations show deviations in their course. This is explained by the mesh that is slightly different due to the different convergence criteria for the mesh deformation. At these low loads between 6.0° and

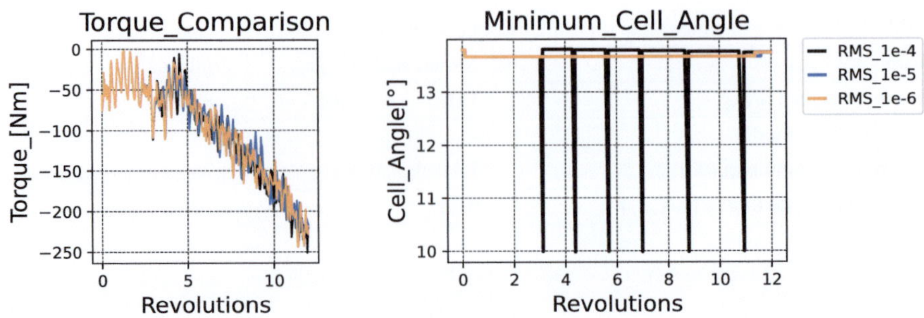

Fig. 5 Torque and minimum cell angle with different convergence criteria for the inner loops of the mesh deformation

12.0° the flow field is highly chaotic, which means slight deviations of any boundary condition influence the course of the fluctuations. Considering this, all three setups deliver similar physical results. However, they behave different in the occurring mesh cell angles and the consequent frequency of the remeshing process. In Fig. 5 it is seen, that the minimum cell angle within the RMS 1e-4 setup decreases suddenly at multiple occasions which initiates the remeshing process to avoid negative cell angles and an abortion of the simulation. Due to this remeshing the RMS 1e-4 setup requires the longest simulation time. Overall, the RMS 1e-5 setup is approximately 10 % faster than the RMS 1e-6 setup and 50 % faster than the RMS 1e-4 setup. According to these findings the RMS 1e-5 configuration is chosen for further simulations. Regarding the presented minimum cell angle of approximately 13.0° it must be mentioned that this angle is only occurring in a very small area within the stay and guide vane area. Except this area the guide vane angles are above 30°. Still this small area appears to affect the mesh deformation and remeshing process.

5.2 Modification of the Used Guide Vane Angle Measurements

As mentioned earlier, the measured guide vane angle needs to be corrected to achieve a better fit between the measured and simulated torque. A correction of −0.75° at angles of 6° and higher is chosen. However, the angles below need to be corrected as well since the measurements start at an angle of around −0.9° which is geometrically unrealistic. The measured value of −0.9° is corrected to be 0° within the CFD. Between this correction and the correction above 6° the guide vane angle is interpolated as displayed in Fig. 6. Moreover, the influence of the high frequency fluctuations of the guide vane angle in the measurements is investigated by compar-

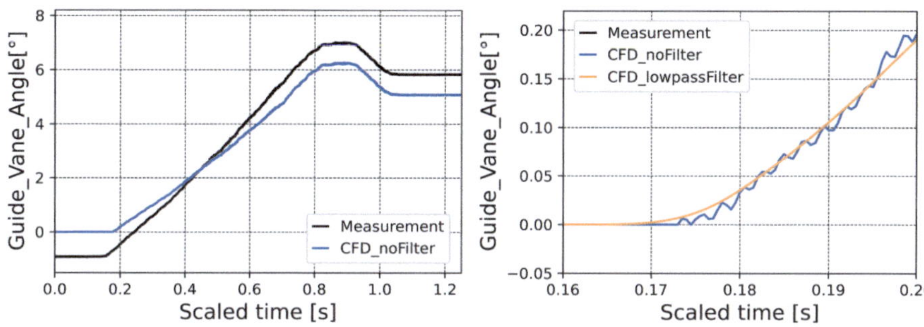

Fig. 6 Measured and modified guide vane angle

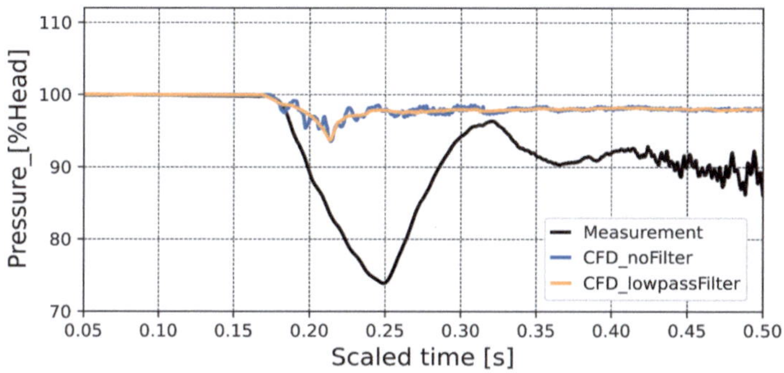

Fig. 7 Influence of the low-pass filtering of the guide vane angle on the simulated pressure upstream of the pump-turbines runner

ing a simulation that includes these fluctuations with a simulation that uses a low pass filter to remove these fluctuations (see Fig. 6).

Figure 7 shows the measured and simulated pressure at sensors located upstream of the turbine at the beginning of the spiral casing. It is seen that the high frequency fluctuations of the guide vane angle lead to high frequency pressure fluctuations that are not visible in the measurement. Consequently, the following investigations will use a low pass filter to remove these fluctuations. The large pressure drop of lower frequency within the measurements between 0.2 and 0.3 s is explained and investigated in Sect. 5.4.

5.3 *Influence of the Remeshing Frequency*

Whereas the influence of remeshing on the simulation time was already presented in Sect. 5.1, the influence of the remeshing and the mesh deformation on the accuracy during the early phase of a startup is presented in Fig. 8. Three setups using a larger timestep equivalent to 3.6° of runner rotation at fixed speed are compared: A base

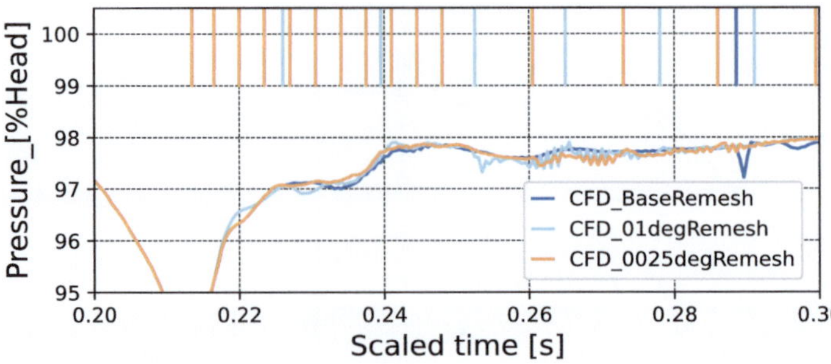

Fig. 8 Influence of the remeshing frequency on the pressure upstream of the pump-turbines runner. The lines above 99% indicate the occurence of a remeshing event

setup with the remeshing triggered by a threshold of the mesh quality, a setup that starts the remeshing process every 0.1° of guide vane rotation and a setup that starts the remeshing process every 0.025° of guide vane rotation in the early beginning. Comparing the results of the simulated pressure values upstream of the pump-turbine runner at the inlet of the spiral casing two unexpected pressure peaks are visible. One at around 0.25 s within the 0.1° remeshing setup and another at 0.29 s for the base setup. Both peaks coincide with the time of the remeshing process. An explanation for this is that if the remeshing occurs after a longer timeframe, the deviation between the old deformed and the new remeshed mesh is larger. Consequently, the meshes lead to different solutions which results in a jump of the simulated quantities. This effect is further intensified as the mesh deformation is made to keep the height of the boundary layer, whereas the remeshing process adjusts the boundary layer to stay within the desirable y^+ values, which results in a jump of the boundary layer resolution as well. For the final simulations the remeshing process starts at every 0.025° of guide vane rotation is then increased to every 0.1° and only for guide vane angles above 1.5° the remeshing is initiated by the mesh quality threshold.

5.4 Influence of Inlet and Outlet Boundary Conditions

To investigate the influence of the boundary conditions at the inlet and outlet during a startup from standstill, three different setups are compared. The first setup is the basic setup with a total pressure inlet and the model tank at the end of the draft tube. In the second setup the model tank is replaced by the prototype outlet geometry including a small part of the tailwater (see Fig. 9). The third setup includes the prototype outlet geometry as well and uses the measured, time dependent static pressure as inlet condition. To reduce the computational effort of this study the mesh size is decreased to around 14 million nodes and the timestep is increased to what would be equivalent to 3.6° of runner rotation at fixed speed.

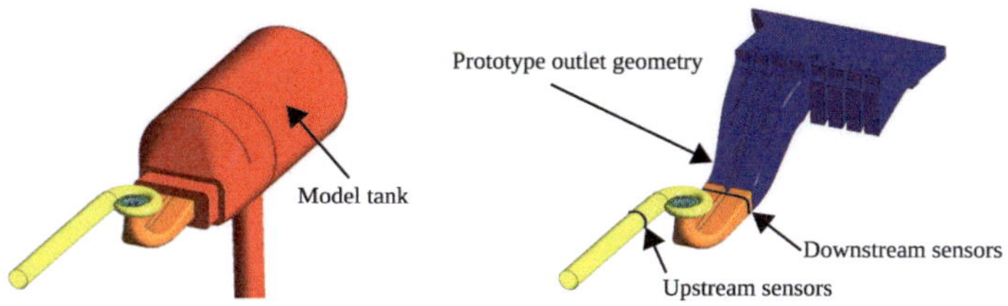

Fig. 9 Model tank and prototype outlet geometry

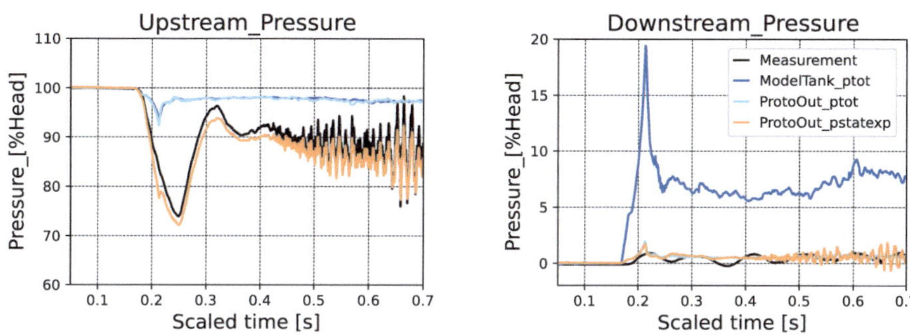

Fig. 10 Influence of the inlet and outlet boundary conditions on the upstream and downstream pressure values

In Fig. 10 the simulated and measured static pressure at the upstream and downstream sensors is displayed. The location of the sensors is shown in Fig. 9. Comparing the upstream pressure values, it is visible, that a constant total pressure inlet is not capable of resolving the pressure drop occuring during the startup. This is as expected, as the pressure drop occurs due to the compressible effects within the pump-turbine and its penstock which are neglected in this setup. Using the measured static pressure as inlet condition obviously results in the pressure drop being visible in the simulation results. The shift between the measured values and the ProtoOut pstatexp setup is present since the inlet of the CFD geometry is not at the same position as the upstream sensors. These results display the necessity of coupling a compressible 1D solver of the penstock with a 3D CFD of the pump-turbine to resolve this phenomenon in cases, where no measurements are available.

The comparison of the downstream pressure values shows, that the model tank leads to different results than the prototype outlet geometry (see Fig. 10). Using the model tank, the pressure increase is significantly larger than with the outlet geometry. Consequently, at least at the beginning of a startup using a model tank as extension of the draft tube is not suitable. In contrast, both simulations with the prototype outlet geometry predict pressure values close to the measured values. However, the sinus-like pressure oscillation in the measurement is not visible in the simulations. Again, this is expected to be a pressure wave that cannot be resolved within an incompressible CFD. Overall the usage of the measured static pressure as inlet boundary condition together with the prototype outlet geometry is the best suited setup to simulate a startup within an incompressible CFD approach that is not coupled with a 1D solver of the penstock.

6 Computational Resources

The numerical simulations have been performed on the HPE Apollo (Hawk) at the High Performance Center Stuttgart. The architecture of the HPE Apollo consists of 5632 compute nodes with 2 AMD EPYC 7742 processors per node with a frequency of 2.25 GHz. Each node comes with a total of 128 cores and 256 GB memory. An InfiniBand HDR200 interconnect with an enhanced 9D-Hypercube topology is used on Hawk.

On the file system ws10 different scaling tests are carried out, that use version 19.5 of ANSYS CFX and the MPI method Open MPIHAWK Distributed Parallel. Whereas simulation results from multiple setups are presented in the previous sections, the scaling tests are only shown for the two setups that overall require the largest amount of computational resources. These setups are.

A 27M mesh within unsteady simulations of constant operating points. This setup is used to investigate the influence of the guide vane angle correction. It is also used in an investigation of multiple deep part load operating points that is still ongoing and therefore not presented within the scope of this paper.

A 29M mesh within the transient simulations including the protoype outlet geometry and moreover the mesh deformation approach leading to further computational effort. This setup includes the findings from Sect. 5 and is used in the final transient simulations of a load increase that will deliver pressure fields for subsequent structural analyses.

The speedup curves as well as the time per timestep for both setups are presented in Fig. 11. The speedup curves are normalized to the simulations with 128 cores. It is visible, that the 29 M setup with mesh deformation shows a better speedup and requires more time per timestep compared with the 27M setup without mesh deformation. According to the results of the speedups the simulations of the 27 M setups are performed on 512 cores with a parallel performance of 75 %, whereas the 29 M setups are using 768 cores with a parallel performance of 76 %. Both values are not ideal but are chosen as a trade-off between performance and overall timeframe to achieve the numerical results.

In addition to the speedups that use all 128 cores per node, the influence of using less cores per node is investigated to identify a possible bottleneck due to bandwidth, which was visible on several workstations used at the IHS. The speedups of simulations using 128, 96 and 64 cores per node are shown for both the 27 and 29 M setup in Fig. 12. No relevant influence of the number of used cores per node is observed for these two setups.

Within the reporting period around 9 million core-h have been used. Approximately 500.000 core-h were needed for the first simulation of the load increase, 2

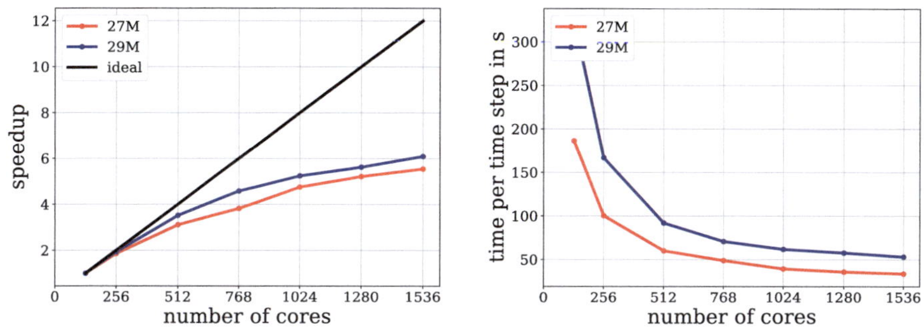

Fig. 11 Speedup and time per time step for the 27 and 29 M setup

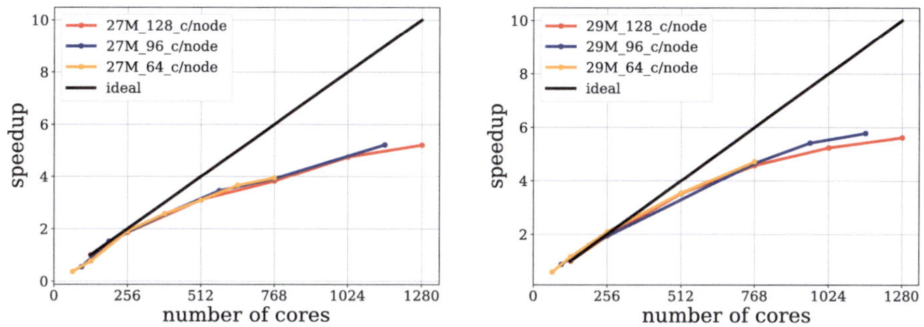

Fig. 12 Speedup for the 27 and 29 M setup using 128, 96 and 64 cores per node

million core-h for the investigations of the runner gap, 2.5 million for the investigation of the guide vane angle correction and 500.000 core-h for the setup investigations. The other around 3.5 million core-h were used within investigations of different deep part load operating points.

7 Conclusion

In this study, a workflow was developed to enable the simulation of transient operations of a pump-turbine. This workflow requires a combination of mesh deformation and remeshing as well as the usage of a porous domain to represent the motion of the guide vanes. A first simulation using this workflow within a load increase from 6° to 15° of guide vane angle demonstrated the capability of this workflow but showed deviations in the comparison between simulated and measured torque. It was found that the measured guide vane angle needs to be corrected by around −0.75° to achieve a better fit between measurement and simulation. Moreover, the influence of the neglected runner gap on the runner torque was investigated to correct this influence as well. Further simulations were carried out to improve parameters of the setup, especially during a startup, where the guide vane angle is starting at 0° and the runner rotational speed is increasing until fixed speed. It was found, that a convergence criterion of a root mean square residual value of 1e-5 is best suited within the solution of the mesh deformation. The high-frequency oscillations within the measured guide vane angle need to be removed by a low-pass filter to avoid oscillations that are not visible within the measured fluid quantities. Remeshing at very low angles needs to be executed more often to avoid jumps within the solution. The prototype outlet geometry must be considered instead of the model tank and the measured static pressure upstream of the pump-turbine must be used as inlet condition if no 1D-3D coupling is implemented.

Future work includes the simulation of a full load increase, using the findings from these investigations, to obtain the time dependent pressure field on the runner and use it within structural analyses to finally estimate the damage of this operation. Moreover, the influence of different mesh sizes, timesteps and the usage of the porous domain will be investigated and a workflow for a 1D-3D coupling to enhance the boundary conditions will be developed and investigated.

Acknowledgements The simulations were performed on the national supercomputer HPE Apollo Hawk at the High Performance Computing Center Stuttgart (HLRS) under the grant number 44047. The Hydropower Extending Power System Flexibility (XFLEX HYDRO) project has received funding from the European Unions Horizon 2020 research and innovation programme under grant agreement No 857832.

References

1. Valero, C., Egusquiza, M., Presas, A., Valentin, D., Bossio, M.: Extension of operating range in pump-turbines. Influence of Head and Load. Energies **10**(12), 2178 (2017)
2. Seidel, U., Mende, C., Hübner, B., Weber, W., Otto, A.: Dynamic loads in Francis runners and their impact on fatigue life. In: IOP Conference Series: Earth and Environmental Science, vol. 22, p. 032054 (2014)
3. Hasmatuchi, V., Decaix, J., Titzschkau, M., Münch, C.: A challenging puzzle to extend the runner lifetime of a 100 MW Francis turbine. In : Proceedings of Hydro 2018, Gdansk, Poland (2018)
4. Zorn, M., Khalfaoui, K., Riedelbauch, S.: Towards accurate numerical fluid dynamics of pump-turbine runners in deep partial load operation: In : Proceedings of the 31st IAHR Symposium on Hydraulic Machinery and Systems, Trondheim, Norway (2022)
5. Khalfaoui, K., Zorn, M., Riedelbauch, S.: Towards accurate structural dynamic modelling of pump-turbine runners in deep partial load operation: In : Proceedings of the 31st IAHR Symposium on Hydraulic Machinery and Systems, Trondheim, Norway (2022)
6. Mössinger, P., Jester-Zürker, R., Jung, A.: Francis-99: Transient CFD simulation of load changes and turbine shutdown in a model sozed high-head Francis turbine. J. Phys.: Conf. Ser. **782**, 012001 (2017)
7. Mössinger, P.: Numerische Untersuchung der Strömung bei instabilen und transienten Betriebszuständen von Francis-Turbinen. Ph.D. thesis, Stuttgart (2019)
8. Nicolle, J., Giroux, A.M., Morissette, J.F.: CFD configurations for hydraulic turbine startup. In: IOP Conference Series: Earth and Environmental Science, vol. 22, p. 032021 (2014)

9. Cherny, S., Chirkov, D., Bannikov, D., Lapin, V., Skorospelov, V., Eshkunova, I., Avdushenko, A.: 3D numerical simulation of transient processes in hydraulic turbines. In: IOP Conference Series: Earth and Environmental Science, vol. 12, p. 012071 (2010)
10. Tholen, J.: Netzgenerierung und CFD Simulation des Schließ- oder Öffnungsvorgangs eines Leitapparats. Master Thesis, Stuttgart(2022)
11. Trivedi, C.: Compressible large eddy simulation of a francis turbine during speed-no-load: rotor stator interaction and inception of a vortical flow. J. Eng. Gas Turb. Power **140**, 112601 (2018)
12. Menter, F.R.: Stress-Blended eddy simulation (SBES)—a new paradigm in hybrid RANS-LES modeling. In: Proceedings of the 6th HRLM Symposium, Strasbourg, France (2016)

Highly Resolved Hybrid RANS-LES Simulations of Different Operating Points of a Model Kaplan Turbine

S. Joßberger and S. Riedelbauch

Abstract Double regulated Kaplan turbines with adjustable guide vanes and runner blades offer a high degree of flexibility and good efficiency for a wide range of operating points. However, this also leads to a complex geometry and flow guidance with, for example, vortices of different size and strength. The flow in a draft tube is especially challenging to simulate mainly due to flow phenomena like swirl, separation and strong adverse pressure gradient and a strong dependency on the upstream flow conditions. The calculation of performance is often inaccurate with standard simulation approaches (i.e. RANS turbulence models, a coarse mesh and large time step size) and can lead to wrong predictions of the operating range. To reveal occurring flow phenomena and physical effects scale resolving hybrid RANS-LES simulations for three different operating points, one on-cam and two off-cam, on a block structured mesh of about 400 million hexahedral elements of a double regulated five blade model Kaplan turbine are carried out. In this paper first results of the ongoing simulations are presented. The major part of the simulation domain is running in LES mode and seems to be properly resolved. The validation of simulation results with experimental data shows good agreement in the global results, i.e. total head and power, and a good visual agreement with three-dimensional PIV measurements of the velocity. For the standard RANS simulations the results are inconsistent and the agreement with the measurement depends on the investigated operating point.

1 Introduction

The increasing use of volatile renewable energies like wind and solar power leads to higher fluctuations of available power in the electricity grid. Hydropower, as a predictable and reliable source of electricity, can be used to compensate the fluctuating

---

S. Joßberger (✉) · S. Riedelbauch
Institute of Fluid Mechanics and Hydraulic Machinery, Stuttgart, Germany
e-mail: simon.jossberger@ihs.uni-stuttgart.de

S. Riedelbauch
e-mail: stefan.riedelbauch@ihs.uni-stuttgart.de

T. Ludwig et al. (eds.), *High Performance Computing in Science and Engineering '23*,
https://doi.org/10.1007/978-3-031-91312-9_25

power output and stabilize the energy grid. Double regulated Kaplan turbines with adjustable guide vanes and runner blades offer a high efficiency over a wide range of operating points. This wide operational flexibility requires a substantial change of the runner blade opening. The gaps on both the tip and hub of the runner blades create flow vortices interacting with the decelerated flow of the draft tube. Because of that the flow in a draft tube is especially challenging to simulate (cf. [1]) and has a strong dependency on the flow conditions at the inlet defined by the quality and fidelity of the simulation upstream (cf. [2]). Typical Kaplan turbines are applied for small to medium heads. Therefore, even small errors in the flow dependent losses can lead to wrong predictions of performance and overall operating range.

The overall goal of this ongoing investigation is to detect and understand occurring flow phenomena and identify their importance to accurately simulate Kaplan turbines. To achieve that, highly resolved simulations for three different operating points with a hybrid RANS-LES turbulence model on a very fine mesh with about 400 million hexahedral elements are currently carried out. Advanced turbulence models demonstrated to improve the results and level of detail (cf. [4–8]), but their numerical requirements, e.g. time step size and mesh resolution, are generally much higher than standard RANS simulation approaches and typically too costly for industrial development processes. A comparison of the high-resolution simulations with different less complex simulations allows to identify important flow phenomena and how standard simulation approaches and development processes can be improved. To validate simulation results, time-averaged experimental values for total head and power of the runner are available, as well as phase-resolved three-dimensional PIV measurements in a plane just downstream of the runner and in a plane in each of the two diffuser channels of the draft tube. Details on the experimental investigation are published by Beck et al. [3]. In this part of the study an estimation of achieved numerical resolution of the high-resolution simulations is described. Furthermore, a comparison to experimental data and standard RANS simulations on a mesh with 20 million hexahedral elements is presented.

2 Model Turbine and Numerical Setup

The investigated turbine is a Kaplan turbine in model scale. The whole simulation domain consists of five subdomains (cf. Fig. 1) and begins with the rectangular semi spiral intake, which includes two piers. The following subdomain consists of 23 stay vanes and 24 adjustable guide vanes. The runner is built of the hub and five adjustable blades. No geometric simplifications are used for the runner blades. Both the tip and hub clearance and the fillets between the runner blades and the hub are modelled. An elbow type draft tube with one pier leads to the tailwater tank. The geometry of the tailwater tank matches the model test arrangement at the hydraulic laboratory. The outlet of the simulation domain is defined at the end of the pipe downstream of the tailwater tank. This minimizes the impact of the boundary conditions on the flow in the draft tube and allows an accurate evaluation of results even at the end of the draft

Fig. 1 Simulation domain. Red: Inlet area and semi spiral casing (SC). Blue: Stay and guide vanes (SVGV). Green: Runner (RU). Orange: Draft tube (DT). Yellow: Tailwater tank (TW)

tube. Due to the fact that the flow in the tailwater tank is of no particular interest, the tailwater tank will be neglected in the following results section.

Three operating points are investigated. The first operating point (OP1) is an on-cam point where the pitch angles of guide vanes and runner blades are matched. The second (OP2) and third (OP3) operating point are off-cam points with the same head and runner angle as the first operating point, but different opening angles of the guide vanes. OP2 has an increased opening angle of the guide vanes of $+2.5°$, leading to an increase in volume flow of about $+2.5\%$. OP3 has an decreased opening angle of the guide vanes of $-2.5°$, leading to an decrease in volume flow of about -3.5%.

ICEM CFD is utilized to generate the block-structured hexahedral mesh with about 20 and 400 million elements for the standard RANS approach and the high-resolution simulation, respectively. The main differences between the 20 million elements (20M) and 400 million elements (400M) mesh are in the resolution of the boundary layer and overall element size, while preserving the distribution of elements as far as possible. To properly resolve the boundary layer without the use of wall functions, a fine mesh at the wall and a small growth rate normal to the wall is required. With an unchanged overall element size, this leads to a mesh with about 50 million elements. For the 400M mesh, the element size is additionally approximately halved in all directions, while maintaining the general block structure as far as possible. The number of cells in the five subdomains intake with semi spiral casing (SC), stay and guide vanes (SVGV), runner (RU), draft tube (DT) and tailwater tank (TW) for both meshes is listed in Table 1.

The simulations are conducted with ANSYS CFX in version 19.5. Automatic wall treatment is activated for the boundary layer (cf. [10]). Depending on the local dimensionless wall distance y^+, either wall functions or direct calculation of the

Table 1 Number of cells of the subdomains in million elements

Sim	Total	SC	SVGV	RU	DT	TW
20M	19.8	2.1	7.0	4.7	4.9	1.1
400M	399.5	29.8	78.6	63.3	218.2	9.6

Table 2 Average, maximum and 95th percentile of time averaged y^+-value in the subdomains for operating point 1

	SC			SVGV		
Sim	Ave	Max	p95	Ave	Max	p95
20M OP1	62.1	145.1	99.9	81.7	267.1	132.9
400M OP1	0.54	1.86	0.89	0.59	69.7	1.07
	RU			DT		
Sim	Ave	Max	p95	Ave	Max	p95
20M OP1	57.0	135.1	88.5	60.5	221.3	105.5
400M OP1	0.96	6.63	3.84	0.42	2.02	0.93

boundary layer are applied. The average and maximum value and the 95th percentile of the time averaged y^+-values of the subdomains neglecting the tailwater tank are listed in Table 2 for operating point 1. In the 20M simulation the average y^+-value ranges between 57 and 82 depending on the subdomain. This leads to the use of wall functions. As indicated by the 95th percentile and the average, most of the elements of the 400M simulation have a y^+-value smaller than 2. This should lead to direct calculation of the boundary layer in most parts of the domain. In all simulations, the same time averaged three-dimensional distributions of velocity and turbulence properties from a preceding unsteady simulation of the flow in the head water tank are specified as inlet boundary conditions (cf. [3, 9]). An average static pressure is set as outlet boundary condition. General Grid Interfaces (GGI) are employed between stationary subdomains. Full 360° transient rotor stator interfaces connect the rotating runner subdomain with the stationary stay vane and draft tube subdomain.

Turbulence is modelled by the RANS $\mathrm{k}-\omega$ Shear Stress Transport model (SST) for the 20M simulations and by the hybrid RANS-LES Stress Blended Eddy Simulation model (SBES) for the 400M simulations. The SST model combines the advantages of the $\mathrm{k}-\omega$ model in wall bounded flows with the advantages of the $\mathrm{k}-\epsilon$ model in free shear flows (cf. [11]). It is a standard RANS model and is commonly applied in simulating hydraulic turbomachinery. The SBES model uses an unpublished shielding function f_S to blend between the stress tensor τ_{ij} modelled by the RANS and LES model (cf. [12]):

$$\tau_{ij}^{SBES} = f_S \tau_{ij}^{RANS} + (1 - f_S)\tau_{ij}^{LES} \tag{1}$$

In this study the SST model and the wall-adapted local eddy-viscosity model (WALE) are used for the RANS and LES part, respectively. Compared to other hybrid RANS-LES approaches, the SBES model has the advantage of a quick transition from RANS to LES mode and a clear distinction between both regions by the shielding function f_S (cf. [13]).

For spatial discretization the high resolution scheme is applied for the SST simulations and a bounded central differencing scheme is used for the SBES simulations (cf. [10]). A second order backward Euler scheme is employed for temporal discretization in all simulations (cf. [13]). A fixed number of five inner loops per time step is used. A time step size of $\Delta t = 2 \cdot 10^{-4}$ s, which is equivalent to 1.8° runner revolution, is used for the 20M simulations. This is a typical order of magnitude for unsteady RANS simulations of hydropower turbines. The unsteady simulations are initialized with a steady state SST simulation on the same mesh. Due to numerical instabilities the 400M simulations can not be initialized with a steady state simulation and results of a preceding unsteady simulation have to be used. In theory, the build up time necessary to obtain statistically valid results can be reduced by a higher quality of initial conditions. Therefore, unsteady simulations with the SBES turbulence model on the 20 million element mesh and a time step size of $\Delta t = 1 \cdot 10^{-4}$ s, equivalent to 0.9° runner revolution are used as initial conditions. After 10 revolutions the time step size is ramped down to $\Delta t = 5 \cdot 10^{-5}$ s over the course of about 100 iterations. After additional five revolutions, the time step size is ramped down to the final value of $\Delta t = 2.5 \cdot 10^{-5}$ s, equivalent to 0.225° of runner revolution.

For all simulations, the averaging of data and evaluation of results is started after a build-up time of 20 revolutions. This is equivalent to 4000 and 15920 time steps in the 20M and 400M simulations, respectively, and corresponds to about two convective flows from the inlet of the semi spiral casing to the outlet of the draft tube. The number of build-up iterations is chosen by experience with similar simulations of hydropower turbines and by the trend of integral values of interest in the subdomains. Due to the unsteady behaviour of the flow in the draft tube for some operating points in the 20M simulation, this subdomain is less suited for the consideration of build-up time.

3 Results

In the following sections, the 20 build-up revolutions are neglected and only results of the subsequent revolutions are presented.

3.1 *Estimation of Achieved Resolution in the 400M Simulation for Operating Point 1*

To derive the RANS equations of the Navier-Stokes equations, the flow variables are decomposed into a mean or time averaged component and a fluctuating turbulent component. Since the latter component and the resulting turbulence energy cascade are almost completely modelled by the turbulence model, the time step size of an unsteady RANS simulation is, in theory, sufficiently small, when it is able to capture

Table 3 Average, maximum and 95th percentile of time averaged Courant Number in the subdomains for operating point 1

	SC			SVGV		
Sim	Ave	Max	p95	Ave	Max	p95
20M OP1	0.07	2.78	0.17	1.21	19.1	3.30
400M OP1 RANS region	0.02	1.01	0.04	1.29	146	4.05
400M OP1 LES region	0.03	0.60	0.08	0.81	123	1.89
	RU			DT		
Sim	Ave	Max	p95	Ave	Max	p95
20M OP1	4.20	358	9.23	0.39	12.8	1.58
400M OP1 RANS region	3.04	423	9.59	0.08	5.25	0.33
400M OP1 LES region	0.87	220	1.77	0.11	3.36	0.26

the unsteady fluctuations of the mean values excited by external influences, such as interactions between the guide vanes and runner blades. In contrast to that, in a LES simulation all but the smallest turbulent scales are directly simulated. To avoid excessive damping of turbulence in the LES region, Menter [13] recommends a Courant number of $CFL \approx 1$ or smaller.

In Table 3 the average and maximum values and the 95th percentile of the time averaged Courant Number for operating point 1 are presented. The values for the 400M simulation are divided into the LES region and the RANS region. In the LES region of all subdomains the average Courant Number is smaller than 1 and most of the elements have a Courant Number smaller than 2 as indicated by the 95th percentile. The left part of Fig. 2 shows the root mean square of about 30 revolutions of the Courant number on a cut through the turbine for the 400M simulation. The areas exceeding the recommendation of $CFL \leq 1$ are mostly in regions with a fine mesh resolution like leading and trailing edges and gaps of the runner blades. The time step size would have to be approximately halved to achieve a Courant number smaller 1 for almost all elements. This would, in turn, approximately double the computing time for one revolution. The selected final time step size of the 400M simulations is a compromise between accuracy and feasibility.

The shielding function f_S of the SBES turbulence model is used to visualize whether the RANS or LES model is used. The black lines in the right picture of Fig. 2 represent a value of $f_S = 0.99$ and the magenta lines of $f_S = 0.01$. The regions outside the black lines are basically completely in RANS mode, while the regions inside the magenta lines are in LES mode. The small regions in-between the lines are the transition zones between the two models. The transition from RANS to LES primarily occurs in the semi spiral casing and stay and guide vanes. Downstream the guide vanes almost all of the simulation domain is in LES mode. In the runner

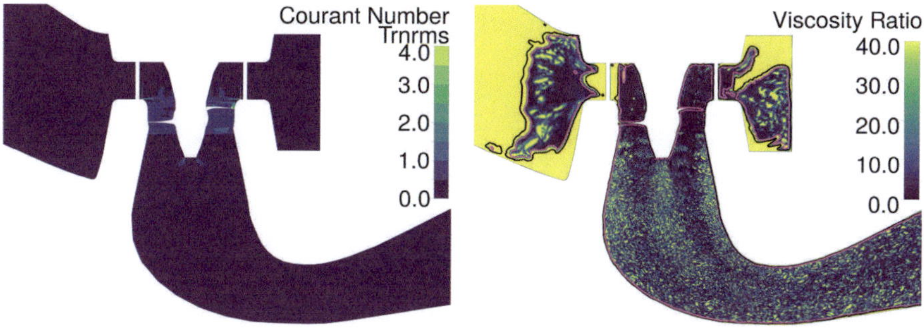

Fig. 2 Cut through the turbine for 400 M simulation of operating point 1. Left: Root mean square of Courant number. Right: Viscosity ratio ν_t/ν and shielding function f_S of the SBES turbulence model (black line: $f_S = 0.99$; magenta line: $f_S = 0.01$)

and draft tube only the boundary layer is in RANS mode, while all other areas are in LES mode. This is the intended state for a hybrid RANS-LES simulation. In the LES region, the amount of modelled turbulence, depicted by the eddy viscosity ratio ν_t/ν built with the turbulent viscosity ν_t and the dynamic viscosity ν, is quite small. In the RANS region, the viscosity ratio is about two orders of magnitude higher. This is also seen in Fig. 3 where the resolved turbulent structures in the draft tube, depicted by an isosurface of the velocity invariant $Q = 50$ and coloured by the viscosity ratio, are visualized. Compared to the 20M simulation, the 400M simulation resolves much finer structures and calculates the energy transfer due to turbulence on a much finer scale.

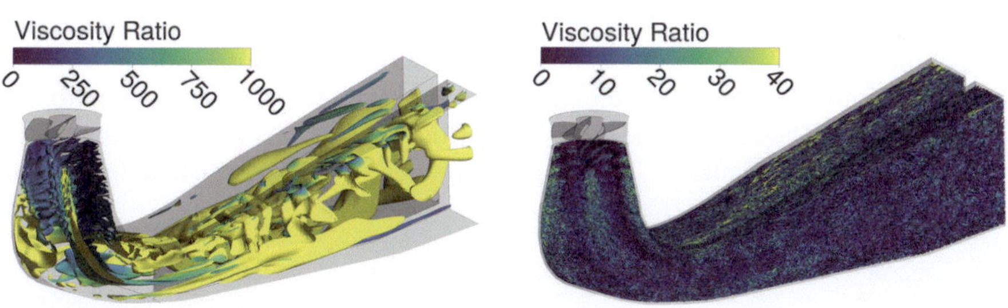

Fig. 3 Instantaneous isosurface of velocity invariant $Q = 50$ to depict resolved turbulent structures coloured by viscosity ratio ν_t/ν. Different scales are used for the viscosity ratio. Left: 20M simulation OP1. Right: 400M simulation OP1

3.2 Comparison of Simulations and Measurements for All Operating Points

At the moment only about 5 revolutions after the build up time for the 400M simulations of operating points 2 and 3 are available. Therefore, the conclusions drawn for these simulations are highly preliminary and must be taken with a grain of salt.

The left part of Fig. 4 shows power and the right part total head of the 20M and 400M simulations for all three operating points relative to the measurement. All simulations match the measured power very good, with a little advantage for the 20M simulations in most cases. The power trends of the 20M simulations exhibit only minor oscillations without high random deflections. In comparison, the power trends of the 400M simulations show higher oscillations and an overall noisier behaviour. This is probably an effect of the higher temporal and spatial resolution and the ability of the SBES model to resolve much smaller turbulent structures. Nonetheless, the power trend of all simulations has reached a quasi steady state. The trend of total head exhibits a different behaviour. The result for all 400M simulations and the 20M simulation of operating point 3 are in good agreement with the measurement and have reached a quasi steady state, which is also observed in the experiment. The trend of the 20M simulations of operating points 1 and 2 are highly unsteady and overestimate the head by more than 12% at the maximum. The trends do not settle over time. The oscillations seem to be no singular event, but the normal behaviour for this simulation approach.

Figure 5 shows the time averaged velocity and Fig. 6 the corresponding standard deviation on the measurement plane in the left diffuser channel of the draft tube. Due to the limited data available in the 400M simulations of operating points 2 and 3, these simulations are not shown. The underlying data used to build the time averaged flow field is quite different between simulations and measurement. In the simulations one revolution consists of 200 and 1600 individual data sets in the 20M and 400M simulations, respectively. But the number of revolutions is limited to less than a hundred. The measurement consists of only 9 data sets per revolution taken with a time lag equivalent to 8° runner revolution between subsequent data sets, but

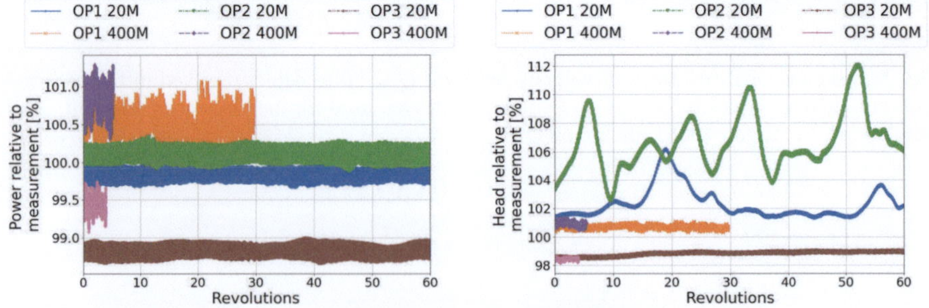

Fig. 4 Power (left) and total head (right) relative to measurements

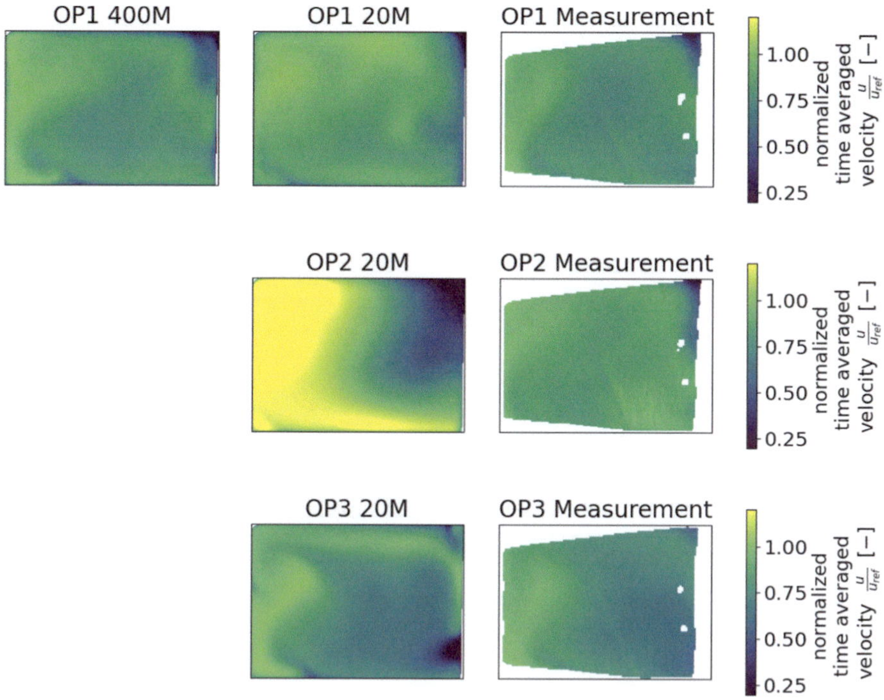

Fig. 5 Time averaged velocity on the measurement plane in the left diffuser channel. Values normalized with the maximum value of the corresponding measurement. Left: 400M simulation. Middle: 20M simulation. Right: Measurement. From top to bottom: Operating point 1, 2 and 3

a total of 1000 revolutions. Despite the different data basis, the velocity distribution and magnitude of the 400M simulation of operating point 1 and the 20M simulation of operating point 3 are in good agreement with the measurement. For the 400M simulation this is also true for the standard deviation, while the 20M simulation of OP3 exhibits a much smaller standard deviation. The velocity distribution of the 20M simulations of operating point 1 and 2 differ compared to the measurement and exhibit a more unsteady behaviour as is seen by the higher standard deviation. This is especially true for operating point 2. The data of the measurement plane in the right diffuser channel exhibit comparable results and are therefore not presented.

To identify the origin of the unsteady trend of total head (cf. Fig. 4) the head losses of the individual subdomains normalized with the corresponding mean of the 400M simulation are shown in Fig. 7. It is evident, that the unsteady behaviour of total head in the 20M simulations of operating point 1 and 2 is mostly determined by the losses in the draft tube. The other subdomains exhibit a more or less constant course similar to the power and have reached quasi steady states. In comparison to the 400M simulations, the 20 M simulations show smaller oscillations and an overall smoother trend.

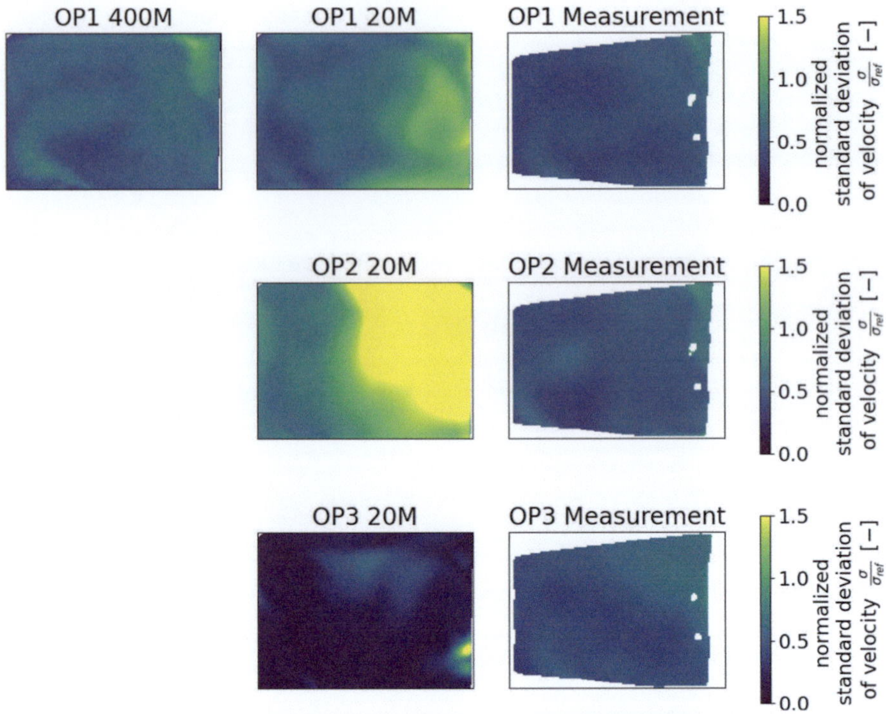

Fig. 6 Standard deviation on the measurement plane in the left diffuser channel. Values normalized with the maximum value of the corresponding measurement. Left: 400M simulation. Middle: 20M simulation. Right: Measurement. From top to bottom: Operating point 1, 2 and 3

To further isolate the origin of the unsteady losses Fig. 8 shows instantaneous velocity fields on multiple monitor planes in the draft tube for different revolutions. For revolution five, the head loss in the draft tube (cf. Fig. 7) has a similar value for both the 20M and 400M simulation for operating point 1. The flow field in the draft tube is largely comparable. Extensive flow features, e.g. higher velocities at the top than the bottom of the elbow, are present in both simulations. In detail there are differences. These are particularly visible on the plane at the end of the cone. The 20M simulation shows a region of low velocity near the wall, that is not present in the 400M simulation. Furthermore, the overall granularity is much finer in the 400M simulation. The 400M simulation does not change drastically with time. Major flow features remain preserved and only spatial limited fluctuations occur. The flow of the 20M simulation exhibits a substantially different behaviour. Over time the region with low velocities at the end of the cone becomes unstable and detaches in the following revolutions from the wall. The detachment is transported downstream and leads to major vortices downstream the cone, especially visible on the monitor plane in the elbow. With increasing distance in downstream direction the impact of the detachment on the flow field weakens. At the outlet of the draft tube only minor

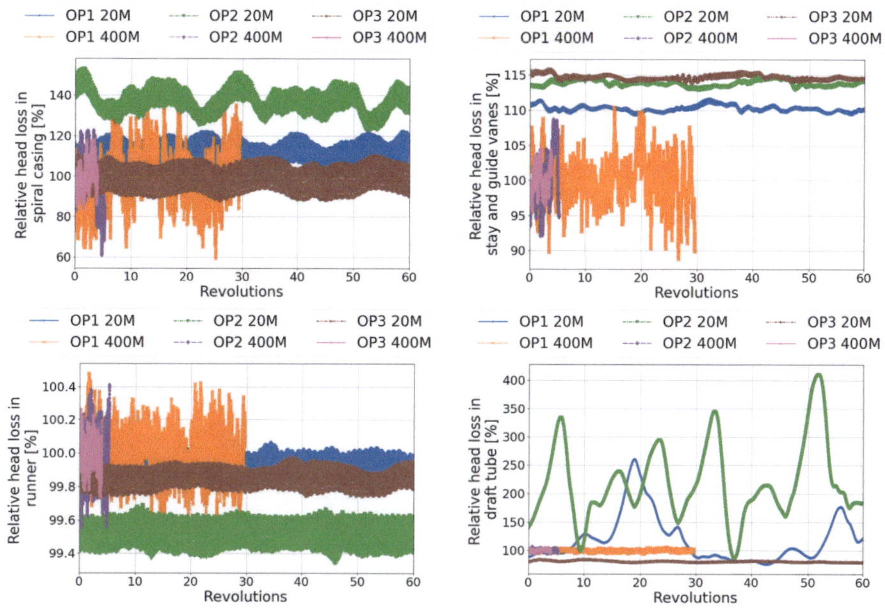

Fig. 7 Head loss in individual subdomains relative to corresponding mean of the 400M simulation. Top left: Spiral casing. Top right: Stay and guide vanes. Bottom left: Runner. Bottom right: Draft tube

changes in the flow field are visible. The 20M simulation of operating point 2 shows a more extensive detachment at the end of the cone with more pronounced vortices in the elbow and an overall more unstable flow field throughout the draft tube. The 20M simulation of operating point 3 shows a very stable flow field without any major changes over time. The shown instability explains the differences in velocity and standard deviation on the measurement planes in the diffuser channels between the 20M simulation and the 400M simulation and measurement shown in Figs. 5 and 6. In all simulations the flow downstream the runner blades at the inlet to the cone is not or only very slightly influenced by the overall flow and behaviour in the draft tube. The shown results lead to the conclusion, that the unsteady behaviour of the draft tube originates in the cone of the draft tube. The instabilities arising in the cone are mainly transported in downstream direction and do not seem to influence the simulation results in upstream direction. This is a possible explanation for the good agreement in power for all simulations (cf. Fig. 4).

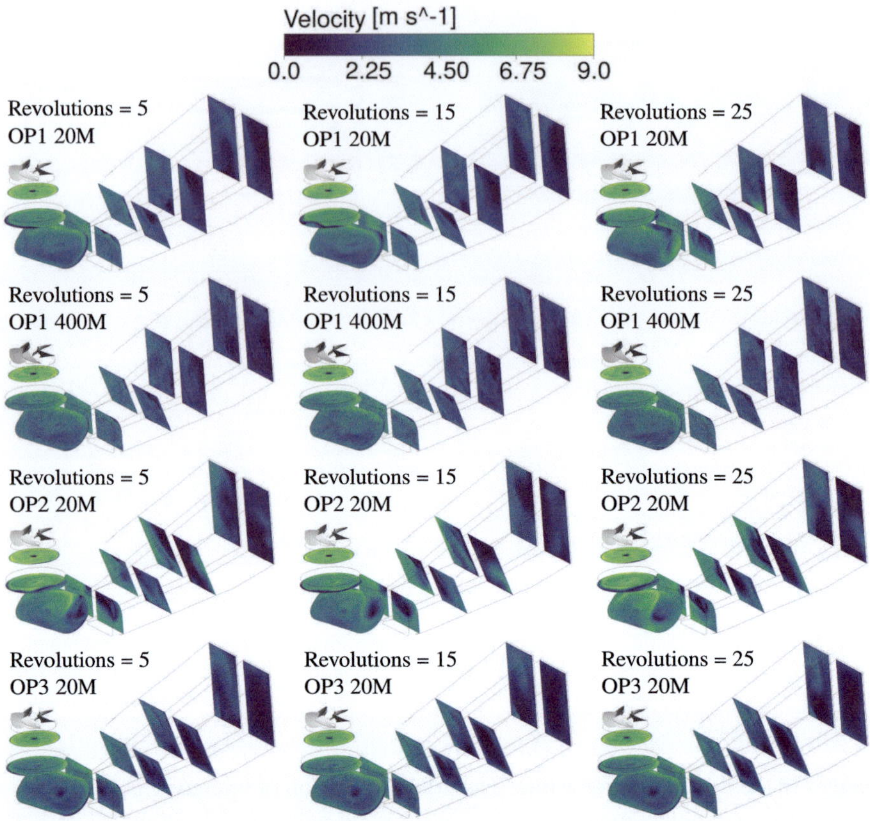

Fig. 8 Instantaneous velocity fields on monitor planes in the draft tube at different revolutions. Row 1: OP1 20M simulation. Row 2: OP1 400M simulation. Row 3: OP2 20M simulation. Row 4: OP3 20M simulation

4 Computational Resources

The numerical simulations have been performed on the Hewlett Packard Enterprise Apollo (Hawk) at the High-Performance Computing Center Stuttgart. The HPE Apollo is built of 5632 compute nodes with 2 AMD EPYC 7742 CPUs with 64 cores each and 256 GB memory. The nodes are connected by InfiniBand HDR200. To run the simulations about 12.4 million core-h were needed. Most of them for the 400M simulations. An additional 5400 core-h were used on post- and preprocessing nodes.

The speedup for a 20M simulation is shown on the left in Fig. 9 relative to one node. Up to 2 Nodes or 256 cores the speedup is almost ideal. For higher core numbers the efficiency decreases. The right side of Fig. 9 shows the speedup for a 400M simulation. Due to the amount of RAM necessary to run the simulation

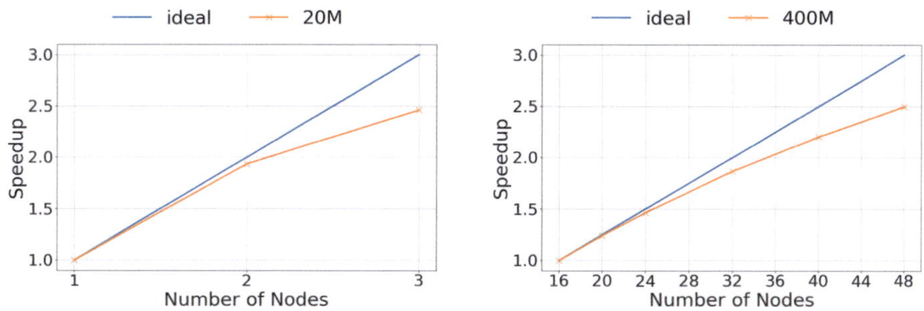

Fig. 9 Speedup for different mesh sizes. Left: 20 million hexahedral elements. Right: 400 million hexahedral elements

a minimum of 16 nodes is necessary. Up to 24 nodes or 3072 cores the speedup is almost ideal. For higher numbers of nodes the efficiency is starting to decrease. Typically 32 nodes or 4096 cores are used to run the simulations. With an efficiency of over 93% 32 nodes are a good compromise between efficiency and wall-clock time. During the optimization workshop at the HLRS in April 2022 an additional performance increase of about $5 - 10\%$ dependent on the number of nodes could be achieved.

5 Conclusion and Outlook

First results of hybrid RANS-LES simulations on a mesh with 400 million hexahedral elements and of standard RANS simulations with 20 million hexahedral elements for one on-cam ond two off-cam operating points are presented and compared to each other as well as to measurements. The results of the 400M simulation of operating point 1 indicate, that most of the simulation domain is running in LES mode and that the regions in LES mode are properly resolved. Furthermore, the results of all 400M simulations are in good agreement with the measurement and the simulations seem to accurately predict the flow in the turbine. A time resolved analysis of the 400M simulations shows, that these simulations have reached a quasi steady state in the integral values, which is also observed in the experiment. This is also true for the 20M simulation of operating point 3. The 20M simulations of the other operating points exhibit a highly unsteady behaviour in the draft tube, leading to an unsteady trend of total head. This instability arises in the cone of the draft tube and is transported primarily in downstream direction. A speedup for both a 20M and 400M simulation is shown. The 20M simulation exhibits a very good scaling up to 2 nodes or 256 cores. The performance of the 400M simulations is almost ideal up to 24 nodes or 3072 cores. For higher numbers of nodes the efficiency is starting to slowly decrease.

Currently, the most crucial flow features and primary influencing factors for precise simulations of Kaplan turbines, such as turbulence model and time-step size, remain unclear. In order to determine these factors, a comprehensive analysis is required, which involves comparing current simulations and experiments and conducting additional simulations with varied settings. Furthermore, a more in-depth analysis of the achieved resolution and quality of the LES simulation has to be performed.

Acknowledgements The authors would like to thank the HLRS Stuttgart for support and supply of computation time.

References

1. Cervantes, M.J., Engstrom, T.F., Gustavsson, L.H.: Turbine-99 III. In: Proceedings of the third IAHR/ERCOFTAC Workshop on draft tube flows, Lulea University of Technology (2005)
2. Nilsson, H., Cervantes, M.J.: Effects of the inlet boundary conditions, on the computed flow in the Turbine-99 draft tube, using OpenFOAM and CFX. IOP Conf. Ser.: Earth Environ. Sci. **15**, 032002 (2012)
3. Beck, J., Joßberger, S., Jester-Zürker, R., Riedelbauch, S.: PIV measurements on a Kaplan turbine and comparison with scale-adaptive numerical analysis. IOP Conf. Ser.: Earth Environ. Sci. **405**, 012024 (2019)
4. Jošt, D., Škerlavaj, A., Lipej, A.: Improvement of efficiency prediction for a kaplan turbine with advanced turbulence models. Strojniski Vestnik/J. Mech. Eng. **60**, I. 2, 124-1 (2014)
5. Krappel, T., Riedelbauch, S.: Scale resolving flow simulations of a francis turbine using highly parallel CFD simulations. In: High Performance Computing in Science and Engineering '16, Springer (2016)
6. Krappel, T., Riedelbauch, S., Jester-Zuerker, R., Jung, A., Flurl, B., Unger, F., Galpin, P.: Turbulence resolving flow simulations of a francis turbine in part load using highly parallel CFD simulations. In: IOP Conference Series: Earth and Environmental Science, vol. 49 (2016)
7. Javadi, A., Nilsson, H.: Detailed numerical investigation of a Kaplan turbine with rotor-stator interaction using turbulence-resolving simulations. Int. J. Heat Fluid Flow **63**, 1–13 (2017)
8. Minakov, A., Platonov, D., Litvinov, I., Shtork, S., Hanjalić, K.: Vortex ropes in draft tube of a laboratory Kaplan hydroturbine at low load: an experimental and LES scrutiny of RANS and DES computational models. J. Hydraul. Res. **55**, 668–685 (2017)
9. Joßberger, S., Riedelbauch, S.: Influence of the Inlet Boundary Conditions on Numerical Flow Simulations of a Model Kaplan Turbine. In: Proceedings of 13th European Conference on Turbomachinery, Fluid dynamics and Thermodynamics
10. ANSYS Inc.: ANSYS CFX-Solver Modeling Guide Release 19.5 (2019)
11. ANSYS Inc.: ANSYS CFX-Solver Theory Guide Release 19.5 (2019)
12. Menter, F.R.: Stress-Blended Eddy Simulation (SBES)—a new Paradigm in hybrid RANS-LES modeling. In: Proceedings of the 6th HRLM Symposium, Strasbourg, France (2016)
13. Menter, F.R.: Best practice: Scale-resolving simulations in ANSYS CFD—Version 2.0. ANSYS Germany GmbH (2015)

A High-Order Framework for Compressible Droplet Dynamics

Daniel Appel, Pascal Mossier, Jens Keim, Steven Jöns, Andrea Beck, and Claus-Dieter Munz

Abstract Due to the multiscale nature of two-phase flows, high-order methods are an essential tool to gain further understanding of droplet dynamics phenomena. However, the construction of robust and accurate high-order schemes is a challenging process due the lack of a stabilization mechanism. This is especially the case when compressible fluids are considered, for which discontinuities in the solution might occur and a complex interaction between fluid flow and thermodynamics is present. In the present work, recent results of our numerical framework for compressible droplet flows are presented. The numerical method is based on the high-order discontinuous Galerkin spectral element method and includes both a sharp and a diffuse interface approach. We investigate the parallel performance of our scheme in the most challenging case and consider several applications that include typical interfacial flow phenomena, e.g., evaporation and droplet collisions.

1 Introduction

Droplet flows have a strong impact on a variety of natural phenomena and engineering applications, e.g., weather prediction, pre-combustion mixing and porous media flows. The multiscale nature of these flows leads to a numerically challenging interaction between the bulk phases and the phase interface. This is especially the case when compressible flows are considered due to the nonlinear interaction between the fluid flow equations and the equation of state.

Two widespread approaches exist to simulate such flows on a continuum scale: the sharp interface method (SIM) and the diffuse interface method (DIM). In the SIM, a two-fluid approach is employed, i.e., the phases are governed by their own set of equations and the phase interface is represented by an infinitesimally thin

D. Appel (✉) · P. Mossier · J. Keim · S. Jöns · A. Beck · C.-D. Munz
Institute of Aerodynamics and Gas Dynamics, Stuttgart, Germany
e-mail: daniel.appel@iag.uni-stuttgart.de

S. Jöns
e-mail: joens@iag.uni-stuttgart.de

T. Ludwig et al. (eds.), *High Performance Computing in Science and Engineering '23*,
https://doi.org/10.1007/978-3-031-91312-9_26

discontinuity. Across the interface, jump conditions govern the exchange of mass momentum and energy between the phases. However, an additional tracking algorithm for the interface is necessary. The DIM on the other hand does not need an additional tracking. It is a one-fluid approach and the two-phase flow is treated as a mixture with one set of governing equations. The interface thus constitutes a region with a smooth transition between the two phases. However, this transition has to be modeled appropriately. Furthermore, the one-fluid formulation necessitates to resolve the interfacial region. Hence, practical applications of the DIM are usually concerned with small-scale applications.

In the present work, we summarize our latest results on compressible droplet flows using a numerical framework that incorporates both modeling approaches. It is based on the high-order discontinuous Galerkin method (DGSEM) and uses a level-set ghost fluid method for the SIM and a phase-field approach based on the Navier-Stokes Korteweg equations (NSK) for the DIM. The paper is structured as follows: We shortly introduce the main building block of our numerical method, the DGSEM, in Sect. 2. Afterwards, Sect. 3 gives an overview of the incorporated SIM and DIM. In Sect. 4, we discuss the latest improvements on the load balancing algorithm necessary for the SIM. In Sect. 5 we demonstrate the capabilities of our framework for three different droplet dynamics test cases. Finally, we give a short summary and outlook in Sect. 6.

2 Discontinuous Galerkin Spectral Element Method

Our numerical framework is based on the open-source DGSEM solver *FLEXI* [17]. It is a high-order numerical method designed to solve advection-diffusion-reaction equations of the form

$$\boldsymbol{U}_t + \nabla \cdot \boldsymbol{F}^a(\boldsymbol{U}) = \nabla \cdot \boldsymbol{F}^d(\boldsymbol{U}) + \boldsymbol{S}, \tag{1}$$

where $\boldsymbol{U}$ represents the conservative state vector, $\boldsymbol{F}^a(\boldsymbol{U})$ an advective flux, $\boldsymbol{F}^d(\boldsymbol{U})$ a diffusive flux and $\boldsymbol{S}$ a source term. The present work will make use of the well-known compressible Navier-Stokes equations, the hyperbolic Godunov-Peshkov-Romenski equations [26], as well as variants of the Navier-Stokes-Korteweg equations [14]. All three are closed via an appropriate equation of state (EOS), for which the framework of [10] is used. The DGSEM tessellates the domain of interest into N_{Elems} non-overlapping hexahedral elements. For each element, the weak form of Eq. (1) is solved on a reference domain $\Omega = [-1, 1]^3$. To this end, an element-local polynomial solution ansatz of degree N is employed, which is based on a tensor product of Lagrange polynomials defined at Gauss-Legendre points. All integrals are evaluated with the Gauss-Legendre quadrature. The solution inside each grid element is continuous but may jump at the element borders. There, Riemann solvers, known from the finite volume methods, couple the elements and ensure a consistent and stable discretization. To obtain solution gradients for the parabolic fluxes, the BR1 scheme [4] is employed. Temporal evolution of the semi-discrete weak form is finally obtained via an appropriate high-order integrator, e.g., low-storage Runge-Kutta schemes [15]. Given the minimal communication effort and the efficient tensor product structure of the spatial operator, the DGSEM is an excellent candidate for highly parallelized large-scale simulations.

A downside of high-order methods, like the DGSEM, is their inherent instability when handling discontinuities. However, these phenomena frequently arise in compressible fluid flow simulations, e.g., shock waves, contact discontinuities and sharp phase interfaces. To handle these discontinuities, different stabilization techniques are found in the literature. We employ the finite volume (FV) shock capturing strategy of Sonntag and Munz [29, 30]. Therein, a troubled grid element, identified with the oscillation indicator of [25], is substituted by a set of $(N+1)^3$ FV sub-cells. The distribution of these cells is equidistant in the reference space. These cells are then treated with a second-order FV method with a minmod limiter. Thus, stability is ensured with a slight loss of accuracy. The efficiency and accuracy of this hybrid DG/FV method can be further improved with a dynamic hp-adaptive refinement, proposed in [21, 22]. It introduces a variable polynomial degree $N \in [N_{\min}, N_{\max}]$ in the DGSEM to obtain a p-adaptive operator and allows for a FV sub-cell resolution independent of the polynomial degree of the DG solution. With a h-refined grid of $N_{\text{FV}} = 2N_{\max} + 1$ sub-cells per direction, the FV operator exploits the maximum possible resolution without constraining the time step compared to a DG element of degree $N_{\max}$ [5]. Crucially, this greatly improves the localization of the phase interface with only a slight increase in the computational costs.

3 Numerical Methods for Compressible Multiphase Flows

Our numerical framework is based on the above described DGSEM. Hence, SIM and DIM are implemented in a similar high-order fashion. In the following, we briefly summarize both approaches: first the SIM, followed by the DIM.

3.1 Sharp-Interface Method

The SIM goes back to the work by Fedkiw et al. [9] and has since been improved by many authors. We follow the ghost fluid approach proposed by Merkle and Rohde [19], who used the solution of the two-phase Riemann problem to define the ghost states at the phase interface. Their idea was later extended to approximate Riemann solvers and applied in a level-set ghost fluid framework [6–8].

The level-set method itself was first introduced by Osher and Sethian [24] and represents an interface as the zero-contour of a scalar function, the level-set field $\phi(\boldsymbol{x})$. The physically relevant normal and mean curvature of the interface, $\boldsymbol{n}$ and κ respectively, are directly given by the derivatives of ϕ according to

$$\boldsymbol{n} = \frac{\nabla \phi}{|\nabla \phi|} \quad \text{and} \quad \kappa = \nabla \cdot \boldsymbol{n} \,. \tag{2}$$

The deformation of the interface under a velocity field $\boldsymbol{v}$ is governed by the level-set equation

$$\phi_t + \boldsymbol{v} \cdot \nabla \phi = 0 \,. \tag{3}$$

For robust and accurate discretizations, the level-set function is chosen to be the signed distance function, satisfying $|\nabla \phi| = 1$. However, since $\phi(t, \boldsymbol{x})$ does not preserve this property as it evolves in time through Eq. (3), Sussman et al. [32] introduced the reinitialization equation

$$\phi_\tau + \text{sign}(\phi_0)(|\nabla \phi| - 1) = 0 \tag{4}$$

to transform a level-set function ϕ_0 into the signed distance function ϕ.

The transport velocity $\boldsymbol{v}(\boldsymbol{x})$ in Eq. (3) is obtained by extrapolating the interface velocity, which emerges from the solution of the two-phase Riemann problem, into the domain. To this end, we solve a similar Hamilton-Jacobi equation for each velocity component v^d, as suggested by Aslam [3]

$$v^d_\tau + \text{sign}(\phi)\boldsymbol{n} \cdot \nabla v^d = 0 \,, \quad d = 1, 2, 3 \,. \tag{5}$$

3.2 *Diffuse-Interface Method*

The alternative approach we employ is a diffuse interface method based on the NSK equations. These originated from the seminal works of van der Waals and Korteweg [16, 35] and provided the first thermodynamic insight into the physics of capillarity. The NSK equations define the properties of the continuous transition zone between the two phases with a Helmholtz free energy composed of two parts: a double well potential that accounts for the two coexisting bulk phases and a gradient term with respect to an order parameter [1]. The latter is directly related to the surface energy of the phase interface and thus models the surface tension effects. Consequently, the total energy per unit mass e includes an additional term compared to the well-known Navier-Stokes equations and is given by

$$e(\rho, \boldsymbol{u}, \eta, \nabla \rho) = \frac{1}{2} \boldsymbol{u} \cdot \boldsymbol{u} + \epsilon(\rho, \eta) + \frac{\gamma_K}{2\rho} \nabla \rho \cdot \nabla \rho, \tag{6}$$

with ρ the density, $\boldsymbol{u}$ the velocity vector, η the entropy and γ_K the so-called capillary parameter. The NSK equations are then given by the Navier-Stokes equations supplemented by an additional stress tensor which models the effect of the interfacial region: the Korteweg tensor. For a detailed derivation of the NSK equations the reader is referred to [1, 35].

The numerical treatment of the NSK equations is relatively simple, as the DGSEM described in Sect. 2 can be applied in a straightforward fashion [14]. Only, the additional second derivative of the density, necessary for evaluating the Korteweg tensor, is in need of an additional lifting step. On the downside, the non-convexity of the Helmholtz free energy may lead to an imaginary speed of sound, which makes an accurate time step estimation for the explicit time discretization difficult and leads to possible instabilities. Hence, we also use a relaxation model of the NSK which circumvents the above described problem and guarantees the hyperbolicity of the advective fluxes. For details on the relaxation method, the reader is referred to [14].

4 Load Balancing for SIM

A central challenge in high-performance computing is an even distribution of the workload to facilitate an efficient utilization of the computational resources. Therefore, a dynamic load balancing (DLB) strategy is vital to respond to temporal changes in the workload distribution among the processor units.

In the current code framework, there are numerous factors that contribute to an uneven load distribution throughout the computational domain. The most prominent factors are the costly interface tracking algorithm, which is confined to a narrow band around the interface [2], and the hp-adaptive discretization with a hybrid DG and FV scheme [22]. Furthermore, the iterative evaluation of the EOS can have varying costs depending on the current state, and the two-phase Riemann problem causes an additional workload for elements adjacent to the phase boundary.

To cope with the highly non-stationary and complex load distribution, an efficient DLB scheme was developed for the present code framework by Appel et al. [2] and Mossier et al. [21]. It is based on a domain decomposition through a space-filling curve and thus reduces the partitioning to a one-dimensional problem. An optimal distribution among a number of P processors is found when the maximum processor load, referred to as bottleneck $\mathcal{B}$, is minimized:

$$\mathcal{B} = \max_{1 \leq k \leq P} \{L_k\} \ , \text{with} \quad L_k = \sum_{i=s_{k-1}+1}^{s_k} \omega_i . \tag{7}$$

The workload of an individual processor k is herein called L_k and obtained as the sum of the weights ω_i, which correspond to the computational costs of the elements. A partitioning is defined as a sequence of separator indices $s_0 \leq s_1 \leq ... \leq s_k \leq s_P$, the first index $s_0 = 0$ and the last index $s_0 = P$ and the number of elements within each sub-domain given as $N_k = (s_k - s_{k-1})$. The DLB scheme within the code framework relies on an bisection algorithm called *ExactBS+PI* [18] to obtain an optimal bottleneck for a given weight distribution $\{\omega_i\}$ and processor count P. The weights of individual elements are obtained dynamically through wall time measurements during the simulation. For a detailed description of the DLB algorithm, the reader is referred to [2, 22].

After a new partitioning is determined, the elements have to be redistributed among the processor units via an MPI interface [20]. During the computation, each processor has to reinitialize the DG and FV operators, e.g., by computing the metric terms for the new elements, thus hiding most of the communication time. After completion of the communication and the operator initialization, the computation can resume on the new partitioning.

To study the parallel scaling of the resulting framework, a strong scaling test is performed on up to 64 compute nodes (8192 CPU cores). As a test problem, a resting droplet within a three-dimensional domain Ω with a resolution of $30 \times 30 \times 30$ elements is considered. A checkerboard-like distribution of the polynomial degree is prescribed in the bulk domain with polynomial degrees $N \in [2, 4]$. For the interface tracking, the DG operator uses a degree $N \in [2, 8]$. The FV operator applies a sub-cell resolution of $N_{\mathrm{FV}} = 9$ and is applied at the interface for the bulk and at the narrow-band cut-off for the level-set. The setup is illustrated in Fig. 1a, b.

To investigate the quality of the load redistribution, a single repartitioning is performed after 5 time steps, followed by 44 time steps to measure the performance index PID:

$$\mathrm{PID} = \frac{\#\text{total wall time} \cdot \#\text{cores } P}{\#\text{DOFs} \cdot \#\text{time steps} \cdot \#\text{RK-stages}}. \tag{8}$$

Computation are performed with $1, 2, 4, ..., 64$ compute nodes to determine the strong scaling behavior of the framework. The results are evaluated in terms of the parallel efficiency

$$\eta = \frac{\mathrm{PID}_{P=128}}{\mathrm{PID}} \tag{9}$$

and the parallel speedup in Fig. 1c, d. Up to 32 compute nodes, the parallel efficiency surpasses 50%. The sudden decline in parallel efficiency at 64 nodes is the result of the most expensive element approaching the bottleneck, $\omega_{\max} \sim \mathcal{B}$, which deteriorates the quality of the partitioning. Aside from that, the suboptimal parallel efficiency can be attributed to the fact that the high costs of the interface tracking cause the processors in the narrow band to host very few elements, which undermines the latency hiding of the MPI communication.

5 Applications

This section finally applies the outlined framework to three scenarios on droplet dynamics in order to demonstrate its capability to handle complex compressible two-phase flow problems. All runs were performed on the HPE Apollo System Hawk at HLRS.

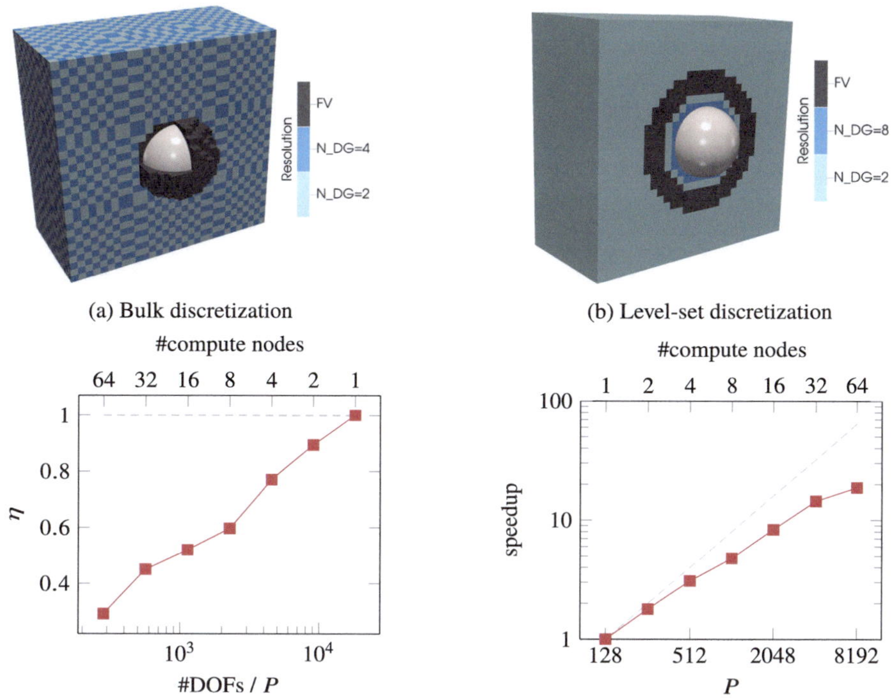

(a) Bulk discretization

(b) Level-set discretization

(c) Parallel efficiency

(d) Parallel speedup

Fig. 1 Static droplet setup with the distribution of DG/FV elements and the element-local degree indicated by color for the bulk discretization (1a) and the level-set interface tracking (1b). The strong scaling test yields a parallel efficiency (1c) and parallel speed up (1d) for a range of $P = 128, \ldots, 8192$ CPU cores

5.1 Evaporation of N-Dodecane

As a first case, we study an evaporation process. This is extremely challenging in compressible flow simulations with real gas EOS due to the complex interaction between the non-linear fluid motion and the thermodynamically driven phase change. In this context, we recently developed a thermodynamically consistent evaporation framework for our sharp-interface method [13, 23]. The key building block of this approach is the solution of the two-phase Riemann problem with evaporation at the interface. The solution to this problem is obtained via the incorporation of a local thermodynamic model for evaporation and condensation based on classical irreversible thermodynamics. The solution of the Riemann problem is then used to couple the individual phases across the interface. The proposed method has been developed for both the Euler-Fourier and GPR equation system and was validated against molecular dynamics data, c.f., [13, 23].

To illustrate the capabilities of the evaporation framework, we consider a shock-droplet interaction. This test case was initially proposed by Fechter et al. [8] and slightly modified in [13], where it was studied for the Euler-Fourier equations in two-dimensions and using a Peng-Robinson equation of state. The setup consists of an n-dodecane droplet which initially rests in its quiescent vapor. A shock is initialized in the domain which moves with a Mach number of $M = 1.5$ towards the droplet. Upon impingement, the droplet is deformed while simultaneously evaporating and condensating on different parts of the surface.

As an extension to [13], we here consider the same test case in three dimensions and with the GPR equations. To handle the massively increased computational effort due to the additional dimension, a quarter droplet with appropriate symmetry boundary conditions was considered and the above described hp-adaptive DGSEM scheme in combination with the load balancing framework was fully exploited. The numerical grid consisted of $90 \times 36 \times 36$ elements with a varying polynomial degree between $N \in [2, 4]$ and a FV sub-cell resolution of $N_{\mathrm{FV}} = 9$, resulting in an effective resolution of approximately 85M degrees of freedom (DOFs). Calculations were performed using 128 nodes for 1.5 MCPUh. Figure 2 shows the droplet surface, local discretization and temperature distribution of the droplet for two different time instances. Impingement of the shock wave leads to instabilities on the droplet surface. This is followed by a flattening of the liquid droplet. The temperature distribution indicates condensation phenomena in the stagnation area, while the rear side of the droplet evaporates. The local discretization distribution shows an excellent capturing of the intricate flow phenomena, i.e., shock waves and vortices, while minimizing the computational effort in areas where the solution only changes sparsely.

(a) $t = 20.0\mu s$

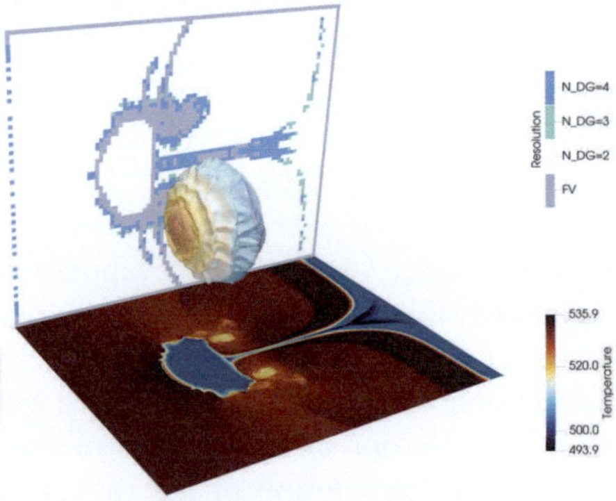

(b) $t = 55.0\mu s$

Fig. 2 3D simulation of a n-dodecane droplet interacting with a shock wave. The projections cut through the symmetry planes and show the temperature distribution and local discretization of the computation. Additionally, the temperature of the droplet surface is indicated by color

5.2 Droplet Grouping

As another application of our sharp-interface framework we consider the grouping behavior of droplets. In general, droplet grouping refers to the tendency of initially distant droplets in gas flows to convene and form clusters. Despite its technical relevance through affecting, for example, evaporation rates and combustion dynamics, this phenomenon is not yet fully understood, especially in the compressible flow regime.

To this end, we consider two equisized droplets aligned in streamwise direction, which constitutes the most fundamental configuration subject to grouping. The first analytical evidence for his phenomenon was presented by Stimson and Jefferey [31] who studied the motion of two solid spheres in this tandem setup in a viscous fluid (Stokes flow, i.e. $\mathrm{Re} \ll 1$). Roth et al. [28] investigated experimentally and numerically the grouping behavior in a monodisperse droplet stream at a moderate Reynolds number ($\mathrm{Re} = 112$). Recently, experiments and direct numerical simulations (DNS) at higher Reynolds numbers ($\mathrm{Re} = 492$) were put forth by Vaikuntanathan et al. [33] and Ibach et al. [12], respectively. Moreover, in [34] they studied the grouping mechanisms in detail based on an analytical model of two droplets in tandem configuration. They revealed that grouping is induced by the lower drag force the trailing droplet experiences in comparison to the leading droplet.

Our configuration is aligned to Ibach et al. [12] and serves as an experimentally validated, incompressible baseline testcase. The computational setup with the two droplets at their initial positions is depicted in Fig. 3a. The streamwise direction was upwards along the x-axis and we neglected the effect of gravity. Two isopropanol droplets of diameter $D = 112\,\mu\mathrm{m}$ were placed in a cuboid domain of edge length $L = 605.5\,\mu\mathrm{m}$ with an initial spacing of $a/L = 0.491$. The droplets were initialized in air at rest with a velocity of $\boldsymbol{u}_{l,0} = [15.07, 0, 0]\,\mathrm{m/s}$, resulting in a Reynolds number of $\mathrm{Re} = (\rho_g |\boldsymbol{u}_{l,0}| D)/\mu_g = 110.6$. Compared to [12], the ambient pressure was reduced to $p_g = 10\mathrm{kPa}$ in order to increase the computational time step and thereby save computational cost. The remaining material parameters and more details on the setup can be found in the cited work. We discretized the droplet uniformly by $D/h = 48$ DOFs per diameter and employed a grid stretching of 1.188 for $y, z \geq 0.154L$. The polynomial degree was set to $N = 3$, implying a grid of $65 \times 25 \times 25$ elements and 2.6M DOFs in total.

Figure 3b shows the obtained velocity field at the exemplary time instance $t = 4.15\,\mathrm{ms}$, where the droplet spacing has dropped to $a/L = 0.234$. While the axial velocity field is almost uniform in streamwise direction for the moderate Reynolds number at hand, the asymmetric radial velocity field reveals a clear interaction of the trailing droplet with the wake of the leading droplet. The continuous decline of the droplet spacing over time is plotted in Fig. 3c and is in good agreement with the reference data by Roth et al. [28]. The simulation results, however, slightly overpredict the grouping tendency and predict an earlier coalescence than observed in the experiment. On the one hand, this difference may be attributed to the minor modifications in the present setup compared to the reference. On the other hand, the included

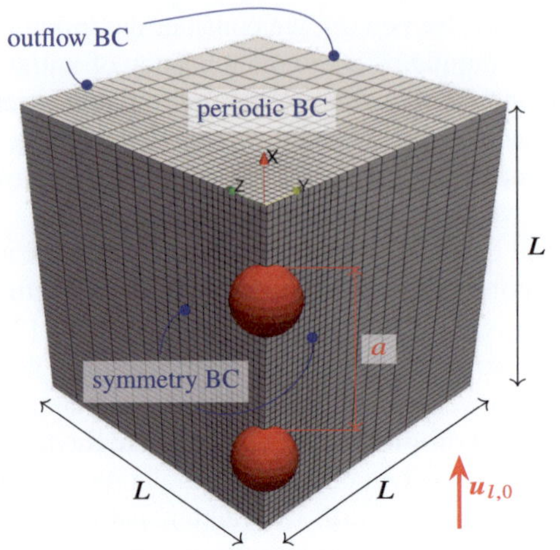

(a) Computational setup with boundary conditions (BC) and initial droplet positions.

(b) Normalized velocity in $x = 0$ plane at $t = 4.15$ ms.

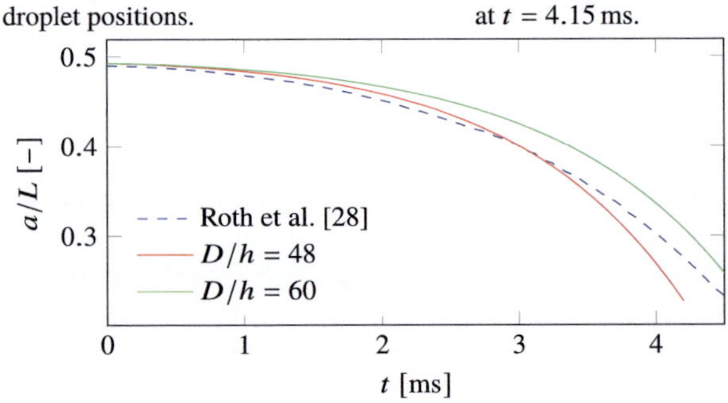

(c) Dimensionless droplet spacing over time.

Fig. 3 Grouping behavior of monodisperse droplet stream with Re = 110.6

trajectory at a higher resolution of $D/h = 60$ indicates that the results are not yet fully grid-converged. Nevertheless, the established setup along with the outlined sharp interface framework provide a promising foundation for upcoming grouping simulations in the compressibleflow regime. A total compute time of approximately 35 000 CPUh on 512 cores and 75 000 CPUh on 1024 cores was spent on the results with a resolution of $D/h = 48$ and $D/h = 60$, respectively.

5.3 Droplet Collision

In a final investigation, the developed relaxation formulation of the NSK equations is compared to the original formulation of the NSK model and reference data from the literature [11, 27]. To this end, we consider a highly dynamic three-dimensional binary head-on collision of two droplets in the stretching separation regime. In contrast to the previously discussed setup of the droplet stream, this test case is in the vicinity of the critical point and features strong compressibility effects and phase transition. An additional difference in the simulation setup is a spatial offset of the initial droplets. Hence, no frontal collision and breakup but a stretching and finally a separation of the droplets is observed after the collision.

The configuration of the test case is based on the setup in [27], which corresponds to a Reynolds number of $Re = 325$, a Weber number of $We = 490$ and a Prandtl number of $Pr = 20$. At $t = 0$, the non-dimensional temperature is chosen as constant with $T = 0.85$ and the droplets, each with a diameter of $D = 0.2$, are positioned at $x_1 = (0.3, 0.5, 0.55)^\mathrm{T}$ and $x_1 = (0.7, 0.5, 0.45)^\mathrm{T}$. With the initial densities of the vapor and the liquid defined as $\rho_{\text{liq}} = 1.8493$ and $\rho_{\text{vap}} = 0.3537$ and based on the specified Reynolds and Weber number, the relative velocity is given as $\boldsymbol{u} = (3, 0, 0)^\mathrm{T}$. In the present simulations, the domain was discretized with $N_{\text{Elems}} = 50^3$ elements and a polynomial degree of $N = 3$. Similar to [14], the relaxation parameters were chosen as $\alpha = 200$ and $\beta = 10^{-3}$ in the case of the relaxation model. Each computation was performed on 8 nodes for overall 33000 CPUh.

The results at different time instances are illustrated in Fig. 4. The two droplets move towards each other and collide. However, due to the spatial offset in the z-direction, first, a stretching of the droplets is observed and finally, a separation. During this process, two ligaments are formed which elongate. At a critical point, the ligaments disconnect from the droplets and eventually merge into a single satellite droplet. At $t = 4$, two droplets and a single satellite droplet in the middle remain. A good qualitative agreement with the results in the literature [11, 27] is obtained.

Additionally, a quantitative comparison of the kinetic and the surface energy of the relaxation formulation, the original model and the data from the literature [11] is given in Fig. 5, where a good agreement is observed. The results of the original NSK model (NSK) and the relaxation formulation (NSKR) match perfectly within the present investigation. Minor deviations of the results to the simulations in [11, 27] are visible in the kinetic energy from $t = 1.5$ and in the surface energy in the time interval $t \in (0.5, 1)$. This difference could be induced by the higher numerical dissipation of the second-order FV discretization applied in [11, 27]. In the present study, the same number of DOFs was utilized however in combination with a fourth-order DGSEM.

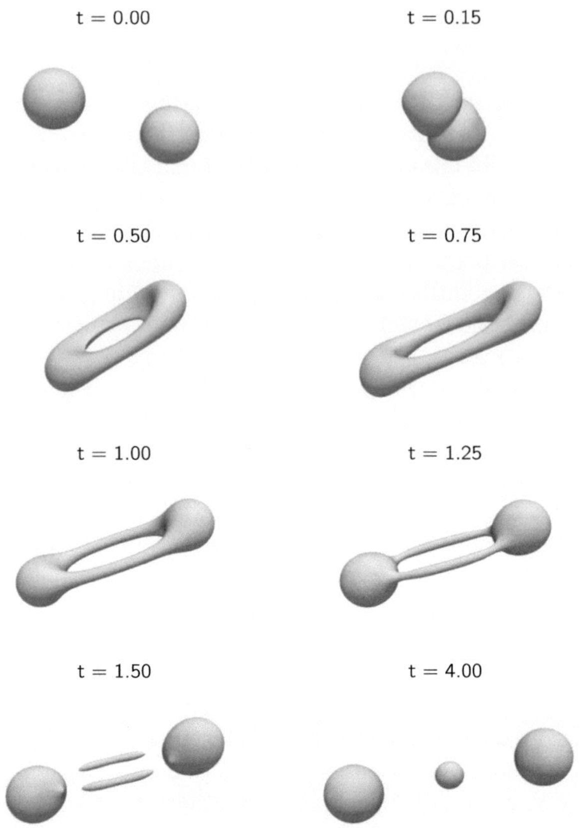

Fig. 4 Temporal evolution of the stretching separation test case visualized by the iso-contour of the mean density

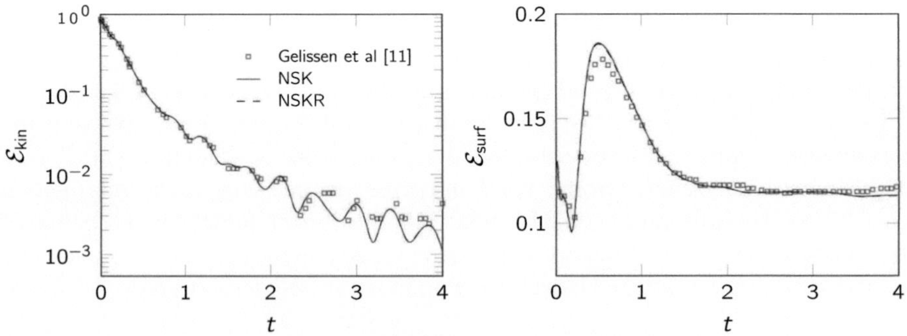

Fig. 5 Comparison of the integral kinetic and surface energy for the stretching separation test cases. The results of the original model are indicated by NSK and those of the relaxation formulation by NSKR. The symbols indicate the reference data in [27]

6 Conclusion

In the present work, a framework to simulate droplet dynamics in compressible fluid flows was presented. It incorporates both a sharp- and a diffuse-interface method based on the Navier-Stokes or GPR equations and the NSK equations, respectively. The numerical scheme is based on the high-order DGSEM with FV shock capturing. To make efficient use of compute power, the scheme is formulated in an hp-adaptive variant that can locally refine the polynomial degree of the solution in order to capture intricate flow features with a minimal amount of DOFs. The adaptivity as well as the necessary interface tracking for the sharp-interface method introduces a significant imbalance in parallelized computations. Hence, a dynamic load balancing scheme is proposed in order to evenly distribute the workload.

Scaling tests for the sharp-interface method indicated a clear speed-up via the load balancing scheme. However, parallel efficiency declined for an increased number of compute nodes. In addition, the framework was used to study a shock-droplet interaction with evaporation, droplet grouping behavior and drop collisions. In all cases, the robustness and flexibility of the numerical method were able to give detailed insights into the droplet dynamics. In future works, we seek to improve the parallel performance in order to further increase the resolved physical details of the simulations. In addition, we plan to extend our sharp-interface method to multi-component simulations, giving the possibility to examine more complex two-phase flow phenomena.

Acknowledgements We gratefully acknowledge the Deutsche Forschungsgemeinschaft (DFG) with their support of D. Appel and P. Mossier through the project GRK 2160/2 "Droplet Interaction Technologies", S. Jöns through the project SFB-TRR 75, Project number 84292822—"Droplet Dynamics Under Extreme Ambient Conditions", J. Keim and C.-D. Munz under Germany's Excellence Strategy—EXC 2075—390740016. The simulations were performed on the national supercomputer HPE Apollo System *Hawk* at the High Performance Computing Center Stuttgart (HLRS) under the grant number *hpcmphas/44084*, of which we used approximately 10.2M CPUh in the current accounting period.

Competing Interests The authors have no conflicts of interest to declare that are relevant to the content of this chapter.

References

1. Anderson, D.M., McFadden, G.B., Wheeler, A.A.: Diffuse interface methods in fluid mechanics. Ann. Rev. Fluid Mech. **30**(1), 139–165 (1998)
2. Appel, D., Jöns, S., Keim, J., Müller, C., Zeifang, J., Munz, C.-D.: A narrow band-based dynamic load balancing scheme for the level-set ghost-fluid method. In: High Performance Computing in Science and Engineering '21 (in press) (2021)
3. Aslam, T.D.: A partial differential equation approach to multidimensional extrapolation. J. Comput. Phys. **193**(1), 349–355 (2004)

4. Bassi, F., Rebay, S.: A high-order accurate discontinuous finite element method for the numerical solution of the compressible Navier-Stokes equations. J. Comput. Phys. **131**(2), 267–279 (1997)
5. Dumbser, M., Loubère, R.: A simple robust and accurate a posteriori sub-cell finite volume limiter for the discontinuous Galerkin method on unstructured meshes. J. Comput. Phys. **319**, 163–199 (2016)
6. Fechter, S., Jaegle, F., Schleper, V.: Exact and approximate Riemann solvers at phase boundaries. Comput. Fluids **75**, 112–126 (2013)
7. Fechter, S., Munz, C.-D.: A discontinuous Galerkin-based sharp-interface method to simulate three-dimensional compressible two-phase flow. Int. J. Numer. Methods Fluids **78**(7), 413–435 (2015)
8. Fechter, S., Munz, C-D., Rohde, C., Zeiler, C.: Approximate Riemann solver for compressible liquid vapor flow with phase transition and surface tension. Comput. Fluids **169**, 169–185 (2018)
9. Fedkiw, R.P., Aslam, T., Merriman, B., Osher, S.: A non-oscillatory Eulerian approach to interfaces in multimaterial flows (the ghost fluid method). J. Comput. Phys. **152**(2), 457–492 (1999)
10. Föll, F., Hitz, T., Müller, C., Munz, C.-D., Dumbser, M.: On the use of tabulated equations of state for multi-phase simulations in the homogeneous equilibrium limit. Shock Waves (2019)
11. Gelissen, E.J., van der Geld, C.W.M., Kuipers, J.A.M., Kuerten, J.G.M.: Simulations of droplet collisions with a Diffuse Interface Model near the critical point. Int. J. Multiphase Flow **107**, 208–220 (2018)
12. Ibach, M., Vaikuntanathan, V., Arad, A., Katoshevski, D., Greenberg, J.B., Weigand, B.: Investigation of droplet grouping in monodisperse streams by direct numerical simulations. Phys. Fluids **34**(8), 083314 (2022)
13. Jöns, S., Munz, C.-D.: Riemann solvers for phase transition in a compressible sharp-interface method. Appl. Math. Comput. **440**, 127624 (2023)
14. Keim, J., Munz, C.-D., Rohde, C.: A relaxation model for the non-isothermal navier-stokes-korteweg equations in confined domains. J. Comput. Phys. **474**, 111830 (2023)
15. Kennedy, C.A., Carpenter, M.H., Michael Lewis, R.: Low-Storage, Explicit Runge–Kutta schemes for the compressible navier–stokes equations. Appl. Numer. Math. **35**(3), 177–219 (2000)
16. Korteweg, D.J.: Sur la forme que prennent les équations du mouvements des fluides si l'on tient compte des forces capillaires causées par des variations de densité considérables mais connues et sur la théorie de la capillarité dans l'hypothèse d'une variation continue de la densité. Archives Néerlandaises des Sciences exactes et naturelles **6**, 1–24 (1901)
17. Krais, N., Beck, A., Bolemann, T., Frank, H., Flad, D., Gassner, G., Hindenlang, F., Hoffmann, M., Kuhn, T., Sonntag, M., Munz, C.-D.: Flexi: a high order discontinuous Galerkin framework for hyperbolic-parabolic conservation laws. Comput. Math. Appl. **81**, 186–219 (2020)
18. Lieber, M., Nagel, W.E.: Highly scalable SFC-based dynamic load balancing and its application to atmospheric modeling. Future Gen. Comput. Syst. **82**, 575–590 (2018)
19. Merkle, C., Rohde, C.: The sharp-interface approach for fluids with phase change: riemann problems and ghost fluid techniques. ESAIM: Math. Modell. Numer. Anal. **41**(06), 1089–1123 (2007)
20. Message Passing Interface Forum. MPI: A Message-Passing Interface Standard, June 2021
21. Mossier, P., Daniel, C.-D.M., Beck, A.: An efficient hp-adaptive strategy for a level-set ghost fluid method (2022)
22. Mossier, P., Munz, C.-D., Beck, A.: A p-adaptive discontinuous Galerkin method with hp-shock capturing. J. Sci. Comput. **91**(1), 1573–7691 (2022)
23. Müller, C., Mossier, P., Munz, C.-D.: A sharp interface framework based on the inviscid Godunov-Peshkov-Romenski equations: Simulation of evaporating fluids. J. Comput. Phys. **473**, 111737 (2023)
24. Osher, A., Sethian, J.A.: Fronts propagating with curvature-dependent speed: algorithms based on Hamilton-Jacobi formulations. J. Comput. Phys. **79**(1), 12–49 (1988)

25. Persson, P.-O., Peraire, J.: Sub-Cell shock capturing for discontinuous Galerkin Methods. In: 44th AIAA Aerospace Sciences Meeting and Exhibit. American Institute of Aeronautics and Astronautics, Jan. 2006
26. Peshkov, I., Romenski, E.: A hyperbolic model for viscous Newtonian flows. Continuum Mech. Thermodyn. **28**(1–2), 85–104 (2014)
27. Rajkotwala, A.H., Gelissen, E.J., Peters, E.A.J.F., Baltussen, M.W., van der Geld, C.W.M., Kuerten, J.G.M., Kuipers, J.A.M.: Comparison of the local front reconstruction method with a diffuse interface model for the modeling of droplet collisions. Chem. Eng. Sci.: X **7**, 100066 (2020)
28. Norbert Roth, H.G., Livne, D.K., Weigand, B.: Comparison between analytical, numerical, and experimental results of grouping effects in droplet streams. In: Proceedings ILASS–Europe 2017. 28th Conference on Liquid Atomization and Spray Systems. Universitat Politècnica València, Sept. 2017
29. Sonntag, M., Munz, C.-D.: Shock capturing for discontinuous Galerkin methods using finite volume subcells. Finite Vol. Complex Appl. **VII**, 78:945–953 (2014)
30. Sonntag, M., Munz, C.-D.: Efficient parallelization of a shock capturing for discontinuous Galerkin methods using finite volume sub-cells. J. Sci. Comput. **70**, 1262–1289 (2017)
31. Stimson, M., Jeffery, G.B.: The motion of two spheres in a viscous fluid. In: Proceedings of the Royal Society of London. Series A, Containing Papers of a Mathematical and Physical Character, vol. 111(757), pp. 110–116 (1926)
32. Sussman, M., Smereka, P., Osher, S.: A level set approach for computing solutions to incompressible two-phase flow. J. Comput. Phys. **114**(1), 146–159 (1994)
33. Vaikuntanathan, V., Amini, K., Arad, A., Katoshevski, D., Greenberg, B., Weigand, B.: Experimental investigations on the grouping behaviour in a mono-disperse droplet stream. In: International Conference on Liquid Atomization and Spray Systems (ICLASS), Aug. 2021
34. Vaikuntanathan, V., Ibach, M., Arad, A., Chu, X., Katoshevski, D., Greenberg, J.B., Weigand, B.: An analytical study on the mechanism of grouping of droplets. Fluids **7**(5), 172 (2022)
35. van der Waals, J.D.: Thermodynamische Theorie der Kapillarität unter Voraussetzung stetiger Dichteänderung. Zeitschrift für Physikalische Chemie **13**(1), 657–725 (1894)

Reactive Flows

Reactive flows are highly challenging to predict accurately due to their multiscale nature and easily require supercomputer resources. The combination of turbulence, transport and mixing as well as highly complex and nonlinear reaction mechanisms poses numerous difficulties for numerical computations. The research presented in this chapter covers this field from various perspectives ranging from different types of burners with focus on pollutant prediction to treating rocket combustion chambers. The numerical approaches encompass stochastic particle methods as well as standard continuum-based approaches.

In their paper "HPC in Fluid-Mechanics—Materials Synthesis and Hydrogen Combustion" the authors from the Chair of Fluid Dynamics, University of Duisburg-Essen, report on their work focusing on nanoparticle generation and a multiregime burner using two different codes, namely PsiPhi developed in the group as well as the open-source code openFOAM. PsiPhi is based on structured meshes and is optimized for fully utilizing memory bandwidth and scalability through asynchronous communication. The application problems addressed in the three individual subprojects clearly demonstrate the need for HPC.

The contribution "IDDES of a LOX/CH4 Model Rocket Combustion Chamber at Real Gas Conditions" by authors from the DLR Institute of Combustion Technology in Stuttgart, the DLR Institute of Space Propulsion in Hardthausen and the Institute of Combustion Technology for Aerospace Engineering at the University of Stuttgart reports on reactive flow simulation for a next-generation rocket combustion chamber using methane/liquid oxygen as fuel. Numerical results are compared with experimental results from a model combustion chamber operated at DLR. Simulations are performed with the code TASCOM3D solving the compressible reactive Navier-Stokes equations on block structured meshes. The paper presents very good strong and weak scalability results on up to 13824 cores and quite good agreement between experiment and simulation is observed.

In their paper "Advancement of Sparse Particle Method for High-Performance Computing" the authors from the Institute for Combustion Technology, University of Stuttgart and the Engler-Bunte-Institute of the Karlsruhe Institute of Technology, present new methodological advances in a reactive turbulent multiphase flow solver based on a multiple mapping conditioning (MMC) approach to compute the averaged chemical source term. MMC is a stochastic particle method requiring a particle pairing with suitable properties in each step. The paper develops a new parallel method for the pairing problem which scales significantly better than the current state-of-the-art at least up to 384 cores.

Peter Bastian
Interdisziplinäres Zentrum für Wissenschaftliches Rechnen
Universität Heidelberg
Heidelberg, Germany
e-mail: peter.bastian@iwr.uni-heidelberg.de

Advancement of Sparse Particle Method for High-Performance Computing

Jan Wilhelm Gärtner, Oliver T. Stein, and Andreas Kronenburg

Abstract A sparse particle CFD solver for reacting, turbulent multiphase flows with multiple mapping conditioning (MMC) is investigated in terms of its scalability and performance in an HPC environment. Previous versions of the solver, called *mmcFoam*, were unable to transfer the significant potential speed-up of the sparse particle method to large cases with hundreds of cores. This work investigates the solver's bottlenecks, and the main issues hindering the scaling are identified and solved. In particular, the selection of particle pairs for the diffusive mixing prevented the scaling of the code due to a required all-to-all MPI communication. A review of the particle pairing model and algorithm led to the development of a new particle pairing method based on a sub-set of processors with a significantly reduced number of required MPI communications. Further, load balancing and particle management have been improved. The overall scaling of the new mmcFoam solver reaches ideal near-linear scaling. Therefore, this work ports the advantages of a sparse particle method for solving turbulent, reacting flows to large-scale HPC environments.

1 Introduction

The simulation of reacting multiphase flows must account for processes such as transport, mixing, chemistry, phase transitions and all their interactions. Particular challenges arise in the presence of turbulence since at elevated Reynolds numbers all relevant scales can no longer be resolved and only the larger scales can be computed

J. W. Gärtner (✉) · A. Kronenburg
Institute for Reactive Flows, University of Stuttgart, Stuttgart, Germany
e-mail: jan-wilhelm.gaertner@irst.uni-stuttgart.de

A. Kronenburg
e-mail: andreas.kronenburg@irst.uni-stuttgart.de

O. T. Stein
Simulation of Reacting Thermo-Fluid Systems, Engler-Bunte-Institute, Karlsruhe Institute of Technology, Karlsruhe, Germany
e-mail: oliver.t.stein@kit.edu

T. Ludwig et al. (eds.), *High Performance Computing in Science and Engineering '23*,
https://doi.org/10.1007/978-3-031-91312-9_27

directly (large eddy simulation, LES). Then, all subgrid scale processes and their interactions require modeling. Various modeling approaches have been explored by the community [4, 7, 11, 12] with their respective advantages and disadvantages discussed intensively in the archival literature. Here, it suffices to state that methods based on the solution of the joint probability density function (JPDF) of the mixture's composition offer some distinct advantages for the modeling of turbulent, reactive flows such as the closure of the chemical source terms. Their standard implementations as stochastic particle methods are, however, costly to use and they have therefore not found their way to mainstream applications in CFD.

Multiple mapping conditioning (MMC) has emerged as a promising alternative since its stochastic implementation is of equal accuracy (see e.g. [10]) as a standard JPDF method but it comes at a significantly reduced computational cost. A reduction of the number of stochastic particles by a factor of around 100 (and sometimes even more [1]) can be realized due to a conditioning of the mixing operator (that emulates molecular and turbulent mixing) on a reference field. MMC thus combines a stochastic PDF method with a conditioning approach such as *conditional moment closure* or *flamelets*, exploiting information on a possible flame structure that allows for such a reduction in particle number and associated computational cost [1, 9, 10]. The MMC approach has demonstrated its suitability as a subgrid model for large-eddy simulations, and for more than 15 years researchers from around the world have developed MMC closures for a wide range of applications.

The first MMC-LES code was developed at the University of Queensland where subroutines for the MMC-specific stochastic particle solution (for the composition field) were coupled to an in-house Fortran solver for the solution of the continuity and momentum equations [1]. Only few years later, MMC libraries were entirely re-coded in C++ and coupled to the open-source computational fluid dynamic (CFD) framework OpenFOAM to increase the method's applicability and code acceptance by other research groups. These new CFD libraries were called mmcFoam.

Until today, the latest mmcFoam version has been implemented for OpenFOAM 5.x and while the MMC method greatly reduces computational costs, it has not been able to fully scale its performance to a large-scale HPC setting. Here, the causes for the scaling problems are identified, discussed and solved, laying the foundation for a more efficient use of mmcFoam on HPC systems such as the super-computer Hawk at HLRS.

2 MMC Modeling

The main challenge of simulating turbulent reactive flows stems from the closure of the averaged chemical source term, $\overline{\dot{w}}$, required for the species transport equation,

$$\frac{\partial \overline{\rho} \widetilde{Y}_k}{\partial t} + \nabla \cdot \left(\overline{\rho} \widetilde{Y}_k \widetilde{\boldsymbol{u}}\right) = \nabla \cdot \left(\Gamma_k \nabla \widetilde{Y}_k\right) + \overline{\dot{w}_k}. \tag{1}$$

Since the reaction rate is highly non-linear, the average reaction rate cannot be computed with the averaged temperature and mass fraction,

$$\overline{\dot{w}_k(T, Y_k)} \neq \dot{w}(\overline{T}, \overline{Y_k}). \quad (2)$$

One method to close the source term without introducing further simplifications is the above-mentioned JPDF approach. Typically this is implemented as a particle based Monte Carlo method, where stochastic particles emulate the evolution of the reactive scalars. This advantage is, however, bought by having to model the molecular diffusion, i.e. the mixing of the stochastic particles. In a direct implementation a sufficient number of particles, up to 100, must be present in each computational finite-volume cell for LES. To model the molecular diffusion, particle pairs close in physical space are selected and mixed with a suitable mixing model, such as the Curl's coalescence/dispersion model [3, 6]. The high particle number density required for this approach leads to a very high computational cost, making large scale simulations very expensive or even unfeasible.

Recently the novel MMC approach has emerged, which significantly reduces the number of stochastic particles required by several orders of magnitude [1, 9]. The concept behind this method is to base the selection of the unique particle pairs not only on their distance in physical space, but to condition it with an additional reference variable. For non-premixed flames the mixture fraction is typically chosen. Hence, for a one dimensional reference space the squared distance between two particles p and q is expressed with,

$$d^2_{(p,q)} = \left[\sum_{i=1}^{3}\left(\frac{d^{p,q}_{x_i}}{r_m}\right)^2\right] + \left(\frac{d^{p,q}_{\xi}}{\xi_m}\right)^2. \quad (3)$$

Here, r_m and f_m are characteristic lengths of the physical and reference space, respectively. Conditioning with the reference variable prevents particle mixing across the flame front, as illustrated in Fig. 1. Here particles A and B are close in physical space but separated by the flame front and thus have a large distance in reference space. In contrast, particle C is further away but within the same range of mixture fraction space, hence A will mix with C instead of B, preventing the mixing across the flame front.

2.1 Particle Pairing Algorithms

A crucial function of the MMC method is the selection of particles to mix and their mixing rate. Particle pairs have to be selected to minimize the distance $d_{p,q}$. For further details on the related theory, see [1, 2]. On a more abstract, mathematical level, the problem described is the assignment problem, or minimum cost bi-partite matching, which can be solved with linear programming. The problem is that n

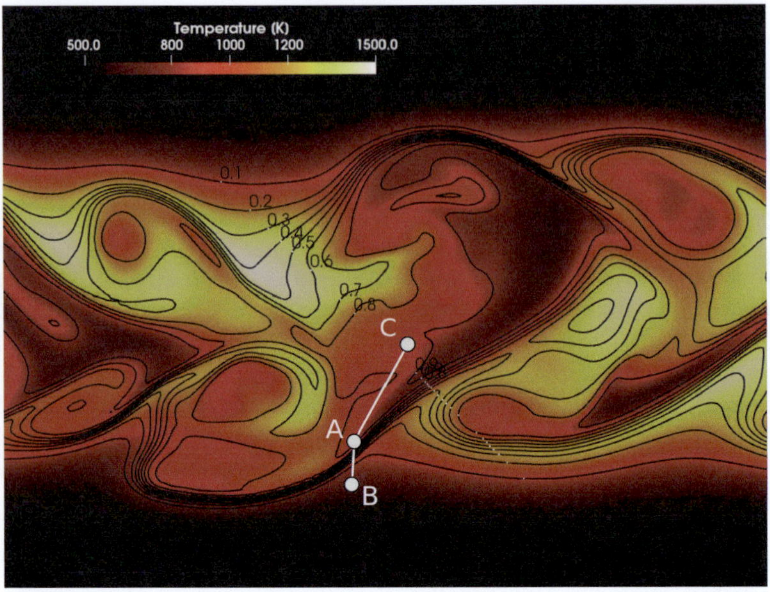

Fig. 1 Sketch of particle pairing based on multiple mapping conditioning. Despite the small physical distance between particle A and B, particle A will mix with the more distant particle C due to their small distance in mixture fraction space. The background image is the temperature profile of a reacting double shear layer, and the black lines represent iso-contours of the mixture fraction

Table 1 Example cost matrix for a four particle matching problem

	Particle 1	Particle 2	Particle 3	Particle 4
Particle 1	Inf	**1.0**	2.0	5.0
Particle 2	**1.0**	Inf	0.8	1.3
Particle 3	2.0	0.8	Inf	**1.2**
Particle 4	5.0	1.3	**1.2**	Inf

workers have to be assigned to n jobs with an associated cost function, f. Finding unique particle pairs is a special case of an assignment problem with the cost function $f(p, q) = d_{p,q}$, as described in Eq. (3). To solve these problems, the cost matrix has to be known, which contains the cost of mixing particle p in row i with particle q in column j. Here an example case with four particles is given. To avoid the pairing of particles with themselves, the distance is set to infinity for $i = j$ (Table1).

In the given example, particle 1 would mix with particle 2, and particle 3 would mix with particle 4. This problem can be solved in polynomial time with, e.g. the Hungarian algorithm [8]. However, the main challenge with this set of problems is that the calculation of the cost function is quite expensive, and when generating the cost matrix, $0.5n^2$ calculations of the distance $d_{p,q}$ are required, where n is the number of particles.

To avoid the calculation of the cost function, the constraint of finding the minimum pairing configuration is relaxed to an approximate minimum mixing distance, which is close to the optimum. For this, an efficient k-d-tree implementation is used, which iteratively divides the particle population until only two particles exist in one tree node which are subsequently mixed. Only log(n) sort operations are required for this operation, and at no point, the cost function has to be evaluated. Further, the implementation of the k-d tree is straightforward, as only the construction of the tree but not the traversing for nearest neighbor search must be considered.

2.2 Parallel Particle Mixing

Most simulations have to be run in parallel on several cores and nodes, to solve the problem in a reasonable amount of time. To run the simulation in parallel, the problem, here the physical domain, is divided into smaller sub-domains and distributed to the cores of the CPU, such that each core solves for one sub-domain. As each sub-domain has its own memory, information must be exchanged explicitly via Message Passing Interface (MPI) commands. Considering particle pairing and mixing in MMC, this means that the particles are distributed to the individual cores and do not know about the others by default. Figure 2a shows a sketch of particle pairing for a domain decomposed into four sub-domains (processors). As long as all processors know about all the other particle positions, the pairing computed individually on each processor is the same. However, this changes if only a subset of particle positions is known to a given processor. For example in Fig. 2a with the information of all processors available to processor 1, none of its particles mixes with any particle on processor 4. If we alternatively communicate only the information from processors 2 and 3 to processor 1, the pairing changes, see Fig. 2b. Hence, a change in the known set of particles results in a different matching, even if the missing particles do not directly contribute to the pairing of the considered processor.

Now, the question is if we can allow different pairings for each processor or if this violates the MMC assumptions. One important rule is that particles should not mix twice within a single mixing step. For parallel runs, this issue would occur if the same particle mixes with two different particles on two different processors. As the processors have no direct access to particle data located on different processors, only copies of the particles can be transferred. This means that in the given example case of Fig. 2b, processor 1 has copies of the particles located and stored on processors 2 and 3. Hence if on processor 1 particle p_1 mixes with particle p_3 as in Fig. 2b instead of p_2 as it would in the fully reconstructed case from Fig. 2a, this does not affect the mixing for the other processors two, three, and four. Therefore, no other particle data is affected by the different mixing partners, which at first appears favorable.

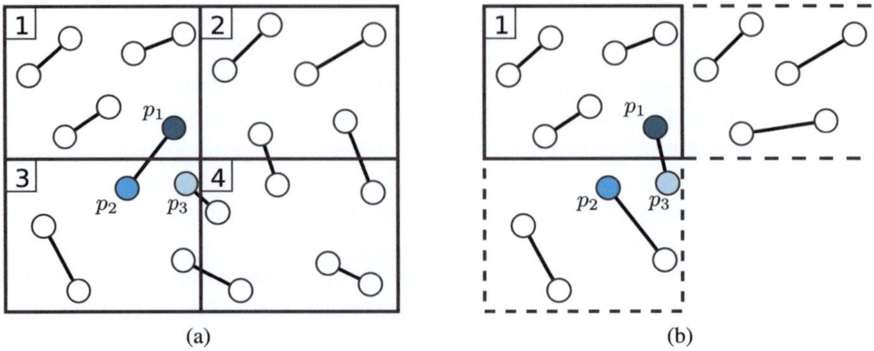

Fig. 2 Sketch of particle pairing for a case decomposed into four sub-domains (processors). In case **a** all particles on all processors are known and considered for pairing. Case **b** shows an alternative pairing result if only the particles of processors 2 and 3 are known to processor 1

However, this leads to problems for the mass balance. As now different particles mix on different processors, depending on their choice of the subset of particles considered, the species mass fraction of the new ensemble of all particles does no longer match,

$$\sum_p \sum_i (Y_i)_p \, m_p \neq \sum_p \sum_i (Y_i)_{p,\text{mixed}} \, m_p. \tag{4}$$

As particle pairing depends on the choice of the particle set, all processors must always know about all other particle positions to ensure mass conservation. However, such *global mixing* requires all-to-all MPI communication which strongly hampers any parallel scaling and has to be addressed before large simulations can be performed on HPC systems with hundreds or thousands of computer cores.

2.3 Review: K-D Tree Pairing

In the introduction is was outlined that finding an exact minimum-cost pairing of particles for MMC is too computationally expensive. As described above, to avoid the computation of the particle-to-particle distance according to Eq. (3), existing MMC implementations construct a k-d tree until only two or maximum three particles remain within a single leaf node. The principle of the k-d tree is to find the dimension with the largest spread of values in each iteration, to sort the particles according to this dimension and to then split the particle population in half. This step is subsequently repeated for the new sub-list of particles until only two particles are left, and then the algorithm terminates, see Fig. 3. It is important to note that this pairing algorithm

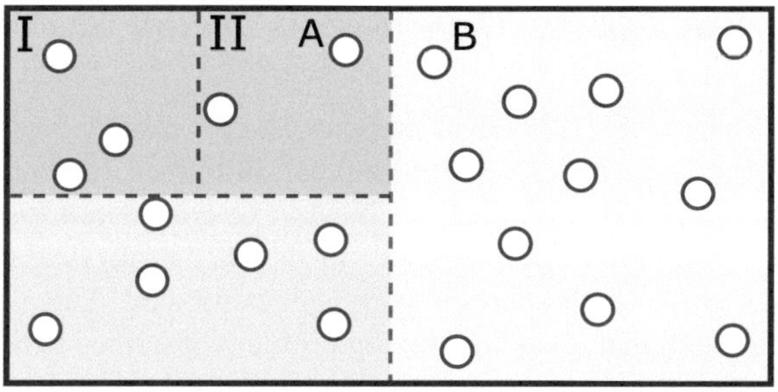

Fig. 3 Example sketch of the first three iterations of the k-d tree pairing algorithm. Due to the first split in the longest dimension (left to right), the particle pair found in the leaf node on the left side, box II, is formed, even though a closer particle for particle A exists on the right side of the first split, marked as B. Hence, the first split prevents any mixing of particles over the barrier marked by the red line

will, in most cases, introduce mixing barriers over which no particles will mix. This behavior is shown with a simple example as follows. Assume the case of a mixing layer which consists of a 3D domain with length, width, and height of $L = 10\,\text{mm}$, $W = 5\,\text{mm}$, $H = 5\,\text{mm}$. Typical normalization parameters r_m and ξ_m (see Eq. (3)) are,

$$r_m = 0.22\,\text{mm}$$
$$\xi_m = 0.03$$

with this, the normalized coordinates become,

$$x^* = 45.45$$
$$y^* = 22.73$$
$$z^* = 22.73$$
$$\xi^* = 33.33$$

Hence in nearly all cases, the k-d tree will split the particle population in the physical x-dimension first. This is illustrated in Fig. 3, which sketches the first three iterations of an example k-d tree construction. In conclusion, this means that no particle on the left side with a low x-coordinate will mix with a particle on the right side with a large x-coordinate, effectively implementing a mixing barrier. This behavior is pronounced if the physical extent of the considered simulation domain is large or severely elongated in one direction, as this will lead to several splits in the same coordinate direction.

2.4 New Particle Pairing Method—Sub-volume Searches

The behavior of the k-d tree algorithm motivates a new pairing method for parallel simulations. As the k-d tree implementation can introduce mixing barriers in the simulation, the condition that all particles have to be considered for pairing can be relaxed. Hence, a new model is proposed, in which the physical domain is split into sub-volumes in which the particle pairing is computed. This avoids the inter-processor communication and double mixing problem described in Sect. 2.2, as only particles within sub-volumes are considered and no particle mixing over sub-volume boundaries is allowed. This idea is sketched in Fig. 4. The challenge is to find these sub-volumes and to avoid the introduction of new mixing barriers. To solve this, not a single set of sub-volumes is computed but rather a multitude of sub-volume sets from which one set is selected in each iteration at random. Further, to generate random sub-volume sets one processor is selected randomly as a starting point from which the sub-volume search is subsequently developed. Therefore, any mixing boundary introduced by the sub-volumes is not fixed in time, and in each time step a different set of particles is considered for mixing. The second parameter of the method is the number of processors considered for a sub-volume. Based on the concept that the k-d tree splits the search domain in the physical dimension if

$$\Delta X \geq \frac{r_m}{\xi_m} \tag{5}$$

holds true, a sufficient number of processors is reached once their extent in one physical dimension satisfies the condition of Eq. (5). Hence, adding more processors to the sub-volume once Eq. (5) is satisfied does not add more particles as potential mixing candidates, as the k-d tree algorithm will split the particle population first, and always, in the same physical dimension.

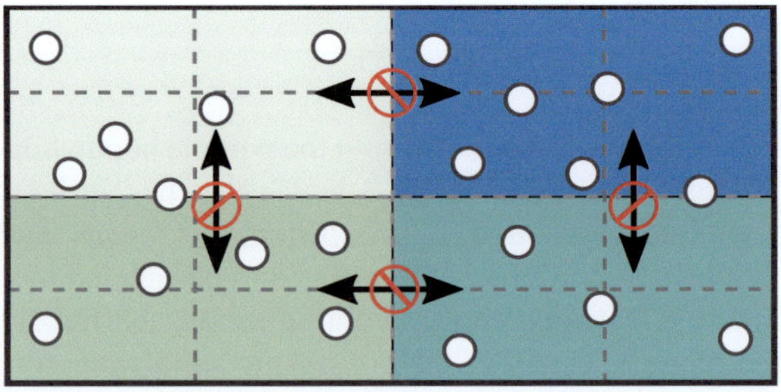

Fig. 4 Sketch of particle pairing based on sub-volumes. The gray dashed lines denote processor boundaries and the background colors the different sub-volumes, consisting of several processor domains. There is no particle pairing or mixing over the sub-volume boundaries. Note that the sub-volume set changes in each iteration and thus mixing barriers are not persistent in time

2.5 Particle Number Management and Load Balancing

Management of the particle number density is a challenging task for any stochastic particle based method. The random movement of particles may lead to depletion of particles within the reaction zone due to the high temperatures and thus fluid expansion, and high diffusivity of the particles. To ensure a sufficient number of particles at each location of the domain, the particle number is controlled in mmcFoam using a *super-mesh*. This super-mesh divides the domain into control regions, in which a fixed particle number is enforced. In the case of locally too many particles, a set of particles with low mass will be combined into a single particle, whereas in regions with a low particle count, heavy particles will be split to form two new particles. In previous versions of mmcFoam, this super-mesh hindered the decomposition of the domain, as one control region could not be split and distributed over several processors. This is fixed in the current mmcFoam version which employs a particle number controller class, based on the same principle as the super-mesh, however, without the constraint of the domain decomposition. To ensure the correct particle number count of control regions distributed over several processors, some remaining MPI communication is still required. However, the benefit of decomposing the domain with any method into any number of processors outweighs the communication costs.

A second challenging task for particle based simulations is the load balance, especially for simulations including chemical reactions. Solving detailed chemical reaction mechanisms takes the most part of the computational time, and particles outside the reaction zone can be computed much faster than particles within the reaction zone. Therefore, to balance the computational load, it is not sufficient to aim for an equal particle count in each processor. Instead the computational time to compute each particle has to be considered. Based on the computational time required to solve each particle for the previous time step, the particles can be redistributed to the processors to achieve an equal load balance, see Fig. 5. A similar concept has already been proposed for mmcFoam and presented in Huo et al. [5].

3 Parallel Performance and Scaling

The performance of the new mmcFoam solver is tested on the HPC system Hawk of HLRS Stuttgart. The test case examined is a non-premixed, turbulent ethylene flame, one of the target flames of the International Sooting Flame (ISF) Workshop, and has been experimentally investigated at Sandia National Laboratories. The mesh comprises of 2.5 million cells with 700 000 stochastic particles. OpenFOAM solvers typically scale well up to a decomposition of 10 000 cells per core, therefore, the solver should scale up to at least 256 cores for this test case. To investigate the effect of the different components of the solver, the scaling is split up into a total scaling of the overall solver, scaling of the particle pairing and mixing, solving chemical reactions on the particles, and coupling the particle cloud with the Eulerian field, i.e. density coupling. Figure 6 depicts the normalized scaling of the solver with and without load

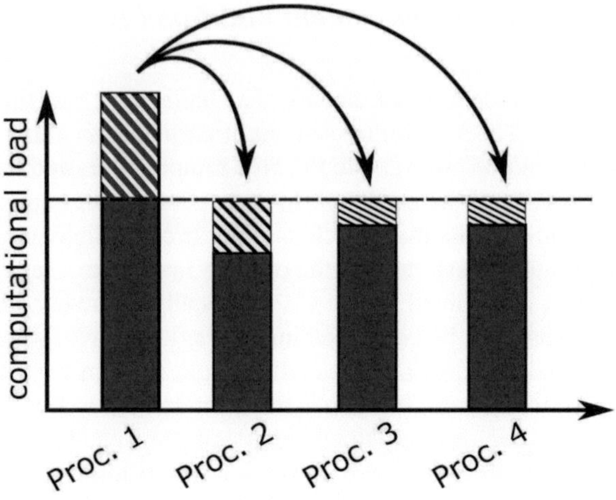

Fig. 5 Sketch of load balancing to solve chemical reactions for different processors and their overheads. Here, the load of the first processor is distributed to the other three to achieve an equal load for all four processors

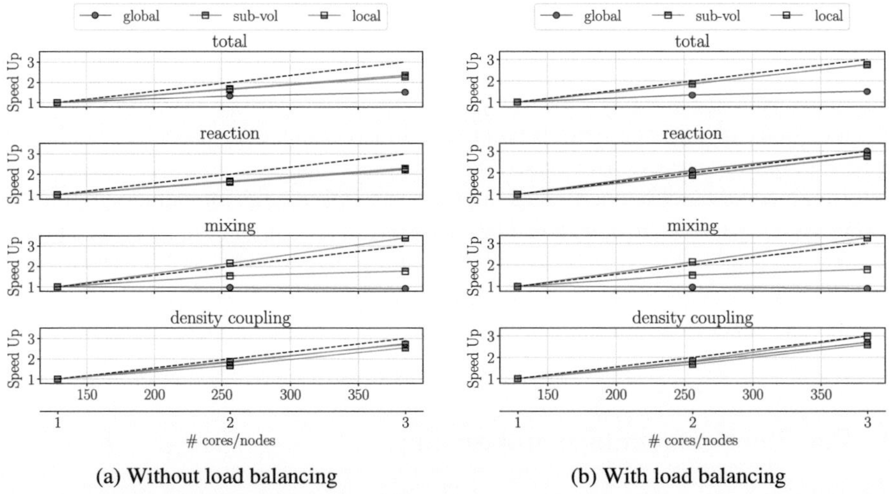

Fig. 6 Total scaling of mmcFoam and scaling of its main components, i.e. solving chemistry, particle pairing and mixing, and density coupling. Results are presented for the three different particle pairing methods without load balancing (**a**) and with load balancing (**b**). The black dashed line represents an ideal linear scaling

balancing, and for the three particle pairing methods, global, local, and sub-volumes. The first pairing method is the existing implementation of mmcFoam, in which the total particle population has to be known to all processors to find a consistent particle pairing for all. The second option selects particle mixing on each processor only; hence other processors are not considered. This can be a potential option for cases that do not use a high decomposition. However, this requires validation by the user to check if the results can be trusted. The last pairing method is the newly implemented sub-volume option. The results emphasize the impact of the load balancing on the overall scaling, as well as the choice of particle pairing. If the particle load balancing is switched off, solving the reaction scales poorly, and for the case of 384 cores, only 66% of the ideal scaling is reached. The reactive part's poor scaling coincides with the solver's overall scaling. The cause for this correlation becomes evident when the percentage of time spent in each of the three main components is investigated in Fig. 7. The reactive part of the solver takes about 75% of the total time, except for the global pairing method. Hence a poor scaling of this part directly affects the total scaling of the solver. It further shows the significance of the particle pairing method, as the global pairing method does not scale and takes up 50% of the time for the case with 256 cores. Due to the inability to scale with more cores, the percentage of time spent on particle pairing and mixing increases with an increasing number of cores. The new sub-volume method also does not scale well, as the number of processors within one sub-volume is dependent on a physical size not on the domain decomposition. However, as only a small subset of all processors is considered for communication, the overall time spent on particle pairing and mixing is much less than for the global method, hence, the impact on the total scaling is negligible. The

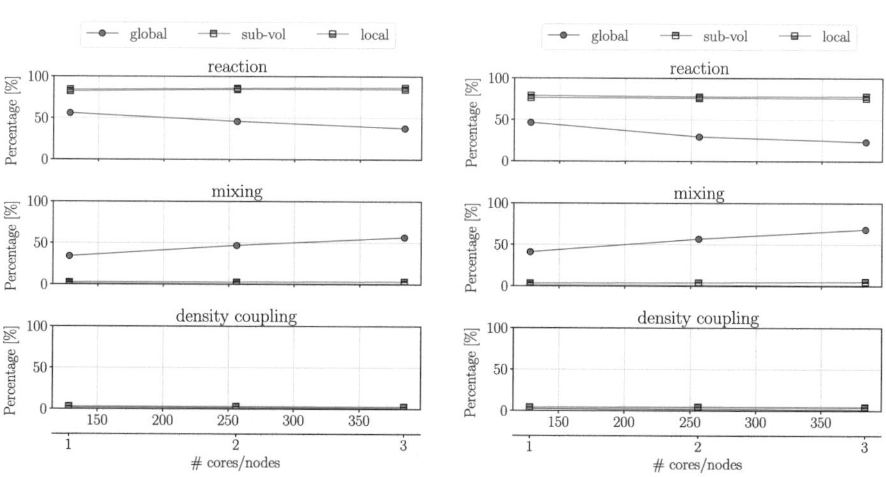

(a) Without load balancing

(b) With load balancing

Fig. 7 Percentage of the total time spent in each component of mmcFoam, solving the reaction, particle pairing and mixing, and density coupling. Results are presented for the three different particle pairing methods without load balancing (**a**) and with load balancing (**b**) local particle pairing method scales perfectly as the problem size decreases linearly with the number of cores. However, to reiterate, the local particle mixing is only valid for large processor domains, hence for a low decomposition, and is not a suitable approach for HPC systems with decompositions into hundreds of cores. On the other hand, the sub-volume method considers multiple processors and scales as well as the local method. For the case of active load balancing, the results scale ideally up to the investigated decomposition of 384 cores.

In conclusion, the new mmcFoam solver combined with the sub-volume particle pairing and particle load balancing shows a near-perfect scaling within the investigated range and can be applied in an HPC setting. During the project reporting period from April 2022 to March 2023, a total of 46 thousand node hours, equaling 5.9 million core hours, have been used on Hawk. Validation, tests, and development of the mmcFoam solver, which is not shown here, took the majority of the core hours, and is attributed to sub-project SP2 "Sparse particle methods for multiphase combustion systems". Simulations required for SP1 "Multidimensional flamelet modelling for LES of pulverised coal flames" take up the remainder of the core hours.

4 Conclusion

The CFD framework *mmcFoam* for turbulent, reacting multiphase flows with detailed chemical reaction mechanisms has been investigated for its use in large-scale HPC environments, and bottlenecks have been identified and solved. Previous versions of mmcFoam were unable to scale very well in HPC environments due to limitations of their methods for particle number management, load balancing, particle pairing and mixing. In particular, the mixing operation for the stochastic particles hindered the use in an HPC context due to its requirement of all-to-all MPI communication to find the particle pairs. Three methods to select particle pairs, global, local and a new *sub-volume method* are discussed. Further, the implications of using a k-d tree algorithm to select particle pairs are highlighted, and based on these, the new sub-volume pairing strategy has been developed. Here, processors are grouped together to form sub-volumes in which particle pairing is calculated, reducing the all-to-all communications of the global pairing method to a significantly smaller number of processors. The reduction of the communication to smaller sub-sets of processors significantly reduces the communicative overhead of the simulations compared to the previously implemented global pairing algorithm. In addition to the particle pairing method, new load balancing and particle management techniques are implemented, enabling the decomposition into any number of processors and with any decomposition method. Overall, the new mmcFoam solver delivers near-ideal linear scaling for several hundred cores, thus porting the sparse particle MMC method to modern HPC systems.

References

1. Cleary, M.J., Klimenko, A.Y.: A generalised multiple mapping conditioning approach for turbulent combustion. Flow Turbul. Combus. **82**(4), 477–491 (2009)
2. Cleary, M.J., Klimenko, A.Y.: A detailed quantitative analysis of sparse-Lagrangian filtered density function simulations in constant and variable density reacting jet flows. Phys. Fluids **23**(11), 115102 (2011)
3. Curl, R.L.: Dispersed phase mixing: I. Theory and effects in simple reactors. AIChE J. **9**(2):175–181 (1963)
4. Ertesvåg, I.S., Magnussen, B.F.: The eddy dissipation turbulence energy cascade model. Combus. Sci. Technol. **159**(1), 213–235 (2000)
5. Huo, Z., Cleary, M.J., Wu, K., Masri, A.R., Fan, X.: A dynamic load balancing model coupled with DAC and ISAT for a stochastic turbulent combustion model. Combus. Theory Model. **27**(3), 317–345 (2023)
6. Janicka, J., Kolbe, W., Kollmann, W.: Closure of the transport equation for the probability density function of turbulent scalar fields. J. Non-Equilibrium Thermodyn. **4**(1) (1979)
7. Klimenko, A.Y., Bilger, R.W.: Conditional moment closure for turbulent combustion. Prog. Energy Combust. Sci. **25**(6), 595–687 (1999)
8. Kuhn, H.W.: The Hungarian method for the assignment problem. Naval Res. Logistics Q. **2**(1–2), 83–97 (1955)
9. Neuber, G., Garcia, C.E., Kronenburg, A., Williams, B.A.O., Beyrau, F., Stein, O.T., Cleary, M.J.: Joint experimental and numerical study of silica particulate synthesis in a turbulent reacting jet. Proc. Combus. Inst. **37**(1), 1213–1220 (2019)
10. Neuber, G., Kronenburg, A., Stein, O.T., Cleary, M.J.: MMC-LES modelling of droplet nucleation and growth in turbulent jets. Chem. Eng. Sci. **167**, 204–218 (2017)
11. Peters, N.: Laminar diffusion flamelet models in non-premixed turbulent combustion. Prog. in Energy Combus. Sci. **10**(3), 319–339 (1984)
12. Pope, S.B.: PDF methods for turbulent reactive flows. Prog. Energy Combus. Sci. **11**(2), 119–192 (1985)

IDDES of a LOX/CH4 Model Rocket Combustion Chamber at Real Gas Conditions

Timo Seitz, Ansgar Lechtenberg, Jan Martin, Justin S. Hardi, and Peter Gerlinger

Abstract The combustion dynamics observed in a sub-scale rocket combustor with optical access and operated with the propellant combination of liquid oxygen and methane at supercritical chamber pressures are investigated numerically. This combustion chamber shows a very intermittent behavior which is extremely difficult to simulate. Temperature and pressure measurements at the wall as well as OH^* radiation images from inside the combustor are used to validate the numerical solver. All simulations are performed with the in-house code TASCOM3D using a spatial discretization up to fifth order. The scalability of TASCOM3D is briefly discussed. It will be shown that TASCOM3D achieves high efficiencies for both weak and strong scaling. Comparison of the time-resolved data with the experiment reveals an excellent agreement concerning the frequencies of the eigenmodes of the combustion chamber. The flame characteristics are similar, too. The presented analysis additionally focuses on the methane combustion process at rocket combustor conditions.

1 Introduction

In Western Europe, mainly hydrogen and hypergolic propellants have been used for liquid rocket propulsion [1]. However, due to its more favorable properties, methane is getting more and more attention, not only in Europe but worldwide. Compared with hydrogen, methane is cheaper and easier to handle and achieves a higher specific impulse than kerosene. Additionally, methane is non-toxic and could theoretically

T. Seitz (✉) · A. Lechtenberg
DLR Institute of Combustion Technology, Stuttgart, Germany
e-mail: timo.seitz@dlr.de

J. Martin · J. S. Hardi
DLR Institute of Space Propulsion, Hardthausen am Kocher, Germany

P. Gerlinger
Institute of Combustion Technology for Aerospace Engineering (IVLR), University of Stuttgart, Stuttgart, Germany
e-mail: peter.gerlinger@dlr.de

T. Ludwig et al. (eds.), *High Performance Computing in Science and Engineering '23*,
https://doi.org/10.1007/978-3-031-91312-9_28

be produced directly on Mars. Well-known rocket engines operated with methane are the Raptor engine of SpaceX or the Prometheus engine [2, 3] which will be used in future developments of the Ariane. But there are also numerous other examples, such as China with its TQ engine family or Japan [4].

The development of rocket engines can be supported by reliable and accurate numerical simulations. Of special interest are wall heat fluxes and pressure fluctuations as well as the flow field itself. Here, time-accurate numerical simulations play an important role, which require high-quality experimental data sets for validation. Due to the extreme conditions in rocket thrust chambers, experimental data are usually limited to a few quantities such as the wall temperature or the unsteady pressure in model combustors. The rocket model combustor BKN is operated with methane as fuel. OH^* and CH^* radiation images are available in addition to temperature and unsteady pressure measurements near the wall. This allows a comparison of experimental and simulated flame shapes. This combustion chamber is experimentally investigated at the German Aerospace Center (DLR) [5].

In this work, the BKN is investigated numerically using the in-house code TASCOM3D. A hybrid Reynolds-Averaged Navier-Stokes (RANS)/Large-Eddy Simulation (LES) method is used here, the improved Delayed Detached Eddy Simulation (iDDES) [6, 7]. This method combines the advantages of RANS close to the wall with the resolution of significant turbulent scales in free shear layers by LES. TASCOM3D has already been used for methane rocket combustors at both ideal [8] (RANS) and real gas conditions [9] (iDDES).

2 Numerical Methodology

Over the past three decades, the in-house code TASCOM3D (Turbulent All Speed Combustion Multigrid 3D) has been developed and successfully applied to various reacting and non-reacting test cases [10–14]. In recent years, the focus has been on the simulation of rocket combustion chambers [8, 9, 15–17]. Flow and combustion processes in thrust chambers are described by the extended compressible Navier-Stokes equations that include equations for turbulence modeling, species transport and turbulence-chemistry interaction (TCI). The applied set of equations is given by

$$\frac{\partial \mathbf{Q}}{\partial t} + \frac{\partial \left(\mathbf{F} - \mathbf{F}_v\right)}{\partial x} + \frac{\partial \left(\mathbf{G} - \mathbf{G}_v\right)}{\partial y} + \frac{\partial \left(\mathbf{H} - \mathbf{H}_v\right)}{\partial z} = \mathbf{S} \tag{1}$$

with

$$\mathbf{Q} = \left[\rho, \rho u, \rho v, \rho w, \rho E, \rho k, \rho\omega, \rho\sigma_T, \rho\sigma_Y, \rho Y_i\right]^\top, \ i = 1, \ldots, N_k - 1. \tag{2}$$

Here, $\mathbf{Q}$ consists of the density ρ, the velocities u, v, w in each coordinate direction, the total specific energy E, the turbulence quantities k and ω, the temperature variance σ_T, the sum of the variances of all species mass fractions σ_Y as well as $N_k - 1$

independent species mass fractions Y_i. N_k is the number of species considered. **F**, **G** and **H** indicate the inviscid fluxes in x-, y- and z-direction, respectively. $\mathbf{F}_v$, $\mathbf{G}_v$ and $\mathbf{H}_v$ are their viscous counterparts. The source term vector

$$\mathbf{S} = \left[0, 0, 0, 0, 0, S_k, S_\omega, S_{\sigma_T}, S_{\sigma_Y}, S_{Y_i}\right]^\top, \; i = 1, \ldots, N_k - 1 \tag{3}$$

includes terms resulting from turbulence and chemistry.

As the propellants are injected cryogenically in the investigated rocket combustion chamber, a real gas equation of state and corresponding thermodynamic and gas properties are required. In this work, the Soave-Redlich-Kwong equation of state [18]

$$p = \frac{R_\mathrm{u} T}{V_\mathrm{m} - b} - \frac{\alpha(T)}{V_\mathrm{m}(V_\mathrm{m} + b)} \tag{4}$$

is used, where R_u is the universal gas constant, while V_m is the molecular volume. $\alpha(T)$ and b are a temperature-dependent and an independent constant, respectively [16, 18].

The above equations are solved with a block-structured, cell-centered finite-volume method that uses an implicit lower-upper symmetric Gauss-Seidel algorithm [19–21]. Temporal discretization is carried out by the implicit backward differentiation formula (BDF) [22], where a dual-time stepping scheme with a couple of inner Newton iterations solves the non-linear set of equations. In addition, an all Mach number preconditioning [14, 23] is used.

The inviscid fluxes are discretized with the SLAU flux vector splitting [24]. The required primitive values at the cell-interfaces are obtained using a multi-dimensional limiting process with low dissipation ($\mathrm{MLP}^{\mathrm{ld}}$) [13, 15], which is up to fifth-order accurate in space. The viscous fluxes are calculated using a central discretization in a cell-oriented coordinate system.

Turbulence is modeled using the hybrid RANS/LES iDDES [6, 7] approach. The version used here is based on the Wilcox k-ω model [25]. While LES is used for free shear layers, the near-wall behavior is treated with unsteady RANS. This avoids the strict grid requirements [26, 27] of pure LES or Direct Numerical Simulation (DNS). The switch between RANS and LES is performed automatically using an empirical function.

In order to model combustion, finite-rate chemistry is applied. An assumed probability density function (APDF) approach [28–31] accounts for TCI. Here, a reduced chemical mechanism with 17 species and 70 steps is used for the combustion of methane.

3 Code Performance

Due to the high numerical effort of the present rocket combustion chamber simulation, it is important to understand and optimize the interaction of the code with the hardware architecture in use. Hence, the strong and weak scaling behavior of TASCOM3D on HPE Apollo (HAWK) at HLRS is investigated. Each node on HAWK consists of 128 cores that are arranged in two sockets. A socket consists of four Non-Uniform Memory Access (NUMA) nodes, thus giving a total of eight NUMA nodes. A further subunit is the so-called Cluster Complex Die (CCD). Two of these form a NUMA node. One CCD is composed of two Cluster CompleXs (CCXs), each consisting of four cores that share a L3 cache. In order to analyze the code, a geometrically simple, three-dimensional fully developed turbulent channel flow is simulated using iDDES.

3.1 Strong Scaling

For the investigation of the strong scaling behavior a fixed number of 96^3 volumes is used. While the problem size is kept constant, an increasing number of CPUs causes a decreasing workload for each CPU. Read and write operations are not considered. For each configuration the runtime for 500 iterations is stated. In order to reduce the influence of varying node performances on the result, the given runtime is averaged over ten single runs. As it is not possible to use a single CPU for the chosen test case, the smallest numbers of CPUs considered is eight. This leads to a blocksize of 48^3 for each CPU. The maximum amount of CPUs considered is 13824 resulting in a blocksize of 4^3. The left side of Fig. 1 depicts the runtime and the efficiency for all considered setups. In addition, Table 1 lists the blocksize and the resulting speedup as well. The efficiency is calculated by

$$\eta_{SS} = \frac{T_{\text{CPU}}^{\text{ref}}}{T_{\text{CPU}}} = \frac{T_{\text{R}}^{\text{ref}} \cdot N_{\text{CPU}}^{\text{ref}}}{T_{\text{R}} \cdot N_{\text{CPU}}}, \tag{5}$$

where T_{CPU} is the computational cost, T_{R} the runtime measured in seconds and N_{CPU} the number of CPUs used. Terms with the superscript *ref* refer to reference values which corresponds to the case with eight CPUs. With an increasing number of CPUs, the runtime decreases from more than 11670 s to less than 6.5 s for the setup with the highest number of CPUs. This is a speedup of approximately 1844. However, the efficiency does not increase monotonically. Instead, the setup with 27 CPUs has a lower efficiency than the reference case. This seems to be hardware related. Thereafter, the efficiency increases until a blocksize of 8^3 is reached. Due to the communication overhead that degrades the performance, efficiency decreases for smaller blocksizes. The observed superlinear speedup is most likely caused by increased cache hits with decreasing block size. For TASCOM3D, a similar behavior

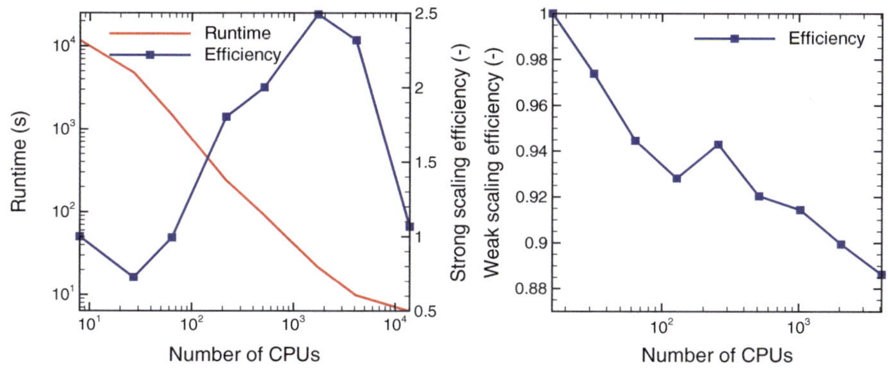

Fig. 1 Strong (left) and weak (right) scaling behavior of TASCOM3D on HAWK at HLRS

Table 1 Strong scaling behavior of TASCOM3D on HAWK at HLRS. This test was performed with a total number of 96^3 volumes

#cores	Blocksize	Absolute timing (s)	Speedup (–)	Efficiency (–)
8	48^3	11673.6	1.0	1.00
27	32^3	4764.2	2.5	0.73
64	24^3	1469.3	8.0	0.99
216	16^3	239.6	48.7	1.80
512	12^3	91.2	128.1	2.00
1728	8^3	21.7	538.0	2.49
4096	6^3	9.8	1186.6	2.32
13824	4^3	6.3	1844.5	1.07

has already been observed on a different system [32]. These results indicate that a blocksize of 8^3 might be optimal for TASCOM3D. However, in order to obtain a high numerical stability, all equations are solved in an implicit manner. Hence, the blocksize should not become too small as this corrupts the implicit nature of the numerical algorithm. Currently, a blocksize of about 16^3 is typical for our combustion chamber simulations. As can be seen on the left side of Fig. 1, this still achieves an excellent efficiency.

3.2 *Weak Scaling*

In the weak scaling tests the workload of each core is kept constant, while the problem size increases. As a workload of 16^3 volumes per CPU is typical for our applications, this size is examined. The reference configuration uses 16 CPUs, which corresponds

Table 2 Weak scaling behavior of TASCOM3D on HAWK at HLRS. This test was performed with a constant blocksize of 16^3. Given are absolute timings per 500 time steps (s) and efficiencies relative to 16 cores

#cores	Absolute timing (s)	Efficiency (–)
16	278.3	1.00
32	285.8	0.97
64	294.7	0.94
128	299.9	0.93
256	295.2	0.94
512	302.4	0.92
1024	304.4	0.91
2048	309.4	0.90
4096	314.0	0.89

to a fully occupied NUMA node. This enables a meaningful comparison as each node should have the same load. The right side of Fig. 1 shows the efficiency

$$\eta_{WS} = \frac{T_{\mathrm{R}}^{\mathrm{ref}}}{T_{\mathrm{R}}} . \tag{6}$$

of the weak scaling tests. Numerical values are given in Table 2. In addition to the efficiency, absolute runtimes are given. As can be seen, the weak scaling behavior of TASCOM3D is excellent as the efficiency with 4096 CPUs is higher than 88%. The occurring losses in efficiency can be explained by the MPI overhead.

4 The Sub-Scale Rocket Combustion Chamber BKN

4.1 Experimental Setup

The sub-scale rocket model combustor BKN was experimentally investigated at the German Aerospace Center (DLR) [5]. It is equipped with various measuring devices in order to provide extensive data for the validation of numerical simulations. The cylindrical chamber has a diameter of 50 mm and a length of 359 mm from faceplate to nozzle throat. Its nozzle throat diameter is 14.5 mm.

A single shear coaxial injector supplies the propellants to the combustion chamber. The injection element has already been used for purely numerical test cases [9, 33, 34]. Its geometrical details can be found in e.g. [5, 33]. Oxygen is introduced through the inner, recessed tube. For the load point considered, the outer annulus supplies methane to the chamber. However, the BKN can also be operated with hydrogen [35] or liquefied natural gas [36]. In addition, film cooling using methane is applied at

Table 3 Experimental inflow conditions for the simulated load point of the BKN

	$\dot{m}$ (g/s)	T (K)
Oxidizer stream (O_2)	362	121
Fuel stream (CH_4)	129	279
Cooling film (CH_4)	298	280

the chamber walls in order to reduce the heat load on the optical access windows. Table 3 summarizes the mass flow rates as well as the corresponding temperatures for all injected streams of the simulated load point. These inflow conditions are similar to the ones typically used in main- and upper stage rocket engines.

One of the characteristic features of the BKN is its large optical access window that can be used for flame visualization via OH* and CH* radiation imaging. Furthermore, the experimental data set includes temperature as well as static and unsteady pressure measurements near the chamber wall.

4.2 Numerical Setup

As TASCOM3D requires block-structured grids, the geometry had to be slightly simplified. This mainly concerns the supply system of the methane cooling film at the outer walls. In the experiment, it consists of two acoustically decoupled plenum chambers that are connected through 62 injection tubes. The second plenum supplies the cooling film to the combustion chamber using an annulus gap. Because it was not possible to resolve the 62 injection tubes by the computational grid, only the latter plenum is considered in the simulation and an annulus is used to model its inflow.

The computational grid consists of approximately 12.3 million volumes. The injector including the recess and the cooling film supply system account for approximately 11.6% of the cells. All other volumes discretize the combustion chamber and the nozzle. In radial direction, the oxidizer post tip is resolved by 20 volumes.

As the experiment provides temperature measurements at the chamber wall, a temperature profile is specified as boundary condition. The injector walls and the faceplate are assumed to be adiabatic. At the inlets, turbulent velocity profiles using the 1/7th power law [37] are specified in a way to fulfill the mass flow rates given in Table 3. It is worth noting that, due to the geometrical simplifications, the temperature of the incoming cooling film had to be slightly adjusted in order to ensure correct temperatures at the combustion chamber entry. A supersonic outflow boundary condition is imposed at the outlet.

Using a time step of $\Delta t = 1 \times 10^{-8} s$, approximately two flow-through times (FTTs) are simulated on 8181 cores. The mean axial velocity obtained from the simulation is $\bar{u} \approx 38.5\ \mathrm{m/s}$ and thus one FTT has a duration of approximately 8.36

ms. Hence, the simulated time period is considerably lower than the sampling period of 0.4 s in the experiment. The relatively large chemical kinetic scheme makes it infeasible to simulate such a long time span.

4.3 Instantaneous Flow Field

The experiment is characterized by intermittent excitation of the first longitudinal mode of the chamber. This can be seen in the high-pass filtered pressure signal at a location 34.5 mm downstream of the faceplate shown in Fig. 2. These bursts of oscillatory behavior are assumed to be caused by the operation of the injector at bistable operating conditions [38]. Thus, small perturbations are sufficient to trigger the switch between different states of combustion, which makes a numerical simulation extremely difficult.

Figure 3 shows simulated instantaneous temperature distributions at two different times. The two snapshots are approximately one FTT apart. The black lines denote stoichiometric mixture fractions and thus indicate the flame position. As can be seen, completely different flames occur at the different times, thus also being an indication for intermittent behavior. In both cases, typical Kelvin-Helmholtz instabilities develop downstream of the recessed oxidizer post tip, inducing small scale vortices. In addition, local flame extinction occurs in the front part due to shear layer instabilities.

At $t = 8.1$ ms, the cold oxygen core breaks down at around $x \approx 60$ mm. In the front part, the developing flame is thin and wrinkled. However, in contrast to similar model rocket combustion chambers (such as the PennState preburner combus-

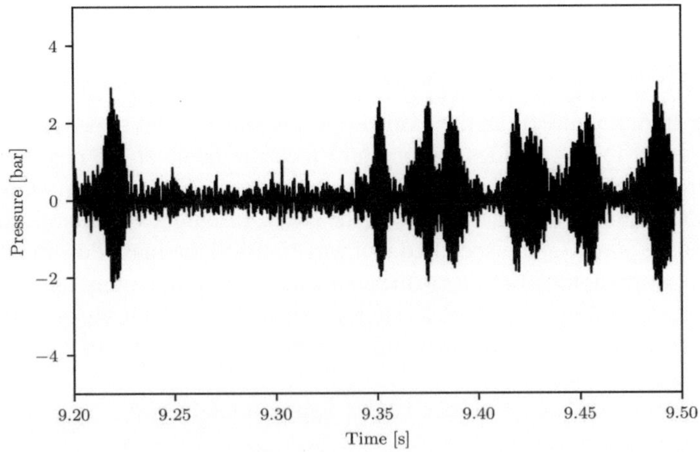

Fig. 2 Experimental high-pass filtered dynamic pressure signal 34.5 mm downstream of the faceplate

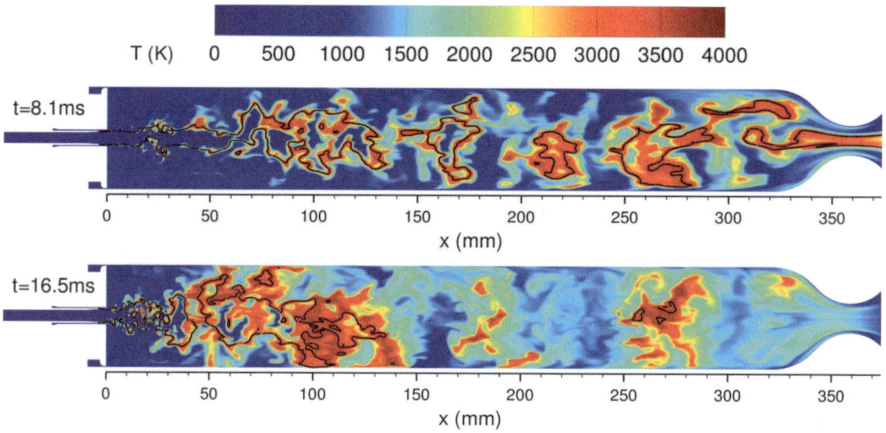

Fig. 3 Instantaneous temperature distribution in the cross section $z = 0$ mm at two different times that are approximately one FTT apart. Black lines denote the stoichiometric mixture fractions, indicating the flame position

tor [17]), no homogeneous temperature distribution is achieved in the rear part. Instead, cold, unburnt regions alternate with hot regions where combustion still takes place. Even in the nozzle, there is combustion and a significant amount of unburnt oxygen leaves the system. This behavior could not be observed in the experiment though.

At $t = 16.5$ ms, the cold oxygen core disintegrates much faster. The flame is mostly in the first half of the combustion chamber and in the rear part there are hardly any regions with stoichiometric conditions. Only traces of oxygen leave the combustion chamber. For $x > 150$ mm, cold and hot regions are still alternating. This is caused by the unburnt methane and the cooling film, which is approximately 70% of the total methane mass flux.

4.4 Flame Characteristics

Using the large optical access window in the experiment the flame topology is investigated using OH^* and CH^* radiation imaging. As the CH radical is not part of the applied chemical mechanism, the following analysis is restricted to OH^* radiation. By using the equilibrium filtered radiation model of Fiala and Sattelmayer [39], the concentration of the OH^* molecule can be determined from the OH radical. In addition, a line-of-sight integration along one axis is performed in order to mimic the experimental imaging. Thereby, absorption was neglected, which somewhat restricts the comparability.

Figure 4 shows a comparison of the instantaneous and the mean OH^* radiation images between experiment and simulation. The upper part of that figure depicts the

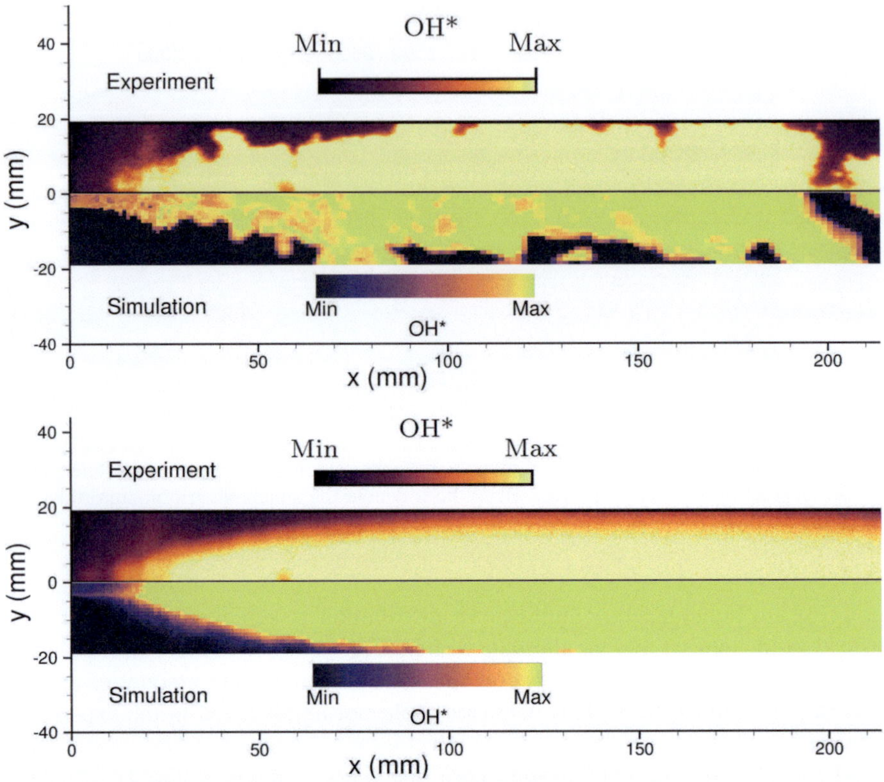

Fig. 4 Comparison of instantaneous (top) and mean (bottom) OH* flame radiation images for the BKN. Experimental data from [5]

instantaneous images. The turbulent nature of the flame is quite similar, especially in the first third of the combustion chamber. At the chosen point in time, the experimental flame front disintegrates at around $x \approx 200$ mm. A similar behavior can also be seen in the simulation.

Images of the averaged OH* radiation are depicted in the lower part of Fig. 4. The simulation correctly predicts the mean opening angle of the flame. Please note that the flame is significantly longer than the optical access window. A good agreement between experiment and simulation exists, both for the instantaneous and the mean OH* radiation.

4.5 Combustion Chamber Pressure

For the given load point, the observed experimental mean pressure is 66.8 bar. In the simulation, a mean pressure of 69.5 bar is obtained. Thus, the relative error is just slightly larger than 4%.

As mentioned above, both the simulation and the experiment show a highly intermittent behavior. The sampling period in the experiment is 0.4 s. Several time intervals that feature an excitation of the first longitudinal mode of the chamber can be identified [5], see also Fig. 2. These alternate arbitrarily with time intervals that are characterized by small amplitudes. This behavior makes it extremely difficult for the numerical simulation to capture all effects, especially as only approximately two FTTs could be simulated. In the first, more stable part of the simulation, small pressure amplitudes occur. This state is similar to the excited state in the experiment. The second part is characterized by high pressure amplitudes. However, compared with the experiment, this more excited part is overestimated in the simulation. Hence, the investigations in this section are based on the more stable part of the simulation. Figure 5 shows the pressure frequency spectra at a location 34.5 mm downstream of the faceplate. This location corresponds to the first unsteady pressure sensor in the experiment [5]. A good agreement with the experiment is achieved as the frequencies of the first and second longitudinal mode, denoted by *1L* and *2L* in Fig. 5, match very well. The frequency of the first longitudinal mode is around 1.3 kHz. On the other hand, the amplitudes are overestimated. At that location, no further eigenmodes could be identified, both in the experiment and in the simulation. Nonetheless, radial modes also exist in this test case. In the simulation, the first radial mode is observed near the faceplate (not shown here), while that mode is not visible in the experimental data. This could be due to the fact that radial (and transverse) modes have their highest amplitudes near the faceplate [40] and are damped quickly thereafter.

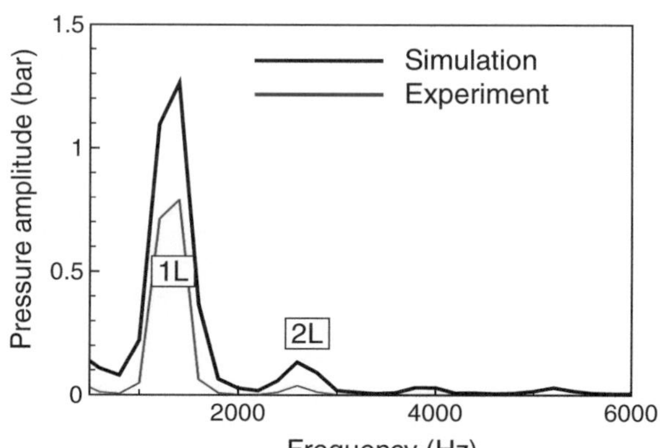

Fig. 5 Comparison of experimental and simulated frequency spectra at a location 34.5 mm downstream of the faceplate at the chamber wall. The terms *1L* and *2L* denote the first and second longitudinal mode, respectively

4.6 Methane Combustion at Elevated Pressures

To focus the analysis on the combustion process, the thermochemical state space is investigated. Figure 6 shows scatter plots of instantaneous temperatures versus mixture fractions taken from volumes in the front ($x < 50$ mm) and the rear part ($x > 50$ mm) of the combustion chamber. The data is extracted at a single snapshot equal to the one depicted in the upper part of Fig. 3. Volumes close to the wall and in the nozzle are blanked. In addition, chemical equilibrium is indicated as a reference by the red line. In the front part, there is a significant number of cells far from equilibrium. At positions with mixed gas but low temperatures the flame is either extinguished or the gas is not heated fast enough to enable ignition. Such finite-rate chemistry effects are enhanced by low injection temperatures, slow chemistry (it is much more pronounced with methane compared to hydrogen) and strongly lean or fuel-rich conditions. Flame extinction is caused by shear layer instabilities near the faceplate and flame stretching. For $x > 50$ mm combustion is completely different. At all positions, chemistry is close to chemical equilibrium, flame extinction does not occur and, especially for lean mixtures, chemistry is in chemical equilibrium. At fuel-rich conditions, there are slight deviations from chemical equilibrium, which may be due to strain or finite-rate chemistry. Especially the results for the front part highlight the necessity for detailed chemistry which is able to reproduce these processes accurately but is computationally expensive. The results presented are in agreement with the results of the purely numerical REST-HF10 test case [9], which uses the same injection head. It demonstrates that, despite the high pressures and the according fast chemistry, non-equilibrium effects occur, in particular under fuel-rich locations.

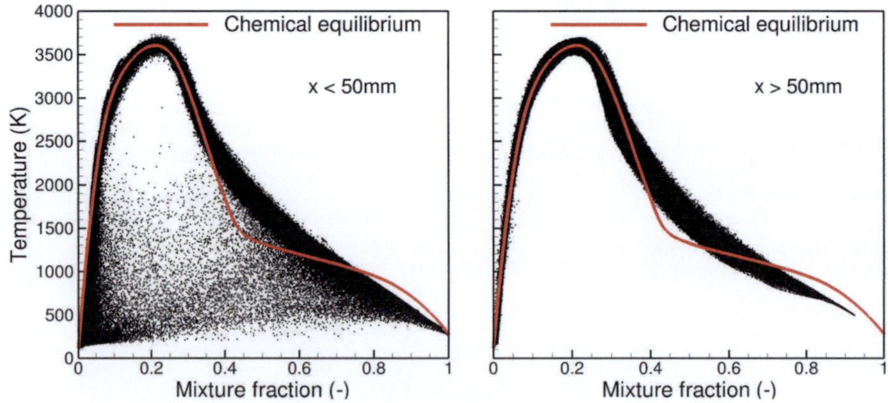

Fig. 6 Scatter plots of instantaneous temperature versus mixture fraction in the front (left) and rear part (right) of the combustion chamber. The red line denotes chemical equilibrium

5 Concluding Remarks

In this paper, a numerical investigation of a sub-scale methane-oxygen rocket combustion chamber with transcritical injection is presented. The simulations are performed using the in-house code TASCOM3D with a hybrid RANS/LES method with a spatial discretization up to fifth order. In addition, finite-rate chemistry is employed.

The investigated test case is highly intermittent, which makes numerical simulations quite challenging. Nevertheless, a good agreement with the experimental data is obtained for the considered load point. On the one hand, the comparison with the experimental OH* radiation images reveal a similar flame shape and opening angle. On the other hand, the frequencies of the first two longitudinal eigenmodes match very well. However, the amplitudes of those modes are overestimated. The same holds for the mean chamber pressure, which is predicted 4% too high. Both in the experiment and in the simulation, the BKN shows an intermittent behavior, which is probably caused by the operation of the injector at bistable operating conditions. Furthermore, flame extinction is observed near the faceplate, demonstrating the necessity for detailed chemistry which is able to account for non-equilibrium effects.

In addition to the simulation of the BKN, the scalability of the used code is discussed. TASCOM3D demonstrates an excellent weak scaling performance which is only limited by an increase of MPI overhead. Also, the strong scaling behavior shows high efficiencies. This is due to the reduced workload per core, while at the same time the MPI communication has a small contribution only. However, as the amount of MPI communication increases, the efficiency decreases as well.

Acknowledgements Part of the present work was conducted within the framework of the DLR project Advanced Methods for reusable Aerospace vehicle DEsign Using artificial intelligence and interdisciplinary numerical Simulation (AMADEUS). The authors gratefully acknowledge the Gauss Centre for Supercomputing e.V. (www.gauss-centre.eu) for funding this project by providing computing time on the GCS Supercomputer HAWK at High Performance Computing Center in Stuttgart (HLRS) (www.hlrs.de).

References

1. Haidn, O.: Advances on propulsion technology for high-speed aircraft **1**, 1 (2008)
2. Iannetti, A., Girard, N., Tchou-Kien, D., Bonhomme, C., Ravier, N., Edeline, E.: In: 7th European Conference for Aeronautics and Aerospace Sciences (EUCASS) (2017). https://doi.org/10.13009/EUCASS2017-537
3. Simontacchi, P., Blasi, R., Edeline, E., Sagnier, S., Ravier, N., Espinosa-Ramos, A., Breteau, J., Altenhoefer, P.: In: 8th European Conference for Aeronautics and Space Sciences (EUCASS) (2019). https://doi.org/10.13009/EUCASS2019-743
4. Asakawa, H., Nanri, H., Aoki, K., Kubota, I., Mori, H., Ishikawa, Y., Kimoto, K., Ishihara, S., Ishizaki, S.: In: De Luca, L.T., Shimada, T., Sinditskii, V.P., Calabro, M. (eds.), Chemical Rocket Propulsion: A Comprehensive Survey of Energetic Materials, pp. 463–487. Springer International Publishing, Cham (2017). https://doi.org/10.1007/978-3-319-27748-6_19

5. Martin, J., Armbruster, W., Suslov, D., Stützer, R., Hardi, J.S., Oschwald, M.: Aerospace **9**(8) (2022). https://doi.org/10.3390/aerospace9080410
6. Shur, M.L., Spalart, P.R., Strelets, M.K., Travin, A.K.: Int. J. Heat Fluid Flow **29**(6), 1638 (2008). https://doi.org/10.1016/j.ijheatfluidflow.2008.07.001
7. Gritskevich, M.S., Garbaruk, A.V., Schütze, J., Menter, F.R.: Flow Turbul. Combus. **88**(3), 431 (2012). https://doi.org/10.1007/s10494-011-9378-4
8. Keller, R., Gerlinger, P.: In: *54th AIAA Aerospace Sciences Meeting (AIAA Paper 2016-2148)* (2016). https://doi.org/10.2514/6.2016-2148
9. Lechtenberg, A., Gerlinger, P.: In: 9th European Conference for Aeronautics and Aerospace Sciences (EUCASS) (2022). https://doi.org/10.13009/EUCASS2022-6075
10. Gerlinger, P., Brüggemann, D.: Int. J. Numer. Methods Fluids **24**(10), 1019 (1997)
11. Gerlinger, P., Möbus, H., Brüggemann, D.: J. Comput. Phys. **167**(2), 247 (2001). https://doi.org/10.1006/jcph.2000.6671
12. Gerlinger, P., Stoll, P., Kindler, M., Schneider, F., Aigner, M.: Aerosp. Sci. Techno. **12**(2), 159 (2008). https://doi.org/10.1016/j.ast.2007.04.003
13. Gerlinger, P.: J. Comput. Phys. **231**(5), 2199 (2012). https://doi.org/10.1016/j.jcp.2011.10.024
14. Simsont, Y.H., Gerlinger, P.: Int. J. Numerical Methods Fluids **82**(7), 417 (2016). https://doi.org/10.1002/fld.4224
15. Lempke, M., Keller, R., Gerlinger, P.: Int. J. Numer. Methods Fluids **79**(9), 437 (2015). https://doi.org/10.1002/fld.4059
16. Seidl, M., Aigner, M., Keller, R., Gerlinger, P.: J. Supercrit. Fluids **121**, 63 (2017). https://doi.org/10.1016/j.supflu.2016.10.017
17. Seitz, T., Gerlinger, P.: In: AIAA Scitech 2021 Forum (AIAA Paper 2021-1543) (2021). https://doi.org/10.2514/6.2021-1543
18. Soave, G.: Chem. Eng. Sci. **27**(6), 1197 (1972). https://doi.org/10.1016/0009-2509(72)80096-4
19. Jameson, A., Yoon, S.: AIAA J. **25**(7), 929 (1987). https://doi.org/10.2514/3.9724
20. Shuen, J.S., Yoon, S.: AIAA J. **27**(12), 1752 (1989). https://doi.org/10.2514/3.10331
21. Shuen, J.S.: J. Comput. Phys. **99**(2), 233 (1992). https://doi.org/10.1016/0021-9991(92)90205-D
22. Gear, C.W.: Numerical Initial Value Problems in Ordinary Differential Equations. Prentice Hall (1971)
23. Yang, V.: Liquid-propellant rocket engine injector dynamics and combustion processes at supercritical conditions. Pennsylvania State University Park Department of Mechanical Engineering, Technical report (2004)
24. Shima, E., Kitamura, K.: AIAA J. **49**(8), 1693 (2011). https://doi.org/10.2514/1.J050905
25. Wilcox, D.C.: AIAA J. **46**(11), 2823 (2008). https://doi.org/10.2514/1.36541
26. Chapman, D.R.: AIAA J. **17**(12), 1293 (1979). https://doi.org/10.2514/3.61311
27. Reynolds, W.C.: In: Lumley, J.L. (ed.) Whither Turbulence? Turbulence at the Crossroads, pp. 313–343. Springer, Berlin, Heidelberg (1990). https://doi.org/10.1007/3-540-52535-1_52
28. Girimaji, S.: Combus. Sci. Technol. **78**(4–6), 177 (1991). https://doi.org/10.1080/00102209108951748
29. Baurle, R.A., Alexopoulos, G., Hassan, H.A.: J. Propuls. Power **10**(4), 473 (1994). https://doi.org/10.2514/3.23797
30. Gerlinger, P.: Combust. Sci. Technol. **175**(5), 841 (2003). https://doi.org/10.1080/00102200302410
31. Gerlinger, P., Noll, B., Aigner, M.: Progress in computational fluid dynamics. Int. J. **5**(6), 334 (2005). https://doi.org/10.1504/PCFD.2005.007066
32. Kindler, M., Gerlinger, P., Aigner, M.: In: Nagel, W.E., Kröner, D.B., Resch, M.M. (eds.) High Performance Computing in Science and Engineering '11, pp. 259–272. Springer, Berlin, Heidelberg (2012). https://doi.org/10.1007/978-3-642-23869-7_20
33. Traxinger, C., Zips, J., Pfitzner, M.: In: 8th European Conference for Aeronautics and Space Sciences (EUCASS) (2019). https://doi.org/10.13009/EUCASS2019-731

34. Schmitt, T., Marchal, D., Ducruix, S., Cuenot, B.: In: 9th European Conference for Aeronautics and Aerospace Sciences (EUCASS) (2022). https://doi.org/10.13009/EUCASS2022-4460
35. Martin, J., Armbruster, W., Stützer, R., General, S., Knapp, B., Suslov, D., Hardi, J.: Case Stud. Thermal Eng. **28**, 101546 (2021). https://doi.org/10.1016/j.csite.2021.101546
36. Martin, J., Armbruster, W., Stützer, R., Suslov, D., Hardi, J., Oschwald, M.: In: 9th European Conference for Aeronautics and Aerospace Sciences (EUCASS) (2022). https://doi.org/10.13009/EUCASS2022-4688
37. De Chant, L.J.: Appl. Math. Comput. **161**(2), 463 (2005). https://doi.org/10.1016/j.amc.2003.12.109
38. Martin, J., Armbruster, W., Stützer, R., Suslov, D., Hardi, J., Oschwald, M.: Int. J. Spray Combust. Dyn. **0**(0) (2023). https://doi.org/10.1177/17568277231172302
39. Fiala, T., Sattelmayer, T.: CEAS Space J. **8**, 47 (2016). https://doi.org/10.1007/s12567-015-0107-z
40. Oefelein, J.C., Yang, V.: J. Propuls. Power **9**(5), 657 (1993). https://doi.org/10.2514/3.23674

HPC in Fluid-Mechanics - Materials Synthesis and Hydrogen Combustion

Seung-Jin Baik, Linus Engelmann, Patrick Wollny, Robert Externbrink, and Andreas M. Kempf

Abstract Large-eddy simulations (LES) of reactive flows are presented, analysing various topics of concern to the combustion community, i.e. multi-regime combustion, solid fuel combustion and nanoparticle synthesis. Regarding multi-regime combustion, a novel burner system is simulated using highly resolved LES and tabulated chemistry. Good agreement with experimental data was observed, while the simulation provides a high-fidelity database for future model development and validation. For solid fuel combustion multi-dimensional flamelet models for LES are developed to represent exhaust gas recirculation and achieve precise pollutant prediction. Relatively new measures to reduce carbon dioxide and nitrogen dioxide emissions, i.e. co-firing of coal and ammonia and the use of a swirl-stabilized low-NO_x burner, are investigated numerically and compared against experimental data. Concerning nanoparticle synthesis, a model of amorphous composite nanoparticles is developed for the use as anode material for lithium-ion batteries. Our novel approach pairs the advantages of a spatial-resolving finite volume simulation with detailed models for formation and growth, known from coarse grained molecular dynamics (CGMD) and grants new insights in the overall process of nanoparticle synthesis. More particular, we developed a new multiscale model in the framework of CFD, which features surface reactions, condensation and phase coupling. The Chair of Fluid Dynamics at the University of Duisburg-Essen is pursuing an approach of bundling computational time requirements and applying for them jointly to reduce review efforts

S.-J. Baik · L. Engelmann · P. Wollny · R. Externbrink (✉) · A. M. Kempf
Chair of Fluid Dynamics, University of Duisburg-Essen, Duisburg, Germany
e-mail: robert.externbrink@uni-due.de

S.-J. Baik
e-mail: seung.baik@uni-due.de

L. Engelmann
e-mail: linus.engelmann@uni-due.de

P. Wollny
e-mail: patrick.wollny@uni-due.de

A. M. Kempf
e-mail: andreas.kempf@uni-due.de

T. Ludwig et al. (eds.), *High Performance Computing in Science and Engineering '23*,
https://doi.org/10.1007/978-3-031-91312-9_29

1 Scaling of the Used CFD-Codes

1.1 Scaling of PsiPhi

The *PsiPhi* CFD code, originally developed by Kempf, has been improved, extended, accelerated, and adapted for efficient computations on tens of thousands of processor cores for this work. The code relies on a Cartesian equidistant cuboid computational domain, where cells inside the geometry are blocked by the immersed boundary approach. The Cartesian grid allows for good vectorization performance by avoiding non-sequential memory access, which often outweighs the additional computational effort of the immersed boundary method. The code is written in Fortan 90/95, which has proven since many years to be a well-suited programming language for engineering purposes due to its simple syntax and the good computational performance of the resulting code. The code is parallelized with the message passing interface (MPI), the overhead of the communication routines is partially compensated by using non-blocking communication overlayed with computational work.

The parallel scaling behavior of the code has been tested on the IBM Blue Gene/Q-System JUQUEEN at the Juelich Supercomputing Centre (JSC) using up to 128,000 cores and HAWK at the (HLRS) using up to 32,768 cores, as shown in Fig. 1. The scal-

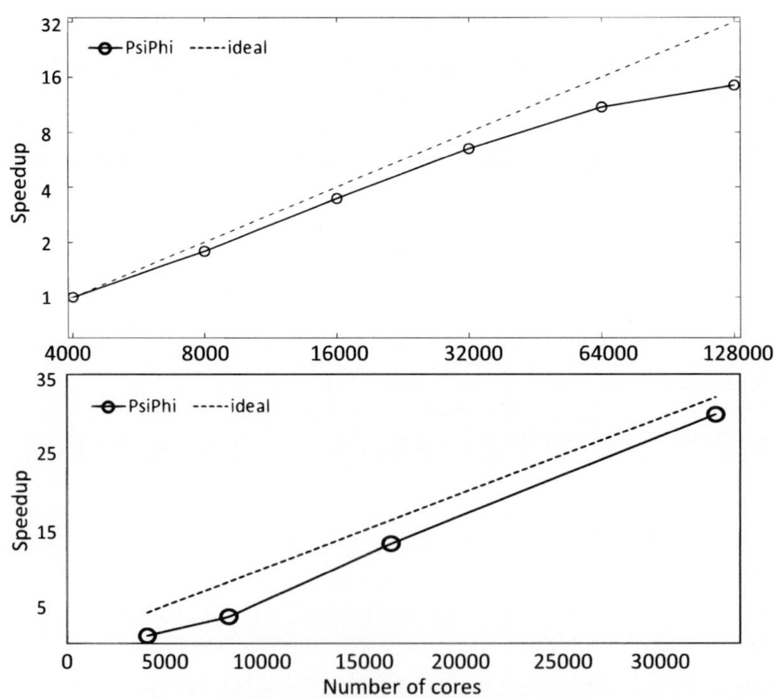

Fig. 1 Results of the parallel scaling tests for PsiPhi from JUQUEEN (top) and HAWK (bottom)

ing test was performed for the 1.6 billion cell computations of the Cambridge/Sandia bluff-body burner, as discussed in Proch et al. [1], and was computed on a progressively increasing number of processor cores. The scaling behavior is comparable to other CFD codes tailored for massively parallel computations, for the 64,000 processor cores that were used for the productive runs the parallel efficiency is around 70% with respect to 4000 processor cores at the scaling test from JUQUEEN. In both tests, cash effects are found until around 8000 cores, and after that, they show a good scaling performance. As the Mach-numbers in the burner are below 0.1, the incompressible version of the code is more efficient overall in spite of its lower parallel efficiency, as the required number of time steps for the density-based solution algorithm would be very large. For cases with higher Mach numbers, the density-based version of the code can be used. This version avoids the communication-intensive Poisson solver and thereby improves the parallel efficiency by approximately 10%.

1.2 Scaling of OpenFOAM

OpenFOAM is an established open-source software framework for CFD with proven performance and scalability. Nevertheless, the solvers used were subjected to a performance test.

Figure 2 shows the results of the strong scaling tests for the compressible FRC solver in OpenFOAM, using an analytically reduced reaction mechanism Lu19. The results were obtained on MagnitUDE in Duisburg. The test case was a simulation of an atmospheric jet flame with a small problem size of 1.77 million cells. For the commonly used range of 20 to 4k cells per core, the scaling of the solver is nearly ideal. The results are mostly representative for the usage of FRC with a reduced

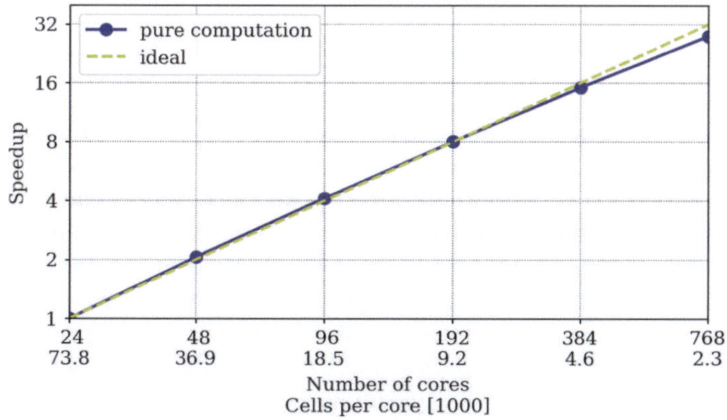

Fig. 2 Strong scaling behavior of finite rate chemistry LES with OpenFOAM on MagnitUDE. This data was obtained with a problem size of 1.77 million cells

mechanism, where the main part of the computational cost is due to the calculation of the chemistry. This requires no MPI communication between subdomains, leading to excellent scaling performance.

2 Large Eddy Simulation of Nanoparticle Synthesis in Ethanol/2EHA Spray Flames

2.1 Introduction

The SpraySyn burner [2], specifically developed within the framework of the DFG priority program SPP 1980 by research groups, is used to explore the synthesis of nanoparticles from spray flames using a combination of diagnostic methods and simulations. In this study, we focus on investigating the production of iron oxide nanoparticles through the utilization of iron nitrate dissolved in a mixture of ethanol and ethyl hexanoic acid. Within the SpraySyn burner, the precursor solution containing the nanoparticles is injected and atomized using an air-blast nozzle, surrounded by a pilot flame. It should be noted that the solvent itself possesses combustible properties, leading to the rapid evaporation of droplets under high heating rates within an extremely reactive environment. To simulate the flame for nanoparticle synthesis, large eddy simulations (LES) are employed. The behavior of liquid droplets is described using Lagrangian particles, while the combustion of the gaseous phase is modeled using the flamelet-generated manifold approach combined with the artificial flame thickening technique. The dynamics of iron oxide nanoparticles are modeled based on their number, surface, and volume concentration using a mono-modal model.

2.2 Numerical Setup

The SpraySyn burner is composed of a slender tube ($D_{in/out} = 0.4/0.7$ mm) for the precursor solution and a nozzle for the dispersion gas ($D = 1.5$ mm), followed by a broad sinter matrix ($D = 70$ mm) that provides hot pilot flame products and sheath gas. The precursor solution consists of 0.05 ml of iron(III)-nitrate nonahydrate per liter of a mixture of ethanol and 2-ethyl hexanoic acid and is injected at a volume flow rate of 2 ml/min at ambient pressure. The entire process can be separated into three stages: (I) the evaporation of the precursor-solution mixture in the shear layer between the dispersion gas and the hot pilot flame, (II) the turbulent combustion arising from the evaporated fuel and oxidizer mixture, and (III) the nucleation, growth, and coalescence of Fe_2O_3 nanoparticles, as depicted schematically in Fig. 3. For further details on the SpraySyn burner and in situ measurements of the nanoparticle synthesis, the previous experiment work [2, 3] should be referred to.

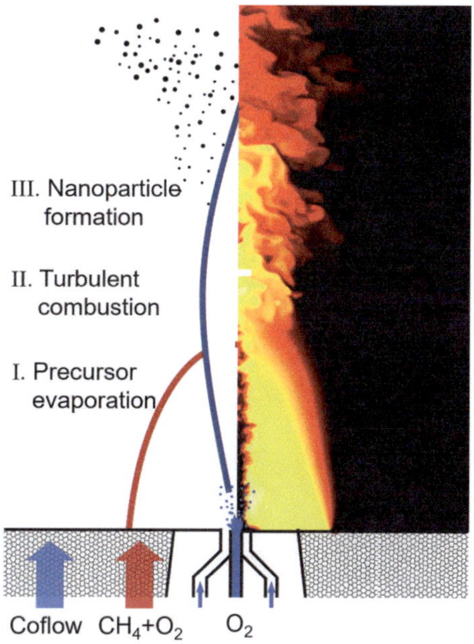

Fig. 3 A schematic drawing of the burner inlet and the main processes

The numerical simulations were carried out utilizing the in-house code *PsiPhi*, which has been successfully validated in various DNS and LES studies. The computational domain $120 \times 40 \times 40$ mm^3 was used for the simulations, consisting of a cartesian grid with an equidistant spacing of $\Delta = 0.125$ mm and a total of $960 \times 320 \times 320$ grid nodes, comprising approximately 100 million cells. The simulation of 0.05 s required approximately 0.4 million cores hours per simulation.

2.3 *Results*

Contour plots in Fig. 4 show the instantaneous (left) and mean (right) axial velocity, temperature, evaporated precursor/solvent mass source, and reaction rate of progress species in the burner cross-section. The region from the nozzle exit up to 50 mm axial distance is referred to as the upstream region. This region represents a zone where the mixing of the hot pilot flame and the dispersion gas containing liquid droplets occurs. The shear layer between fast dispersion gas and the co-flow is responsible for the creation of turbulent structures, which result in increased heat penetration. The liquid droplets evaporate in a long, narrow region up to 50 mm above the burner. The average evaporation rate $d\dot{M}$ indicates that most of the evaporation occurs near this region as shown in Fig. 4c.

Figure 5 shows contour plots of the predicted particle number concentration *NN*, surface area concentration *NA*, and volume concentration *NA* in a burner cross section.

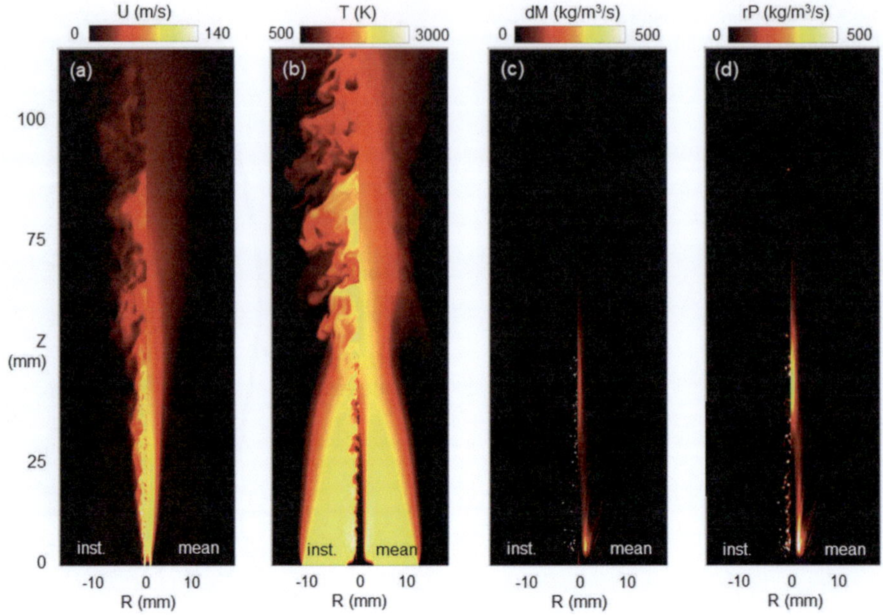

Fig. 4 Instantaneous (left) and mean (right) contour plots of **a** the gas phase velocity u, **b** temperature T, **c** evaporated precursor/solvent dM, and **d** reaction rate of progress species rP in a burner cross-section

The particle number concentration *NN* initially increases due to the formation of monomer particles. The velocities in the downstream region are lower, resulting in longer residence times, which give the particles more time to coagulate, leading to a decrease in the number concentration. The surface area concentration and volume concentration rise sharply in the shear layer between dispersion and hot pilot flame due to the nucleation of monomers, but the surface area concentration *NA* decreases in the high temperature and long residence time region due to sintering, as depicted in Fig. 5b, c.

Figure 6 shows the comparison between the volume-averaged particle diameter obtained by simulation (lines) and the count median diameters obtained by various measurements (dots), along the centerline of the flame. SMPS, TEM, and PMS stand for scanning mobility particle sizing, Transmission electron microscopy, and particle mass spectrometry respectively. Readers can refer to the recent work of Suleiman et al. [2, 3] for more detailed information on measurements. Until $Z = 60$ mm from the burner surface, the simulation shows an underestimated nanoparticle diameter compared to measurements. However, further downstream, the simulated particle diameter shows faster growth than the measurements.

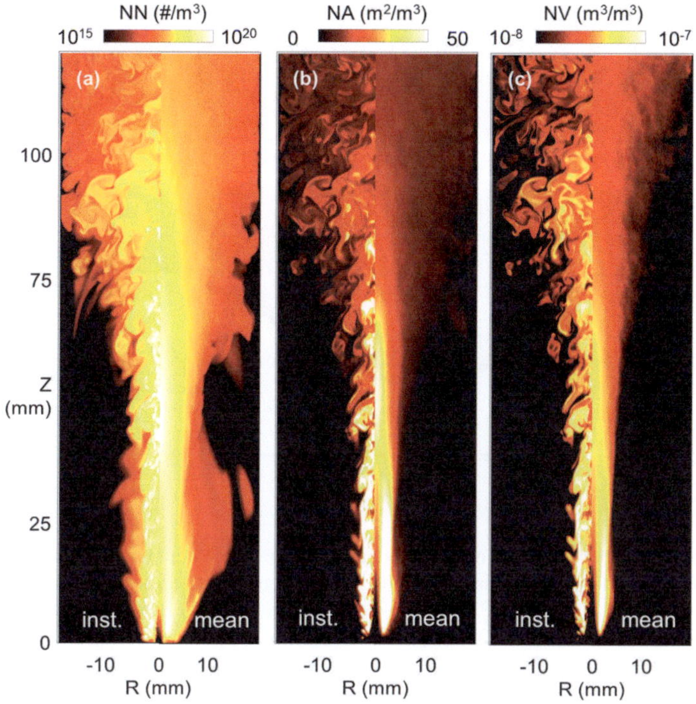

Fig. 5 Instantaneous (left) and mean (right) contour plots of **a** the particle number concentration NV, **b** surface area concentration NA, and **c** volume concentration NV in a burner cross-section

2.4 Conclusion

This study utilized a large eddy simulation (LES) technique to investigate nanoparticle synthesis with SpraySyn burner. A cost-effective monodisperse model was employed in conjunction with the LES approach to study the formation of Fe_2O_3 nanoparticles in a spray flame. The results demonstrated the ability of the model to predict nanoparticle generation at a suitable computational cost. Iron(III)-nitrate nonahydrate precursor, dissolved in ethanol/2-ethyl hexanoic acid solvent mixture, was observed to evaporate primarily in the shear layer between the hot pilot flame and dispersion gas. The primary source of Fe_2O_3 monomers was pre-calculated and tabulated with other gas-phase reactions using the premixed flamelet generated manifolds (PFGM) approach. The simulated volume averaged nanoparticle size was compared with the in-situ measurements performed by Suleiman [3]. The comparison demonstrated good agreement between the simulation and the experiment for nanoparticle size after 70 mm from the burner surface while the simulated particle diameter was under-predicted compared to measurements in the upstream region (0–50 mm from

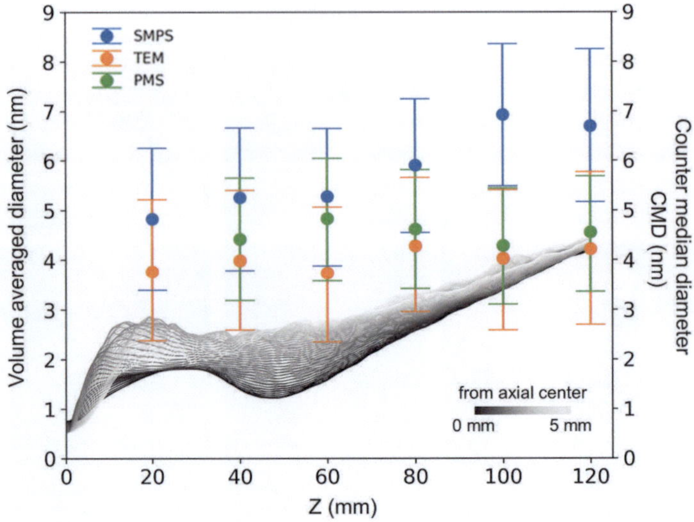

Fig. 6 Comparison between the volume-averaged particle diameter obtained by simulation (lines) and the count median diameters obtained by various measurements (dots) [3], along the centerline of the flame. SMPS, TEM, and PMS stand for scanning mobility particle sizing, Transmission electron microscopy, and particle mass spectrometry respectively

the burner surface). The results have been presented at the European Combustion Meeting 2023. We have published selected results to Applications in Energy and Combustion Science [4].

3 Large-Eddy Simulation of a Labscale Multi-Regime Burner System

3.1 Introduction

In practical applications, partial premixing of fuel and oxidizer, as well as recirculation of combustion products, result in complex combustion scenarios where multi-regime effects arise and a numerical representation of local reaction zones by purely premixed or purely non-premixed flame structures may not hold. A novel burner system is being simulated to investigate the fundamental characteristics of multi-regime combustion and to provide a basis for validating numerical models. This multi-regime burner (MRB) is specifically designed to produce flames with multi-regime characteristics while maintaining well-defined boundary conditions [5]. The importance of different flame zones, indicated by their population fractions, are found to evolve with downstream distance and show distinct differences between the two selected flames.

3.2 Numerical Setup

The simulation was performed with the in-house code *PsiPhi* that solves the governing equations in a low-Mach number finite-volume formulation. The spatial discretization relies on a central-differencing scheme for momentum and a total variation diminishing (TVD) scheme for the scalars, with the non-linear CHARM limiter applied. The solution is advanced in time with an explicit low-storage third order RungeÃ¢â‚¬â€œKutta scheme using a CFL number of 0.3. The geometry was described by an immersed boundary technique, continuity is ensured by a pressure-correction method embedded in a predictor-corrector scheme. A successive over-relaxated GaussÃ¢â‚¬â€œSeidel solver was applied for the solution of the Poisson-equation for the pressure. The code is parallelized using the distributed memory message passing interface (MPI), with non-blocking communication used to overlay the communication with computation. Equidistant cells with an edge-length of 100 μm were used. The last 25 mm of the burner geometry were included into the computational domain as immersed boundaries. This led to a total amount of 1.6 billion control volumes [1].

3.3 Results

3.3.1 Coarse Grid Study

"The burner in question was first examined using a grid of 0.4 mm size to investigate the case sensitivity on the grid. It was found that while overall the behavior near the burner cap was captured sufficiently, the downstream behavior was not predicted in a satisfying manner by the coarse grid, so that a refined study had to be conducted, requiring significant HPC resources. Figure 7 gives qualitative and quantitative insight into the simulation by showing the mixture fraction and temperature fields on the center plane (cross-section), as well as showing radial average and rms profiles at different axial locations. The coarse grid revealed to produce an excessive contraction, which affects the flame quantities notably after a distance larger than 30 mm from the burner cap.

3.3.2 Fine Grid Study

HPC resources are used to investigate the phenomenon of multi-regime combustion, the most dominant combustion mode in most technical applications, which is still only marginally studied. A partially flame-resolved calculation was performed using a cubic-isotropic mesh of 0.1 mm size. The simulation was performed on 65,536 cores using around 7 mio. CPUh. High order numerics using a employing a similar framework as mentioned above were used and the combustion was modeled using PFGM tables. The presented fields in Fig. 8 show the fuel mass fraction and the

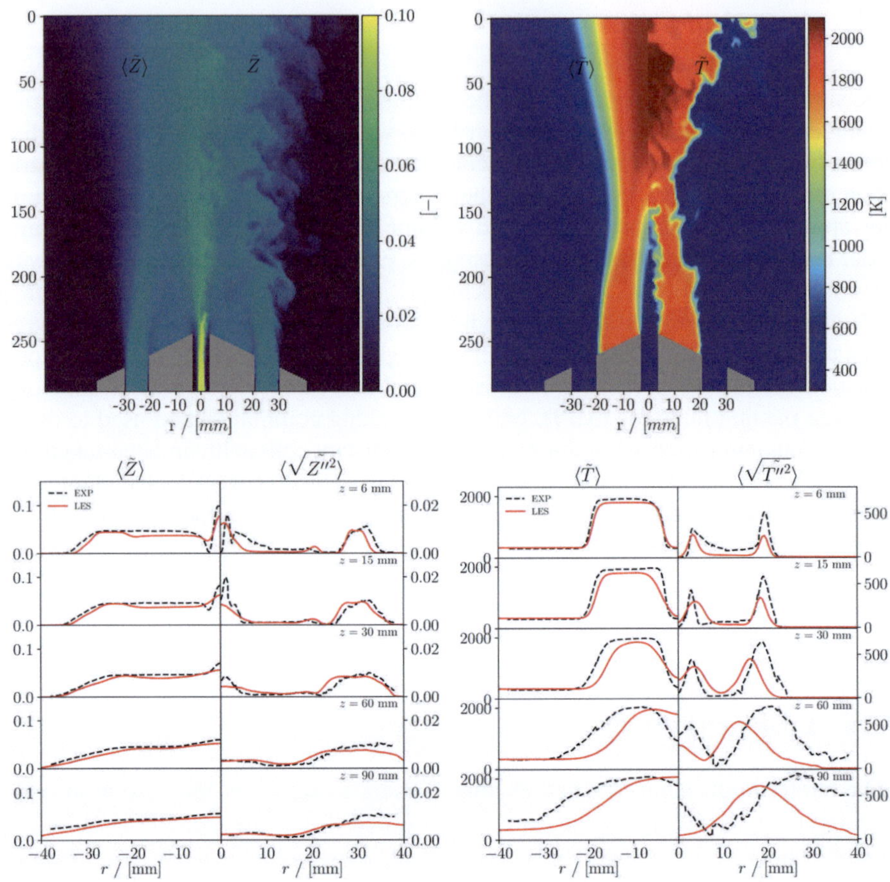

Fig. 7 Results of the coarse grid calculation. Top row: Mean and instantaneous mixture fraction and temperature fields, Bottom row: Radial profiles of mixture fraction and temperature at different axial locations

temperature for the burner cross-subsection. Both images reveal the finely resolved structures of the flame. Due to the high velocity of the inner jet, the inner flame features very small turbulent structures, which characterize the flame up to 80 mm away from the burner. These fine structures are missing in the coarse grid calculation, which leads to the strong difference of the inner-flame resolution, hence, stressing the need for HPC.

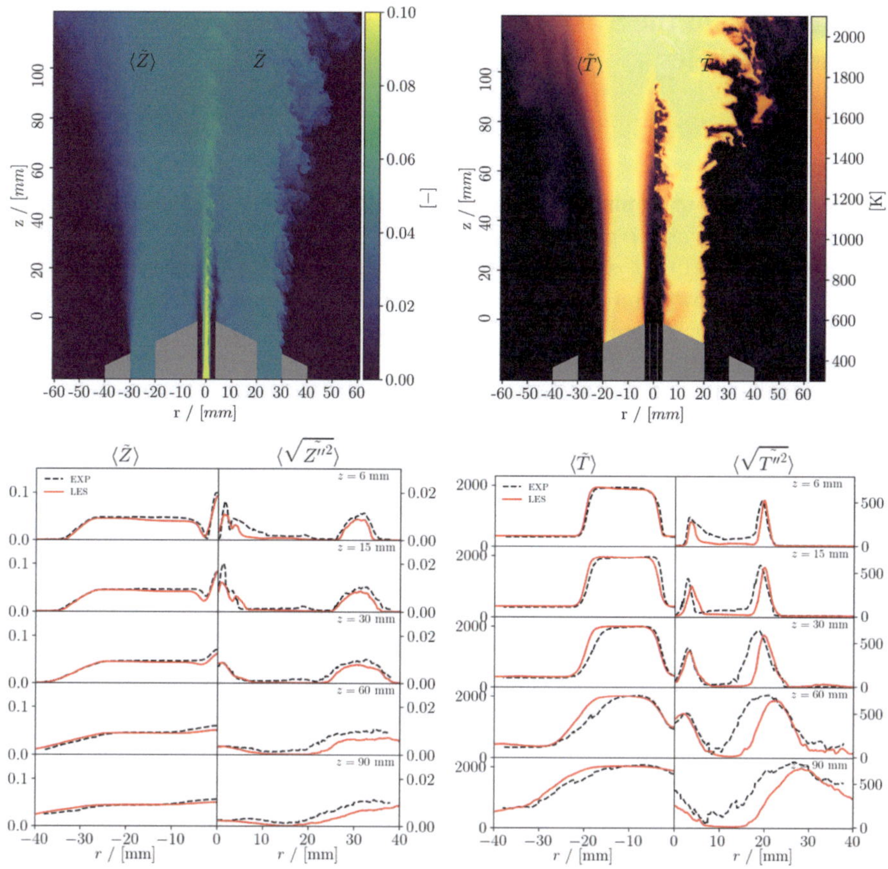

Fig. 8 Results of the fine grid calculation. Top row: Mean and instantaneous mixture fraction and temperature field, Bottom row: Radial profiles of mixture fraction and temperature at different axial locations

3.4 Conclusion

The calculation produced a high-fidelity database of a novel burner, which will be the basis for future model development and validation in the field of multi-regime combustion. Good agreement with the experimental database was observed, providing evidence for the power of the setup. Further diagnostics such as the flame index and the temperature locus diagram were employed to highlight the difference between the combustion modes, especially between the inner and the outer flame. Selected results have been presented at the TNF-workshop 2022, and published at Combustion and Flame [6].

4 Modeling of the Synthesis of Amorphous Composite Nanoparticles

4.1 Introduction

This project aims to model the synthesis of amorphous composite nanoparticles, intended for use as anode material in lithium-ion batteries. Our innovative approach combines the benefits of spatially-resolved finite volume simulation with detailed formation and growth models, derived from coarse-grained molecular dynamics (CGMD). This provides new insights into the overall process of nanoparticle synthesis. More specifically, we developed a new multiscale model within the framework of CFD, incorporating surface reactions, condensation, and phase coupling. It is based on the open source CFD library *OpenFOAM* and allows to simulate the complex formation of amorphous silicon nanoparticles in hot wall reactors.

4.2 Numerical Setup

The calculations are used to understand the process and to fine-tune the phase transition model, for which no comparable data exist. Since the reactor shows particle residence times of up to 20 s, an affordable resolution was chosen (about 6.5 million cells), in order to yield results with reasonable computational effort. For the decomposition kinetics of the precursor gas, a 10-step mechanism introduced by Ho was employed and a set of ordinary differential equations (ODEs) were obtained by applying a finite rate chemistry approach. A second set of ODEs was defined for the particle phase, represented by a bivariate moment method, including the phase transition models by surface reaction, condensation and nucleation. Both sets of equations were solved using the Rosenbrock solver for stiff ODEs and is of order O(2–3). Buoyancy effects are considered by volume forces in the momentum conservation and by hydrostatic pressure. The spatial derivatives are approximated by a second order central differencing scheme for the momentum equations and by total variation diminishing (TVD) schemes for the other equations. Time derivatives are discretized using the Crank-Nicolson scheme. The resulting partial differential equations were solved by a preconditioned bi-conjugate gradient solver with diagonal incomplete-LU preconditioner. The code is by default parallelized utilizing the distributed memory message passing interface (MPI) in a non-blocking manner. Simulations were performed using 12 nodes with 128 cores each, in order to reach a value of around 4230 cells per rank, which we found to be close to the maximum for linear scaling for this setups using OpenFOAM. (see Sect. 1.2).

4.3 Validation

Two experimentally measured cases with different silane mass fractions, wall temperatures and flow rates were simulated to validate the new model on the basis of the overall reactor system. As the reactor efficiency highly depends on the wall heat transfer, the temperature distribution in the reactor was given by an experimentally measured profile for both cases. For this reason, the temperature determined in each experiment was transferred to the simulation as a function of the axial distance. The instantaneous simulation results are shown on 2D cutting planes (z-normal) along the central axis for setup 1 (Fig. 9) and setup 2 (Fig. 10). The chaotic (laminar) flow field reveals two recirculation zones (see Ua and T in Figs. 9 and 10), which are formed by buoyancy force and differ in the area of heating (axial distance <1.5 m) and active cooling (axial distance >1.5 m). Due to the the fact that particles with longer overall residence time are captured within the lower recirculation zone, the biggest particle diameters (collision diameter dc and primary particle diameter dp) are formed here. A comparison of the BET-surface and Ã¢â,¬âŁœarea for setup 1 is exemplary shown by Fig. 11, where a good agreement with the experiments could be achieved by the model.

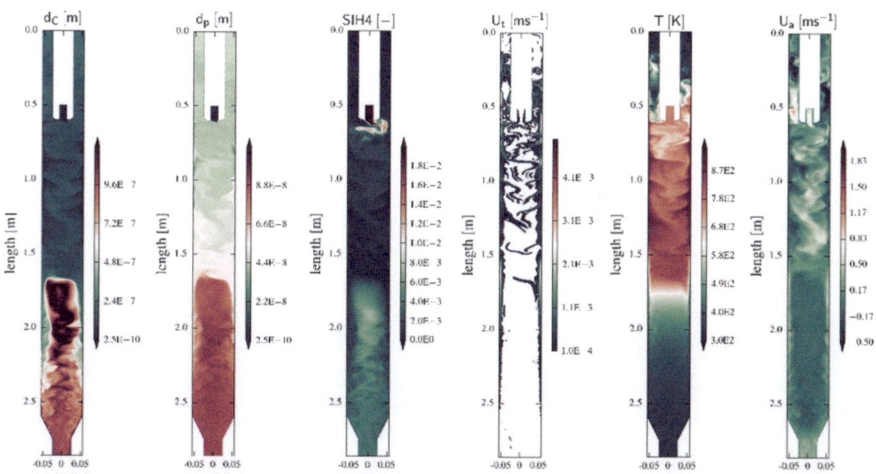

Fig. 9 Cutting planes of 3D domain of setup 1; from left to right: instantaneous collision diameter d_c, primary particle diameter d_p, silane mass fraction Y_{SiH4}, thermophoretic velocity U_t, temperature T, axial velocity U_a

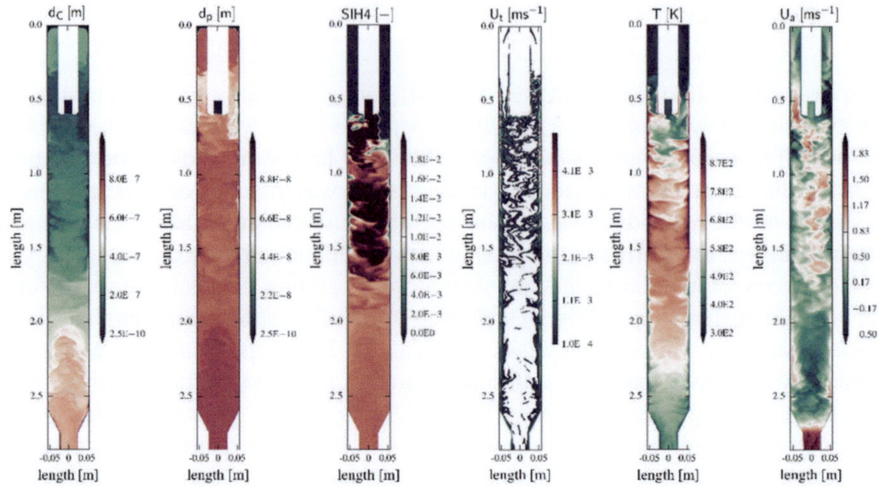

Fig. 10 Cutting planes of 3D domain of setup 2; from left to right: instantaneous collision diameter d_c, primary particle diameter d_p, silane mass fraction Y_{SiH4}, thermophoretic velocity U_t, temperature T, axial velocity U_a

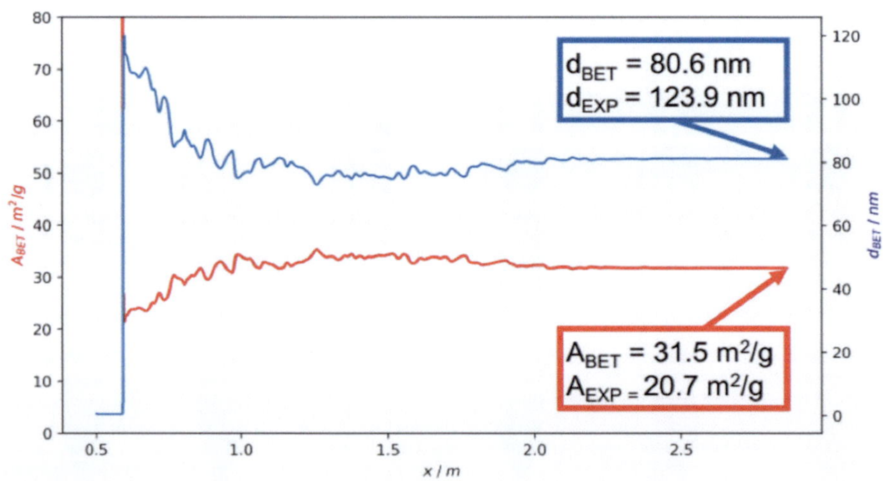

Fig. 11 Comparison of the BET-diameter and BET-surface along the centerline of the calculation (setup 1). The final state of the calculation data (d_{BET}, A_{BET}) is compared with the data measured in the experiment (d_{EXP}, A_{EXP})

4.4 Conclusion

A novel model approach for predicting the synthesis of amorphous nanoparticles has been applied in order to simulate the synthesis within a semi -industrial hot wall reactor. Two experimental setups with different silane mass fractions, wall temperatures, and flow have been considered for simulation at high-performance computing level. The simulations have been taken out at moderate temperatures (665 to $715°C$), while setup 1 has been proven to deliver a good agreement in comparison to the measured BET-surface and BET-area, and verifies the capabilities of the developed, computationally affordable model. Further simulations have to be performed to adjust the model parameters further and check the grid dependency of the dispersed phase. Selected results are published in Chemical Engineering Journal [7].

5 Conclusion and Future Work

The presented research results of the ongoing project DUEHAWK2021 in the domain of fluid dynamics relied on the resources and performance provided by the HPC HAWK. With the HPC, the investigation of complex problems in the realm of combustion is feasible, enabling the development of better models for cost restricted simulations and furthering the understanding of processes in cutting-edge technologies. The analyses carried out in this project rely on the computational resources of HPC for several reasons. The prediction of mixing, combustion, thermophysical state, and nanoparticle formation in various simulations has been compared to available experimental data. The results of DNS investigation shed light on the effect of turbulence on nanoparticle synthesis in flames. To create a high-fidelity database for multi-regime combustion and achieve the necessary accuracy, a problem size of 1.6 billion was required, necessitating high parallelization for affordable wall times. For the investigation of solid fuel combustion processes and nanoparticle syntheses, long physical times and for the latter a complex reaction mechanism and particle syntheses had to be simulated resulting in a high number of needed iterations and computational complexity.

The bundling of resource requirements in a joint project application has once again shown its effectiveness. Compared to submitting several applications, this not only reduced the time and effort required for the review process, but also generated synergy effects in the coordination and handling of the individual subprojects. In the current reporting period, a total of 26.6 million core-h were used.

As part of the ongoing project, additional research into current topics of interest in combustion will be carried out, concerning thermoacoustics. Thermoacoustic combustion instabilities remain a concern for the development of highly efficient gas turbines since they limit the operating range. To investigate the causes and possible prevention mechanisms of high-frequency azimuthal-mode thermoacoustics, a combustor showing self-excited thermoacoustics is planned to be developed.

Acknowledgements The authors gratefully acknowledge support through the Gauss Centre for Supercomputing e.V. (www.gauss-centre.eu) by funding this project (grant DUEHAWK2021) by providing computing time on the GCS Supercomputer HAWK at Hoechstleistungsrechenzentrum Stuttgart (www.hlrs.de). We would also like to express our profound gratitude for the computing time provided at UDE, FZ-Jülich, and LRZ Munich, and we are immensely thankful to the DFG for their financial support.

References

1. Proch, F., Domingo, P., Vervisch, L., Kempf, A.: Combust. Flame **180**, 321–339 (2017)
2. Schneider, F., Suleiman, S., Menser, J., Borukhovich, E., Wlokas, I., Kempf, A., Wiggers, H., Schulz, C.: Rev. Sci. Instrum. **90**, 085108 (2019)
3. Suleiman, S., Nanjaiah, M., Skenderovic, I., Rosenberger, T., Kunze, F., Wlokas, I., Kruis, F.E., Wiggers, H., Schulz, C.: J. Aerosol Sci. **158** (2021)
4. Baik, S.-J., Wollny, P., Nanjaiah, M., Wlokas, I., Kempf, A.: Appl. Energy Combust. Sci. **15** (2023)
5. Popp, S., Hartl, S., Butz, D., Geyer, D., Dreizler, A., Vervisch, L., Hasse, C.: Proc. Combust. Inst. **38**(2), 2551–2558 (2021)
6. Engelmann, L., Wollny, P., Breicher, A., Geyer, D., Chakraborty, N., Kempf, A.: Combust. Flame **251** (2023)
7. P. Wollny, L. Engelmann, M. Loewenich, I. Rahinov, H. Wiggers, C. Schulz, A. Kempf, I. Wlokas, Chem. Eng. J. **463** (2023)

Transport and Climate

The three progress reports in transport and climate tackle problems relating to the Atmosphere, aiming to understand its dynamics, facilitate model improvements, and model-based extrapolation for climate change impact assessment.

Geophysical flows in atmosphere and ocean are crucial in the Earth system, transporting momentum, heat and matter. The complexity of the three-dimensional flow is described by the Navier-Stokes equations. Yet, until today the existence and nature of smooth solutions remain a million-dollar question, and approximations are used to investigate and model properties of the atmospheric boundary layer. The scales relevant to the boundary layer range from millimeters (where viscous dissipation occurs) to kilometers (where the interaction with the large-scale flow provides external forcing). Over the planetary scale atmospheric flow is driven by the planetary vorticity, energy redistribution and interactions with other compartments of the system. High-Performance Computing, together with advances in model formulation, is a fundamental prerequisite to improve understanding of the climate system.

Ansorge and Kostelekcy draw on HAWK to investigate properties of turbulence in an idealized setup of a rotating plate in their compute project on the "Direct Representation of Roughness in the Atmospheric Boundary Layer". They use an open-source tool-suite for turbulent flow to represent incompressible flow and interactions with obstacles.

Bauer, Branch and Körner draw on the Weather Research and Forecast (WRF) model and the Parallelized Large Eddy Simulation Model (PALM) with resolutions from km-scale down to meter-scale. Weather Research and Forecasting model (WRF) in their project on "WRF Simulations to Investigate Processes Across Scales".

Feldmann and coauthors aim to take this understanding further and report on the Flagship pilot studies on dynamical downscaling from the largest scales of the circulation to the km-scale with a regional climate model, that is three-dimensional and non-hydrostatic and represents the compressibility of air. This allows them to contribute "High-Resolution Climate Change Information for Extremes and Land Use Change".

Ansorge and Kostelekcy draw on a massively parallel and highly optimized research framework based on direct numerical simulation of turbulent Ekman flow over a rotating plate. They analyze the properties of the flow to assess the potential for parameterization of the turbulent properties towards larger-scale models for the large-scale boundary layer. Ansorge and Kostelekcy demonstrate that the scaling of their code depends on the domain size. For a cubic domain size of $(256)^3$ grid points the algorithm shows superlinear scaling from 32 to 512 cores, where the code benefits from local matrix transposition), and saturation at the switch-level (1024 parallel tasks on 2048 cores). Due to the complexity of the problem this is where the network load becomes large and reduces the parallelization benefits. Carrying out iterations of the code for the $(768)^3$ grid points shows slightly sublinear scaling beyond the node size and saturates beyond the rack size, which implies benefits from a larger network size. Due to the computational power available Ansorge and Kostelekcy were able to achieve a higher level of accuracy in determining the Reynolds-number dependency of higher-order features in the idealized flow.

Bauer, Branch und Körner investigate the next larger set of spatio-temporal scales spanning from the boundary layer to land surface modification over larger regions with nested regional models, driven by high-resolution weather forecasting output. They investigated cases for Southern Oklahoma (targeting weather dynamics at the Land Atmosphere Feedback Experiment, LAFE), Southern Germany (Land Atmosphere Feedback Observatory, LAFO) and the United Arab Emirates (targeting the potential modification of weather patterns from smaller or larger plantations). The nested domains in WRFscale bridge from 2.5 km to 100 m resolution, resolving the finer features of the Southern Great Plains topography. Detailed output allowed intercomparison of the model output with local observations showed that dynamics are approximately represented, but that simulated turbulent eddies are larger than observed, extend too far upward, and move too much, which lead the team to conclude that the horizontal and vertical resolution was initially too coarse. A refinement to 20 m resolution appears more reasonable to resolve these statistics. At LAFO, a similar threefold nesting approach down to 50 m resolution was applied, whereas for the UAE a single domain at 2.7 km was used, demonstrating the impact of convection and precipitation modification by convection-permitting simulations. Data from high-resolution turbulence 'Large Eddy Simulations' with the PALM model were also stored to train a neural network in order to identify functional interrelations for a better parameterization of turbulence over heterogeneous domains. WRF requires a minimum of 768 cores and scales approximately linearly, with the radiation code as the strongest limiting factor. PALM scales sublinearly with a minimum of 2048 cores.

Feldmann and coauthors aim to bridge from the results out of large-scale simulations for future projections with General Circulation Models of the Atmosphere and Ocean to local climate change information, so to the km-scale. They draw on the numerical weather prediction model COSMO to (1) downscale GCM output to the km-scale for Germany, and specifically the Stuttgart region, and test (2) the role of land-use characteristics (e.g., afforestation with darker and shallow-root coniferous versus lighter deciduous forest) on energy partitioning at the land surface. The shortwave radiative absorption is locally reduced, leading to lower heating in an experiment where all coniferous forests are replaced by deciduous forests. In COSMO the number of sub-domains for domain-decomposition is limited because a halo of three grid-boxes from neighbouring processors is required for time integration. Tests on HAWK suggested a good scalability of the model up to 40 nodes, although variability of execution time was noted. Due to the high amount of input/output required for atmospheric computation with 3D boundary conditions, and high frequency of output, in realistic applications a computing time increase of 9% is expected based on initial tests.

Kira Rehfeld
Geo and Environmental Research Center
University of Tübingen
Tübingen, Germany
e-mail: kira.rehfeld@uni-tuebingen.de

High-Resolution Climate Change Information for Extremes and Land Use Change

Hendrik Feldmann, Marie Hundhausen, Regina Kohlhepp, Marcus Breil, Armenia Franco-Diaz, and Joaquim G. Pinto

Abstract This report provides an overview of the IMKTRO's contributions to the Flagship pilot studies (FPS) within the CORDEX initiative, specifically focusing on the Convective Phenomena over Europe and the Mediterranean (FPSC) and Land Use and Climate Across Scales (LUCAS) projects. FPSC found improvements in the representation of small-scale processes from the analysis of kilometer-scale simulations. However, there are still open research questions and problems with this type of climate simulations. FPS-LUCAS examines the impact of land use on climate. The study found that targeted land use changes, such as afforestation, have the potential to mitigate the impacts of climate change on a regional scale in Europe. Additionally, the report highlights the use of the KIT-KLIWA ensemble for climate adaptation strategies in the Stuttgart region and to examine the climate change signal of heat stress and extreme precipitation.

1 Introduction - Overview

Within the CORDEX (Coordinated Downscaling Experiment) initiative of the World Climate Research Programme (WCRP) several so called Flagship pilot studies (FPS) have been endorsed, where scientists from multiple international institution tackle pioneering tasks with respect to regional climate research. The IMKTRO contributes to several of these initiatives based on simulations performed at the HLRS. Among them is the FPS: Convective Phenomena over Europe and the Mediterranean (FPSC, [1]) which investigates the representation of convective processes in a first of its kind multi-model ensemble of kilometre scale simulations. At such scales the resolution

H. Feldmann (✉) · M. Hundhausen · A. Franco-Diaz · J. G. Pinto
Institute for Meteorology and Climate Research—Troposphere Research (IMKTRO), Karlsruhe Institute of Technology (KIT), Karlsruhe, Germany
e-mail: hendrik.feldmann@kit.edu

R. Kohlhepp
LUBW Landesanstalt für Umwelt Baden-Württemberg, Karlsruhe, Germany

M. Breil
Institute of Physics and Meteorolgy, University of Hohenheim, Stuttgart, Germany

T. Ludwig et al. (eds.), *High Performance Computing in Science and Engineering '23*,
https://doi.org/10.1007/978-3-031-91312-9_30

is sufficient to represent deep convection explicitly without parametrizations, which reduces the uncertainties due to parametrizing convective processes. Therefore, these simulations are called "convection permitting modelling (CPM)". In addition, local meteorological phenomena like e.g. orographic effects and channelling, the heterogeneity of the land-cover, or coastal wind systems are resolved better than in global or coarser regional climate simulations. A large number of studies – with contributions from IMKTRO – are currently published. This report will give an overview of the main findings.

Secondly, we contribute to another CORDEX-FPS, namely the FPS: "Land Use and Climate Across Scales (LUCAS; [2])", which is dedicated to study the effects of land use and land use change on the climate via exchange processes. Several experiments have been performed by the group and have subsequently been analyzed and published.

In addition to the CORDEX related activities, we analyzed the KIT-KLIWA CPM ensemble, generated at HLRS to provide high resolution information on future heat stress and extreme precipitation under climate change. These studies, performed in the framework of the BMFTR research program RegIKlim (regional information for climate action) subproject ISAP (Integrative city-regional adaptation strategies for a growing polycentric region: Region Stuttgart), will be used to develop adaptation strategies to climate change and to be included in an information portal for the region of Stuttgart.

2 The Regional Climate Model COSMO-CLM

We performed our own simulation within this study with the regional climate model (RCM) COSMO-CLM (CCLM) is the climate version of the former operational weather forecast model COSMO (Consortium for Small-scale MOdeling; www.cosmo-model.org; [3]). It is a three-dimensional, non-hydrostatic, fully compressible numerical model for the atmosphere and includes a land surface model. The model solves prognostic equations for wind, pressure, air temperature, different phases of atmospheric water, soil temperature, and soil water content. The standard land surface model of CCLM is TERRA-ML [4]. Alternatively, the sub-model VEG3D [5], which includes an explicit treatment of the vegetation layer, can be applied. CCLM is a community model maintained and developed for climate applications by the CLM Community (www.clm-community.eu; [6, 7]). It has been applied in multiple projects with resolution down the kilometer scale [8] and for domains across the globe [9].

Table 1 Convection permitting scale climate simulations performed on the HLRS supercomputers

Project	Domain	Scenario	Period	Resolution	Dimension
KLIWA	SGER-28	Evaluation	1971–2000	50, 7/2.8 km	322*328*40 GP
		Historical	1971–2005		25 s
		rcp8.5	2006–2100		for Nest3
FPSC	ALP-3	Evaluation,	2000–2016	12.5(25)/3 km	522*490*50 GP
		Historical,	1996–2005		25 s
		rcp8.5	2041–2050		for Nest2
			2090–2099		

3 Extreme Events in the KIT-KLIWA Ensemble

The KIT-KLIWA Ensemble is a model ensemble of high-resolution climate simulations consisting of four ensemble members driven by four global climate models (GCMs; Tables 1 and 2). Using the regional climate model COSMO-CLM, the driving simulations are downscaled over a domain covering Germany to a resolution of 2.8km, which is on the convection permitting scale. The simulations and post processing were partly performed at HLRS. The KIT-KLIWA ensemble is used in the ISAP project (Integrative stadt-regionale Anpassungsstrategien in einer polyzentrischen Wachstumsregion: Modellregion – Region Stuttgart). As part of the project, the data is processed for end users and presented in the form of climate parameters (for example, hot days, tropical nights, and frost days). The results will be made available in an online tool that is developed in the course of the project. Within the framework of the ISAP project, additional detailed evaluations of extreme events in the ensemble were carried out. The analysis focused on heat waves and heavy rainfall, which are of particular importance for urban regions.

A comprehensive analysis of heatwaves within the ensemble, accepted in NHESS [11], revealing the following key findings: Compared to coarser simulations with parametrized convection, the convection-permitting simulations exhibited a reduced cold bias, particularly in summer. The projected temperature increase favored the development of longer and hotter heat waves, especially in late summer and early autumn. Heatwaves are projected to become longer and more intense, resulting in a significant increase in the magnitude of heatwave events (Fig. 1). In a 2°C (3°C) warmer world, the Heat Wave Magnitude Index is expected to increase by 26% (100%).

Additionally, another study investigates the climate change signal of extreme precipitation [16]. The data from the KIT-KLIWA Ensemble demonstrated added value in the precipitation intensities compared to coarser simulations, especially for hourly resolution. The strongest climate change signal of extreme precipitation in the ensemble was found for events of short duration and with high intensity/long return period. For these events, the climate change signal ranges from 6 to 8.5%, which is in

Table 2 Name, realization, Equilibrium Climate Sensitivity (ECS; cf. [10] supplementary), 30-year periods corresponding to the analysed global warming levels (GWLs) +2 and +3 degrees relative to pre-industrial conditions and main reference for the CMIP5 GCMs downscaled for the ensemble (additional information in [11])

GCM	Realization	ECS in °C	GWL2	GWL3	References
CNRM-CM5	r1i1p1	3.28	2029– 2058	2052–2081	[12]
MPI-ESM-LR	r1i1p1	3.66	2029–2058	2052–2081	[13]
EC-EARTH	r12i1p1	4.18	2026–2055	2051–2080	[14]
HadGEM2-ES	r1i1p1	4.64	2016–2045	2037–2066	[15]

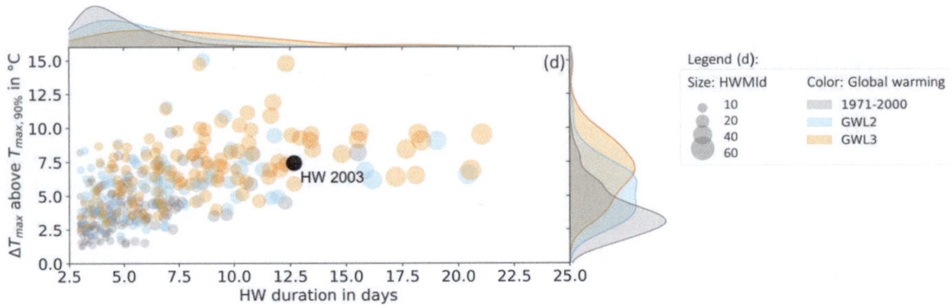

Fig. 1 The bubble plots show the strongest heat wave in each summer half-year (May-Oct) in every projection run with respect to duration on the abscissa and excess temperature on the ordinate. Bubble size indicates mean Heat Wave Magnitude Index (HWMId) over all grid point results affected by the HW. Marginal plots show the distribution of duration of the heat waves in days (abscissa) and the distribution of the excess temperature, which is the maximum temperature excess within the heat wave compared to the climatological mean temperature in 1971-2000 (ordinate). Colors indicate different global warming levels: 1971-2000, GWL2 and GWL3. The black data point corresponds to the HW in 2003 derived from the observation data set HYRAS (Figure and analysis by [11])

the range of Clausius-Clapeyron scaling—the increase of the water-holding capacity of the atmosphere for rise in temperature.

4 Summary of the Results from the CORDEX Flagship Pilot Study on Convection

The representation of extreme precipitation events strongly depends on the resolution of the underlying simulation. Figure 2a shows an example of such an event from the GCM MPI-ESM-LR historical simulation with 200 km grid spacing and from a downscaling simulation with COSMO-CLM to 3 km from the FPSC ensemble. This event represents quite typical condition for a summertime flooding event in Central Europe, roughly similar to e.g. the Elbe and Donau flooding in August 2002 or the Central European flooding in May/June 2013. Both simulations indicate an

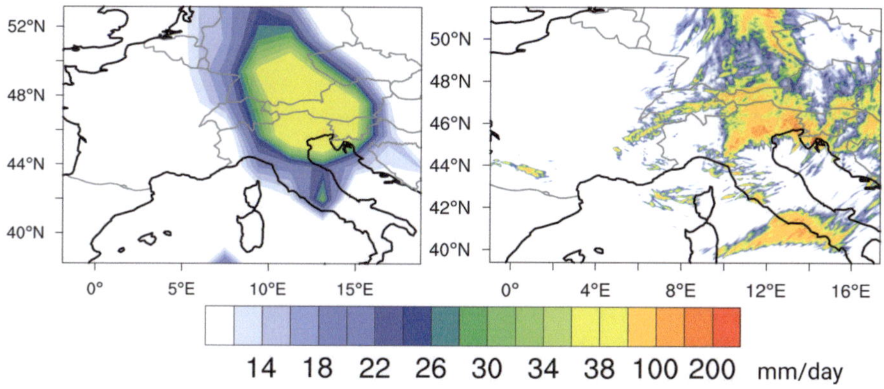

Fig. 2 Representation of an extreme precipitation [mm/day] event from **a** the MPI-ESM-LR historical climate simulation from CMIP5 with 200km resolution and **b** a FPSC downscaling simulation with COSMO-CLM to 3km for the same day: 31 August 2002

extreme precipitation event on the day within the top five extremes within the 10-year CPM simulation period. The results for the GCM display a large quite homogeneous precipitation field on the eastern side of the Alps from Northern Italy to Eastern Germany with heavy precipitation of more than 38 mm/day. The high-resolution CPM simulation shows far more realistic differentiation with extreme precipitation and embedded convective cells over the Eastern part and the northern slope of the Alps, but as well over Eastern Germany with peak intensities well above 100 mm/day, affecting for instance the Donau and Elbe catchments. This high-resolution data provide a far more meteorologically plausible representation of such events.

In the reporting period several papers have been published within a special issues for FPSC and another one from BMFTR project ClimXtreme.

Within The ClimXtreme sub-project SEVERE (Scale Dependent Process Representation and Sensitivity Analysis for Most Extreme Events) Caldas-Alvarez et al. [17] analyzed the scale dependent representation of the relevant processess in FPSC reanalysis driven simulations performed by IMKTRO with CCLM at 25 and 3 km resolution. They applied the Precipitation Severity Index PSI to quantify the extremeness of events and used a principle component analysis to identify the most frequent associated weather pattern. The CPM simulation showed higher peak intensities and a better representation of the extremeness of precipitation events than the coarser 25 km CCLM simulation (here RCM). The peak intensities were overestimated by the CPM simulation in mountainous regions compared to the HYRAS observations [18] on a 5 km grid. The paper assesses the different representation of the events in both resolutions. the coarse resolution data indicate a higher integrated water vapour and higher equivalent potential temperature at 850 hPa due to a wetter mid-troposhere conditions and a higher latent heat flux over the Mediterranean Sea. On the other hand is latent heat flux over land north of the Alps higher in the CPM simulation causing a higher convective available potential energy (CAPE).

The differences in the surface fluxes as a main distinction between CPM and RCM simulations was confirmed for the larger FPSC ensemble by Sangelantoni et al. [19], where the different representations of heat-waves were addressed. The findings indicate that the higher latent heat-fluxes in Northern Italy, caused a drying of the soil in summer, which led to higher heat wave intensities there compared to the CPM simulations and observations. This shows that the differences in the surface fluxes are a common feature over multiple regional models and forcing data and not a specific feature of CCLM.

Nevertheless, Soares et al. [20] found some added value for the representation of temperatures in the CPM ensemble with respect to local atmospheric circulation especially in region with strongly changing spatial conditions as coastal areas.

Belusic Vozila et al. [21] analyzed this specifically for the Adriatic region and found a good and detailed representation of the small-scale orographic features and the resulting dynamical and thermal circulations in this region, which are not well depicted in coarser scale simulations. In particular, a clear added value of CPM was found for The northern Adriatic for the Bora maximum wind-speed and the frequency of Scirocco in the southern Adriatic.

These collective results indicate an added value of high-resolution simulation for multiple purposes as process understanding, assessment of extremes and potentially as forcing data for impact modelling. However, further studies are needed to understand and reduce deficits of CPM simulations to make more broadly usable for adaptation planing.

5 The Impact of Land-Use Characteristics on the Earth's Energy Balance

The land surface characteristics decisively influence earth's energy balance by controlling the way solar radiation is absorbed, transformed and released again as turbulent heat into the atmosphere. Thus, changes in the land surface characteristics due to land use changes can have a strong impact on the climate conditions, especially on the regional scale (e.g. [22]). However, land use changes are generally not implemented in RCMs, constituting a potential reason for biases in regional climate simulations.

In order to investigate the impact of land use changes on the regional climate conditions in Europe in detail, therefore, the LUCAS (Land Use and Climate Across Scales; [2]) project was initiated by WCRP-CORDEX. In the framework of this project, an RCM ensemble is established, consisting of several RCMs, coupled to different Land Surface Models (LSMs). The Institute of Meteorology and Climate Research at the Karlsruhe Institute of Technology contributes to this multi-model ensemble by performing RCM simulations with CCLM coupled to the LSM VEG3D [5].

During the year 2021, first experiments with CCLM-VEG3D were conducted to quantify the effects of land use changes on the European climate conditions (see report 2022). The results of these preliminary studies indicated that targeted land use changes indeed have the potential to mitigate the burdens of man-made climate change on the regional scale. A particularly promising mitigation strategy in this regard is afforestation.

Forests are able to transpire large amounts of water and are additionally characterized by a high surface roughness [23, 24]. Absorbed solar radiation is therefore efficiently transformed and released into the atmosphere without intensively warming the surface [25].

However, European forests are currently dominated by coniferous forests, due to forestry reasons. Coniferous forests are in comparison to deciduous forests very dark, and consequently absorb more solar radiation. In addition, coniferous forests are generally characterized by a shallow root system, limiting evapotranspiration particularly during long-lasting heat periods. Thus, coniferous forests are potentially intensifying such extreme events [26]. Therefore, an additional experiment was conducted to investigate whether an increase in the proportion of deciduous forests in Europe has the potential to reduce the intensity of severe heat periods [27].

For this purpose, an idealized CCLM-VEG3D simulation was performed in which all coniferous forests in Europe were replaced by deciduous forests (called DECID). The results of this simulation were compared with a simulation in which the actual deciduous forest fraction was used (REF). Maps of the actual and the adapted forest distribution in Europe are shown in Fig. 3. Both CCLM-VEG3D simulations were performed for the period 1986-2015, driven by the ERA5 reanalysis [28] on a spatial resolution of 0,11°.

Figure 4 shows the climatological changes in regions, in which coniferous forests were replaced by deciduous forests. In all these regions, net short-wave radiation is reduced (Fig. 4a), since deciduous forests have a higher albedo than coniferous forests. As a consequence, less radiative energy is absorbed and available to heat up the land surface. In central and southern Europe, additionally, latent heat fluxes of deciduous forests are increased (Fig. 4b), while sensible heat fluxes are reduced

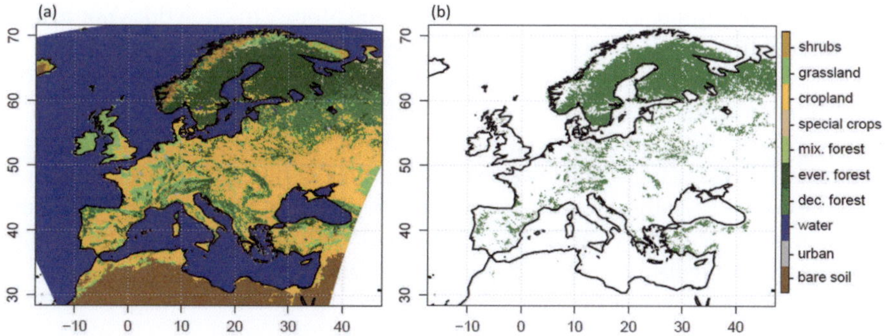

Fig. 3 **a** CCLM-VEG3D land use classes. **b** grid cells in which coniferous forests were replaced by deciduous forests in the DECID simulation. From: [27]

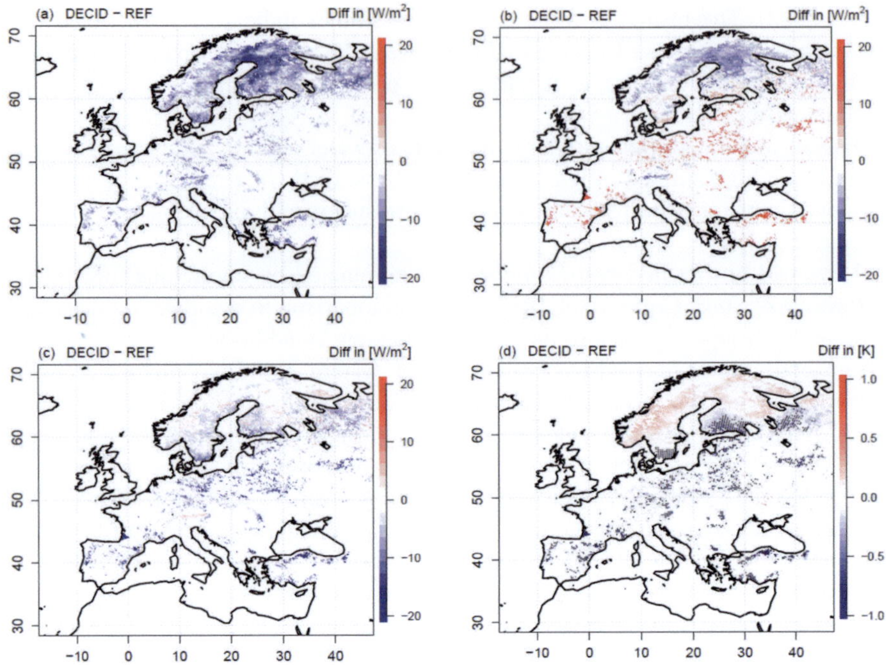

Fig. 4 Local differences between DECID and REF for **a** net short-wave radiation, **b** latent heat fluxes, **c** sensible heat fluxes, and **d** daily maximum 2 m temperatures during heat periods. The black circles in (**d**) indicate significant results calculated with a Wilcoxon-Rank-Sum-Test at a 95 % level. From: [27]

(Fig. 4c). Thus, the reduced radiative energy input during heat periods is transformed into latent heat instead of warming the land surface. A reduction of the daily maximum 2 m temperatures during heat periods is the consequence in central and southern Europe (Fig. 4d). This reduction of heat period intensities is in mean about 0.2 K.

In northern Europe, an increase in the deciduous forest fraction does not lead to such a reduction in the heat period intensities. Particularly in the northern part of Scandinavia, an increase in the daily maximum 2 m temperatures is simulated. This intensification of heat periods can be explained by reduced evapotranspiration rates in this area, and thus an increased amount of radiative energy that is available to heat up the land surface (Fig. 4b). However, although the cooling effect on heat period intensities is statistically significant in central and southern Europe, the temperature reduction is comparatively small. Therefore, the potential of an increased deciduous forest fraction to reduce heat period intensities is only limited [27].

Table 3 Test setup for scaling tests with CCLM on HPE Apollo

Grid points x-,y-direction	522 × 490
Vertical layers	50
Number of grid points	12.789.000
Horizontal resolution	3 km
Time-step	25 s
Simulation period	48 h

6 Computational Aspects and Scaling Properties

The numerical weather prediction mode of CCLM, the COSMO model, has been applied to operational weather forecasts for over a decade. It is highly optimized for super computer architectures. It uses a 2-dimensional domain decomposition method for parallelization via MPI on distributed memory machines. Each sub-domain requires a halo of three grid-boxes from neighbouring processors to provide the necessary data for the Runge-Kutta time-integration scheme. This limits the number of sub-domains for the decomposition. The recommendation by the DWD (U. Schättler, COSMO developing group, personal communication) is that for fast and efficient computations the ratio between the number of grid points in x- and y-direction divided by the number of sub-domains in both directions should be in the order of 10. Within the range, we performed scalability tests with different settings for stride (st) and block-size (bs). Additional experiments on the new HLRS computing system HPE Apollo “Hawk” were executed with respect to the I/O strategy. The test environment was taken from the CORDEX FPS Convection setup (Table 3).

The results for test series with strides of 1 and 2 and a block size of 1 are presented. Tests with other settings provided a lower efficiency. Giving the restrictions on the number of sub-domains the computational domain can be split in to about 50 × 50 processes using about 2500 CPUs. Compared to the single node (128 CPUs, st = 1,

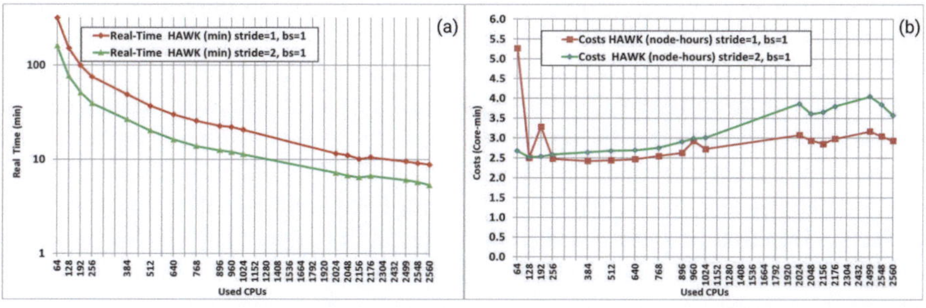

Fig. 5 **a** Real-time execution duration for the CCLM scaling experiments versus the number of CPUs used. **b** Node hour costs for the CCLM scaling experiments versus the number of CPUs used

bs = 1) execution time (ET) of about 150 minutes for the test simulation the model provides a very good scalability up to 40 nodes with an ET of 5.36 min (st = 2, bs = 1, 2560 CPUs, Fig. 5a). This means a speed-up of a factor 30 by applying a factor of 40 more CPUs and about 33.6% higher node-hours (Fig. 5b). Please note, that the ET and node-hours for single node test were higher than expected from experiments with more nodes. Repeated tests showed some variability of ET. Some of the tests have therefore been repeated several times.

The stride=2 and block-size=1 case provides the shortest ET and is up to a factor of 2 faster than the case with st = 1. With increasing number of nodes being used the factor is slightly decreasing. In the test with 2560 CPUs (for st = 1: 20 nodes, respectively 40 nodes for st = 2) the st = 2, bs = 1 case is still faster, needing about 65% of the ET compared to the st = 1 test. The costs for the st = 2, bs = 1 case increase moderately by about 22 %.

Atmospheric applications typically require large amounts of I/O, since they require 3-dimensional input data and produce data with high spatio-temporal frequency. The results presented so far were generated with low amounts of output fields and no specifically dedicated CPUs for the output (synchronous I/O). The test with 52×49 CPUs has been repeated with 12 additional CPUs ($52 \times 49 + 12 = 2560$ CPUs) dedicated just for I/O (asynchronous I/O). This reduced computing time and costs by ca. 9%.

7 Outlook

For more than ten years the global forcing data for regional downscaling stem from the Coupled Model Intercomparison Project Phase 5 (CMIP5) global models. The simulations presented within this report still belong to this era, which is now going to end. Currently, CORDEX finalized the plans for a downscaling of the CMIP6 simulations (e.g. [29] for Europe. This new generation of global models generally show improvements compared to the older one, due model improvements and higher resolution (e.g. [30]. Within the upcoming BMFTR project "Updating the data basis for adaptation to climate change in Germany (UDAG)" an ensemble of CMIP6 downscaling simulation will be performed within the next years based in the new regional climate model ICON CLM [31]. The ensemble will consist of simulations with forcing from different CMIP6 GCMs and different emission scenarios. The nesting steps will include a European domain with 12.5km resolution and a further nesting step to 3 km covering Germany, Switzerland, Austria and it's surroundings. The simulations and subsequent analysis from FPSC and KLIWA paved the way for such convection permitting scale ensembles in the future. The UDAG project is expected to start in September 2023, starting with a phase of testing and optimization. Parts of the simulations will be performed at HLRS. Regarding the contributions to the CORDEX Flagship Pilot Studies there will be a shift in the work by IMKTRO: The FPS on convection recently ended and the simulation data will be provided to the public via the Earth System Grid Foundation (ESGF) portal. Its successor will focus on the representation of urban areas in high-resolution models. IMKTRO will contribute to these efforts. The FPS LUCAS will continue into a next phase also with a focus on very high resolution. The analysis of the regional climate ensembles will continue in the second phase of RegIKlim projects NUKLEUS and ISAP.

Acknowledgements The work presented in this work has been funded by the BMFTR project ISAP (FKZ 01LR2007B) and NUKLEUS (FKZ 01LR2002B) which are projects within the RegIKlim ("Regionale Informationen zum Klimahandeln") research program and to the BMFTR research program ClimXtreme ("Climate Change and Extremes": FKZ: 01LP1901A).

References

1. Coppola, E., co authors.: A first-of-its-kind multi-model convection permitting ensemble for investigating convective phenomena over europe and the mediterranean. Sci. Rep. **55**, 3–34 (2020)
2. Davin, E.L., co authors.: Biogeophysical impacts of forestation in europe: first results from the LUCAS (Land Use and Climate Across Scales) regional climate model intercomparison. Earth Syst. Dyn. **11**, 183–200 (2020)
3. Baldauf, M., Seifert, A., Förstner, J., Majewski, D., Raschendorfer, M., Reinhardt, T.: Operational convective-scale numerical weather prediction with the COSMO model: description and sensitivities. Mon. Wea. Rev. **139**(12), 3887–3905 (2011)
4. Schrodin, R., Heise, E.: The multi-layer version of the DWD soil model terra-lm. COSMO Technical report (2001). Available at http://www.cosmo-model.org/content/model/documentation/techReports/cosmo/docs/techReport02.pdf
5. Breil, M., Schädler, G.: The reduction of systematic temperature biases in soil moisture–limited regimes by stochastic root depth variations. J. Hydrometeorol. **27**(7), 1897–1911 (2021)
6. Rockel, B., Will, A., Hense, A.: The regional climate model COSMO-CLM (CCLM). Meteorologische Zeitschrift **17**(4), 347–348 (2008)
7. Steger, C., Bucchignani, E.: Regional climate modelling with cosmo-clm: history and perspectives. Atmosphere **11**(11) (2020)
8. Ban, N., co authors.: The first multi-model ensemble of regional climate simulations at kilometer-scale resolution, Part I: evaluation of precipitation. Clim. Dyn. **57**, 275–302 (2021). https://doi.org/10.1007/s00382-021-05708-w
9. Sørland, S.L., co authors.: COSMO-CLM regional climate simulations in the CORDEX framework: a review. Geosci. Model Dev. Discuss. **2021**, 1–44 (2021)
10. Nijsse, F.J.M.M., Cox, P.M., Williamson, M.S.: Emergent constraints on transient climate response (TCR) and equilibrium climate sensitivity (ECS) from historical warming in cmip5 and cmip6 models. Earth Syst. Dyn. **11**(3), 737–750 (2020)
11. Hundhausen, M., Feldmann, H., Laube, N., Pinto, J.G.: Future heat extremes and impacts in a convection permitting climate ensemble over Germany. Natural Hazards Earth Syst. Sci. **23** (8), 2873–2893 (2023). https://doi.org/10.5194/nhess-23-2873-2023
12. Voldoire, A., co authors.: The CNRM-cm5.1 global climate model: description and basic evaluation. Clim. Dyn. **40**(9), 2091–2121 (2013)
13. Giorgetta, M.A., co authors.: Climate and carbon cycle changes from 1850 to 2100 in mpi-esm simulations for the coupled model intercomparison project phase 5. J. Adv. Model. Earth Syst. **5**(3), 572–597 (2013)
14. Prodhomme, C., Doblas-Reyes, F.J., Bellprat, O., Dutra, E.: Impact of land-surface initialization on sub-seasonal to seasonal forecasts over Europe. Clim. Dyn. **47**(3), 919–935 (2016)
15. Collins, W.J., co authors.: Development and evaluation of an earth-system model – hadgem 2. Geophy. Mod. Dev. **4**(4), 1051–1075 (2011)

16. Hundhausen, M., Feldmann, H., Kohlhepp, R., Pinto, J.G.: Climate change signals of extreme precipitation return levels for germany in a transient convection-permitting simulation ensemble. Int. J. Climatol. **44** (5), 1454–1471 (2024). https://doi.org/10.1002/joc.8393
17. Caldas-Alvarez, A., Feldmann, H., Lucio-Eceiza, E., Pinto, J.G.: Convection-parameterized and convection-permitting modelling of heavy precipitation in decadal simulations of the greater alpine region with cosmo-clm. Weather Clim. Dyn. **4**(2), 543–565 (2023)
18. Razafimaharo, C., co authors.: New high-resolution gridded dataset of daily mean, minimum, and maximum temperature and relative humidity for Central Europe (HYRAS). Theor. Appl. Climatol. **142**(3), 1531–1553 (2020)
19. Sangelantoni, L., Sobolowski, S., Lorenz, T., Hodnebrog, O., Cardoso, R.M., Soares, P.M.M., Ferretti, R., Lavin-Gullon, A., Fernandez, J., Goergen, K., Milovac, J., Katragkou, E., Kartsios, S., Coppola, E., Pichelli, E., Adinolfi, M., Mercogliano, P., Berthou, S., de Vries, H., Dobler, A., Belusic, D., Feldmann, H., Tölle, M.H., Bastin, S.: Investigating the representation of heatwaves from an ensemble of km-scale regional climate simulations within cordex-fps convection. Clim. Dyn. **62**(6), 4635–4671 (2024). https://doi.org/10.1007/s00382-023-06769-9
20. Soares, P.M.M., Careto, J.A.M., Cardoso, R.M., Goergen, K., Katragkou, E., Sobolowski, S., Coppola, E., Ban, N., Belusic, D., Berthou, S., Caillaud, C., Dobler, A., Hodnebrog, O., Kartsios, S., Lenderink, G., Lorenz, T., Milovac, J., Feldmann, H., Pichelli, E., Truhetz, H., Demory, M.-E., de Vries, H., Warrach-Sagi, K., Keuler, K., Raffa, M., Tölle, M., Sieck, K., Bastin, S.: Investigating the representation of heatwaves from an ensemble of km-scale regional climate simulations within cordex-fps convection. Clim. Dyn. **62**, 4491–4514 (2024). https://doi.org/10.1007/s00382-022-06593-7
21. Belusic Vozila, A., Belusic, D., Telisman Prtenjak, M., Güttler, I., Bastin, S., Brisson, E., Demory, M., Dobler, A., Feldmann, H., Hodnebrog, O., Kartsios, S., Keuler, K., Lorenz, T., Milovac, J., Pichelli, E., Raffa, M., Soares, P.M.M., Tölle, M.H., Truhetz, H., de Vries, H., Warrach-Sagi, K.: Evaluation of the near-surface wind field over the adriatic region: local wind characteristics in the convection-permitting model ensemble. Clim. Dyn. **62**(6), 4617–4634 (2024). https://doi.org/10.1007/s00382-023-06703-z
22. Jia, G., co authors.: Climate change and land: an IPCC special report on climate change, desertification, land degradation, sustainable land management, food security, and greenhouse gas fluxes in terrestrial ecosystems, chapter Land–climate interactions. Cambridge University Press, in Press (2019)
23. Bonan, G.B.: Forests and climate change: forcings, feedbacks, and the climate benefits of forests. Science **320**(5882), 1444–1449 (2008)
24. Pielke, R., Sr, co authors, Land use/land cover changes and climate: modeling analysis and observational evidence. WIREs: Clim. Change **2**(6), 828–850 (2011)
25. Breil, M., co authors.: The opposing effects of reforestation and afforestation on the diurnal temperature cycle at the surface and in the lowest atmospheric model level in the European summer. J. Clim. **33**(21), 9159–9179 (2020)
26. Schwaab, J., co authors.: Increasing the broad-leaved tree fraction in european forests mitigates hot temperature extremes. Sci. Rep. **10**(14153) (2020)
27. Breil, M., Weber, A., Pinto, J.G.: The potential of an increased deciduous forest fraction to mitigate the effects of heat extremes in Europe. Biogeosciences **20**(12), 2237–2250 (2023)
28. Hersbach, H., co authors.: The era5 global reanalysis. Q. J. R. Meteorol. Soc. **146**(730), 1999–2049 (2020)
29. Gutowski, W.J. Jr., co authors.: WCRP Ccordinated regional downscaling experiment (CORDEX): a diagnostic MIP for CMIP6. Geosc. Model Dev. **9**(11), 4087–4095 (2016)
30. Fernández-Granja, J.A., Casanueva, A., Bedia, B., Fernández, J.: Improved atmospheric circulation over Europe by the new generation of CMIP6 earth system models. Clim. Dyn. **56**, 3527–3540 (2021)
31. Pham, T.V., co authors.: ICON in climate limited-area mode (ICON release version 2.6.1): a new regional climate model. Geosci. Model Dev. **14**(2), 985–1005 (2021)

Direct Representation of Roughness in the Atmospheric Boundary Layer

Cedrick Ansorge and Jonathan Kostelecky

Abstract Geophysical flow is generally characterized by huge Reynolds numbers—which limits our ability to directly represent these systems on a computer. Most practical applications such as weather forecasting or climate projection rely on the representation of small-scale processes, one of which is turbulent mixing, by parameterizations. Sometimes, however, an explicit representation is inevitable to further process-level understanding, for instance for informed representation of mixing processes in parameterizations. While an explicit representation of turbulence is not possible across the entire geophysical range of scales, hydrodynamic/Reynolds-number similarity can be exploited to quantitatively extrapolate the behavior at reduced scale to the geophysically relevant limit of large scale separation. In previous work, we established a research framework based on direct numerical simulation of turbulent Ekman flow to proceed this path. Here, we outline results from the explicit representation of roughness in turbulent Ekman flow. Preliminary results indicate that roughness acts as a geometric reduction of the scale separation in the flow. This potentially allows us to narrow the gap in turbulence scale separation between the numerical simulations possible on modern HPC systems on the one side and the real atmospheric system on the other.

1 The Challenge of Atmospheric Turbulence and Its Practical Manifestations

Atmospheric turbulence affects humans as we spend most of our lifetime directly or indirectly exposed to the mostly turbulent atmospheric boundary layer (ABL). Turbulence enhances the vertical transport of momentum, heat, vapor, and pollutants by orders of magnitude in comparison to a non-turbulent flow. It is therefore

C. Ansorge (✉) · J. Kostelecky
Institut für Meteorologie, Freie Universität Berlin, Berlin, Germany
e-mail: c.ansorge@fu-berlin.de

J. Kostelecky
Institut für Geophysik und Meteorologie, Universität zu Köln, Köln, Germany

T. Ludwig et al. (eds.), *High Performance Computing in Science and Engineering '23*,
https://doi.org/10.1007/978-3-031-91312-9_31

an important mechanism for the evolution of the bulk state within the ABL, which impacts on both the compartments of the earth system coupled to it (land surface, ocean surface, man-made surface, land or ocean ice) and the free atmosphere aloft. Hence, an accurate representation of the ABL is key to correctly represent the earth system, which also manifests in large forecast errors when the models used to represent ABL physics become inaccurate or fail entirely [40]. Taking a closer look at the development of the forecasting skill for short- and medium-range weather forecasts, we find a tremendous increase over the past decades [22, 45] over all. At the same time, some common regimes, for example certain instances of the stable atmospheric boundary layer (SABL), remain problematic—and indeed winter-time, polar night and transitional regimes (for instance the morning and evening transition, [44]) pose severe challenges for modeling the SABL [31, 40]. It turns out that a breakdown of the classical homogeneous and equilibrium Kolmogorov-type turbulence [25] due to stratification and associated coexistence of turbulent and non-turbulent flow patches inside the atmospheric boundary layer (ABL) poses the major hurdle towards an improved understanding [2, 23, 31]. Systematic insight to the super-critical regime of stratification, also known as weak or intermittent turbulence is lacking from a number of perspectives.

This lack of process-level understanding and the associated inaccurate representation of turbulence translates into long-standing problems of meteorological applications [17, 42]: (i) Night-time and polar-night minimum temperatures are a major challenge for numerical models of the earth system [40, 42], in particular with respect to food, aviation and transport safety. (ii) Turbulent transfer of humidity from the surface throughout the ABL determines availability of moisture for cloud formation, a key source of multi-model uncertainty in current climate models; better turbulence models will improve numerical weather forecasting and climate projection [31, 40, 42]. (iii) With the increasing role of wind energy for power production, the accurate now- (up to 8 h into the future) and fore-casting (up to 72 h into the future) of turbulence and wind conditions in elevated regions of the SABL gains major economic relevance. Effects of stable stratification and their accurate representation are of major importance for future advances [14]. (iv) Both accurate flux modeling at scales in the so-called *terra incognita* or gray zone [46] and scale-aware turbulence closures for the ABL require quantitative consideration of large-scale turbulence intermittency as the vertical ($\sim$10 m) and horizontal resolution ($\sim$1 km) in such simulation overlaps with the length scale of large-scale intermittency on the order of the ABL depth.

This compute project is driven by the need for an improved process-level understanding in the globally intermittent ABL, crucial to address standing issues in the representation of turbulent mixing in large-scale models of the earth system. These difficulties are related (i) to the breakdown of continuous turbulence, a fluid mechanics problem at its very core, and (ii) to the interaction of the turbulent flow with surface roughness–which on the one had acts to enhance turbulence and thus vertical transport, but on the other strengthens gradients in the roughness sub-layer by the increase of vertical transport and the associated buoyancy effects. We will advance here the fluid mechanical understanding of intermittency that develops when the external intermittency in a turbulent flow extends towards the surface and eventually transfer the results to the geophysical limit. To do so, we use Ekman flow as a virtual laboratory for the ABL and introduce in this paper an approach for the explicit representation of roughness.

2 Conceptual Minimalism

2.1 A Virtual Lab Focusing on Turbulent Processes in the Atmospheric Boundary Layer

The real-world geophysical problem of the atmospheric boundary layer is not only characterized by its very broad range of scales—ranging from the sub-millimeter scale to planetary waves spanning the entire globe—the ABL is also an intrinsically multi-physical system where many processes interact. On the one hand, this interactive nature may be understood as an imperative towards incorporation of ever more physical processes as done in weather forecasting and climate modelling towards a more and more complete representation of the earth system. On the other hand, a quantitative process understanding is sometimes hampered by the interaction of too many processes as mechanistic attribution becomes more and more of a problem. We follow here a minimal approach in which we try to break down existing geophysical problems to their core—with a focus on cases where the intricate non-linearity of turbulence acts. While this does not imply other processes do not act, they are understood as external forcing or modulators of the intrinsic turbulence dynamics in the flow. This enables unique attribution and quantification of some mechanisms active in a turbulent flow. On the other hand, it comes with limitations regarding strong, local forcing for instance due to microphysical processes, interaction with or strong heterogeneity at the surface, or synoptic-scale systems, all of which may mask the turbulent effects due to large tendencies.

In the quest for a minimal system that represents the turbulence dynamics and peculiarities of turbulence in the atmospheric boundary layer, previous work has shown that the Ekman boundary layer over a flat plate serves as a very good physical model of the system. It represents the main features of ABL turbulence, namely its interaction with non-turbulent external fluid aloft including external intermittency [2], its exposure to background rotation, and to different levels of density stratification. This set-up reproduces all regimes of stably stratified turbulence observed in the ABL–without a change in the modelling paradigm [1]. The system simulated is sketched in Fig. 1: a flow over a flat plate that rotates around the wall-normal axis at a rate $\omega = f\hat{e}_z$ is exposed to a no-slip boundary condition and a spanwise pressure gradient ∇p along the direction $\hat{\mathbf{e}}_y$. In absence of interaction with the surface, i.e. far away from the surface, the flow is in geostrophic balance and the geostrophic wind blows along $\hat{\mathbf{e}}_x$; in vicinity of the surface, the Coriolis force is reduced due to the hampering effect of surface friction on the mean wind and the wind vector

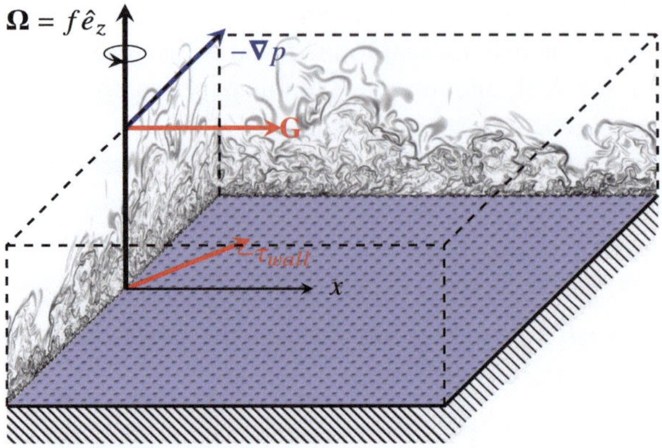

Fig. 1 **a** Sketch describing the Ekman layer that develops from the interaction between a flow in geostrophic balance $2\Omega\mathbf{k} \times \mathbf{G} = -\nabla p$ with the no-slip condition $\mathbf{u}(z = 0) = 0$ at the lower boundary. Density stratification may be imposed by a negative surface buoyancy $B_{\text{wall}} = -B_{\text{ref}}$. Cuboids on the ground illustrate roughness

turns towards the low pressure $-\nabla p$ [15]. In absence of buoyancy, i.e. in a neutrally stratified flow, this system has a statistical equilibrium that is defined by a single parameter, namely the Reynolds number, which is the scale separation between the largest scales in the system and the smallest (viscous) scale; a second parameter characterizing the strength of stratification (Richardson, Froude number or Obukhov length) and a time dependency enter the system once stratification is allowed.

For some questions in the ABL, the rotation and externality of the flow are of limited importance—for instance, the corner stone of surface-layer similarity, namely Monin–Obukhov similarity theory [35, 36] considers the surface layer in analogy to a non-rotating, non-external flow and has proven extremely useful in doing so.

2.2 *The Problem and Forcing in Spectral Space Under Different Modelling Paradigms*

The multiphysical nature of the atmosphere complicates process understanding. Further, the sheer size of geophysical problems normally prohibits assessing the scales involved without a break of modelling paradigms: an easy estimate for a boundary layer of one kilometer depth with ordinary turbulence down to 1 mm yields 10^6 degrees of freedom per spatial direction and a similar number of degrees of freedom in time. The broad-scale nature is best illustrated in spectral space as in Fig. 2, where different scaling hypotheses are used for different scales:

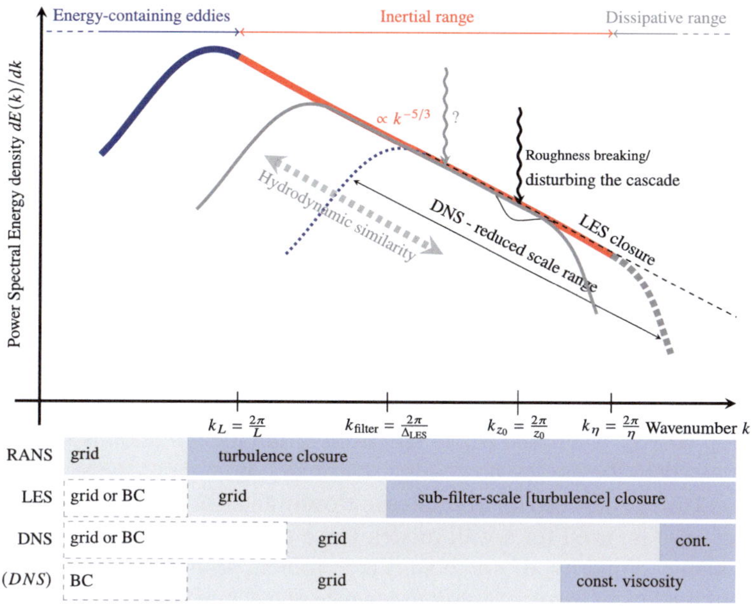

Fig. 2 Schematic of power spectral energy density in a turbulent flow. The full spectrum of turbulent motion for a real-world geophysical flow is shown as thick solid lines in blue color for the scales corresponding to the energy-containing eddies, red for the inertial cascade and gray for the viscous region. The gray bars on the bottom indicate the resolved range of scales; subgrid or sub-filter scale closures are indicated in blue. L refers to the large scale set by the geometry of the flow (for instance the boundary-layer depth in the ABL), Δ_{LES} to the effective grid spacing of an LES, η to the Kolmogorov scale indicative of the onset of the dissipative range. The hydrodynamically similar configuration of a direct numerical simulation is labelled as $\sim (DNS)$ and the corresponding spectrum is indicated by a thin, solid gray line

(i) The **energy-containing eddies** at the large-scale end of the spectrum ($k_L \to 0$) are set by the geometry and forcing of the problem; these eddies are anisotropic and provide the flow with energy.
(ii) In the **dissipative range**, the viscous forces are sufficiently large such that—given the large-scale forcing—turbulent eddies cannot survive.
(iii) The **inertial range** spans in between the energy-containing range of scales and the dissipative range. This is where the turbulence cascades according to the seminal works by [25, 37] and the eddies become more isotropic along the cascade of energy down the range of scales.

Modeling paradigms differ—depending on the physical focus. In Renolds-averaged Navier–Stokes (RANS) models, turbulence is treated as a (mostly linear) diffusive agent. Large-Eddy simulation (LES, for a review of atmospheric applications, see [43]) acknowledges that in some conditions turbulent mixing needs to be resolved; only beyond the filter scale turbulent mixing is parameterized by the subfilter-scale (SFS) closure assuming the scale is well within the inertial range of turbulence [39]. The spectrum of turbulence is then extrapolated to very high scale separation not taking into account of viscous dissipation, i.e. viscous dissipation is neglected in relation to turbulent dissipation; it is only due to the coupling with the wall through Monin–Obukhov similarity theory (or a different wall model) that the LES learns about the finite Reynolds number of the real-world problem.

Under certain conditions, for instance strong stability [1, 31], strong heterogeneity, or strong non-stationarity, LES becomes problematic: the underlying paradigms, in particular the assumptions on existence of resolved large scales of turbulence and their separation from the viscous scaling regime, are challenged. Non-existence of an LES closure that successfully covers this range is hard to prove, and there is indeed ongoing effort to overcome this challenge [8–10, 23]. However, an approach independent of turbulence closure may pave the way for process studies in these challenging regimes. Such an approach is available by means of direct numerical simulation (DNS), where the whole spectrum of turbulent motion is resolved [34]; in that case, both the energy-containing scales and the viscous range are explicitly defined and there is no need for a turbulence closure and–for resolution of all turbulent structures–also no need for a wall model. Once these scales are explicitly defined, hydrodynamic similarity allows to shift the problem arbitrarily along the range of scales without loss of accuracy—a consequence of Buckingham's Π theorem [6]—as shown in Fig. 2.

By shifting the representation along the range of wavenumbers, the size of the problem can be altered by orders of magnitude, similar to the truncation of the problem at wavenumber k_{filter} in LES; but this comes at the price of a Reynolds number that is not exactly matched. Here, the routinely employed concept of Reynolds-number similarity can help: It postulates the existence of fully developed turbulence for a sufficiently large–but finite–Reynolds number [5]. Certain statistics of fully developed turbulence, such as dissipation [13] or profiles of the mean velocity [4], become independent of the Reynolds number when appropriately scaled; other statistics, such as the near-wall maximum in velocity fluctuation depend on Re [3]. For such quantities the externality of the flow may impact on the near-wall scaling behavior [12]. It appears that for certain statistics in Ekman flow, fully-developed turbulence is reached with the Reynolds numbers that became possible due to an increase of computing capabilities over the past decades.

While the exact Reynolds number with respect to the geophysical problems is not matched, in principle, a physically analogous configuration exists; it is, however, most likely not realizable on earth. We hence, employ the concept of Reynolds-number similarity to rescale results to the relevant scales. In trainABL, we exploit the robust features of direct simulations across a range of Reynolds numbers to quantify these scaling laws and thus allow for a quantitative extrapolation of the insight from DNS, whereas so far we are mostly limited to qualitative comparison between DNS and the outdoor ABL.

In a dimensional representation of the turbulent flow, a forcing or process at a particular wavenumber k_{forc}, for instance the forcing due to roughness at a scale related to the roughness parameter labelled as k_{z_0}, is either resolved (and interactions with the turbulent motion need to be taken into account explicitly), or it is unresolved (and needs to be represented as part of the subgrid closure where a zero-order approximation is the trivial neglect of a process). In any case, such forcing modulates the energy content—here around the wavenumber k_{z_0}—and thus also below k_{z_0}. If a non-dimensional representation is chosen, i.e. one makes use of Buckingham's Π theorem, an additional aspect is the scaling of the forcing: It is a priori unclear how k_{z_0} changes when the absolute scales of motion are changed, which may be due to mechanisms of a viscid, i.e. $k_{z_0}/k_\eta \propto \mathbf{Re}_{z_0} = const.$ (cf. Eq. (1)), or inviscid nature, i.e. $k_{\mathrm{forc}}/k_\eta = f(Re)$.

2.3 *The Computational Burden*

Given the potential strength of the concept outlined above, the question remains: Why cannot we reduce the degrees of freedom so much that we do not need access to TIER-0 compute infrastructure? (In fact, the first DNS of Ekman flow have been carried out in the early 1990s [11] on supercomputers with less power than modern laptops.) First, these early simulations were in what was called an intermediate Reynolds-number regime that can be understood as turbulent but not far away from the transition to turbulence. This means that low-order statistics, such as velocity profiles, quantitatively match those of a turbulent flow in an approximate manner; convergence to an Re-independent limit can, however, neither be understood nor quantified. Second, consideration of additional processes may sometimes require a wider range of scales. An illustrative example here is the Buoyancy Reynolds number introduced by [18]: if this Reynolds number is larger than that of the flow, it does not have much of an effect in the simulated flow whereas it may substantially impact the dynamics at much larger Reynolds number [1, 2].

In the context of our rough problem, suppose we want to model a rough flow with a roughness Reynolds number

$$Re_{z_0} = z_0 u_\star / \nu = z_0^+ \tag{1}$$

on the order of 50. If now, we let the friction Reynolds number $Re_\tau = 500$, then roughness elements would span more than 10% of the ABL, i.e. the surface layer would cease to exist in favor of an extremely deep roughness layer. On the other hand, a roughness element spanning about 1 m in a 1 km thick ABL at $Re_\tau = 500$ would extend over 0.5 wall units when scaled geometrically, but the flow would be hydrodynamically smooth. This problem is alleviated when $Re_\tau = 5000$: in that case, we obtain a height of the roughness elements of 1% of the ABL depth (allowing, at least in principle, for the surface layer to exist) and 5 wall units in geometric scaling (corresponding to a rough wall indeed). Hence, clearly, the consideration of roughness (as an additional process) also requires a wider range of scales to be represented in the numerical solution–manifest here in the need for a higher Reynolds number.

3 Research Objective and Computational Strategy

Research questions. As outlined above, the objective of this work is to use the HAWK system for targeted high-performance computing of taylored fluid-mechanical systems that resemble specific problems in atmospheric science. Of the issues identified above, we used the current allocation on HAWK to tackle the problem of roughness: We have implemented an immersed boundary method and could obtain a first direct validation against data from a wind tunnel in the rough boundary layer. This paves a way towards the quantitative investigation of roughness effects that does not rely on the specification of the bulk roughness parameter z-nought (z_0); Preliminary results that became available in our project indicate that roughness parameters and the exact drag partitioning in vicinity of the surface can be established to surprising accuracy.

Simulations. Simulations used here are shown in Table 1. An existing set of simulations for $Re_\tau = 1400$ (gray shading, data from [1, 2]) is extended in a horizontal domain extent up to 8×10^3 viscous units. This precluded developing an immersed boundary representation in `tLab` that now allows for a direct comparison of results with wind-tunnel data over rough surfaces thus facilitating a validation of the code with experimental reference data. The immersed boundary condition explicitly represents three-dimensional roughness elements as bluff bodies and thus requires us to resolve the viscous layer around these three-dimensional elements. Two unexpected complications required us to spend a larger share than originally expected on the rough problem:

1. Direct representation of bluff bodies imposes additional constraints on the numerical stability and consistency. It turned out that `tLab` needed to move to a staggered pressure solver and which required major development efforts and heavy testing, also for turbulent problems at scale, to ascertain that the modifications of

Table 1 DNS data used in this work; the gray-shaded set of data is reused from a previous compute project at Jülich Supercomputing Centre. The friction Reynolds number Re_τ is based on the boundary layer depth scale $\delta = u_\star/f$ for Ekman flow and the channel half-height in channel flow. The mean obstacle height H for the cases with a rough wall is expressed here in wall units of the smooth case

Configuration	Mean obstacle height $\overline{H}^+$	Re_τ	L_{xy}/δ	$N_x \times N_y \times N_z$	wall-clock / it	#nodes	node-hours
Smooth Ekman	0	1403	20.4	$3072 \times 6144 \times 512$			
Smooth Ekman	0	1403	5.1	$3072 \times 3072 \times 656$	22	64	23,500
Rough Tests	0-30	1000-2300	5.1	$3072 \times 3072 \times 656$	22	64	30,000
Rough Ekman	10	1700		$3072 \times 3072 \times 656$	22	64	35,200
	20	2000		$3072 \times 3072 \times 656$	22	64	35,200
	30	2300		$3072 \times 3072 \times 656$	22	64	35,200
							Σ 159,100

the numerical schemes do not affect the quality of the solution. Thus, a substantial amount of the allocation was spent for testing and development.

2. The representation of bluff bodies changes the topology of the boundary from a one-dimensional topology (characterized by a single vector normal to the wall) to a three-dimensional surface where boundaries can be oriented normal to any of the cartesian directions in the problem; this comes with the formation of viscous sub-layers not only in the vertical, but also in horizontal directions. These viscous sub-layers need to be explicitly resolved, which changes the resolution requirements of the simulations from $\Delta(x^+, y^+, z^+)_{\text{smooth}} \approx (8, 4, 1)$ (the resolution in the streamwise, spanwise and wall-normal direction) to $\Delta(x^+, y^+, z^+)_{\text{rough}} \approx (1, 1, 1)$. This is a factor of $\mathcal{O}(10)(\approx 32)$ in the number of grid points and a factor of $\approx 4 - 8$ for the time step, which makes the simulations up to two orders of magnitude ($\approx 64 - 128$) more expensive. We extensively tested what grid is needed to actually resolve the horizontal viscous sub-layers, and could limit the grid to $\approx \Delta(x^+, y^+, z^+)_{\text{rough}} \approx (2, 2, 1)$ and a time step is reduced to $\approx \frac{1}{3}$ requiring about 3 times more iterations to run the problem into a statistically steady state. The increase in number of grid points is approximately compensated by a smaller domain of $d^2 \approx (1/4 \times 1/4)$ of the original horizontal extent. This is still large in comparison to many other studies of the problem, but it sacrifices the accuracy on interaction between large outer-scale phenomena in Ekman flow with the viscous and rough dynamics close to the surface and to some extent the statistical convergence of the simulated problem. However, the reduced time step proved as a major hurdle for the simulations of the rough problems—and other than originally expected, we were not able to finish the proposed parameter study within the framework of the current allocation.
3. The introduction of bluff bodies causes the mean profiles to change which introduces a background unsteadiness and drift related to the inertial oscillation into the problem that comes with a new time scale and slows down the convergence of the problem to statistical equilibrium.

Algorithm. The problem is solved numerically by `tLab`,[1] an open-source toolsuite to simulate and analyze turbulent flow. We use here a fourth-order–five-stage Runge–Kutta integration and sixth-order compact schemes for spatial derivatives in all directions. The incompressibility constraint is enforced by a fractional step approach where the Poisson equation for the pressure field is solved to machine accuracy using a combined spectral/compact approach as described in [32].

`tLab` is routinely run on large compute systems and has been shown to scale up to 256 000 physical CPU cores using 128 000 MPI tasks which each two threads on juqueen at Jülich supercomputing centre. The operational settings used on hawk routinely employ 8192 to 16384 MPI tasks. Scaling of the code up to this configuration is demonstrated in Fig. 3. We find excellent scaling on the level of a single node

[1] https://github.com/turbulencia/tlab.

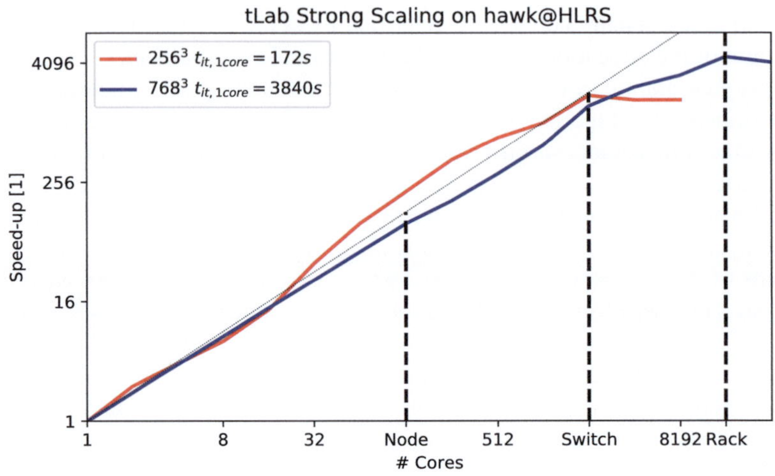

Fig. 3 Strong scaling of tLab on the supercomputer HAWK at Hochleistungsrechenzentrum Stuttgart (HLRS). We show the averaging for carrying out a single iteration of the code for two benchmark cases, one using $256 \times 256 \times 256$ grid points (labelled as 256^3) and the analogous cube using 768^3 grid points

with even a super-linear scaling behavior up to 512 MPI tasks for the smaller case—despite only using 64 MPI tasks on the 128 physical cores available on the two AMD chips per node. This indicates that the code heavily benefits from reduced number of cache misses (which is easily explained by frequent local matrix transpositions) and that the MPI implementation is performing very well and nicely exploits the network infrastructure up to about the level of a single switch (16 nodes/2048 cores). Speed-up for the smaller cases saturates at the switch-level, i.e. using 1024 MPI tasks on 2048 physical cores. In that case, there are only 8192 grid points per processor such that the network-load, in particular in terms of the number of message to be sent, becomes disproportionately large with respect to the computing benefit involved in using more nodes. In fact, for the larger case, the performance only saturates beyond the Rack level, i.e. using 256 and more nodes. This is consistent with the simulation size. The sub-linear growth of the speed-up with increasing number of nodes used indicates a higher network cost due to both a larger number of hops involved in the communication and a growing number of messages that need be transferred across the blocking dimensions of the network topology.

Implementation of an immersed boundary condition for rough surfaces. The consideration of roughness effects in turbulent flow requires an appropriate boundary condition, which was implemented and tested in `tLab` using the hawk system. Several options compatible with a structured finite difference mesh exist in the form of immersed boundary methods among which we chose to follow [19, 20] and implement an alternating direction forcing to impose the boundary condition within the simulation cuboid. With an alternating direction forcing, the physical flow field is kept at zero whenever a fluid point falls underneath the surface; however, when a derivative is calculated, the interior of the solid region is treated as if it would contain values that are artificially generated to ascertain smooth transition across the boundary and thus avoid artificial oscillations or ringing. This yields higher (second-order) accuracy than direct forcing or simple restore methods which normally cause a degradation to first order. In the `tLab` implementation of the algorithm, we, however use an unstaggered grid for the velocity; pressure is staggered along the periodic horizontal directions only. In the vertical, pressure is calculated on the unstaggered grid to avoid the definition of a computational boundary condition or ghost cells. Instead, the staggering is mimicked by a compact filtering operation of the pressure forcing in the vertical direction. The interpolation of the forcing term for the pressure Poisson equation is carried out by a combined sixth-order compact interpolatory derivative operation following [28]. This saves parallel transposes, operation counts, and it increases the accuracy of the interpolation. This procedure eliminates pressure oscillations which tend to occur when pressure is computed on an unstaggered grid. Further, the virtual flow fields needed when derivatives are calculated in the time-stepping procedure of the simulation are generated by a cubic spline interpolation rather than Lagrange polynomials. These splines have less of a tendency to overshoot in comparison with the Lagrange polynomials. We have validated this based on wind-tunnel data and ascertained that the results of turbulent flow agree with that of other solvers and our own solver without these modification.

4 Explicit Representation of Roughness

Roughness ubiquitously characterizes the surface underneath the ABL. Nonetheless, our understanding of roughness is still squeezed into a single bulk parameter, termed z-nought, a reference height in which all effects of roughness are lumped together. From the perspective of the logarithmic law for the mean velocity, the roughness height z_0 is a divisor of the height above the surface measure in viscous units. This suggests that roughness elements limit the range of scales available for turbulence according to their vertical extent. In other words, aloft the roughness sub-layer there should exist an effective Reynolds number Re_{eff} that is defined from Re_τ and Re_{z_0}, better characterizing the turbulent state of the ABL as a whole. We are working towards identifying the exact formulation of this Reynolds number which should be

$$Re_{\text{eff}} = f(Re_\tau, Re_{z_0}) \propto Re_\tau / Re_{z_0}, \tag{2}$$

and want to quantify modifications of higher-order statistics in the surface layer under the reduced scale separation Re_{eff}. Further questions regarding roughness relate to the triggering of intermittency events by roughness elements and related dynamics and the limits of validity for Monin–Obukhov similarity theory when roughness elements become very large, sparse, or dense. When resorting to the z-nought-concept, the properties of roughness are lumped into a single parameter, but an explicit representation warrants appropriate choice of key roughness parameters to determine the mechanical heterogeneity of the boundary condition.

4.1 *Inertial Oscillations and Transition Towards New Equilibrium Profiles*

The inertial oscillation in Ekman flow is tightly linked to its hodograph and a key manifestation of rotation in the flow: Perturbations of the velocity profile inevitably cause oscillations about the long-term statistically stead quasi-equilibrium state of the flow. This also happens when the flow over a smooth wall is altered by introduction of roughness (which causes a new equilibrium profile due to the additional drag introduced to the problem. The shift of the velocity profiles due to roughness can be well-understood by studying the hodograph (Fig. 4): First, we find that larger obstacles cause lower velocity and stronger rotation, the former being a manifestation

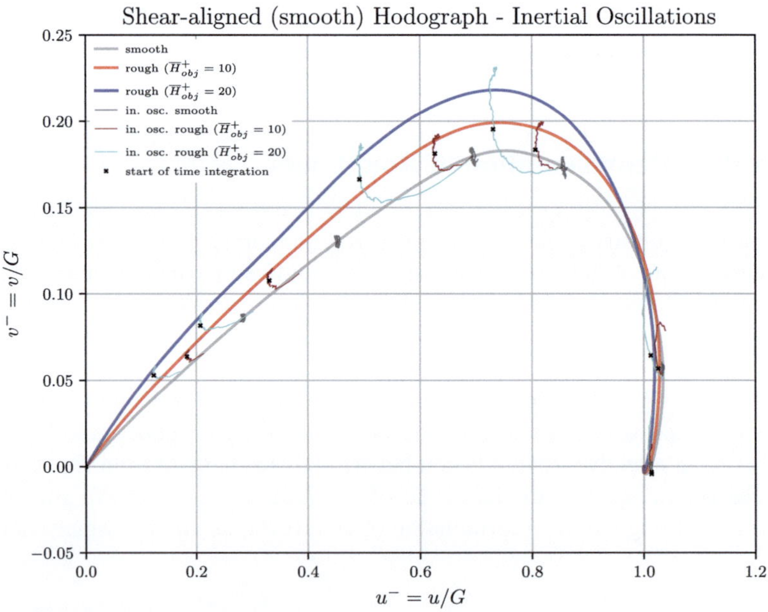

Fig. 4 Characteristic Ekman-layer hodograph of a smooth simulation in equilibrium (gray) and two rough simulations with obstacle heights $\overline{H}^+ = 10$ and $\overline{H}^+ = 20$ corresponding to roughness elements of a size of 0.6 and 1% of the boundary layer depth. The thick lines show the hodograph data and are averaged over approximately one eddy-turnover period $\delta/u_\star = 1/f$. Thin lines show the time evolution of the vector (u, v) at six heights corresponding to the layers driven by different dynamics within the boundary layer from the viscous sub-layer (left; near the origin) to the top of the boundary layer at $(u^-, v^-) = (u/G, v/G) = (1, 0)$

of increased coupling of the turbulence to the surface while the latter is due to increased coupling of the outer layer with the inner layer of the flow in the presence of roughness. Second, the damping of the inertial oscillation is stronger close to the surface, a consequence of increased turbulent mixing (which damps the inertial oscillation) close to the surface: While we see an over-damped transition to the new equilibrium profiles in the viscous and roughness sub-layers (lowest to sampling heights), the outer sampling height shows rather clean inertial oscillations that do not appear damped over the relatively short period of time sampled here—future simulations will complete these inertial cycles and allow us to estimate damping coefficients from Poincaré intersects of the inertial oscillation.

4.2 Visualizations of the Turbulent Flow in the Boundary Layer and Within the Roughness Sub-layer

Instantaneous visualizations of the flow (Fig. 5) illustrate the local dynamics and give an illustration of the problem setting. In Fig. 5, we see that the roughness elements are placed approximately equidistant, but their horizontal and vertical size is varied randomly to avoid singular forcing of the problem in spectral space (which a regular

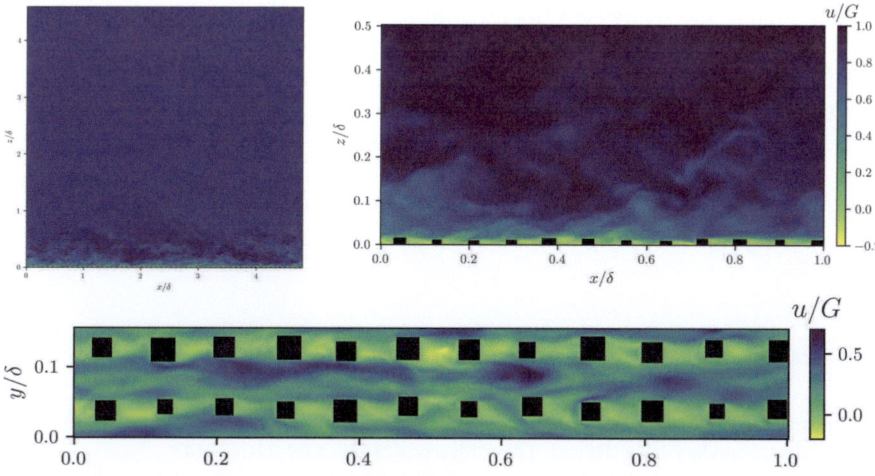

Fig. 5 *Instantaneous snapshots of velocity fields illustrate the resolved dynamics for the case with obstacles (black rectangles) of height* $\overline{H}^+ = 20$. **(Top left)** The full domain in a streamwise-vertical intersect. **(Top right)** The same intersect as in a but zoomed in to a portion with the extent of the boundary-layer depth-scale δ in the horizontal direction and 0.5δ in the vertical direction—corresponding to about 80% of the boundary extent measure in terms of the turbulent stress criterion δ_{95} **(Bottom)** Top-down view (streamwise-spanwise intersect) onto a plane within the roughness layer

roughness pattern would introduce). On the lee-side of obstacles, the velocity is close to zero, and locally even recirculating flow can be found. In between the obstacles, the flow is channeled and accelerated. A viscous layer is found on the lateral sides (bottom panel of Fig. 5) and on top of the obstacles (top right panel of Fig. 5). Further, it appears from the top and top left panels of Fig. 5) that ejections are forming within the displacement flow ahead of an obstacle. Despite the fact that we aligned our coordinate system for simulations such that the smooth simulation has a wall shear stress approximately parallel to the direction Ox, a clear diversion of flow features to the right is found on the bottom panel. This indicates that—as a consequence of changing drag in the flow, the rotation also changes as is well seen in Fig. 4 and discussed above.

5 Conclusions

The use of TIER-0 HPC in combination with modern computing approaches and application of scale similarity arguments allows us to tackle important issues in boundary-layer meteorology. Already during the first year of our compute project on the HAWK-system, we were able to extend our existing set of benchmark simulations reaching a scale separation of $Re_\tau \approx 3000$. We thus reach a new level of accuracy for estimation of the Reynolds-number dependency of some higher-order features in the flow such as the spanwise velocity profile close to the ground. We further show the use of a massively parallel and highly optimized immersed-boundary method and were able to validate our algorithm against data from a wind tunnel. In the current allocation phase, we could open a new horizon for direct numerical simulation in the planetary boundary layer by allowing the explicit representation of roughness. After some 70 years were roughness could–across all scales and paradigms of modelling–only be considered via the buulk roughness parameter z-nought (z_0), this allows quantitative insight to the dynamics within the roughness sub-layer.

References

1. Ansorge, C., Mellado, J.P.: Global intermittency and collapsing turbulence in the stratified planetary boundary layer. Bound.-Layer Meteorol. **153**(1), 89–116 (2014)
2. Ansorge, C., Mellado, J.P.: Analyses of external and global intermittency in the logarithmic layer of Ekman flow. J. Fluid Mech. **805**, 611–635 (2016)
3. Baars, W.J., Marusic, I.: Data-driven decomposition of the streamwise turbulence kinetic energy in boundary layers. Part 1. Energy spectra. J. Fluid Mech. **882** (2020)
4. Barenblatt, G.I.: Scaling laws for fully developed turbulent shear flows. Part 1. Basic hypotheses and analysis. J. Fluid Mech. **248**, 513–520 (1993)
5. Barenblatt, G.I., Goldenfeld, N.: Does fully developed turbulence exist? Reynolds number independence versus asymptotic covariance. Phys. Fluids **7**(12), 3078–3082 (1995)
6. Buckingham, E.: On physically similar systems; Illustrations of the use of dimensional equations. Phys. Rev. **4**(4), 345–376 (1914)

7. Calaf, M., Meneveau, C., Meyers, J.: Large eddy simulation study of fully developed wind-turbine array boundary layers. Phys. Fluids **22**(1), 015110 (2010)
8. Chinita, M.J., Matheou, G., Miranda, P.M.A.: Large-eddy simulation of very stable boundary layers. Part I: Modeling methodology. Q. J. R. Meteorol. Soc. **148**(745), 1805–1823 (2022)
9. Chinita, M.J., Matheou, G., Miranda, P.M.A.: Large-eddy simulation of very stable boundary layers. Part II: Length scales and anisotropy in stratified atmospheric turbulence. Q. J. R. Meteorol. Soc. **148**(745), 1824–1839 (2022)
10. Chung, D., Matheou, G.: Large-eddy simulation of stratified turbulence. Part I: a vortex-based subgrid-scale model. J. Atmos. Sci. **71**(5), 1863–1879 (2014)
11. Coleman, G.N., Ferziger, G.N., Spalart, P.R.: A numerical study of the turbulent Ekman layer. J. Fluid Mech. **213**(-1), 313 (1990)
12. da Silva, C.B., Hunt, J.C.R., Eames, I., Westerweel, J.: Interfacial layers between regions of different turbulence intensity. Annu. Rev. Fluid Mech. **46**(1), 567–590 (2014)
13. Dimotakis, P.E.: Turbulent mixing. Annu. Rev. Fluid Mech. **37**(1), 329–356 (2005)
14. Dörenkämper, M., Witha, B., Steinfeld, G., Heinemann, D., Kühn, M.: The impact of stable atmospheric boundary layers on wind-turbine wakes within offshore wind farms. J. Wind. Eng. Ind. Aerodyn. **144**, 146–153 (2015)
15. Ekman, V.W.: On the influence of the earth's rotation on ocean currents. Ark. Mat. Astron. Fys. **2**, 1-53 (1905), **2**, 1–53 (1905)
16. Etling, D.: Theoretische Meteorologie, 2nd edn. Eine Einführung. Springer, Berlin, Heidelberg (2002)
17. Fernando, H.J.S., Weil, J.C.: Whither the stable boundary layer?: a shift in the research agenda. Bull. Am. Meteorol. Soc. **91**(11), 1475–1484 (2010)
18. Flores, O., Riley, J.J.: Analysis of turbulence collapse in the stably stratified surface layer using direct numerical simulation. Bound.-Layer Meteorol. **139**(2), 241–259 (2011)
19. Gautier, R., Laizet, S., Lamballais, E.: A DNS study of jet control with microjets using an immersed boundary method. Int. J. Comput. Fluid Dyn. **28**(6–10), 393–410 (2014)
20. Giannenas, A.E., Laizet, S.: A simple and scalable immersed boundary method for high-fidelity simulations of fixed and moving objects on a Cartesian mesh. Appl. Math. Model. **99**, 606–627 (2021)
21. Gryning, S.-E., Batchvarova, E., Brümmer, B., Jørgensen, H., Larsen, S.: On the extension of the wind profile over homogeneous terrain beyond the surface boundary layer. Bound.-Layer Meteorol. **124**(2), 251–268 (2007)
22. Haiden, T., Janousek, M.: Update on ECMWF forecast performance (2014)
23. Jiménez, M.A., Cuxart, J.: Large-eddy simulations of the stable boundary layer using the standard Kolmogorov theory: range of applicability. Bound.-Layer Meteorol. **115**(2), 241–261 (2005)
24. Kelly, M., Troen, I.: Probabilistic stability and 'tall' wind profiles: theory and method for use in wind resource assessment. Wind Energy **19**(2), 227–241 (2016)
25. Kolmogorov, A.N.: Dissipation of energy in locally isotropic turbulence. Dokl Akad Nauk SSSR **434**(1890), 15–17 (1941)
26. Laizet, S., Lamballais, E.: High-order compact schemes for incompressible flows: a simple and efficient method with quasi-spectral accuracy. J. Comput. Phys. **228**(16), 5989–6015 (2009)
27. Lee, M., Moser, R.D.: Direct numerical simulation of turbulent channel flow up to. J. Fluid Mech. **774**, 395–415 (2015)
28. Lele, S.K.: Compact finite difference schemes with spectral-like resolution. J. Comput. Phys. **103**(1), 16–42 (1992)
29. Li, Q., Gentine, P., Mellado, J.P., McColl, K.A.: Implications of nonlocal transport and conditionally averaged statistics on Monin–Obukhov similarity theory and Townsend's attached eddy hypothesis. J. Atmos. Sci. **75**(10), 3403–3431 (2018)
30. Llaguno, M., Bou-Zeid, E., Hultmark, M.: The influence of building geometry on street canyon air flow: validation of large eddy simulations against wind tunnel experiments. J. Wind. Eng. Ind. Aerodyn. **165**, 115–130 (2017)

31. Mahrt, L.: Stably stratified atmospheric boundary layers. Annu. Rev. Fluid Mech. **46**(1), 23–45 (2014)
32. Mellado, J.P., Ansorge, C.: Factorization of the Fourier transform of the pressure-Poisson equation using finite differences in colocated grids. ZAMM - J. Appl. Math. Mech./Zeitschrift für Angewandte Mathematik und Mechanik **92**(5), 380–392 (2012)
33. Mellado, J.P., van Heerwaarden, C.C., Garcia, J.R.: Near-surface effects of free atmosphere stratification in free convection. Bound.-Layer Meteorol. **159**(1), 69–95 (2016)
34. Moin, P., Mahesh, K.: Direct numerical simulation: a tool in turbulence research. Annu. Rev. Fluid Mech. **30**, 539–578 (1998)
35. Monin, A.S.: The atmospheric boundary layer. Annu. Rev. Fluid Mech. **2**, 225–250 (1970)
36. Monin, A.S., Obukhov, A.M.: Osnovnye zakonomernosti turbulentnogo peremeshivanija v prizemnom sloe atmosfery (Basic Laws of Turbulent Mixing in the Atmosphere Near the Ground). Trudy geofiz. inst. AN SSSR **24**(151), 163–187 (1954)
37. Obukhov, A.M.: O Raspredelenii energii w spektre turbulentnowo potoka. Isvestija Akademii Nauk SSSR **1941**(4–5), 453–466 (1941)
38. Optis, M., Monahan, A., Bosveld, F.C.: Moving beyond Monin-Obukhov similarity theory in modelling wind-speed profiles in the lower atmospheric boundary layer under stable stratification. Bound.-Layer Meteorol. **153**(3), 497–514 (2014)
39. Pope, S.B.: Turbulent Flows. Cambridge University Press (2000)
40. Sandu, I., Beljaars, A., Bechtold, P., Mauritsen, T., Balsamo, G.: Why is it so difficult to represent stably stratified conditions in numerical weather prediction (NWP) models?: stable conditions in NWP models. J. Adv. Model. Earth Syst. **5**(2), 117–133 (2013)
41. Spalart, P.R.: Theoretical and numerical study of a three-dimensional turbulent boundary layer. J. Fluid Mech. **205**(-1), 319 (1989)
42. Steeneveld, G.-J.: Current challenges in understanding and forecasting stable boundary layers over land and ice. Front. Environ. Sci. **2** (2014)
43. Stoll, R., Gibbs, J.A., Salesky, S.T., Anderson, W., Calaf, M.: Large-eddy simulation of the atmospheric boundary layer. Bound.-Layer Meteorol. **177**(2–3), 541–581 (2020)
44. Antoon van Hooft, J., Baas, P., van Tiggelen, M., Ansorge, C., van de Wiel, B.J.H.: An idealized description for the diurnal cycle of the dry atmospheric boundary layer. J. Atmos. Sci. **76**(12), 3717–3736 (2019)
45. Vitart, F.: Evolution of ECMWF sub-seasonal forecast skill scores: evolution of the ECMWF sub-seasonal forecast skill. Q. J. R. Meteorol. Soc. **140**(683), 1889–1899 (2014)
46. Wyngaard, J.C.: Toward numerical modeling in the "Terra Incognita". J. Atmos. Sci. **61**, 11 (2004)

WRF Simulations to Investigate Processes Across Scales (WRFSCALE)

Hans-Stefan Bauer, Oliver Branch, and Benjamin Körner

Abstract Several scientific aspects ranging from boundary layer research and land modification experiments to turbulence research were addressed in this project with the Weather Research and Forecast (WRF) model and the Parallelized Large Eddy Simulation Model (PALM) applying resolutions from km-scale down to meter-scale. Case study simulations in as different regions as the central United States, The United Arab Emirates and southwestern Germany were performed to investigate the evolution of the convective boundary layer. The multi-nested WRF-setup driven by the operational analysis of the European Centre for Medium-range Weather Forecasts (ECMWF), high-resolution terrain, land cover and soil data sets simulated a realistic evolution of the turbulent internal structure of the boundary layer including realistic transitions between nighttime and daytime conditions and interactions between the surface and the overlying atmosphere. Land modification simulations in the United Arab Emirates showed that plantations of $10 \times 10\,\text{km}^2$ have only a small influence, but larger plantations clearly modify the weather patterns in a way that more precipitation reaches the desert. Monin-Obukhov Similarity Theory (MOST), based on simplyfied assumptions is commonly used for parameterizing turbulent fluxes. However, several processes are ignored that are the more important the finer the resolution is. Therefore, the implementation of MOST in the models will be improved with very high-resolution idealized PALM simulations, providing training data for neural networks.

H.-S. Bauer (✉) · O. Branch · B. Körner
Institute of Physics and Meteorology, University of Hohenheim, Stuttgart, Germany
e-mail: Hans-Stefan.Bauer@uni-hohenheim.de

T. Ludwig et al. (eds.), *High Performance Computing in Science and Engineering '23*,
https://doi.org/10.1007/978-3-031-91312-9_32

1 Introduction and Motivation

Numerical models provide the full 4D representation of the atmosphere and a consistent state with respect to the 3D thermodynamic atmospheric fields, cloud water, cloud ice, and diagnostic variables such as precipitation. Therefore, they are excellent tools to improve our understanding of atmospheric processes across scales and complement to observation-based approaches.

Commonly-used mesoscale models with resolutions on the km scale do not permit the explicit simulation of smaller-scale processes such as large turbulent eddies (∼100–2000 m), and the whole turbulence spectrum is parameterized. Further improvements are expected if a chain of model runs is performed down to the turbulence-permitting (TP) scale (100 m resolution and below), as further details of land-surface atmosphere (LA) interaction are resolved. Nevertheless, processes between the atmosphere and the land surface are still to date not completely understood and subject of ongoing research.

The WRF model system [1], described with more detail in earlier reports, provides the opportunity to perform LES simulations under realistic conditions because of the wide range of scales it can be applied over. Using a nesting strategy that covers synoptic and mesoscale circulations will ensure a forcing of the LES that is as close as possible to the real weather situation.

The Parallelized Large Eddy Simulation Model (PALM) [2] is used to run simulations of the atmospheric boundary layer. In order to resolve a large part of the atmospheric turbulence spectrum, a resolution of only a few meters is required. A special focus is on land surface heterogeneities and the secondary circulations that develop in response. Their impact on the turbulent fluxes of momentum and heat is of interest, as this mechanism is not represented in weather and climate models. The goal is to generate a dataset that allows the training of Neural Networks (NN) capable of identifying functional interrelationships between surface heterogeneities and turbulent fluxes.

2 Work Done Since March 2022

2.1 LES-PROC

2.1.1 LAFE Cases

The Land Atmosphere Feedback Experiment (LAFE) took place in August 2017 at the Southern Great Plains site of the Atmospheric Radiation Measurement Program (ARM) in Oklahoma. Many different instruments were brought to the site and the measurement strategy was optimized in order to derive as much information as possible. More details about the campaign, the applied measurement strategy and its aims can be found in [3].

Simulations and analyses for the clear sky LAFE intensive observation period 6 on the 23rd of August 2017 were continued and the optimized model setup described in the 2021 report of WRFSCALE was applied. The simulations were performed using 128 nodes, applying 32 MPI tasks with 4 OpenMP threads per node. Two nodes were reserved for IO-quilting to accelerate the I/O performance. A scaling diagram for the simulations is shown in Fig. 14 at the end of this report.

WRF was set-up using three domains with 2500, 500 and 100 m grid resolution. The outer domain was driven by the operational analysis of the European Centre of Medium Range Weather Forecasting (ECMWF). The size of the domains is 1000×1000 grid cells in the outer two domains and 1201×1001 grid cells in the innermost 100 m domain. The vertical resolution is the same in all domains with 100 vertical levels up to a pressure height of 50 hPa with 30 levels in the lowest 1500 m of the troposphere. Figure 1 illustrates the 3-domain setup. The location of the ARM site is marked in the different domains.

In addition to the standard WRF output in NetCDF, ASCII data from selected variables was written at each model time step and selected grid cells (e.g. the locations of the instruments). This so-called time series output was applied for direct comparison with the ARM and Hohenheim lidar systems. Figure 2 shows time-height cross sections of water vapor mixing ratio for the grid cell the ARM Raman lidar was located. The latter were continuously operated in vertical staring mode and offers the possibility to directly compare the model results with lidar observations.

The figure illustrates the capability of WRF to reasonably represent the internal structure of the daytime convective boundary layer. The turbulent eddies transporting moist air upwards in the boundary layer and drier air downwards are simulated properly. However, the size of the simulated eddies is larger than observed and the simulated boundary layer height (height of the sharpest vertical moisture gradient) is overestimated and rising during the three-hour period whereas it remains constant in the observation. Cause for the differences are probably a too coarse horizontal and vertical resolution applied in the simulation.

The differences can also be explained by the systematically overestimated surface temperature (see comparison WRF and Eddy-Covariance data in [4]. A higher surface temperature is associated with a stronger surface sensible heat flux which rises the boundary layer top. In addition, the model cannot reproduce the strong drying above the boundary layer, indicating that some important processes may not be captured properly e.g. due to the too coarse vertical resolution not capturing the observed strength of the inversion at the top of the boundary layer properly.

To investigate the influence of an increase in the horizontal and vertical resolution, another nest was included into the 100 m domain. To do this the NDOWN utility calculated the initial and boundary conditions for a 20 m domain from the output of the 100 m domain and the high-resolution lower boundary from a 20 m wrfinput file. During the calculation, a vertical refinement with a factor of two is done, so that the final 20 m domain applies 200 vertical layers. Figure 3 illustrates the influence of this refinement on the horizontal representation of turbulence.

It is clearly seen that much more detail is simulated with the 20 m resolution. Small filaments of updrafts and subsidence are much better represented. The stochastic

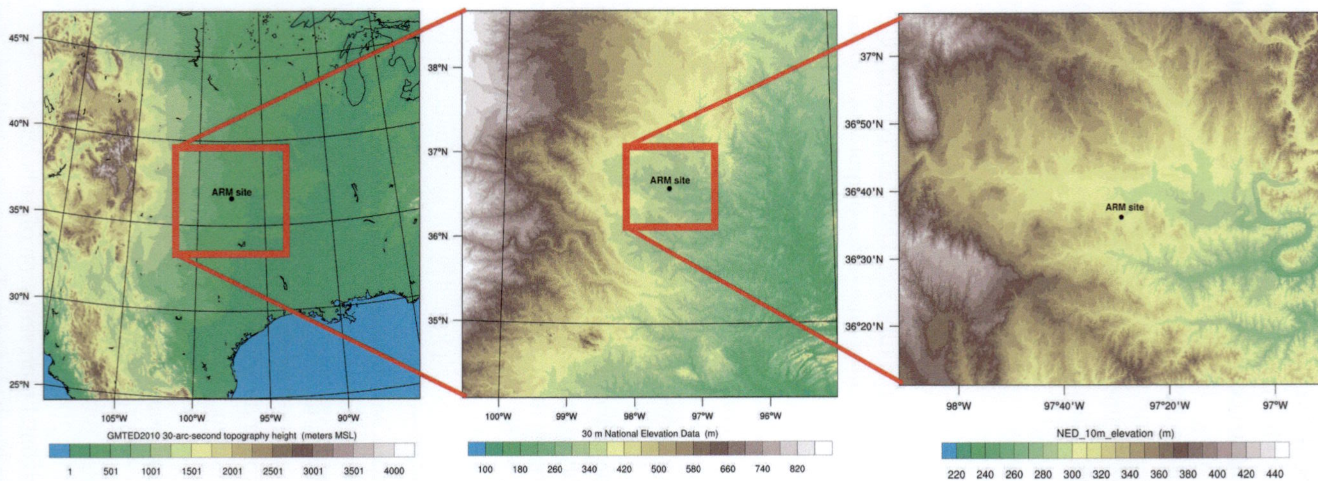

Fig. 1 Domain configuration for the LAFE simulations. From left to right the domains with 2500, 500 and 100 m resolution. The outer and middle domain have sizes of 1000×1000 grid cells and the inner domain consists of 1201×1001 points

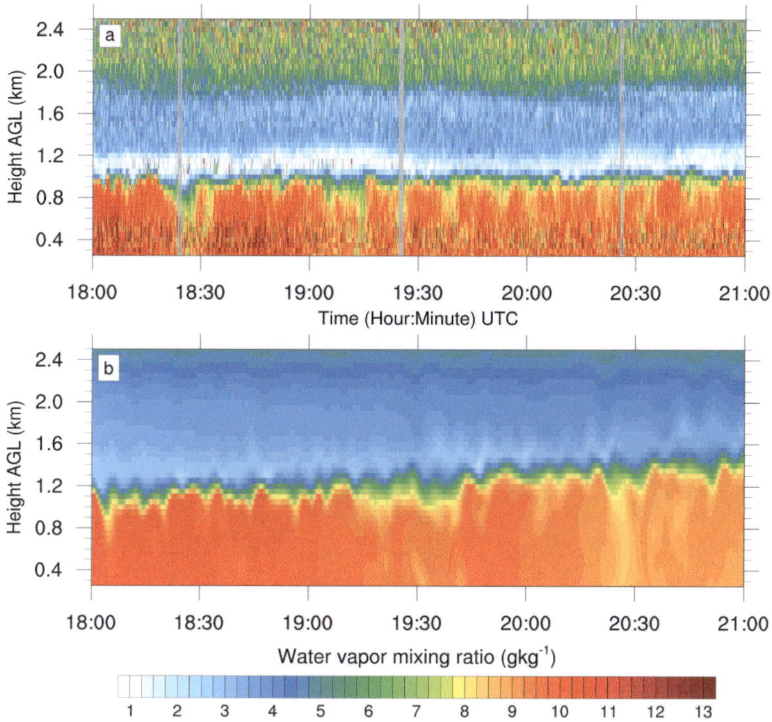

Fig. 2 Time-height cross sections ranging from 250 to 2500 m above ground of the water vapor mixing ratio (g/kg) for the 3-h period 18 to 21 UTC, 23 August 2017 for the ARM Raman lidar (panel a) and the WRF simulation with 100 m horizontal resolution (panel b)

nature of turbulence leads to similar features in some parts of the region and different developments in other parts. The influence of the higher resolution on the vertical structure of the boundary layer is currently investigated in detail.

2.1.2 Simulations at the Land Atmosphere Feedback Observatory (LAFO)

In spring 2023, the focus area changed to southwestern Germany. At the University of Hohenheim, the Land Atmosphere Feedback Observatory (LAFO) is operated. To investigate the temporal evolution of high-impact weather, turbulence evolution as well as land atmosphere interactions, a multi-nested configuration focusing on south-western Germany was set up. Figure 4 shows the new configuration for WRF version 4.4.1., applying domains with 1250, 250 and 50 m horizontal resolution with 100 vertical levels up to 60 hPa (approx. 19 km). A fourth nest with 10 m resolution will be added on purpose when detailed comparisons with the LAFO instrumentation will be performed. As for the LAFE cases, the simulations are initialized with data

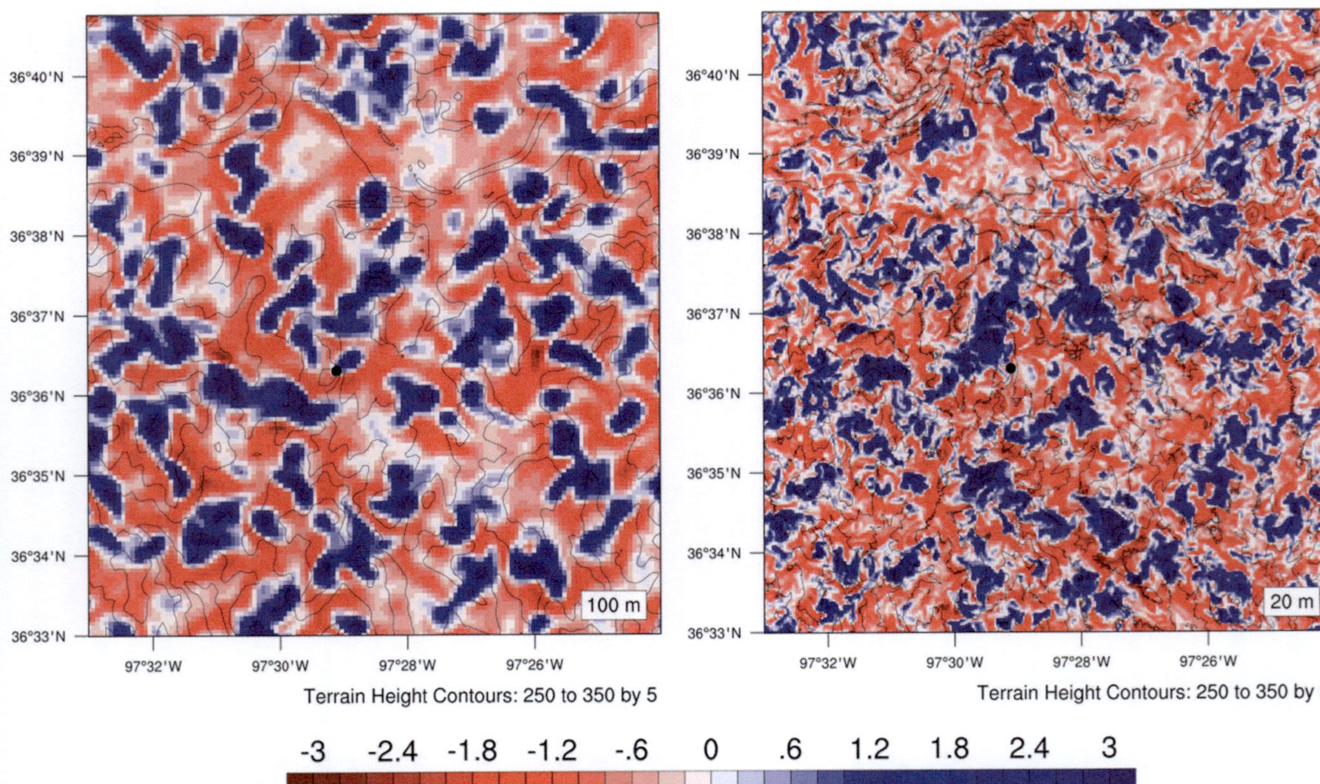

Fig. 3 Vertical wind velocity 1000 m above sea level (m/s) for one time step during the 20 m simulation. The time and area shown on both panels is the same—to the left the 100 m resolution and to the right the 20 m resolution

from the ECMWF analysis without additional data assimilation. The impact of the latter will be investigated in a future study.

Applying this setup, the simulation for a first test case was performed for 28 June 2021. The large-scale weather pattern was dominated by a strong low-pressure system located over the Bay of Biscay and western France. Along its eastern flank it transported hot and humid air into southwestern Germany. The day started mostly sunny. However, convergence of northerly winds east of the Black Forest and northeasterly winds over and south of the Swabian Jura triggered first convective cells to the west of lake Constance in the afternoon, before a line of severe convection moved from southwestern/western part of Baden-Württemberg to the east and northeast in the evening. The severe thunderstorms were accompanied by heavy rain (50 mm within one hour in Hohenheim), hail and wind gusts of more than 100 km/h. The evolution of the boundary layer from after sunrise until the passage of the line of thunderstorms in the evening is illustrated with the vertical velocity and the water vapor mixing ratio approx. 100 m above the surface in Fig. 5.

At 08 UTC (upper line), the surface heats up after sunrise, initiating the development of turbulence. The weak vertical velocities and "streak-like" appearance of the updrafts and downdrafts indicate that the turbulence is still weak and that it is organized in horizontal convective rolls. At 13 UTC (middle row), the turbulence is fully evolved. Stronger updraft cores are surrounded by larger-scale and weaker compensating downdrafts. Interaction with the underlying orography impedes the development of more regular structures. At 17:50 UTC, the surface is already cooling, reducing the general turbulent activity, clearly seen in the eastern part of the domain. However, the approach of the line of thunderstorms dominates the western part of the domain. Much stronger updrafts even close to the surface are seen (further up the values are much larger) and the approach of clearly moister air is seen.

To improve the capability to investigate the evolution of turbulence in the simulation, many additional variables are calculated in the simulation and can be written as additional output fields. As an example, Fig. 6 shows the 5-min averaged deviations of the vertical wind velocity w' and the water vapor mixing ratio q' at 13 UTC when the convective boundary layer is fully evolved.

As expected the vertical velocity deviations have the same order of magnitude as the absolute values. For water vapor, the deviations are approx. 10% of the absolute values. Combining the two values provides the turbulent vertical moisture flux w'q' which can also be directly measured by the lidar systems. This allows detailed comparison and therefore process evaluation of the model results. This will be done in future studies for this as well as other synoptically interesting cases.

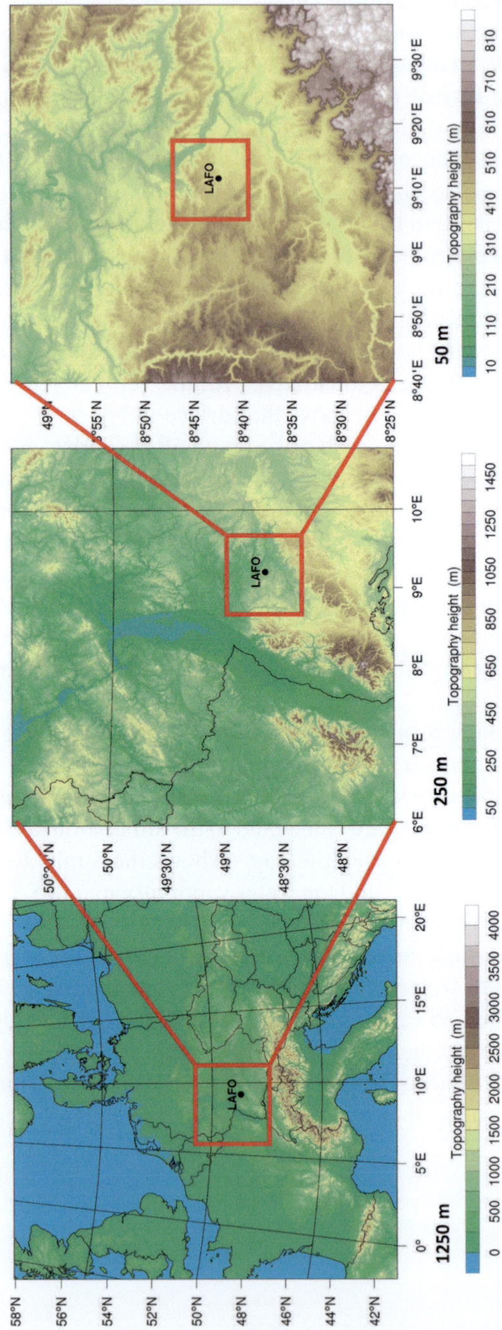

Fig. 4 Domain configuration focusing on southwestern Germany. The three domains with 1250, 500 and 50 m horizontal resolution consist of 1500 × 1500 grid cells

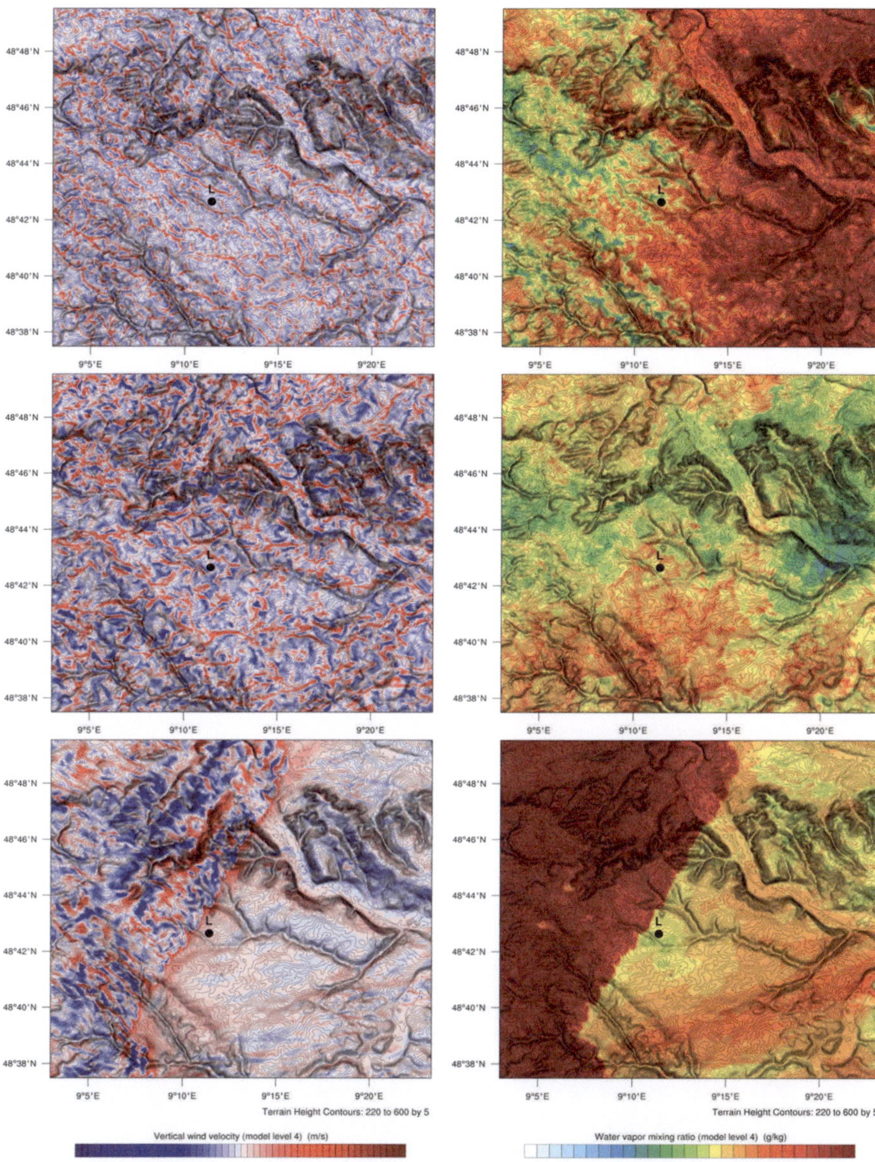

Fig. 5 Simulated vertical velocity (m/s) (left) and water vapor mixing ratio (g/kg) (right) approx. 100 m above the surface at 08 UTC (top row), 13 UTC (middle row) and 17:50 UTC (bottom row). Shown is a sub-region of the 50 m domain centered around the location of the LAFO site (marked by the black dot)

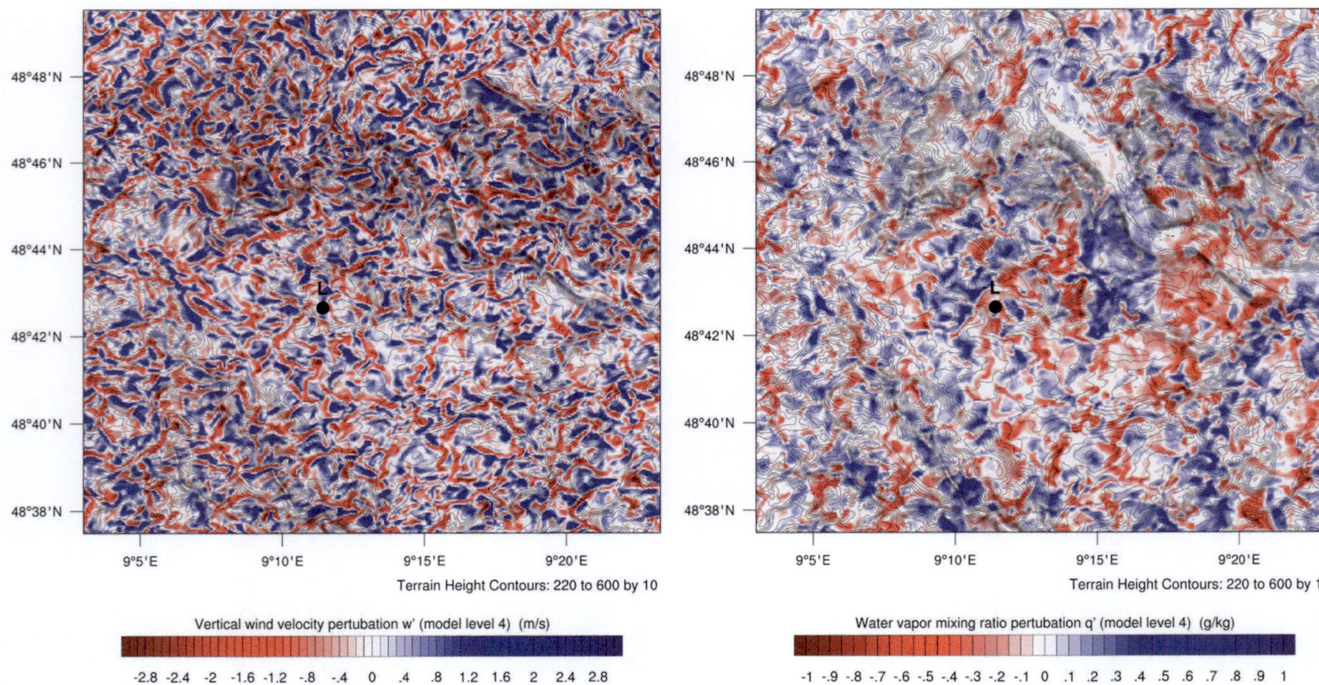

Fig. 6 5-min averages of the turbulent vertical velocity perturbation w' (m/s) (left) and the turbulent water vapor mixing ration deviation q' (g/kg) approx. 100 m above the surface at 13 UTC, 28 June 2021. The same domain as in Fig. 5 is shown. As in Fig. 4, the black dot indicates the location of the LAFO site

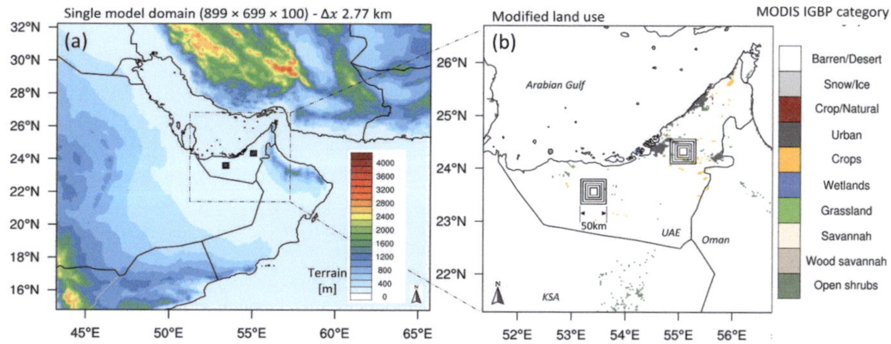

Fig. 7 Domain configuration for the single-domain WRF CP scale runs showing the location of the domain over the Arabian Peninsula (left panel). The right hand inset shows the position of the 10, 20, 30, 40, and 50 km^2 CPR scenarios to be compared with a scenario with unchanged land use. Adapted version of Fig. 1 of [7]

2.2 *Seasonal Land Surface Modification Simulations over the United Arab Emirates (UAE-1)*

The sub-project UAE-1 has now been executed via an ensemble of WRF simulations of 'devices' called cloud and precipitation reactors (CPRs), developed by the authors to induce convective rainfall in hot dry regions (e.g. see [5] for a description of the effect). Previously, these were conceived as patches of vegetation. Now we are simulating the use of artificial black surfaces with very low albedo to achieve the same effect. This design has now been the subject of a European patent (Europäische Patentanmeldung 20 174 927.2). Based on the ensemble, analyses of atmospheric impacts and scale dependency were carried out to test the smallest scale whereby impacts occur, and the underlying processes at high resolution.

For our land surface modification simulations over the United Arab Emirates a single nest configuration was used on a 2.7 km grid resolution (see Fig. 7). This model chain was evaluated previously in a year-long validation study [6].

Our goal is to investigate the lower limits of patch size on atmospheric impacts, down to 10×10 km^2 dimensions. Using a 2.7 km, or convection-permitting (CP) grid scale facilitates rapid computing and the ability to run many case studies. WRF "CP runs also allow" (not allows) for the satisfactory reproduction of convective processes, important when assessing rainfall enhancement potential. A range of scenarios were chosen to test the impact of several sizes of CPRs in the UAE for four one-day case studies in conditionally unstable weather conditions (18th, 24th, 25th, and 27th July 2015). Square CPR sizes of 10, 20, 30, 40, and 50 km^2 were executed (see Fig. 7). The Figure shows all sizes on one plot, with the land use data shown in the background. However, each size was actually simulated separately, as well as a 'control' simulation with no land use change—making a total of eight scenarios. Additionally, several extra tests were made on other scenarios which are not included

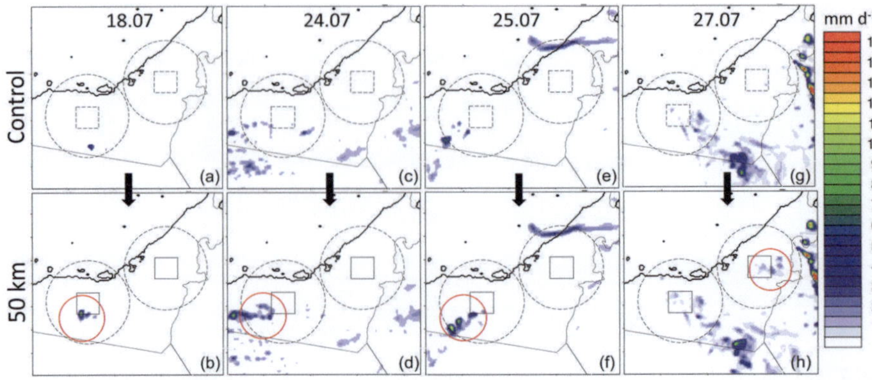

Fig. 8 Total precipitation (mm) over 24 h for each case study (00:00–23:30 UTC). The top row shows the Control simulation (panels a–d). The bottom row shows the 50 km^2 scenario (panels b-h), other scenarios not shown here). The black squares on the bottom row indicate the 50 km black surfaces. The dotted squares in Control show hypothetical surfaces for visual reference. The dotted black circles are 150 km diameter circles in which the rainfall volume was calculated (m^3). This limitation was imposed to show only those impacts due to change in land use, and discard those arising from numerical differences related to initial conditions. Red circles show areas with highly visible impacts compared to the Control. Adapted version of Fig. 4 of [7]

in the analyses. Note that two CPRs were simulated for each scenario in different parts of the UAE.

For the model execution, 64 nodes were used with 64 MPI procs and 1 OMP thread. Additionally, for CPU/domain allocation efficiency, the WRF quilting facility was used, utilizing 4 groups and 64 tasks per group. Each 24-h simulation required approximately 1.5 h to run.

For the UAE simulations, the first variable analyzed was precipitation to assess the impact of CPR size on the desired effect—rainfall enhancement. The one-day total rainfall is shown in Fig. 8. For visual clarity, only three out of the five CPR scenarios are shown (plus the Control). Note that in the Control panel a CPR outline is also marked on for visual reference, although it was not actually simulated. Also marked on the figure are 70 km diameter circles, which were used as search 'windows' for the purpose of calculating the total mass of water falling in the vicinity of each CPR. These 'total water volumes' (m^3) are shown at the top of each panel as well as the maximum rainfall intensity. In Fig. 8, it is clear that the 50 km^2 scale CPR has a large impact on the precipitation. In all Control simulations a small amount of precipitation falls, reflecting the slightly unstable conditions of the day. However, the amounts are significantly enhanced in all other scenarios. It is clear that the rainfall enhancement tends to scale with increasing CPR size.

This is quantified further in Fig. 9. The line plots (panels (a-d)) show total rainfall volume accumulating over the 24-h period (sum of east and west amounts). The curves indicate that the approximate time of convection initiation (CI) for all cases is somewhere between 09:00 and 10:30 UTC aligning with the arrival of the daily sea breeze. There is some precipitation in the Control cases too, but the CPRs consistently enhances these amounts, and the enhancement increases with scale. This holds at least from around a 20 km scale and upwards. The impact of the 10 km CPR appears to be more or less negligible, except for the 24 July where there is an enhancement of 0.4 million m^3 (panel (e)). On the 25 July, the amount is even lower than in the Control. Therefore, it is likely that the 10 km scale CPR has a negligible impact. To provide a benchmark for these amounts, we have also expressed the volumes in terms of per capita supply at regional consumption rates. The UAE has one of the highest per capita water consumptions in the world at ~500-liter per capita and day [8]. The blue right-hand scale in Fig. 9 (panel (e)) shows how many extra people the surplus volumes would supply for each of the cases. Assuming that these amounts are plausible, this would represent a considerable enhancement. There is some variability, both between the cases and between scales. The mean of the four cases is plotted (panel (f), right-hand scale), along with the total mean volumes (left-hand scale). Here we see a relationship between scale and the amounts. This is not quite linear in appearance, but slightly exponential as one might expect, given that areal size increases as the square of the side length. Here we also see that 20 km appears the lower boundary for impacts. But even this scenario could provide a mean increase of >0.5 million m^3 (>2500 people per year) per rainfall event. Extrapolating this, if such rainfall impacts occurred just 10 times over a summer period this would mean an extra 5 million m^3 of water, enough to supply 250,000 people for their annual supply.

In order to investigate the processes involved we examined buoyant motions, as seen in the vertical velocity field. Figure 10 shows all instantaneous [updrafts only] vertical velocity values against height at 08:00 UTC (27 July), along with a line of best fit. These velocities are all taken from within the 50 km CPR boundaries, so there are the same number of values for each scenario. Firstly, we can see a clear increase in velocity values with scale. We can also see that the maximum velocities increase with height as evidenced by the slope of the respective lines.

What is interesting is that the slope angle changes from negative to positive as we go from the 10 to 20 km scale indicating a strong transition in impact between 10 km and the larger scales.

Our results, from four moderately-unstable cases demonstrate that rainfall impacts could be generated by smooth low-albedo surfaces down to scales of approximately $20 \times 20\ km^2$. Smaller scales are likely to have negligible impacts due to slow onset times and limited extent. Impacts are likely predicated on the passing of the sea breeze and high temperatures, radiation and moderate humidity. Increasing scale is associated with increasing vertical velocities and the heights of the maximum velocities and associated mean surface convergence. As expected, the increase in scale increases PBL height and lowers the LFC. Assuming the microphysics of the model produces a good representation of rainfall amounts, then the volumes are considerable, with increased annual amounts equivalent to the supply for 3000–15,000 people (20–50 km scales). These estimated amounts would be of huge benefit to hyper-arid regions like the UAE, which are in desperate need of water resources.

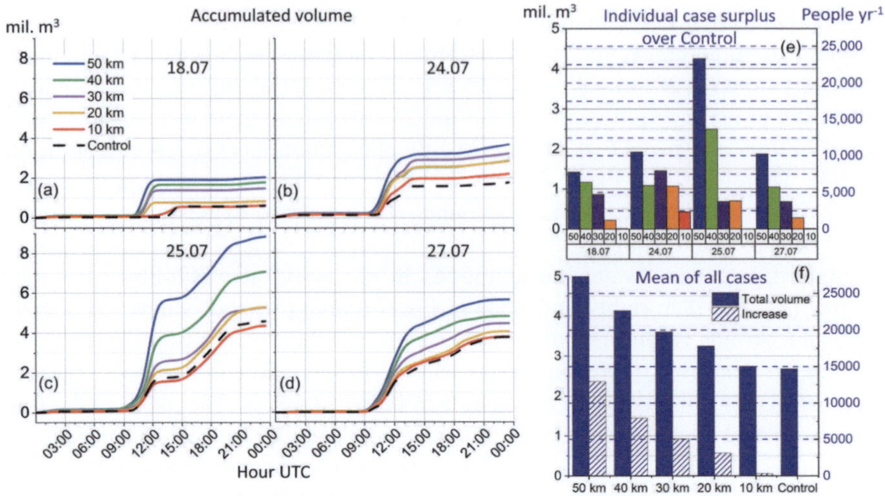

Fig. 9 Spatially-integrated precipitation volume within the perimeter of the 180 km diameter circles shown in Fig. 8. Panels **a–d** show the six scenarios on each of the four cases. Panel **e** shows the volumetric increase over Control (mil. m^3, left axis), and also the number of extra people this would supply annually in the UAE (people per year, right axis)—based on published figures for UAE water consumption of 500 liter per day and capita. Panel **f** shows the mean values for the total volume (left) and annual supply (right). Adapted version of Figs. 5 and 6 of [7]

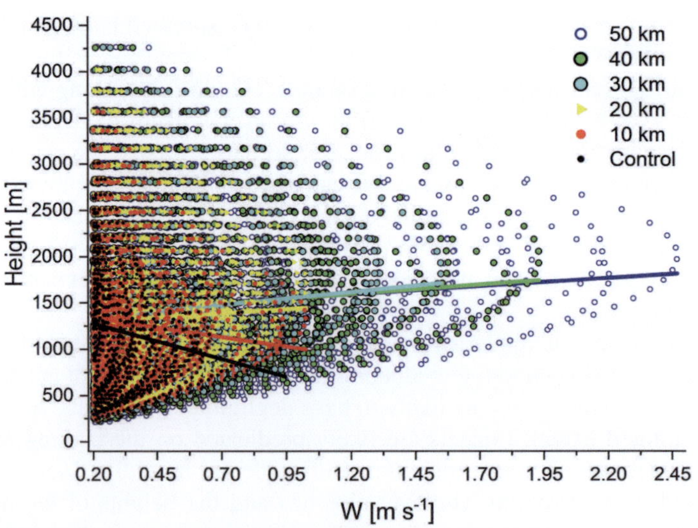

Fig. 10 All vertical velocity values and their heights on 27 July at 08:00 UTC within the 50 km footprint. Also shown is a line of best fit for each scenario to show the relationship of velocities with height. Adapted version of Fig. 11 of [7]

2.3 *Improving the Turbulent Flux Parameterization by Extending the Classic Monin-Obukhov Similarity Theory (MOSTex)*

Monin Obhukov Similarity Theory (MOST) is commonly used for parameterizing turbulent fluxes. However, MOST is based on simplifying assumptions, for example windy and stationary conditions over a flat terrain and low vegetation homogeneous surface [9]. In reality, surface heterogeneities can cause secondary circulations that are not accounted for in MOST but influence the turbulent fluxes. The objective of MOSTex is to develop an approach for calculating turbulent fluxes that is based on MOST, but produces robust results for heterogeneous land surfaces and low wind speeds. The strategy is to provide a Neural Network with training data of turbulent fluxes over homogeneous and heterogeneous surfaces, so that it is able to find functional interrelations. The PALM model [2] will be used to create this dataset.

After the installation, several debugging processes had to be completed to ensure a reliable production of data. This included problems with the MPI and the writing of output and restart data.

MOST is used as parametrization method for the bottom boundary condition in weather and climate models, which have a vertical resolution of several tenths of meters. The parametrized value is therefore applied to this height. To verify LES output as realistic and therefore usable as training data, the shares of explicitly resolved

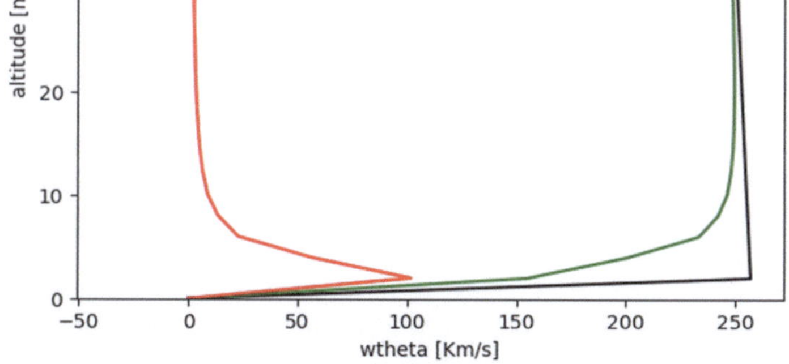

Fig. 11 The black curve denotes the total flux sensible vertical heat flux with height in a simulation of daytime convective boundary layer. The green curve shows the resolved and the red curve the parametrized part of it

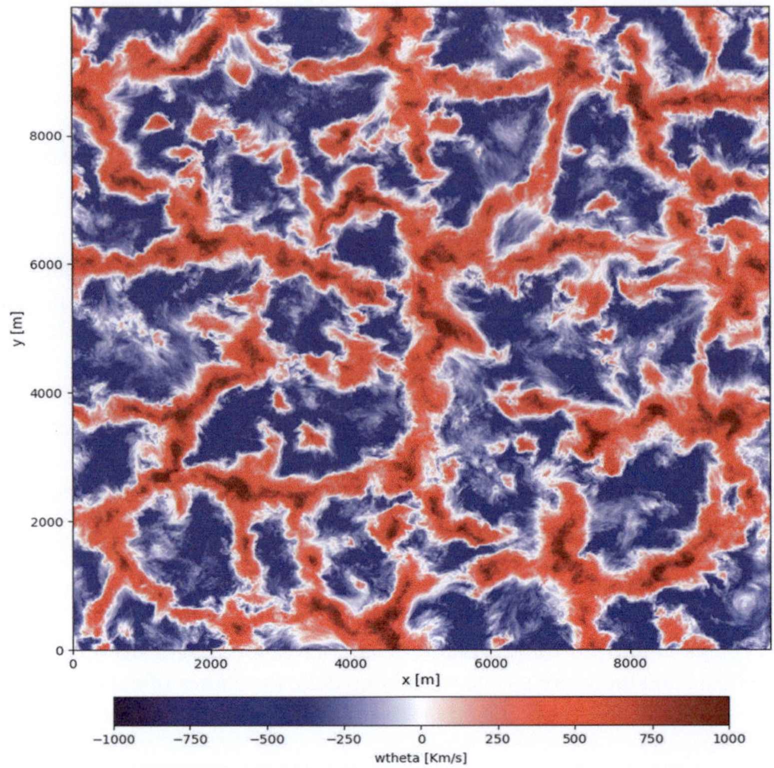

Fig. 12 Sensible vertical heat flux in daytime convective boundary layer at a height of 400 m

sensible heat flux and parametrized fluxes with height is plotted in Fig. 11. The simulation of a daytime convective boundary layer has a horizontal resolution of 5 m and vertical resolution of 2 m. The resolved flux is controlling the total sensible heat flux at a height of 10 m and higher, therefore output from a height larger than 10 m can be used as training data.

Figure 12 shows the horizontal vertical heat flux at 400 m above the surface. Organized structures (convective cells) of positive and negative sensible heat flux are clearly visible. Their horizontal diameter extends up to 2 km. The horizontal extent of the domain is therefore required to be around 10 km, so that interaction between multiple convective cells is possible.

Figure 13 shows the horizontal and vertical distribution of total sensible vertical heat flux in the same simulation, but at a height of 40 m for the horizontal plot. Convective cells are clearly visible; however they are smaller compared to that at 400 m height. In the vertical plot one can see that regions of positive heat flux extend through most of the boundary layer, but there are more regions of positive and negative flux at the bottom.

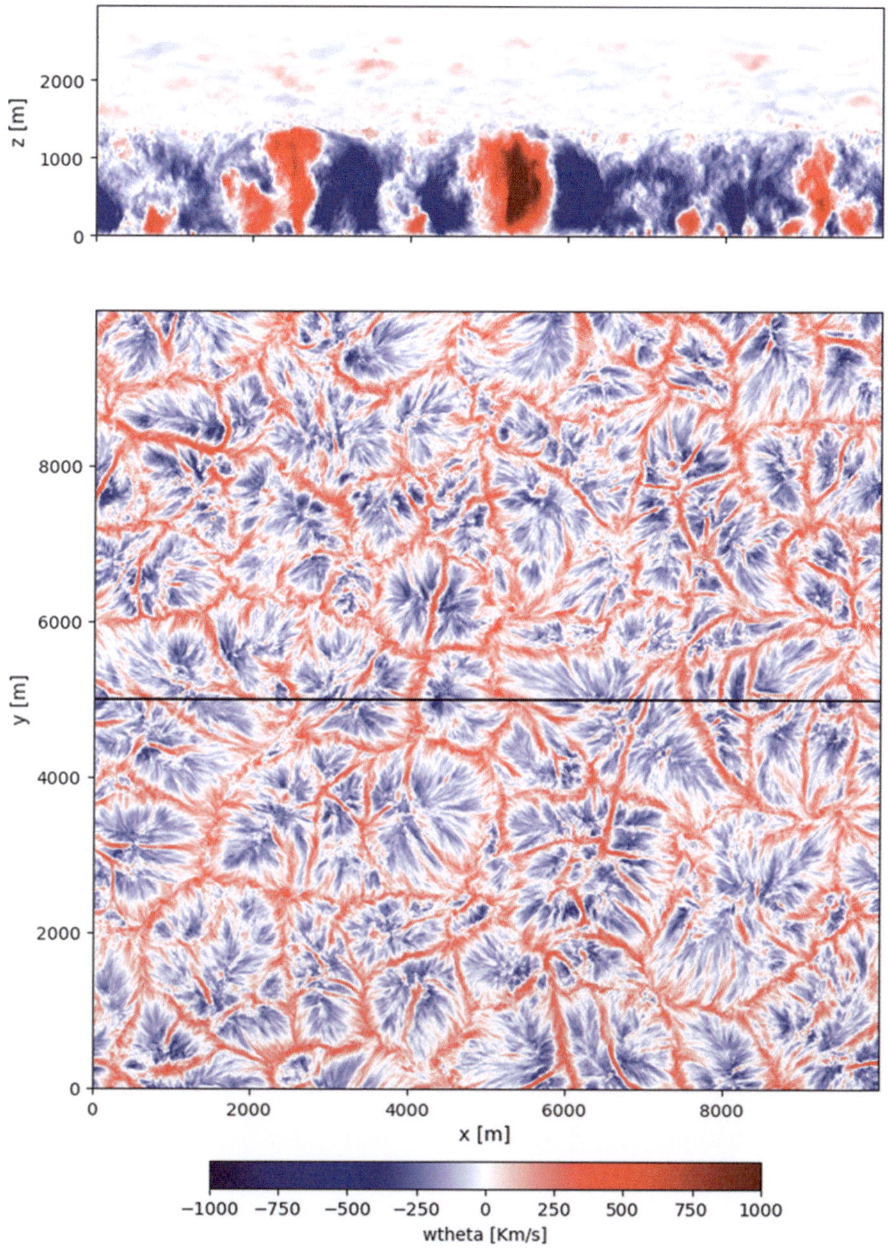

Fig. 13 Sensible vertical heat flux in daytime convective boundary layer. Top panel showing the xz-plain, bottom panel the xy-plain in a height of 42 m. The black line in the bottom panel indicates the location of the xz-plain

3 Scaling and Used Resources

Figures 14 and 15 illustrate the scaling results of the WRF (representative for the two sub-projects LES-PROC and UAE-1—here the example of the configuration applied in the LAFE case study simulations of LES_PROC was used) and PALM.

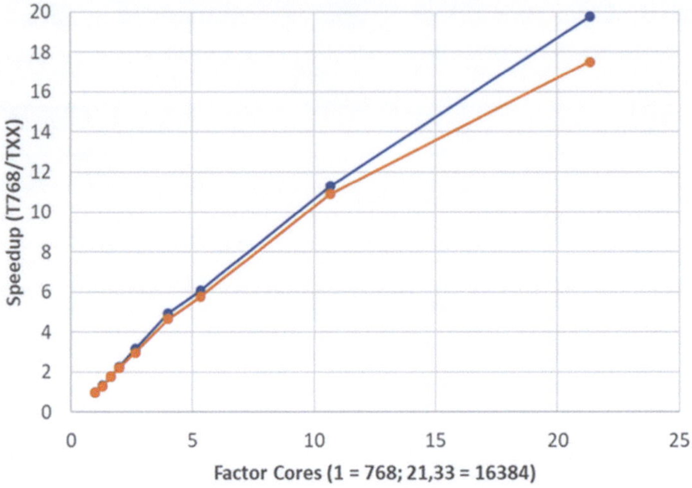

Fig. 14 Scaling of different WRF configurations. T768 is the time needed with 768 cores, the smallest possible configuration, and TXX is the time needed by the larger configurations. See WRFSCALE report of 2021 for more details. Blue line: time for regular time step; orange line: time for radiation time step

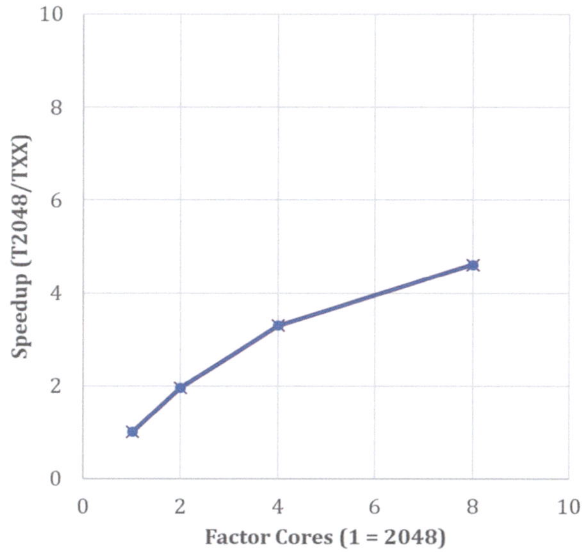

Fig. 15 Scaling of different PALM configurations applied in the MOSTex subproject

Table 1 Resources (RTh.) used by the WRFSCALE project between March 2022 and May 2023

Subproject	User	Computing time (Rth.)	Purpose
LES_PROC, Sect. 2.1	Hans-Stefan Bauer	41635.87	Case study simulations, analyses and tests
UAE-1, Sect. 2.2	Oliver Branch	89065.64	Case study simulations, analyses and tests
MOSTex, Sect. 2.3	Benjamin Körner	24774.85	Case study simulations, analyses and tests
Total		155476.36	

Table 1 lists the resources used during the report period March 2022 to May 2023 for the simulations, analyses and tests.

References

1. Skamarock, W.C., Klemp, J.B., Dudhia, J., Gill, D.O., Liu, Z., Berner, J., Wang, W., Powers, J.G., Duda, M.G., Barker, D.M., Huang, X.Y.: NCAR Tech. Note **NCAR/TN-556+STR**, 145pp. (2019). https://doi.org/10.5065/1dfh-6p97
2. Maronga, B., Banzhaf, S., Burmeister, C., Esch, T., Forkel, R., Fröhlich, D., Fuka, V., Gehrke, K.F., Geletič, J., Giersch, S., Gronemeier, T., Groß, G., Heldens, W., Hellsten, A., Hoffmann, F., Inagaki, A., Kadasch, E., Kanani-Sühring, F., Ketelsen, K., Khan, B.A., Knigge, C., Knoop, H., Krč, P., Kurppa, M., Maamari, H., Matzarakis, A., Mauder, M., Pallasch, M., Pavlik, D., Pfafferott, J., Resler, J., Rissmann, S., Russo, E., Salim, M., Schrempf, M., Schwenkel, J., Seckmeyer, G., Schubert, S., Sühring, M., von Tils, R., Vollmer, L., Ward, S., Witha, B., Wurps, H., Zeidler, J., Raasch, S.: Geosci. Model. Dev. **13**(3), 1335 (2020). https://doi.org/10.5194/gmd-13-1335-2020
3. Wulfmeyer, V., Turner, D.D., Baker, B., Banta, R., Behrendt, A., Bonin, T., Brewer, W.A., Buban, M., Choukulkar, A., Dumas, E., Hardesty, R.M., Heus, T., Ingwersen, J., Lange, D., Lee, T.R., Metzendorf, S., Muppa, S.K., Meyers, T., Newsom, R., Osman, M., Raasch, S., Santanello, J., Senff, C., Späth, F., Wagner, T., Weckwerth, T.: Bull. Amer. Meteorol. Soc. **99**(8), 1639 (2018). https://doi.org/10.1175/BAMS-D-17-0009.1
4. Bauer, H.S., Späth, F., Lange, D., Thundathil, R., Ingwersen, J., Behrendt, A., Wulfmeyer, V.: J. Geophys. Res. Atmos. **128**(8), e2022JD037212 (2023). https://doi.org/10.1029/2022JD037212
5. Branch, O., Wulfmeyer, V.: Proc. Natl. Acad. Sci. U. S. A. **116**(38), 18841 (2019)
6. Branch, O., Behrendt, A., Gong, Z., Schwitalla, T., Wulfmeyer, V.: Meteorologische Zeitschrift **29**, 67 (2020). https://doi.org/10.1127/metz/2019/0997
7. Branch, O., Jach, L., Schwitalla, T., Warrach-Sagi, K., Wulfmeyer, V.: EGUsphere (preprint) (2023). https://doi.org/10.5194/egusphere-2023-1771
8. Albannay, S., Kazama, S., Oguma, K., Hashimoto, T., Takizawa, S.: Water **13**(20), 2073 (2021). https://doi.org/10.3390/w13202827
9. Monin, A., Obukhov, A.: Contrib. Geophys. Inst. Acad. Sci. USSR **151**, e187 (1954)

Particle-Based Simulations

In this chapter, we collect research projects utilizing particle-based computational approaches. High-fidelity simulations typically require the use of huge particle numbers and, thus, the resources of the largest supercomputers. Applications of these methods cover explaining the function of large molecules in biology, unveiling thermodynamic properties of pure fluids and mixtures as well as investigating in detail the processes at molecular-continuum interfaces.

In particular, modeling and simulating thermodynamic and interfacial phenomena using atomistic approaches enable a profound understanding of these phenomena in fields such as chemistry, biology, materials science, and engineering. With the possibilities of high-performance computing, the team of the Chair of Thermodynamics of the TU Berlin conducts such molecular dynamics (MD) simulations and delivers impressive results in high temporal and spatial resolution. In their contribution "Unveiling Thermodynamic Properties and Interfacial Phenomena in Pure Fluids and Mixtures", they examine three main aspects: First, MD simulations are used to investigate the evaporation process of different fluids, varying not only the anisotropy and quadrupole of the molecular species itself but also boundary conditions such as the temperature of the fluid. Second, MD simulations are used to explore the properties of supercritical carbon dioxide mixtures close to the Widom line, where the phase transitions between liquid and gas become blurred. And thirdly, MD simulations are used to investigate the interfacial properties of hydrocarbon propellant-oxygen mixtures, which shows the wide range of applications of MD simulations.

In their paper "Regulation of Substrate Transport in the Early Stages of Fatty Acid Synthesis" the authors from the Max-Planck Institute for Multidisciplinary Science in Göttingen investigate the multifunctional enzyme complex "fungal type 1 fatty acid synthase (FAS)" which plays an important role in de novo fatty acid synthesis. All-atom molecular dynamics (MD) simulation is used in this novel work to elucidate the precise molecular mechanism governing the initiation of fatty acid synthesis. The average size of the fully solvated system was 2.8 million atoms, and the computations were done with the software GROMACS on typically 8192 cores.

In the paper "Extensions to the MaMiCo Software Stack for Flexible, Massively Parallel Molecular-Continuum Flow Simulation", the authors present an efficient approach for coupled molecular-continuum flow simulations, thus extending the feasible spatial domain of molecular dynamics simulations. A massively parallel approach in space and time and across molecular and continuum dynamics models is used to yield acceptable time to solution. The approaches exploit different scales in time and space in a sophisticated way, thus providing an exemplarily solution that might be generalizable to other multi-scale simulations.

Anita Schöbel
Fraunhofer-Institut für Techno- und Wirtschaftsmathematik
Kaiserslautern, Germany
e-mail: anita.schoebel@itwm.fraunhofer.de
Peter Bastian
Interdisziplinäres Zentrum für Wissenschaftliches Rechnen
Universität Heidelberg
Heidelberg, Germany
e-mail: peter.bastian@iwr.uni-heidelberg.de
Miriam Schulte
Institut für Parallele und Verteilte Systeme (IPVS)
Universität Stuttgart
Universitätsstrasse 38
Stuttgart, Germany
70569
e-mail: Miriam.Schulte@ipvs.uni-stuttgart.de

Unveiling Thermodynamic Properties and Interfacial Phenomena in Pure Fluids and Mixtures

Simon Homes, Isabel Nitzke, Denis Saric, and Jadran Vrabec

Abstract High-performance computing has revolutionized engineering by enabling investigations across multiple scales. Molecular-level simulations offer an exceptional temporal and spatial resolution, which is important for engineering thermodynamics. The present works employ molecular dynamics (MD) simulations to explore a range of thermodynamic and interfacial phenomena. The evaporation process is extensively investigated for a wide range of fluids. Not only the properties of the molecular species are varied, but also the boundary conditions, i.e. magnitude of evaporation or the bulk liquid temperature. Moreover, a comprehensive analysis of the Widom line of supercritical CO_2 mixtures is conducted. This study discusses the effect of temperature and solvent species on relevant thermodynamic properties, like the density and speed of sound. Leveraging the power of high-performance computing and advanced simulation techniques, another work focuses on hydrocarbon propellant-oxygen mixtures, leading to a better understanding of their thermodynamic and interfacial properties as well as their implications for engineering applications.

1 Influence of Molecular Anisotropy and Quadrupole on Evaporation

Evaporation processes are of great importance in various fields, including process engineering, materials science and environmental studies. Understanding the fundamental mechanisms and factors influencing evaporation can contribute to the development of more efficient processes and applications. Despite being investigated with a wide range of experimental and numerical methods, the evaporation process was mainly studied with atomistic approaches only for simpler molecular

S. Homes · I. Nitzke · D. Saric · J. Vrabec (✉)
Thermodynamik, Technische Universität Berlin, Berlin, Germany
e-mail: vrabec@tu-berlin.de

S. Homes
e-mail: homes@tu-berlin.de

T. Ludwig et al. (eds.), *High Performance Computing in Science and Engineering '23*,
https://doi.org/10.1007/978-3-031-91312-9_33

Table 1 Properties of the six investigated 2CLJQ models. The reduced critical temperature T_c^* was taken from Ref. [36]. The column on the right contains real molecules that are similar to the selected 2CLJQ models in terms of reduced elongation and reduced squared quadrupolar moment [40]

	L^*	Q^{*2}	T_c^*	Substance
Fluid 1	0	0	5.236	Argon
Fluid 2	0	4	6.398	–
Fluid 3	0.4	0	3.163	Oxygen
Fluid 4	0.4	4	3.692	Acetylene
Fluid 5	0.8	0	2.049	Ethane
Fluid 6	0.8	4	2.408	Carbon dioxide

species [1, 5, 8, 13, 14, 16, 21]. In this work [15], molecular dynamics (MD) simulations were employed to investigate the evaporation process of six different fluids, see Table 1. The selected molecular models were chosen to represent a wide range of real fluids [24, 34, 35, 40, 41], incorporating variations of molecular anisotropy in terms of the reduced elongation L^* and reduced squared quadrupolar moment Q^{*2}. Additionally, the influence of boundary conditions, i.e. bulk liquid temperature and evaporation magnitude, was examined to gain insights into their impact on the evaporation process.

A refined method for specifying the evaporation magnitude in terms of hydrodynamic velocity was developed. A total of 48 simulation runs was carried out using the massively parallelized software *ls1 mardyn* [31]. Spatially resolved profiles of the key quantities were sampled and analyzed to gain a detailed understanding of the evaporation process. The molar flux and energy flux were identified as crucial parameters for describing the process of evaporation and for assessing the impact of different factors.

The results confirm that the interface temperature plays a major role in the evaporation process. Both the molar flux and the energy flux were observed to remain constant across the simulation volume, aligning with the expectations for stationary evaporation processes. Notably, the molecular properties exhibit significant effects on the fluxes. Spherical molecules are associated with the highest fluxes, while the anisotropy of the molecules also demonstrates a pronounced influence. Furthermore, the role of the quadrupole moment was analyzed, revealing its impact on the resulting molar flux especially for higher bulk liquid temperatures. At lower bulk liquid temperatures, the molar flux remains quite similar for molecules with the same elongation and evaporation magnitude, while the quadrupole moment primarily affects the interface temperature. However, the quadrupole moment raises the energy flux regardless of the bulk liquid temperature, as it transports more energy compared to equally structured non-polar molecules. Consequently, a higher quadrupolar moment led to a larger energy flux under the same boundary conditions.

To gain further insights into evaporation, the average molecular orientation was analyzed using an order parameter. It was found that molecules from different fluid species exhibit varying preferred orientations in the interface region, but no clear linkage to the resulting fluxes could be made. The evaporation coefficient, calculated on the basis of the Hertz flux, was studied as well. The evaporation coefficient displayed a limited dependence on the properties of the molecular species and bulk liquid temperature when using the interface temperature as a reference. Nevertheless, the evaporation magnitude and interface temperature were identified as major factors of influence.

To provide further insights into the evaporation process, the thermodynamic states experienced by molecules during evaporation were analyzed, see Fig. 1. When moving from the bulk liquid toward the interface, the molecules follow the saturated liquid line. At the interface, evaporation occurs, resulting in a significant decrease of the local density. The molecules then traverse states along the saturated vapor line before ultimately ending up in the far field of the vapor. These analyses underscore the crucial importance of the interface temperature throughout the entire evaporation process and contribute to a better understanding of the fundamental mechanisms and factors influencing fluid evaporation.

Present simulations were performed on a high-performance computing system using the open-source software *ls1 mardyn* [31]. The computational resources comprised up to 32 nodes, each equipped with 128 cores. To effectively utilize such a large number of cores, parallelization techniques including Message Passing Interface (MPI) and Open Multi-Processing (OpenMP) were employed. Additionally, the vector intrinsics of the processors were leveraged to further enhance the simulation speed. Despite these optimization efforts, the simulations required a significant amount of computational resources, totaling approximately $1.8 \cdot 10^7$ core-hours. To

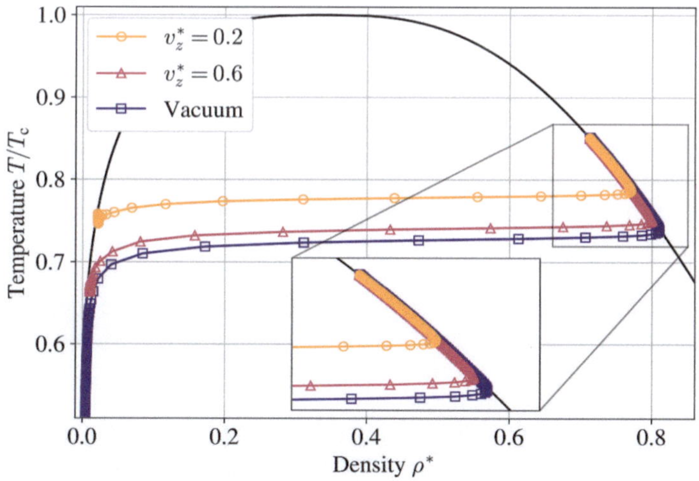

Fig. 1 Reduced temperature T/T_c over density. Three simulation runs of fluid 2 with the reduced bulk liquid temperature $T_{liq}/T_c = 0.85$ are shown, where the evaporation magnitude was varied. The black line shows the saturated liquid and vapor states of the fluid

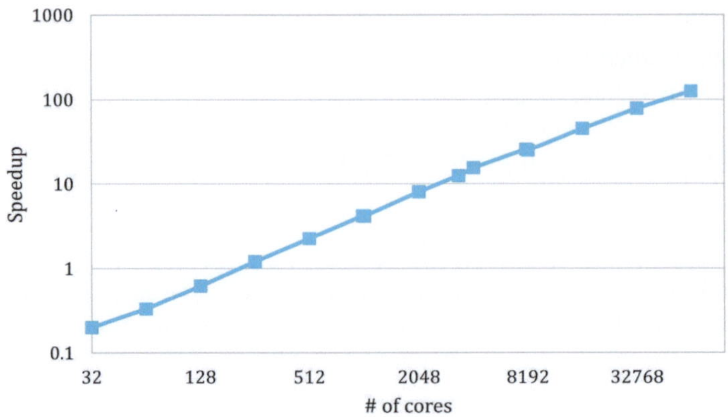

Fig. 2 Strong scaling behavior of the software *ls1 mardyn* [31] on Hawk

evaluate the scalability and performance of the software, various tests were conducted. Figure 2 illustrates the strong scaling behavior observed in a test case which was similar to the production runs of the aforementioned work. The simulations were conducted on the most advanced supercomputer available at the High Performance Computing Center Stuttgart (HLRS), namely Hawk.

2 Widom Line of Supercritical CO_2 Mixtures

Data on thermophysical properties of supercritical carbon dioxide ($scCO_2$) mixtures are important for many industrial processes. However, such data are hardly available, especially close to the critical point of the fluid. In fact, thermodynamic properties like density or enthalpy change strongly over a narrow temperature range in the supercritical state. These anomalous changes are related to the transition between liquid-like and gas-like regions of supercritical fluids, i.e. the Widom line. Many definitions of the Widom line have been proposed in the literature [3, 29]. Most commonly, the maximum of the isobaric heat capacity has been used to determine the Widom line temperature at a specified pressure [6, 42]. Furthermore, extrema of other thermodynamic response functions, e.g. the minimum in the speed of sound or the inflection point of density, can also be used to evaluate the Widom line temperature [17]. Different Widom lines span the Widom region [18], which contains a set of temperatures at a given supercritical pressure.

Pure $scCO_2$ and seven of its binary mixtures with hydrogen (H_2), methane (CH_4), ethane (C_2H_6), isobutane (C_4H_{10}), benzene (C_6H_6), toluene (C_7H_8) or naphthalene ($C_{10}H_8$) were studied over a large range of thermodynamic states close to the Widom line [30]. These mixtures were evaluated at three compositions (0.5, 1.0 or 1.5 mol% of the solute) in the temperature range from $T = 290$ to 350 K and along the isobar

$p = 9$ MPa. Furthermore, the mixtures of $scCO_2$ with benzene, toluene and naphthalene were studied along two additional isobars, i.e. $p = 10$ MPa and 12 MPa. In total, more than 1000 thermodynamic state points were simulated and evaluated to generate insight into the thermodynamic, transport and structural properties of these mixtures. Additionally, these simulations were compared to the most precise equation of state (EOS) models available.

Six thermodynamic properties (see Fig. 3) were calculated from the derivatives of the Helmholtz energy, which are accessible with canonic ensemble simulations through the Lustig formalism [7, 22, 23]. Here, the ideal contributions were taken from the reference EOS and added to the residual contributions sampled by simulation. The mathematical relations between the Helmholtz energy derivatives and the remaining thermodynamic properties can be found, e.g. in Thol et al. [38].

Figure 3 shows the temperature dependence of six thermodynamic properties of $scCO_2$ with 1.5 mol% ethane or toluene along the isobar $p = 9$ MPa. Molecular simulation results agree quantitatively with EOS data. Further, simulation data scatter close to the Widom line temperature due to strong density fluctuations that are characteristic for this region. Despite that, the location of the extrema of the thermodynamic properties was predicted well by MD simulation. Density and enthalpy resemble a sigmoidal function with respect to the temperature along the isobar $p = 9$ MPa. The isothermal compressibility β_T, isobaric heat capacity c_p and volume expansivity α exhibit significant peaks within this region. The slope of the speed of sound changes from negative to positive, indicating different liquid-like and gas-like regions of the supercritical fluid. The strongest changes in these thermodynamic properties occur close to the Widom line temperature, which was determined from the maximum of the isobaric heat capacity. In fact, these results underline the existence of distinct liquid-like and gas-like supercritical regions, as well as to the Widom line crossover between these regions.

Furthermore, the addition of a second component to $scCO_2$ shifts the Widom line temperature of the mixture, which can be related to the displacement of the critical point of the mixture [2]. Components with a higher critical temperature than pure CO_2, like toluene, increase the critical temperature and consequently the Widom line temperature of their mixture. For instance, traces of toluene with 0.1 mol% can shift this temperature by $\Delta T = +0.3$ K, as calculated by molecular simulation. This shift was replicated well by both molecular simulation and EOS data, cf. Fig. 3. The knowledge of the Widom line temperature shift is important for the design of supercritical fluid units [43], as $scCO_2$ is unlikely to be present in its pure form in industrial applications. Although not described here, transport and structural properties of these $scCO_2$ mixtures were also evaluated.

Two consecutive equilibrium MD simulations were carried out with *ms2* [7] to sample thermodynamic, transport and structural properties at a given state point for the specific $scCO_2$ mixture. First, a simulation in the isothermal-isobaric ensemble was carried out at specified conditions of temperature, pressure and composition to obtain the mixture density. These simulations were equilibrated for $3 \cdot 10^4$ time steps and followed by a production run of $2 \cdot 10^6$ steps. Second, the obtained mixture density was used for the subsequent simulation run in the canonic ensemble

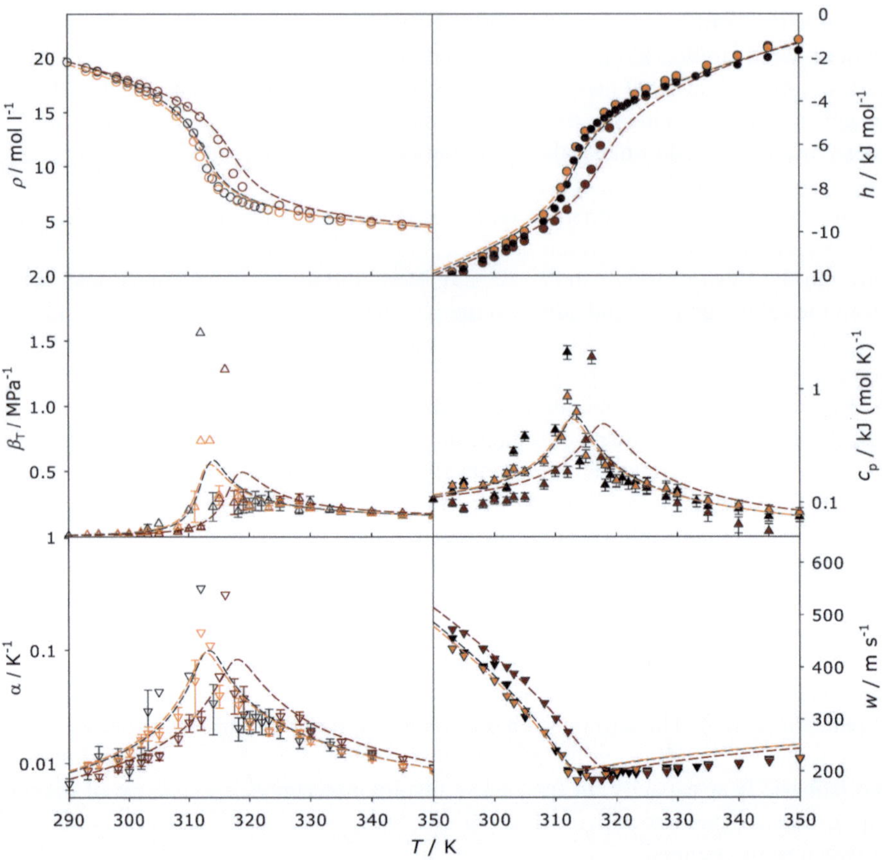

Fig. 3 Six thermodynamic properties of $scCO_2$ (black) and its mixtures with 1.5 mol% of ethane (orange) or toluene (dark red) at $T = 290$ to 350 K at $p = 9$ MPa. Symbols represent the molecular simulation results and dashed lines show the EOS by Span and Wagner for CO_2 [32], the GERG-2008 EOS [19] for CO_2 + ethane and the mixture model by Blackham and Lemmon [20] for CO_2 + toluene. Density and enthalpy are shown at the top, isothermal compressibility and isobaric heat capacity in the center, volume expansivity and speed of sound at the bottom. Error bars were omitted when they are either within symbol size or lead to visual clutter

to determine the thermodynamic, transport and structural properties of interest. The simulations in that ensemble were equilibrated for $5 \cdot 10^5$ time steps and followed by a production run of approximately $24 \cdot 10^6$ steps. To obtain adequate statistics, the number of molecules in the cubic simulation volume with periodic boundary conditions was 5000 throughout. Due to relatively large system sizes and extensive memory demand, between 1536 and 3072 cores were utilized for the multi-ensemble simulations. Around 24 simulations were performed per mixture composition and pressure. This multiplies to over 1000 single simulations that were performed in each

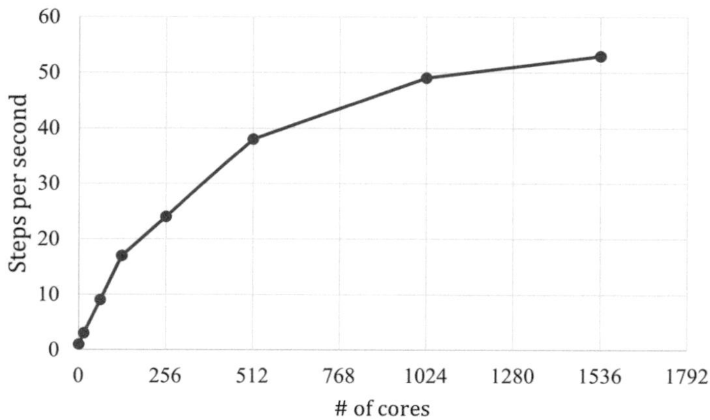

Fig. 4 Strong scaling behavior of the software *ms2* [7]

of the two ensembles. One short isothermal-isobaric simulation required a production time of 4 h and one long canonic ensemble simulation required a production time of approximately 60 h. In total, about $8.2 \cdot 10^6$ core-hours were used.

Extensive scaling and performance tests were conducted to evaluate the software *ms2* [7]. Figure 4 illustrates the strong scaling behavior observed in a simulation of pure CO_2, which comprised 16,000 molecules and utilized up to 2,048 cores. It is important to note that this simulation represents a single state point. In a typical study, numerous other state points need to be considered concurrently, necessitating a proportional increase in the number of cores required.

3 Phase Equilibria and Interface Properties of Hydrocarbon Propellant-Oxygen Mixtures

Energy conversion machines and process technology heavily rely on the processes of high-pressure injection, mixing and combustion, among others. They have applications in rocket engines, diesel fuel injection and nano-particle generation. Increasing pressure improves performance, such as specific impulse in rockets and engine efficiency [37]. However, operating pressures often exceed the critical pressure of individual components in fuel-oxygen mixtures, leading to local vapor-liquid phase transitions. When molecular weight differences are significant, coexisting phases resemble liquid-liquid equilibria, referred to as liquid-supercritical liquid-like equilibrium (LSCE). Numerical simulations using computational fluid dynamics (CFD) are vital for optimizing these processes, requiring accurate thermodynamic data on density, heat capacity, vapor-liquid equilibria (VLE), shear viscosity, thermal conductivity and surface tension. Flow structures are influenced by the thermodynamic properties of multicomponent and multiphase systems, which differ from those of single-phase, pure fluids [4, 44, 45]. While experimental data are often constrained to pure fluids near ambient conditions, molecular simulations offer a convincing route to predict those missing data.

In this work [27], the interface properties of mixtures containing methane (CH_4) or cyclohexane (C_6H_{12}) as hydrocarbon propellant and oxygen (O_2) as oxidant were investigated using MD simulation, as well as density functional theory (DFT) and density gradient theory (DGT) on the basis of the PC-SAFT equation of state (EOS) [9–12, 39]. Since nitrogen (N_2) is often used as a substitute for O_2 in experiments [26, 28, 33], it was also included. Spatial density profiles were sampled and analyzed to determine the interface thickness, enrichment and relative adsorption. In addition, the surface tension and equilibrium bulk properties were compared to empirical EOS data.

The interface properties predicted with MD simulations as well as the theoretical methods DFT and DGT are fully independent. However, the excellent agreement of DFT and DGT is not surprising due to the consistent use of a Helmholtz energy PC-SAFT model in both methods. The findings revealed that the oxygen-nitrogen analogy holds fairly well for mixtures containing higher-carbon fuels like C_6H_{12}, but exhibited discrepancies for lower-carbon fuels like CH_4. In the case of CH_4, significant differences among the compositions of the bulk phases resulted in higher enthalpy of vaporization and surface tension of mixtures containing O_2, see Fig. 5. Conversely, mixtures containing N_2 displayed higher values for enrichment and relative adsorption.

The choice of the low-boiling component has little influence on the properties of $C_{64}H_{12}$ mixtures. Atypical behavior was observed in both cases, indicating a transition from liquid-liquid equilibrium to a LSCE. This was evident from the increase of molar density along the saturation line and the presence of a maximum interfacial thickness. Overall, this thorough study demonstrated that fluid interface properties in the transcritical regime can be accurately predicted with MD, DFT and DGT methods, although the models were parameterized solely on the basis of pure component data and bulk VLE mixture data [25, 40].

For future research, performing a comprehensive investigation into the sensitivity of enrichment to the Helmholtz energy functional used in DFT/DGT or the force field employed in MD, would be a captivating avenue for research. Further investigations could determine whether the maximum interface thickness that was observed in C_6H_{12} systems is a local or global phenomenon.

MD simulations were conducted using both the software *ms2* [7] and *ls1 mardyn* [31]. Each simulation was equilibrated over $7.5 \cdot 10^5$ time steps before a production run over $10 \cdot 10^6$ time steps was carried out. Since a large total number of particles was used with *ls1 mardyn*, up to 1024 cores were needed for each run. For approximately 50 simulations including test runs, this led to a total amount of about $2.4 \cdot 10^6$ core-hours.

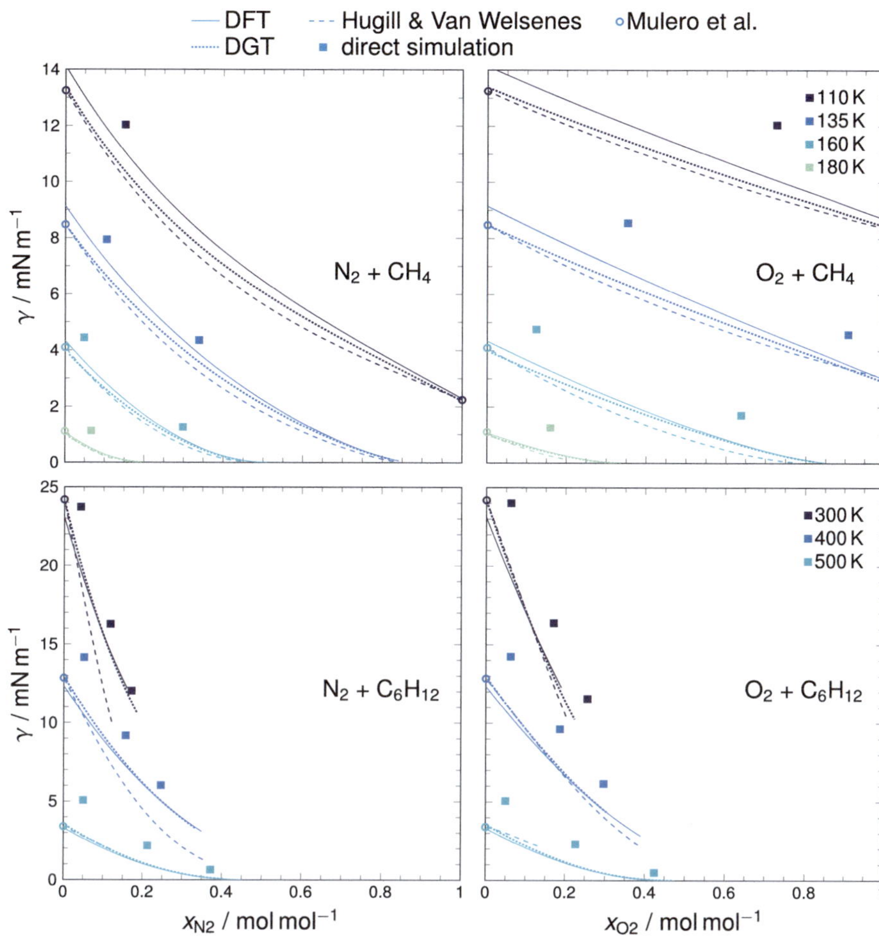

Fig. 5 The surface tension is shown over the composition for the four binary mixtures

Acknowledgements The authors acknowledge funding in the scope of the project *MaST: Macro/Micro-simulation of Phase Decomposition in the Transcritical Regime* which is funded by *dtec.bw—Digitalization and Technology Research Center of the Bundeswehr. dtec.bw* is funded by the European Union—NextGenerationEU. This work was carried out under the auspices of the Boltzmann-Zuse Society of Computational Molecular Engineering (BZS), and it was facilitated by activities of the Innovation Centre for Process Data Technology (Inprodat e.V.), Kaiserslautern. The simulations were performed on the *HPE Apollo (Hawk)* at the High Performance Computing Center Stuttgart (HLRS).

References

1. Anisimov, S.I., Dunikov, D.O., Zhakhovskii, V.V., Malyshenko, S.P.: Properties of a liquid-gas interface at high-rate evaporation. J. Chem. Phys. **110**(17), 8722–8729 (1999)
2. Artemenko, S., Mazur, V., Krijgsman, P.: Phase Behavior of Biomass Components in Supercritical Water. In: Near-critical and Supercritical Water and Their Applications for Biorefineries, pp. 41–68. Springer Netherlands, Dordrecht (2014)
3. Banuti, D., Raju, M., Ihme, M.: Similarity law for Widom lines and coexistence lines. Phys. Rev. E **95**, 052120 (2017)
4. Bellan, J.: Future Challenges in the Modelling and Simulations of High-pressure Flows. Combust. Sci. Technol. **192**(7), 1199–1218 (2020). https://doi.org/10.1080/00102202.2020.1719404.
5. Bird, E., Plascencia, J.G., Keblinski, P., Liang, Z.: Molecular simulation of steady-state evaporation and condensation of water in air. Int. J. Heat Mass Trans. **184**, 122285 (2022)
6. Corradini, D., Rovere, M., Gallo, P.: The Widom line and dynamical crossover in supercritical water: popular water models versus experiments. J. Chem. Phys. **143**, 114502 (2015)
7. Fingerhut, R., Guevara-Carrion, G., Nitzke, I., Saric, D., Marx, J., Langenbach, K., Prokopev, S., Celný, D., Bernreuther, M., Stephan, S., Kohns, M., Hasse, H., Vrabec, J.: ms2: A molecular simulation tool for thermodynamic properties, release 4.0. Comput. Phys. Commun. **262**, 107860 (2021)
8. Frezzotti, A., Barbante, P., Gibelli, L.: Direct simulation Monte Carlo applications to liquid-vapor flows. Phys. Fluids **31**(6), 062103 (2019)
9. Gross, J.: An equation-of-state contribution for polar components: quadrupolar molecules. AlChE J. **51**(9), 2556–2568 (2005)
10. Gross, J., Sadowski, G.: Perturbed-Chain SAFT: an equation of state based on a perturbation theory for chain molecules. Ind. Eng. Chem. Res. **40**(4), 1244–1260 (2001). https://doi.org/10.1021/ie0003887
11. Gross, J., Sadowski, G.: Application of the perturbed-chain SAFT equation of state to associating systems. Ind. Eng. Chem. Res. **41**(22), 5510–5515 (2002)
12. Gross, J., Vrabec, J.: An equation-of-state contribution for polar components: dipolar molecules. AlChE J. **52**(3), 1194–1204 (2006)
13. Heinen, M., Vrabec, J.: Evaporation sampled by stationary molecular dynamics simulation. J. Chem. Phys. **151**(4), 044704 (2019)
14. Heinen, M., Vrabec, J., Fischer, J.: Communication: evaporation: influence of heat transport in the liquid on the interface temperature and the particle flux. J. Chem. Phys. **145**(8), 081101 (2016)
15. Homes, S., Heinen, M., Vrabec, J.: Influence of molecular anisotropy and quadrupolar moment on evaporation. Phys. Fluids **35**(5) (2023)
16. Homes, S., Heinen, M., Vrabec, J., Fischer, J.: Evaporation driven by conductive heat transport. Mol. Phys. **119**(15–16), e1836410 (2020)
17. Imre, A., Deiters, U., Kraska, T., Tiselj, I.: The pseudocritical regions for supercritical water. Nucl. Eng. Des. **252**, 179–183 (2012)
18. Imre, A.R., Groniewsky, A., Györke, G., Katona, A., Velmovszki, D.: Anomalous properties of some fluids with high relevance in energy engineering in their pseudo-critical (Widom) region. Period. Polytech. Chem. Eng. **63**, 276–285 (2019)
19. Kunz, O., Wagner, W.: The GERG-2008 wide-range equation of state for natural gases and other mixtures: an expansion of GERG-2004. J. Chem. Eng. Data **57**(11), 3032–3091 (2012)
20. Lemmon, E., Bell, I.H., Huber, M., McLinden, M.: NIST Standard Reference Database 23: Reference Fluid Thermodynamic and Transport Properties-REFPROP, Version 10.0, National Institute of Standards and Technology (2018). Gaithersburg, Maryland 20899
21. Lotfi, A., Vrabec, J., Fischer, J.: Evaporation from a free liquid surface. Int. J. Heat Mass Transf. **73**, 303–317 (2014)
22. Lustig, R.: Direct molecular NVT simulation of the isobaric heat capacity, speed of sound and Joule-Thomson coefficient. Mol. Simul. **37**, 457–465 (2011)

23. Lustig, R.: Statistical analogues for fundamental equation of state derivatives. Mol. Phys. **110**(24), 3041–3052 (2012)
24. Madin, O.C., Boothroyd, S., Messerly, R.A., Fass, J., Chodera, J.D., Shirts, M.R.: Bayesian-inference-driven model parametrization and model selection for 2CLJQ fluid models. J. Chem. Inf. Model. **62**(4), 874–889 (2022)
25. Merker, T., Vrabec, J., Hasse, H.: Molecular simulation study on the solubility of carbon dioxide in mixtures of cyclohexane+cyclohexanone. Fluid Ph. Equilibria **315**, 77–83 (2012). https://doi.org/10.1016/j.fluid.2011.11.003
26. Niño-Amézquita, O.G.: Interfacial properties and phase equilibria for mixtures relevant in the oil industry. Ph.D. thesis, TU Berlin (2013)
27. Nitzke, I., Stierle, R., Stephan, S., Pfitzner, M., Gross, J., Vrabec, J.: Phase equilibria and interface properties of hydrocarbon propellant-oxygen mixtures in the transcritical regime. Phys. Fluids **35**(3), 032117 (2023)
28. Pereira, L.M., Chapoy, A., Burgass, R., Tohidi, B.: Measurement and modelling of high pressure density and interfacial tension of (gas + n-alkane) binary mixtures. J. Chem. Thermodyn. **97**, 55–69 (2016)
29. Ploetz, E.A., Smith, P.E.: Gas or Liquid? The supercritical behavior of pure fluids. J. Phys. Chem. B **123**, 6554–6563 (2019)
30. Saric, D., Guevara-Carrion, G., Vrabec, J.: Thermodynamics of supercritical carbon dioxide mixtures across the Widom line. Phys. Chem. Chem. Phys **24**, 28257–28270 (2022)
31. Seckler, S., Gratl, F., Heinen, M., Vrabec, J., Bungartz, H.J., Neumann, P.: AutoPas in ls1 mardyn: massively parallel particle simulations with node-level auto-tuning. J. Comput. Sci. **50**, 101296 (2021)
32. Span, R., Wagner, W.: A new equation of state for carbon dioxide covering the fluid region from the triple-point temperature to 1100 K at pressures up to 800 MPa. J. Phys. Chem. Ref. Data **25**, 1509–1596 (1996)
33. Stephan, S.: Vapor-liquid interfaces: molecular simulation,density gradient theory, and experiments. Ph.D. thesis, TU Kaiserslautern (2020). https://doi.org/10.26204/KLUEDO/6559
34. Stöbener, K., Klein, P., Horsch, M., Küfer, K., Hasse, H.: Parametrization of two-center Lennard-Jones plus point-quadrupole force field models by multicriteria optimization. Fluid Phase Equilib. **411**, 33–42 (2016)
35. Stoll, J., Vrabec, J., Hasse, H.: Vapor-liquid equilibria of mixtures containing nitrogen, oxygen, carbon dioxide, and ethane. AIChE J. **49**(8), 2187–2198 (2003)
36. Stoll, J., Vrabec, J., Hasse, H., Fischer, J.: Comprehensive study of the vapour-liquid equilibria of the pure two-centre Lennard-Jones plus pointquadrupole fluid. Fluid Phase Equilib. **179**(1–2), 339–362 (2001)
37. Sutton, G.T., Biblarz, O.: Rocket Propulsion Elements. Wiley, Minneapolis, Minnesota (2016)
38. Thol, M., Rutkai, G., Köster, A., Lustig, R., Span, R., Vrabec, J.: Equation of state for the lennard-jones fluid. J. Phys. Chem. Ref. Data **45**(2), 023101 (2016)
39. Vrabec, J., Gross, J.: Vapor-Liquid equilibria simulation and an equation of state contribution for dipole—quadrupole interactions. J. Phys. Chem. B **112**(1), 51–60 (2008)
40. Vrabec, J., Stoll, J., Hasse, H.: A set of molecular models for symmetric quadrupolar fluids. J. Phys. Chem. B **105**(48), 12126–12133 (2001)
41. Werth, S., Horsch, M., Hasse, H.: Surface tension of the two center Lennard-Jones plus quadrupole model fluid. Fluid Phase Equilib. **392**, 12–18 (2015)
42. Xu, L., Kumar, P., Buldyrev, S.V., Chen, S.H., Poole, P.H., Sciortino, F., Stanley, H.E.: Relation between the Widom line and the dynamic crossover in systems with a liquid-liquid phase transition. Proc. Natl. Acad. Sci. USA **102**, 16558–16562 (2005)
43. Yu, A., Su, W., Lin, X., Zhou, N.: Recent trends of supercritical CO2 Brayton cycle: bibliometric analysis and research review. Nucl. Eng. Technol. **53**, 699–714 (2021)
44. Zhang, T., Sun, S., Bai, H.: Thermodynamically-consistent flash calculation in energy industry: from iterative schemes to a unified thermodynamics-informed neural network. Int. J. Energy Res. **46**(11), 15332–15346 (2022). https://doi.org/10.1002/er.8234
45. Zhang, T., Zhang, Y., Katterbauer, K., Al Shehri, A., Sun, S., Hoteit, I.: Phase equilibrium in the hydrogen energy chain. Fuel **328**, 125324 (2022). https://doi.org/10.1016/j.fuel.2022.125324

Regulation of Substrate Transport in the Early Stages of Fatty Acid Synthesis

Florian Leidner and Helmut Grubmüller

Abstract The fungal type 1 fatty acid synthase (FAS) is a complex multienzyme system that plays a crucial role in the synthesis of saturated fatty acids. The progression of the intricate reaction cycle relies on the transfer of reaction intermediates between catalytic domains, facilitated by an acyl carrier domain. The initial step of the reaction involves the transfer of acetyl to the carrier, necessitating a substantial conformational change that is stabilized by the binding of a regulatory γ-subunit. This study aims to elucidate the molecular mechanism underlying the induction of the conformational change and the stabilization of the initiation complex. To achieve this, all-atom molecular dynamics simulations were performed on the FAS in the initiation conformation. A comprehensive model of the full-length 2.6 MDa FAS complex was reconstructed, serving as the basis for the simulations conducted in the presence and absence of the γ-subunit, as well as by relocating the carrier to different catalytic domains. The results demonstrate that the binding of the γ-subunit is essential for stabilizing the conformational change, whereas the binding of the carrier protein did not exert a significant effect. These findings support the hypothesis that the initiation of fatty acid synthesis is regulated through conformational selection, rather than solely relying on the binding and dynamics of the carrier protein.

1 Introduction

De novo fatty acid synthesis is a central pillar of metabolism. It provides essential molecules for energy storage, cell wall integrity, and signal transduction [21]. The synthesis of C16- and C18-fatty acids from acetyl- and malonyl-CoA proceeds in an iterative sequence of enzymatic reactions extending the nascent fatty acid by two carbons at a time (Fig. 1A). Although the reactions are conserved, enzyme organization differs significantly between species [22]. Plants and bacteria utilize mono-functional enzymes, available in high concentrations in the cytosol [3].

F. Leidner · H. Grubmüller (✉)
Max Planck Institute for Multidisciplinary Sciences, Göttingen, Germany
e-mail: hgrubmu@mpinat.mpg.de

T. Ludwig et al. (eds.), *High Performance Computing in Science and Engineering '23*,
https://doi.org/10.1007/978-3-031-91312-9_34

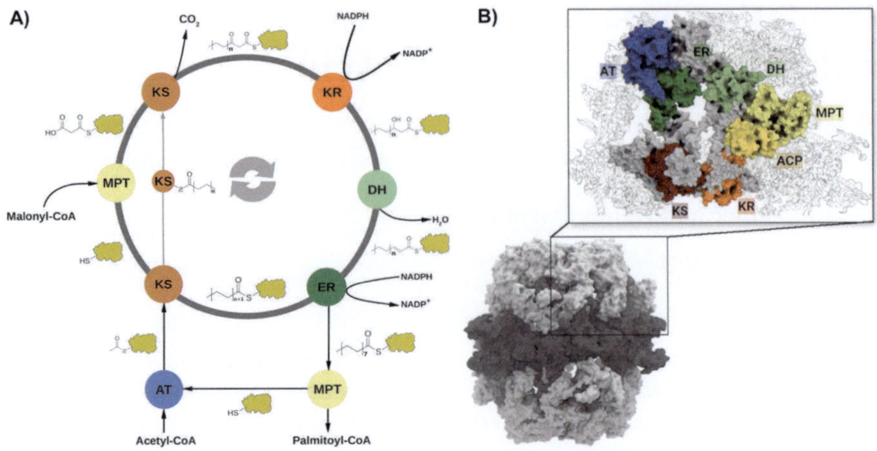

Fig. 1 Synthesis of fatty acids in the fungal FAS **A** Diagram depicting the progression of fungal fatty acid biosynthesis. The component enzymes (illustrated as colored circles) are: Acetyltransferase (AT), Ketoacyl Synthase (KS), Malonyl Palmitoyl Transferase (MPT), Ketoacyl Reductase (KR), Dehydrogenase (DH), Enoyl Reductase (ER). Reaction intermediates are shown above the connecting lines, tethered to the ACP (shown in yellow) **B** Structure of the fungal fatty acid (PDBID: 6QL9) showing the $\alpha_6\beta_6$ heterododecamer. The equatorial α-subunits are colored black, and the polar β-subunits are shown in gray. The inset depicts the $\alpha\beta$-dimer with catalytic enzymes colored in accordance with Fig. 1A

In contrast, animals, fungi, and the CMN group bacteria[1] combine all enzymes in a single, multifunctional enzyme complex; The type 1 Fatty Acid Synthase (FAS).

The fungal fatty acid synthase is a paradigm for microbial FAS systems. Its constituent enzymes are encoded by two genes and form a barrel-shaped $\alpha_6\beta_6$ heterododecamer (Fig. 1B). This oligomeric complex contains two hemispheric reaction chambers where fatty acids are synthesized. The active sites of the enzymatic domains face the lumen of these reaction chambers. To facilitate transfer of reaction intermediates between catalytic domains, nascent fatty acids are covalently linked to an integral Acyl Carrier Protein (ACP). This carrier protein is an essential component of the FAS system and, in conjunction with the spatial clustering of enzymatic domains, contributes to its high turnover efficiency [11]. The exceptional catalytic efficiency of the yeast FAS not only establishes it as a model for multifunctional enzymes but also positions it as a promising platform for the biotechnological production of biofuels and fine chemicals [11].

Despite its vital role in metabolism and potential industrial applications, the molecular mechanism underlying FAS fatty acid synthesis remains elusive. While experimental structural biology has shed light on the composition of fungal FAS, it is challenging to resolve the numerous transient intermediates required for fatty acid synthesis through experimental means. Additionally, until recently, it was believed that substrate transport solely depends on the stochastic motion of ACP, with the FAS

[1] Corynebacteria, Mycobacteria, Nocardia.

core acting as a rigid scaffold. However, the recent discovery of a regulatory FAS γ-subunit suggests that processive steps may not only involve the relocation of the carrier protein but also the rearrangement of the FAS scaffold [23]. This regulatory γ-subunit modulates the catalytic activity of FAS in response to the concentration of NADPH. In the γ-subunit bound state, the β-subunits undergo significant structural rearrangement concurrent with the binding of ACP to the Acetyltransferase (AT) domain. This implies that the progression of fatty acid synthesis is linked to the conformational state of FAS and highlights the ability of regulatory factors to influence the transfer of reaction intermediates and catalytic activity.

In this project, we employed all-atom molecular dynamics (MD) simulations to investigate how the binding of the regulatory γ-subunit and subsequent rearrangement of the β-subunits affect the dynamics of ACP and, consequently, the catalytic activity of fungal FAS. To conduct these simulations, we developed an all-atom model of the FAS complex. The oligomeric complex comprises both folded domains and disordered regions. Although experimentally unresolved, these disordered regions play an important role in the function and regulation of the complex. We reconstructed a total of 1080 unresolved amino acids, resulting in the first all-atom model of FAS. This model enabled us to explore the molecular mechanism governing the initiation of fatty acid synthesis. The binding of ACP to the acetyltransferase domain requires a significant conformational change in the β-subunit. It is unclear whether this conformational change is induced by the binding of the carrier protein or represents a minor component of the room temperature ensemble. Since the rotated conformation can only be observed in the presence of the γ-subunit, we conducted simulations where we either removed the subunit or relocated the carrier protein. Our simulations led us to the conclusion that the γ-subunit stabilizes the rotated state and that it is not induced by ACP binding. This supports the model in which conformational sampling of FAS determines the accessible catalytic states. Although we did not observe a complete transition between the rotated and non-rotated conformations, this work provides an important foundation for future investigations into the kinetics of the conformational transition.

2 Results and Discussion

All simulations were based on the 2.8 Angstrom structure of the *S. Cerevisiae* FAS in complex with the γ-subunit (PDBID: 6ql5) [23]. To account for missing regions, we used a combination of MD simulations, Alphafold2 predictions, and the protein design tool implemented in Rosetta [16]. The missing regions are summarized in Table 1.

First, we modeled the intrinsically disordered regions flanking the ACP domain in the α-subunit. The 45 amino acid linker connecting ACP to the N-terminus of the α-subunit is characterized by a high abundance of alanine and proline amino acids. In contrast, the 25 amino acid linker connecting ACP to the C-terminal part of the α-subunits is enriched in alanine and glycine residues. Due to the putative functional

Table 1 Unresolved amino acids in the *S. Cerevisiae* FAS (PDBID:6ql5)

Name	Subunit	Missing residues	Methods
N-Terminal linker	α	95–139 (45)	MD + Rosetta
C-Terminal linker	α	303–327 (25)	MD + Rosetta
DM2-domain	α	540–603 (64)	Alphafold2 + Rosetta
SSS-loop	β	1111–1120 (10)	Rosetta
γ-loop1	γ	60–76 (17)	Rosetta
γ-loop2	γ	114–132 (19)	Rosetta

significance of these regions, we ran molecular dynamics simulations of the linker peptides in solution and modeled the linker based on the generated ensemble.

The second region that was missing in the starting structure consisted of a 64-amino acid stretch located at the interface between adjacent α-subunits. This region, known as Dimerization Motif 2 (DM2), had been partially resolved in the crystal structure of *T. Lanuginosus* FAS and was found to adopt a helix-turn-helix motif [15]. Although experimental structures of *S. Cerevisiae* FAS do not provide full resolution of this region, certain parts of its sequence were predicted to have a high propensity for forming an α-helix, suggesting a similar fold. Therefore, we used AF2 to predict the structure of the DM2 region. Although the helix-turn-helix motif was predicted with high confidence, we were unable to predict the dimeric form. In all AF2 predictions, the helical region was predicted to dimerize in an antiparallel helical bundle rather than the parallel helical bundle that had been observed experimentally in other species. To exclude that this antiparallel fold is unique to *S. Cerevisiae* FAS, we performed short (500 ns) MD simulations of the antiparallel model, during which we observed partial unfolding of the helical bundle. Therefore, we modeled the dimeric structure based on the *T. Lanuginosus* FAS structure. The disordered regions flanking the α-helical motif were connected to the FAS core using the kinematic loop closure algorithm.

The experimental structure provided resolution for most of the β-subunit. Only a 10-amino acid region between residues 1110 and 1121 remained unresolved. This region, here termed the SSS-loop due to its triple serine motif, is located on the extraluminal side of the FAS and is enriched in polar amino acids. It has been identified as a potential phosphorylation target [13]. To complete the missing residues, we employed kinematic loop closure methodology. In the γ-subunit, two regions were found to be missing. The first region corresponds to a loop that connects the first and second α-helices of the γ-subunit. The second region is a stretch of 19 amino acids that bridges the α-helical region with the short peptide. Binding to the malonyl-palmitoyl transferase domain. Both regions were modeled using kinematic loop closure techniques.

In total, we reconstituted 1080 (180×6) missing residues or 4.4% of the total residues of the FAS $\alpha_6\beta_6\gamma_6$ octadecamer. This is a relatively small amount, considering that up to 30% of the human proteome is predicted to be disordered regions [19].

Two of the disordered regions are located on the extraluminal surface of the FAS. The SSS-loop is a target for CDK1, and the flexibility likely contributes to recognition by the kinase. DM2 was initially hypothesized to contribute to the folding and stability of FAS; however, experimental studies suggest that it is not essential for FAS folding [10]. Instead, there is mounting evidence that the DM2 motif acts as an anchor for the recruitment of other enzymes and the formation of higher-order metabolomic structures [17, 25]. The reconstructed structure of the DM2 motif would support the latter function. The region is not completely disordered but folds into an α-helical bundle flanked by disordered segments. Its interactions with the core of the FAS scaffold are limited, and in subsequent MD simulations, it showed a high degree of flexibility, but notably, the helical bundle remained stable over multiple microseconds.

The remaining disordered regions face the intraluminal space, encapsulated by the FAS scaffold. This is where fatty acid biosynthesis takes place. Based on our all-atom model, we estimated the volume of the reaction chambers with and without the disordered regions. In total, the linker regions flanking the ACP and the extended loop of the γ-subunit reduce the volume of the reaction chamber by 58.3 nm^3. It was hypothesized that volume exclusion effects play an important role in regulating the progression of fatty acid synthesis.

Based on our all-atom model, we investigated the dynamics of the type 1 Fatty Acid Synthase and how binding of the regulatory γ-subunit alters both the conformational and catalytic state of the multienzyme complex. In principle, these conformational changes can be triggered either by binding of the γ-subunit or by binding of the carrier protein. Therefore, we ran all-atom simulations of the complex under different conditions (Table 2). For all conditions, we simulated three independent replicates for at least 5 μs. The average size of the fully solvated system was 2.8 million atoms.

Previous experimental results showed that the γ-subunit stabilizes the rotated conformation of the *S. Cerevisiae* FAS and recruits ACP to the Acetyltransferase (AT) domain. However, it is unknown whether rotation proceeds ACP binding or happens after the carrier binds to the (AT) domain. To understand the interdependence between β-subunit rotation, γ-subunit binding, and ACP recruitment, we ran molecular dynamics simulations of the FAS in the rotated state, where we either removed

Table 2 Simulation overview. Catalytic state corresponds to the location of the ACP

Name	Catalytic state	γ-subunit bound	Replicates	Cumulative simulation time	Platform
FAS_{AT}	Acetyltransferase	Yes	3	15	MPCDF—Cobra
$FAS_{AT}\Delta\gamma$	Acetyltransferase	No	3	15	MPCDF—Cobra
FAS_{KS}	Ketoacyl synthase	Yes	3	15	HPE Apollo (Hawk)
FAS_{AT}	Ketoacyl synthase	No	3	15	HPE Apollo (Hawk)

the γ-subunit or modeled ACP bound to the Ketoacyl synthase domain ($FAS_{AT}\Delta\gamma$, FAS_{KS}). As a control, we ran simulations of FAS in complex with the γ-subunit with ACP bound to the AT domain (FAS_{AT}), as well as simulations of FAS without the γ-subunit and with ACP bound to the KS domain ($FAS_{KS}\Delta\gamma$). Our expectation was that FAS_{AT} would maintain the rotated conformation, while $FAS_{AT}\Delta\gamma$ would gradually relax towards the non-rotated conformation. By analyzing the dynamics of the $FAS_{AT}\Delta\gamma$ and FAS_{KS} systems, we aimed to elucidate whether the rotated conformation primarily results from the binding of the γ-subunit or the recruitment of ACP to the AT domain.

The rotated conformation of FAS is characterized by a rigid body rotation of the β-subunit, leading us to expect significant conformational rearrangements upon destabilization of this conformation. To assess the overall structural changes, we aligned the β-subunit trimers with the experimental structure and calculated the root mean squared deviation (RMSD) of the Cα atoms. In all simulations, we initially observed a sharp increase in the RMSD between the simulated and experimentally observed structures of the β-subunit trimer, reaching approximately 0.6 nm (Fig. 2A). This rapid increase can be attributed to the weak coupling between the dynamics of the domains and subunits, resulting in non-equilibrium dynamics that do not maintain the C3 symmetry observed in the experimental structure of FAS. After 2 μs, the RMSD of the FAS_{AT} simulation plateaued at approximately 0.6 nm, indicating that the simulations were sampling a local minimum. In contrast, the differences between the experimental structure and simulations of the modified FAS continued to increase over 5 μs, suggesting that all perturbations to the system resulted in long-timescale relaxation processes. The average RMSD of the β-subunit trimer was 0.66 and 0.67 nm in the FAS_{KS} and FAS_{AT} simulations, respectively, while it was 0.76 and 0.71 nm in the $FAS_{KS}\Delta\gamma$ and $FAS_{AT}\Delta\gamma$ simulations. This indicates that the γ-subunit makes the largest contribution to stabilizing the rotated conformation.

To determine if the observed changes in the structure of the β-subunit trimer are driven by changes in the orientation of the subunits, domain rearrangement, or changes in the local protein structure, we performed alignment and RMSD calculations for all β-subunits and their constituent domains.

To define these domains, we employed a community detection approach based on mutual information of the displacement vectors, enabling us to identify regions with strongly coupled internal dynamics within the β-subunit. In total, we identified six regions corresponding to both structural and enzymatic domains (Fig. 2B). The structural integrity of these domains remained largely unchanged compared to their fold in experimental structures, with an average RMSD of 0.198 ± 0.049 nm. The most notable changes were observed in the regions corresponding to the AT domain and the regions corresponding to the MPT domain (Fig. 2C). Previous experimental studies had already indicated the relative flexibility of the MPT domain, which is further supported by our analysis, as it was subdivided into two weakly coupled domains. As for the AT domain, which binds both ACP and the γ-subunit, we anticipated that its flexibility would increase upon their removal. However, we found that the average RMSD did not show significant changes across the variants, suggesting

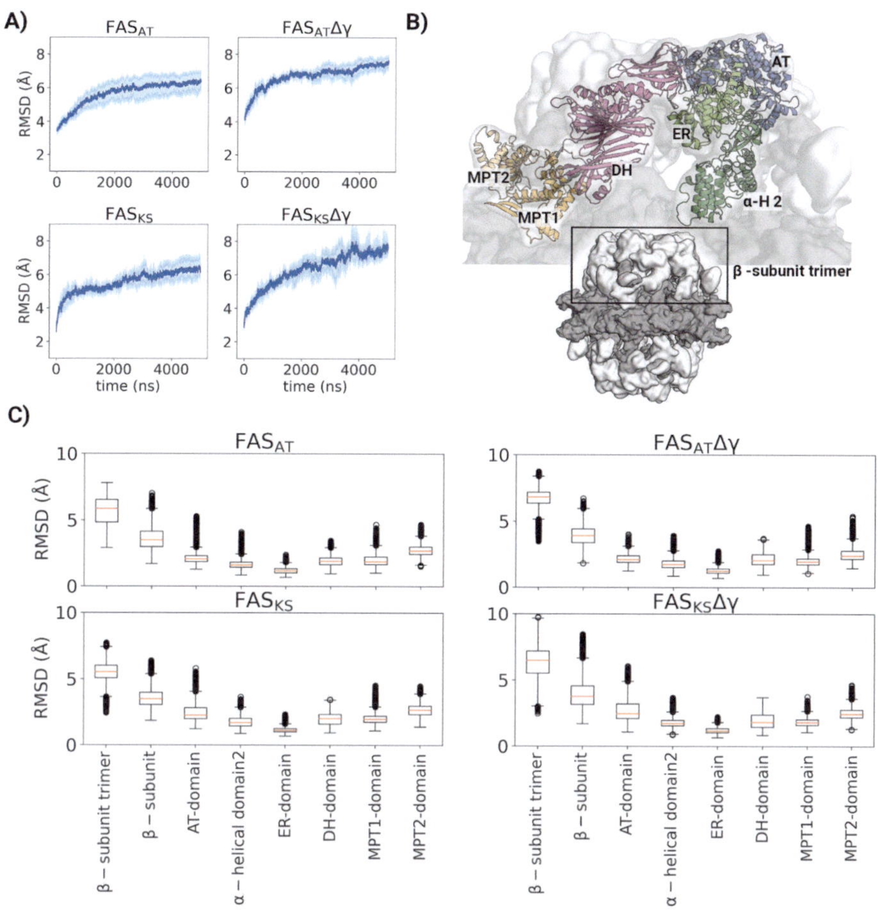

Fig. 2 Structural heterogeneity observed in MD-Simulations **A** The RMSD of the β-trimer relative to the experimental structure in the rotated conformation is shown. The average over three replicates with two trimers per replicate is represented by the bold line. The shaded area indicates the $\pm 1\sigma$ band, representing the standard deviation. **B** The structure of the β-subunit trimer is depicted in the surface representation. The upper panel provides a detailed view of one of the β-subunit trimers, shown in cartoon representation. The coloring highlights the regions with strongly coupled dynamics, as identified by community analysis **C** The RMSD distributions for different FAS variants are presented as boxplots. The interquartile range between the 25th and 75th percentiles is depicted by the boxed area. The whiskers extend to all data points falling within 1.5 times the interquartile range, while any additional points beyond that range are shown as dots

that the structural heterogeneity of the AT domain is an intrinsic property rather than one induced by the removal of either the γ-subunit or the ACP.

By comparing simulations with different variants of the complex, we found that the γ-subunit is the primary factor stabilizing the rotated conformation. Its removal led to a significant destabilization of the rotated conformation (Fig. 2C). Binding of the ACP did not stabilize the rotated conformation in our simulations; in fact, in our simulation, the RMSD of the β-trimer was slightly higher when ACP was bound to the AT domain. This can be attributed to the increased capacity of the β-subunit to form intersubunit interactions when in complex with ACP.

Finally, we aimed to determine whether the β-subunits transitioned from the rotated conformation to the experimentally observed non-rotated conformation. However, the RMSD towards the non-rotated conformation (PDBID: 6ql9) remained consistently high across all simulations. This indicates that a complete transition between the two conformations occurs on longer timescales than those accessible through the current all-atoms simulations. Given that the fully solvated FAS system consists of 2.8 million atoms, sampling a full transition between the rotated and non-rotated conformations may be computationally impractical. Nevertheless, alternative methods such as path sampling simulations and enhanced sampling techniques offer a promising approach that maintains atomistic resolution while facilitating the sampling of transition states [7, 26]. To conduct such simulations, a reaction coordinate must be defined to effectively sample the transition between states.

Our simulations revealed that the changes observed in the β-subunit were primarily driven by rearrangements of the subunits, with domain rearrangements making a lesser contribution. Interestingly, in simulations with the γ-subunit, we observed the stabilization of the rotated conformation, while ACP binding had no significant effect. These findings support the conformational selection model, which suggests that the rotation of the β-subunit is a prerequisite for ACP binding rather than being induced by it. According to this model, the kinetics of the conformational transition play a critical role in regulating the progression of the reaction cycle. To gain further insights, future investigations will focus on studying the kinetics of the transition between the rotated and non-rotated conformations.

3 Methods

3.1 Atomic Model Building

The atomic models of the *S. Cerevisiae* FAS were based on the 6ql5 cryoEM structure. The Rosetta macromolecular modeling platform (version 2020.08) was used to model unresolved regions. Different protocols were used for different regions (Table 3), however, all protocols used the default Rosetta energy function [2]. The SSS-loop and γ-loops 1/2 were modeled using the next generation kinematic closure (NGK) algorithm [24]. Only the missing residues were reconstructed during

Table 3 Scaling behavior of GROMACS 2022.2 on HLR-Hawk. For this test the 3 mil. atom FASγ was used. The combination of MPI ranks and OpenMP threads, that gave the best performance is reported

Nodes	Number of cores	Speed (ns/day)	Speedup	Ideal speedup
1	128	3.1	1.0	1
2	256	6.1	2.0	2
4	512	12.0	3.9	4
8	1024	22.2	7.2	8
16	2048	39.9	13.0	16
32	4096	65.7	21.4	32
64	8192	91.6	29.9	64
128	16384	122.0	39.8	128
256	32768	111.6	36.4	256

loop closure. Subunits adjacent to the missing regions were kept frozen. Subunits that were not adjacent to the missing region were not included in the process to reduce memory requirements. Following loop closure, all sidechains within 0.1 nm of the reconstructed region were repacked. 10,000 models were generated for each missing region. The models were ranked based on the Rosetta energy function, and the final model was chosen from the ten best models based on visual inspection.

For 64 missing residues corresponding to dimerization domain 2, an initial model was generated using Alphafold2 (version 2.3.1). Because we were unable to model the dimer with Alphafold2, we used the *T. Lanuginosus* structure (PDBID: 4V58) as a template to reconstitute the dimer. The dimer interface was repacked using the relax protocol implemented in Rosetta [20]. The unstructured regions flanking the helical region were connected to the core of the FAS using the NGK algorithm. To this end, we used the same protocol that had been described earlier, however, only the final 20 amino acids were reconstructed during loop closure, keeping the helix bundle intact.

For the unresolved linker-regions flanking the acyl carrier protein, we first generated a conformational ensemble based on $3 \times 5\mu s$ molecular dynamics simulation. The simulation protocol for the linker peptides is identical to the one used for the full-length FAS, which is described below. For each of the six α-subunits, linker structures were drawn from the simulated ensemble and modeled into the full-length FAS using the NGK algorithm. Any resulting clashes were resolved using the relax protocol implemented in Rosetta.

3.2 Molecular Dynamics Simulations

Molecular dynamics simulations were carried out using GROMACS 22.3 compiled in double precision mode [1]. The simulation system was parameterized using the CHARMM36M forcefield and the TIP3P water model [14]. Forcefield parameters for the phosphopantetheinylated serine on the ACP domain were based on the CHARMM General Force Field, but dihedral and electrostatic were optimized to better fit Ab Initio calculation [8, 27]. Default protonation states at pH 7 were assigned for all protonatable groups. Virtual sites were used for all non-polar hydrogens, which allowed us to use a longer integration time step in the production simulation.

The system was solvated in a rhombic dodecahedral box, leaving at least 1.6 nm between the box boundaries and the nearest protein heavy atom. Sodium and chloride ions were added to neutralize the charge of the system and set a total ion concentration of 150 mM. Subsequently, the system was minimized using the steepest descent algorithm, while applying position restraints (1000 kJ/mol nm^2) to the protein heavy atoms. This was followed by a second minimization without position restraints. Both minimizations were carried out to machine precision.

Subsequently, the system was equilibrated in a series of short simulations. The temperature was kept constant at 300 K using the velocity rescale thermostat [6]. The temperature of the solute and the solvent were maintained separately. In the first simulation, initial velocities were drawn from a Maxwell distribution; in subsequent simulations, initial velocities were assigned based on the previous simulation. Lennard-Jones interactions were truncated at 1.2 nm and smoothly switched to 0 between 1.0 and 1.2 nm. Electrostatic interactions were calculated using the particle mesh Ewald method with a real-space cutoff of 1.4 and a 0.12 nm grid spacing [9]. The P-LINCS algorithm was used to constrain all bonds [12]. During equilibration, a 2 fs integration time step was used for the equations of motion; in the production simulation, the integration timestep was increased to 4 fs.

First, the system was simulated for 1 ns in the NVT ensemble. Position restraints of 1000 kJ/mol nm^2 were applied to the protein. This was followed by a 1 ns simulation in the NPT ensemble. Pressure was kept at 1 bar using the Berendsen thermostat, position restraints were applied to the protein backbone atoms [4]. This was followed by a second simulation in the NVT ensemble but without position restraints. Two simulations were repeated without position restraints. In the second NPT simulation, exponential relaxation pressure coupling was used to keep a constant pressure of 1 bar [5].

3.3 Community Analysis

To identify subsets of residues with coupled dynamics, we employed a method based on the generalized correlation measure [18]. Initially, all β-subunits were aligned to a common reference, minimizing the deviation between the Cα atoms. We selected chain G from the experimental structure (PDBID: 6ql5) as the initial reference. After the initial alignment, we calculated the average structure, which served as the reference for a second alignment. Using this aligned structure, we computed the generalized correlation between all pairs of Cα atoms in the β-subunit. This measure quantifies the pairwise mutual information of the displacement from the reference coordinates, normalized to a range of 0–1. A high generalized correlation indicates a strong correlation in the displacement of two atoms, while a low value indicates low or no correlation.

Next, we employed an agglomerative clustering protocol to identify subsets of residues with strong internal correlations. We first calculated the generalized correlation measure for each β-subunit trajectory separately. Then, we assigned all residues to one of 10 clusters based on the initial clustering. Using the initial clustering results, we constructed a relationship matrix that indicated the probability of a pair of residues appearing in the same cluster. This matrix was utilized as input for a second round of agglomerative clustering. This bootstrap approach offers the advantage of identifying regions that couldn't be unambiguously assigned to a cluster in the first round of clustering. After visually inspecting the dendrogram, we decided to cluster the residues into 20 clusters. It is worth noting that 14 clusters had fewer than 20 members and were combined into a single noise cluster. Mapping the residues onto the structure confirmed that the residues in the noise cluster corresponded to flexible regions between folded domains.

4 Code Performance

4.1 Scaling

At the beginning of this project, we evaluated the performance and scalability of GROMACS on HLRS - Apollo HAWK. To this end, we performed a comprehensive set of benchmark simulations (Table 3). For the benchmark, we used the software version GROMACS 2022.3 compiled with the double precision option, the precompiled external fftw-library, and AVX2_256. Double precision was necessary to avoid numeric instabilities due to the size of the simulation system. To evaluate performance, we used the same system and settings that have been used before to generate the preliminary results (FASγ; 3,001,945 atoms). The system was simulated for 20,000 steps, and the performance was calculated from the last 10,000 steps. Performance was measured in nanoseconds of simulation time per day (ns/day). Overall, GROMACS showed close to ideal scaling up to 2048 cores (16 nodes) and acceptable scaling up to 8192 cores (64 nodes), whereafter the performance dropped significantly (Fig. 3A).

Because GROMACS uses mixed MPI/OpenMP parallelization, we first evaluated how the ratio of MPI ranks to OpenMP threads affected performance. Optimal performance was achieved when using up to 2 OpenMP threads per MPI rank (Fig. 3B).

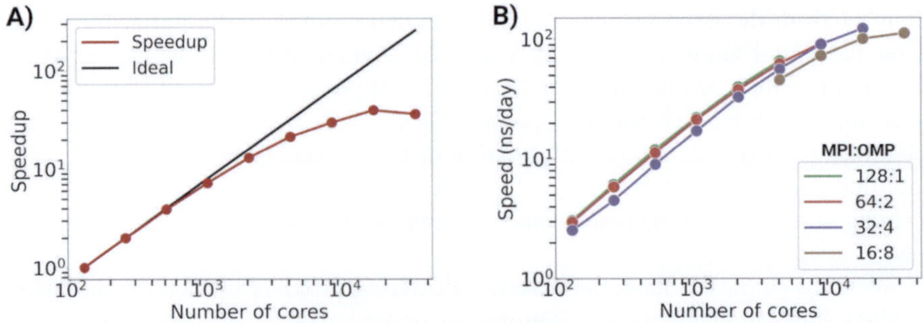

Fig. 3 **Performance of GROMACS 2022.3 on HLRS-Hawk**. **A** Scaling of FASγ simulations on HLRS-HAWK. For each datapoint, the combination of MPI ranks and OpenMP threads that gave the best performance, was used. **B** Speed of FASγ simulations with different combinations of MPI ranks and OpenMP threads

The highest overall performance (122 ns/day) was achieved with 32 MPI ranks and 4 OpenMP threads, albeit at a relatively low efficiency (ns/day/core). We also evaluated simulations utilizing all logical cores, but for highly parallel runs, we found that performance decreased significantly when hardware threading was enabled (data not shown). The memory requirements of the simulations were modest; on average, the benchmark jobs required 275 MB of memory per core ($\sigma = 28$ MB), or 34.375 GB per node, respectively.

We further evaluated and optimized performance during the "Optimization of Scaling, I/O, and Node-level Performance on Hawk" Workshop (07.11.22–11.11.22). Because it was infeasible to modify the software, optimization focused on the compute protocol. It was decided to run simulations on 4096 MPI processes and 2 OMP threads, due to the comparatively high efficiency. To further optimize parallel efficiency, two simulations were bundled together, using the multidir utility implemented in GROMACS.

4.2 Resource Utilization

The results presented in this report were achieved utilizing multiple HPC platforms. The simulations of the FAS complex in the FAS$_{KS}$ and FAS$_{KS}\Delta\gamma$ state were carried out on HPE Apollo Hawk, while simulations in the FAS$_{AT}$ and FAS$_{AT}\Delta\gamma$ state were carried out on the Cobra Cluster of the Max Planck Computing and Data Facility. In addition, we used HPE Apollo (Hawk) to perform simulations of the ACP linker, which were required to model the full-length FAS.

During the reporting period spanning from April 2022 to March 2023, a total of 58.6 million Core-h was utilized on HPE Apollo (Hawk), accounting for approximately 33% of the allocated project resources. The project was initiated in November 2022, with the first production simulations initiated by December 2022.

Around 46 MCore-h were dedicated to simulations of the FAS_{KS} and $FAS_{KS}\Delta\gamma$ systems. For each system, three independent simulations were initiated. To enhance parallel efficiency, two simulations were combined using the GROMACS multidir utility, resulting in the submission of three distinct jobs. Each job used 8192 cores, automatically assigned to the simulations. Additionally, 3 MCore-h were used for simulations of the peptide linker, which connects the ACP domain with the core of FAS. These simulations were crucial for generating the initial models. The remaining resources were allocated to various post-processing tasks.

Acknowledgements The author F.L. acknowledges funding from the European Union under the Marie Sklodowska-Curie grant agreement No. 101065616 – TRISAFA – HORIZON-MSCA-2021-PF-01. Simulation were performed on the national supercomputer HPE Apollo at the High Performance Computing Center Stuttgart under the grant number: GCS-ReSTFAS/44243.

References

1. Abraham, M.J., Murtola, T., Schulz, R., Páll, S., Smith, J.C., Hess, B., Lindahl, E.: Gromacs: high performance molecular simulations through multi-level parallelism from laptops to supercomputers. SoftwareX **1**, 19–25 (2015)
2. Alford, R.F., Leaver-Fay, A., Jeliazkov, J.R., O'Meara, M.J., DiMaio, F.P., Park, H., Shapovalov, M.V., Renfrew, P.D., Mulligan, V.K., Kappel, K., et al.: The Rosetta all-atom energy function for macromolecular modeling and design. J. Chem. Theory Comput. **13**(6), 3031–3048 (2017)
3. Arike, L., Valgepea, K., Peil, L., Nahku, R., Adamberg, K., Vilu, R.: Comparison and applications of label-free absolute proteome quantification methods on Escherichia coli. J. Proteomics **75**(17), 5437–5448 (2012)
4. Berendsen, H.J., Jv, Postma, Van Gunsteren, W.F., DiNola, A., Haak, J.R.: Molecular dynamics with coupling to an external bath. J. Chem. Phys. **81**(8), 3684–3690 (1984)
5. Bernetti, M., Bussi, G.: Pressure control using stochastic cell rescaling. J. Chem. Phys. **153**(11) (2020)
6. Bussi, G., Donadio, D., Parrinello, M.: Canonical sampling through velocity rescaling. J. Chem. Phys. **126**(1) (2007)
7. Chong, L.T., Saglam, A.S., Zuckerman, D.M.: Path-sampling strategies for simulating rare events in biomolecular systems. Curr. Opin. Struct. Biol. **43**, 88–94 (2017)
8. Croitoru, A., Park, S.J., Kumar, A., Lee, J., Im, W., MacKerell, A.D., Jr., Aleksandrov, A.: Additive charmm36 force field for nonstandard amino acids. J. Chem. Theory Comput. **17**(6), 3554–3570 (2021)
9. Essmann, U., Perera, L., Berkowitz, M.L., Darden, T., Lee, H., Pedersen, L.G.: A smooth particle mesh Ewald method. J. Chem. Phys. **103**(19), 8577–8593 (1995)
10. Fischer, M., Joppe, M., Mulinacci, B., Vollrath, R., Konstantinidis, K., Kötter, P., Ciccarelli, L., Vonck, J., Oesterhelt, D., Grininger, M.: Analysis of the co-translational assembly of the fungal fatty acid synthase (FAS). Sci. Rep. **10**(1), 895 (2020)
11. Gajewski, J., Pavlovic, R., Fischer, M., Boles, E., Grininger, M.: Engineering fungal de novo fatty acid synthesis for short chain fatty acid production. Nat. Commun. **8**(1), 14650 (2017)
12. Hess, B.: P-LINCS: a parallel linear constraint solver for molecular simulation. J. Chem. Theory Comput. **4**(1), 116–122 (2008)
13. Holt, L.J., Tuch, B.B., Villén, J., Johnson, A.D., Gygi, S.P., Morgan, D.O.: Global analysis of Cdk1 substrate phosphorylation sites provides insights into evolution. Science **325**(5948), 1682–1686 (2009)

14. Huang, J., Rauscher, S., Nawrocki, G., Ran, T., Feig, M., De Groot, B.L., Grubmüller, H., MacKerell, A.D., Jr.: CHARMM36m: an improved force field for folded and intrinsically disordered proteins. Nat. Methods **14**(1), 71–73 (2017)
15. Jenni, S., Leibundgut, M., Boehringer, D., Frick, C., Mikolásek, B., Ban, N.: Structure of fungal fatty acid synthase and implications for iterative substrate shuttling. Science **316**(5822), 254–261 (2007)
16. Jumper, J., Evans, R., Pritzel, A., Green, T., Figurnov, M., Ronneberger, O., Tunyasuvunakool, K., Bates, R., Žídek, A., Potapenko, A., et al.: Highly accurate protein structure prediction with alphafold. Nature **596**(7873), 583–589 (2021)
17. Kastritis, P.L., O'Reilly, F.J., Bock, T., Li, Y., Rogon, M.Z., Buczak, K., Romanov, N., Betts, M.J., Bui, K.H., Hagen, W.J., et al.: Capturing protein communities by structural proteomics in a thermophilic eukaryote. Mol. Syst. Biol. **13**(7), 936 (2017)
18. Lange, O.F., Grubmüller, H.: Generalized correlation for biomolecular dynamics. Proteins: Struct., Funct., Bioinform. **62**(4), 1053–1061 (2006)
19. Necci, M., Piovesan, D., Tosatto, S.C.: Critical assessment of protein intrinsic disorder prediction. Nat. Methods **18**(5), 472–481 (2021)
20. Nivón, L.G., Moretti, R., Baker, D.: A pareto-optimal refinement method for protein design scaffolds. PLoS ONE **8**(4), e59004 (2013)
21. Paiva, P., Medina, F.E., Viegas, M., Ferreira, P., Neves, R.P., Sousa, J.P., Ramos, M.J., Fernandes, P.A.: Animal fatty acid synthase: a chemical nanofactory. Chem. Rev. **121**(15), 9502–9553 (2021)
22. Schweizer, E., Hofmann, J.: Microbial type I fatty acid synthases (FAS): major players in a network of cellular FAS systems. Microbiol. Mol. Biol. Rev. **68**(3), 501–517 (2004)
23. Singh, K., Graf, B., Linden, A., Sautner, V., Urlaub, H., Tittmann, K., Stark, H., Chari, A.: Discovery of a regulatory subunit of the yeast fatty acid synthase. Cell **180**(6), 1130–1143 (2020)
24. Stein, A., Kortemme, T.: Improvements to robotics-inspired conformational sampling in Rosetta. PLoS ONE **8**(5), e63090 (2013)
25. de Teresa-Trueba, I., Goetz, S.K., Mattausch, A., Stojanovska, F., Zimmerli, C.E., Toro-Nahuelpan, M., Cheng, D.W., Tollervey, F., Pape, C., Beck, M., et al.: Convolutional networks for supervised mining of molecular patterns within cellular context. Nat. Methods **20**(2), 284–294 (2023)
26. Tiwary, P., Parrinello, M.: From metadynamics to dynamics. Phys. Rev. Lett. **111**(23), 230602 (2013)
27. Vanommeslaeghe, K., Hatcher, E., Acharya, C., Kundu, S., Zhong, S., Shim, J., Darian, E., Guvench, O., Lopes, P., Vorobyov, I., et al.: CHARMM general force field: a force field for drug-like molecules compatible with the CHARMM all-atom additive biological force fields. J. Comput. Chem. **31**(4), 671–690 (2010)

Extensions to the MaMiCo Software Stack for Flexible, Massively Parallel Molecular-Continuum Flow Simulation

Louis Viot, Amartya Das Sharma, and Philipp Neumann

Abstract Molecular-continuum flow simulations apply computationally intensive molecular dynamics (MD) simulations in localized regions of a geometry under consideration, and classical, computationally cheaper computational fluid dynamics (CFD) solvers are employed for the remainder of the vast computational domain. The macro-micro coupling tool MaMiCo handles the coupling between a MD solver and a CFD solver. It is highly parallelized and provides interfaces to couple various MD and CFD solvers. In this report, we describe, test and evaluate the performance of new couplings to the MaMiCo framework. On the MD side, we present new developments for the coupling between the MD solver ls1 mardyn and MaMiCo allowing to use an ensemble of ls1 mardyn instances distributed on different MPI ranks for averaging of MD data. On the CFD side, we introduce a new coupling between MaMiCo and the preCICE library allowing to use different time coupling schemes and data interpolation functionalities of preCICE within MaMiCo as well as many CFD solvers already coupled with preCICE.

1 Introduction

Computational fluid dynamics (CFD) solvers [1] can be used to model the fluid behavior at the continuum scale. However, they are not always able to capture fluid phenomena at a molecular scale. This holds for instance when modeling the fluid behavior in nano channels [19] or modeling the fluid close to surfaces, implying inter-molecular surface-fluid interactions [12]. A remedy would be to use molecular dynamics (MD) methods [18], modeling the molecular interactions in the fluid.

L. Viot (✉) · A. Das Sharma · P. Neumann
Chair for High Performance Computing, Helmut Schmidt University, Hamburg, Germany
e-mail: viotl@hsu-hh.de

P. Neumann
e-mail: philipp.neumann@hsu-hh.de

T. Ludwig et al. (eds.), *High Performance Computing in Science and Engineering '23*,
https://doi.org/10.1007/978-3-031-91312-9_35

Then, these interactions can be passed to the CFD solver, allowing to model fluid phase separation or fluid-surface interactions, resolving aforementioned challenging situations.

However, since typically millions of molecules (or significantly more) might be easily required, computing all their trajectories is, compared to continuum-based CFD solvers, inherently computationally intensive and often results in the use of supercomputers [21, 22].

In these cases, hybrid molecular-continuum methods [11, 14] represent one way to remain computationally relatively cheap while remaining precise enough to capture fine-scale phenomena in some part of the domain. Molecular-continuum methods are based on a domain decomposition of the CFD domain, coupling computationally affordable inaccurate (at the nano scale) CFD methods with computationally intensive, yet molecular interaction-resolving MD methods.

Nevertheless, MD still typically dominates computational cost by far, see for example [2]. Besides, due to inherent Brownian motion of the molecules (i.e., thermal noise), sampling and averaging MD quantities (such as flow velocities) are required to provide smooth quantities to the CFD solver. For the latter reasons, molecular-continuum and related simulation approaches still often rely on the use of HPC systems [7, 15].

In this report, a software stack for molecular-continuum simulations is presented. Considerations are based on the macro-micro-coupling tool (MaMiCo) [16], which is extended by the use of the preCICE library [3] for partitioned coupling, data interpolation and coupling with many CFD solvers, as well as by the use of the highly parallelized MD software ls1 mardyn [17] and the CFD solver OpenFOAM [5]. While prior couplings to both ls1 mardyn and OpenFOAM have already been published in the past [16, 25], both required to be revisited. Significant software changes in ls1 mardyn such as augmentation of the underlying MD algorithms by node-level auto-tuning [6] or re-structuring of distributed-memory communication patterns [20], resulted in the need of a re-implementation of the existing interface structures in MaMiCo. Besides, with MaMiCo supporting multi-instance MD simulation sampling [15], using this feature with ls1 mardyn required a revision of the existing interfaces and communication structure in ls1 mardyn. Concerning OpenFOAM, with preCICE being incorporated into MaMiCo, its pre-existing, more evolved and powerful preCICE-OpenFOAM adapter [4] can now be used for a coupling, making the existing OpenFOAM-MaMiCo interfaces less important (but still potentially helpful in the future, for example for rapid or more simplified, direct testings). Using preCICE, fully partitioned simulations with separate executables for CFD and MD solver can now be executed, which is deemed helpful in the future for using commercial (closed-source) CFD software together with MaMiCo, or for simply running the CFD solver on one's desktop while executing the computationally intensive MD simulations on a supercomputer; the feasibility of the latter was demonstrated recently by the authors [24]. The full new software stack of MaMiCo is validated on a Couette flow and its scalability is discussed.

In Sect. 2, we introduce the full software stack and its different components. Validation and scalability results are provided in Sect. 3. We close with a discussion and outlook to future work in Sect. 4.

2 Software Stack

2.1 The Macro-Micro-Coupling Tool MaMiCo

MaMiCo [15, 16] is an open source C++ highly parallelized, yet not vectorized, framework for coupling MD and continuum (e.g. CFD) solvers. It tackles various challenges arising from molecular-continuum simulations: it supports 2D and 3D simulations; the respective coupling algorithm is hidden from the solvers; interfaces are provided for different CFD and MD solvers such as LAMMPS [23], ESPResSo [13]; it supports steady state and transient simulations; massively parallel MPI-based execution is possible; multi-instance MD simulation sampling (ensemble averaging) can be activated.

The latest v2.0 release of MaMiCo [10] introduced new functionalities such as noise filtering [9], dynamic handling of MD simulation samples [8], python bindings, and coupling to the community CFD code OpenFOAM [5, 25].

In this report, we mainly focus on the new coupling between MaMiCo with the highly efficient MD software ls1 mardyn [17] and with the preCICE library [3]. The latter allows to use data interpolation functionalities of preCICE for MaMiCo-based molecular-continuum coupling as well as time coupling schemes available in preCICE in the future. Moreover, every CFD solver that has been coupled with preCICE[1] becomes available to MaMiCo.

2.2 The PreCICE Library

preCICE [3] is a free/open-source coupling library. It allows partitioned multi-physics simulations by coupling different solvers. Each solver needs to implement the interface provided by preCICE but is then seen as a black-box by preCICE allowing a high level of software modularity.

An overview of the preCICE library is given in Fig. 1. Different so-called participants such as a CFD solver or a FEM structural solver participate in a coupled simulation steered by preCICE. At its core, the preCICE library is divided into four parts: coupling schemes, data mapping, communication, and time interpolation.

preCICE offers different *coupling schemes*. They can be uni-directional or bi-directional. Uni-directional coupling requires data transfer only from one participant

---

[1] See https://precice.org/adapters-overview.html for a list of available solver adapters; accessed 31 May 2023.

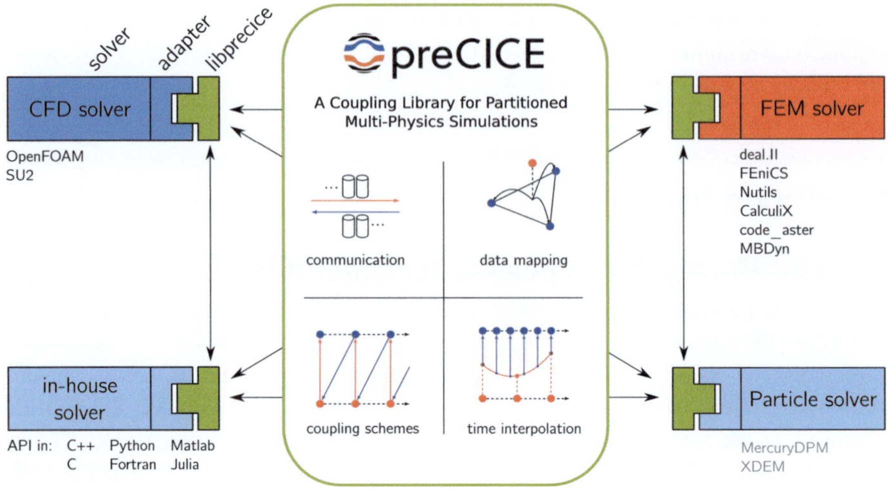

Fig. 1 Overview of preCICE, cf. https://precice.org/docs.html; accessed 31 May 2023

to the other. In this aspect, only an explicit coupling scheme makes sense, which can be serial or parallel. Bi-directional coupling schemes can be serial- or parallel-explicit or serial- or parallel-implicit schemes.

The different participants of a coupled simulation share their coupling meshes with preCICE. The latter then takes care of data interpolation in case of non-matching meshes between the participants. Different *data mapping* methods such as nearest-neighbour, nearest-projection or radial basis functions can then be used.

The third important feature of preCICE's core is *communication* between the participants. This is of high importance for large-scale simulations within a massively parallel high performance computing system. preCICE employs a parallel point-to-point communication where each solver calls its own instance of the preCICE library. Each preCICE parallel instance of one participant then connects to another parallel preCICE instance of another participant. preCICE supports MPI and TCP/IP backends for the communication.

preCICE's fourth feature is about *interpolation in time*. Currently, only sub-cycling between the participants is supported. However, higher-order interpolation schemes such as waveform relaxation methods are under development. They will be available in future releases.

2.3 *The Molecular Simulation Software ls1 Mardyn*

The MD simulation software ls1 mardyn is a flexible, scalable and vectorized MD program capable of simulating small rigid molecules at extreme scales and evolved from a long-term collaboration of numerous computer scientists, engineers and physicists. To satisfy the variety of use cases from the different disciplines, ls1 mardyn features a highly modular code base [17]. A salient feature of ls1 mardyn is the capability of load balancing across ranks, with different load balancing schemes supported. Its capabilities were showcased amongst others in a twenty trillion particle simulation [22], executed on Hazel Hen, the predecessor of Hawk at HLRS.

AutoPas is a node-level particle simulation library, acting as a black-box data container through which particles can be accessed using interfaces. AutoPas uses auto-tuning to select combinations of MD algorithms, data structures and shared-memory parallelization schemes at the node level at runtime to continuously monitor and improve particle simulation performance [6].

Over the last years, ls1 mardyn has been extended to use AutoPas at node level, and this was shown to yield significant speedup in several applications on tested hardware [20].

3 Results

In the following, reported simulations were made on the Hawk computer. The total accumulated run time in terms of node hours spent over the reporting period (April 2022–March 2023) accumulates to 58423 node hours.

3.1 Description of Test Cases

We validate the molecular-continuum software stack on a Couette flow scenario. The ls1 mardyn MD domain is embedded in the OpenFOAM domain representing the channel. We consider two different setups, namely MD30 and MD60, described in Table 1.

In the following, all values are given in MD units. The macroscopic cell size used in MaMiCo for mesh-particle coupling remains constant and equal to 2.5. The OpenFOAM channel domain is of height 50 (resp. 100) for MD30 (resp. MD60), discretized into $20 \times 20 \times 20$ cells of size 2.5 (resp. 5.0) for MD30 (resp. MD60). Hence, for MD60, both meshes don't match and the interpolation methods of preCICE to map data between CFD and MaMiCo grids are needed.

Table 1 MD30 and MD60 Couette flow scenario description

Name	No. macroscopic cells	No. molecules	Channel height	No. CFD cells
MD30	$12 \times 12 \times 12$	21952	50	$20 \times 20 \times 20$
MD60	$24 \times 24 \times 24$	175616	100	$20 \times 20 \times 20$

On the CFD side, OpenFOAM boundary conditions are as follows: the bottom wall is moving in the x-direction with a wall velocity of 0.5 (resp. 1.5) for MD30 (resp. MD60) while the top wall is fixed; left, right, front and back walls are periodic. The OpenFOAM domain is overlapping with the macroscopic cells of MaMiCo. In this overlapping part, three layers of macroscopic cells are covered the OpenFOAM domain. On the MD side, periodic boundary conditions are applied on all six boundaries of the MD domain.

From CFD to MD, particle velocities from the inner cells of the MD domain, which are not in the overlapping part, are averaged cell-wise and imposed on the OpenFOAM mesh at the boundary with the inner MD domain. From MD to CFD, momentum and mass flux are imposed by the CFD solver to the MD solver on the three layers of overlapping cells. For more details about the coupling between CFD and MD, see [15].

The fluid particle density is 0.81, fluid dynamic viscosity is 2.14 and a thermostat is applied on each macroscopic cell with a temperature of 1.1.

Simulations are run for 1000 coupling cycles, each coupling cycle (i.e., CFD time step) comprises of 50 MD time steps of 0.005.

All MD simulations are using 64 MPI ranks with a $4 \times 4 \times 4$ domain decomposition. OpenFOAM is executed sequentially. We tested both strong and weak scaling of the aforementioned MD30 and MD60 Couette flow scenario. We do not use hyperthreading. For the strong scaling, 128 MD instances were used with 1, 2, 4, 8, 16, 32 and 64 nodes. For the 64 nodes case, this leads to one rank handling exactly one subdomain and one MD instance (since Hawk is equipped with two sockets, each with 64 cores). For the weak scaling, the simulation with 1 node uses 2 MD instances, then, as the number of nodes is doubled (until 64 nodes), we double the number of MD instances, i.e. No. MD instances = No. nodes $\times$ 2 for No. nodes $\in \{1, 2, 4, 8, 16, 32, 64\}$.

3.2 Velocity Profile Validation

Figure 2 shows exemplarily velocity profiles for the cases MD30 and MD60 using 64 nodes and 128 MD instances. For both MD30 and MD60 cases with 128 MD instances, the calculated values by the molecular-continuum full software stack seem to follow the Couette flow analytical solutions which validates the software stack for the simulation of coupled molecular-continuum flows. Due to the larger macroscopic cell sizes in MD60 (Fig. 2b), less fluctuations are observable. Note that the visible deviations from the analytical profile may arise from the periodic boundary handling in ls1 mardyn (which is subject to current work). A further reduction of computational efforts can be achieved via noise filters for the MD data, see, e.g., [9] for details.

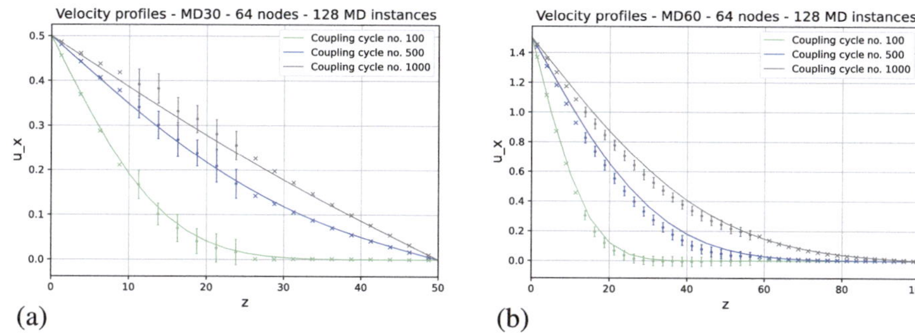

Fig. 2 Couette flow velocity profiles along the z-direction for the cases **a** MD30 and **b** MD60 with 64 nodes and 128 MD instances. Continuous lines are the analytical solution. Crosses are the values extracted from OpenFOAM while dots with error bars are values extracted from ls1 mardyn

3.3 Strong Scaling of the Software Stack

Figure 3 shows the strong scaling results of the full software stack in terms of time-to-solution (also indicating the consumed node hours per run) for the cases MD30 and MD60 with 128 MD instances and No. nodes $\in \{1, 2, 4, 8, 16, 32, 64\}$. Both strong scaling results show that the software stack scales properly until ca. 16 nodes but reaches a plateau for ≥ 32 nodes.

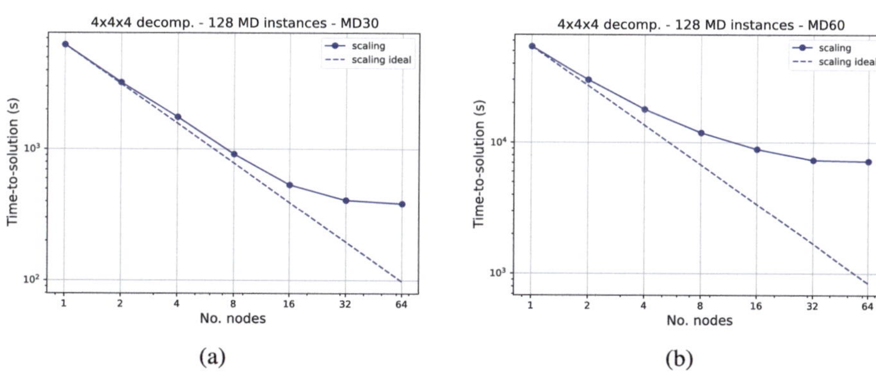

Fig. 3 Strong scaling of the full software stack for the cases **a** MD30 and **b** MD60 with 128 MD instances and No. nodes $\in \{1, 2, 4, 8, 16, 32, 64\}$

3.4 *Towards Weak Scaling*

Currently, OpenFOAM is coupled sequentially to preCICE/MaMiCo. While this can be sufficient for interactive desktop-to-supercomputer molecular-continuum simulation [24], it implies limited weak scalability. This is confirmed by Fig. 4 which shows the weak scaling results of the full software stack for the cases MD30 and MD60 with No. nodes $\in \{1, 2, 4, 8, 16, 32, 64\}$ and No. MD instances $=$ No. nodes $\times$ 2. The time-to-solution increases rapidly for both MD30 and MD60. We therefore conducted further experiments on the Hawk cluster to measure which part of the software stack was causing this behavior. Time measurements are reported in Table 2. We measured the time spent each coupling cycle to solve the MD solver (named 'MD solving' in the table) and the time spent for all ranks of all MD solvers to send their data to MaMiCo (named 'MD to macro comm.' in the table).

As indicated in the table, the communication overhead becomes important when increasing the number of nodes and MPI ranks. Currently, MaMiCo takes care of gathering the data on the rank on which OpenFOAM is executed. An alternative approach would be to use preCICE to gather all the data from MaMiCo's ranks, which is under current investigation.

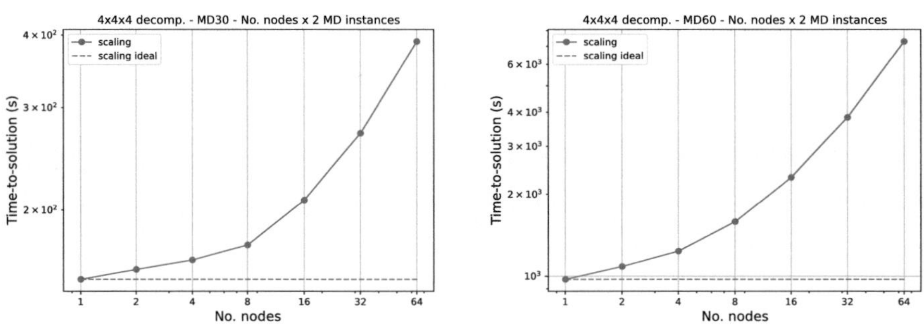

(a) (b)

Fig. 4 Weak scaling of the full software stack for the cases **a** MD30 and **b** MD60 with No. nodes $\in \{1, 2, 4, 8, 16, 32, 64\}$ and No. MD instances $=$ No. nodes $\times$ 2

Table 2 Time measurements captured from MaMiCo for the MD60 case with the weak scaling setups. Values are given as a percentage of the total runtime

No. nodes	1	2	4	8	16	32	64
MD solving (%)	79	73	64	51	37	25	17
MD to macro comm. (%)	7	13	23	37	51	64	72

4 Conclusion

In this report, we have introduced a new software stack for the simulation of molecular-continuum flows. It is constituted by the highly parallelized and vectorized molecular dynamics solver ls1 mardyn, the highly parallelized framework MaMiCo, the coupling library preCICE and the CFD solver OpenFOAM.

Validation tests and scaling experiments have been conducted on the Hawk supercomputer. The full software stack has been validated on a Couette flow with different sizes for the CFD and MD domain, different number of MD instances and different number of nodes used. Next steps comprise, amongst others, the parallelization of the preCICE-MaMiCo coupling so that parallelism can also be exploited on CFD side.

Acknowledgements We thank HLRS and the Gauss Centre for Supercomputing for the provision of computational resources (project *GCS-MDDC*). We further acknowledge funding for MaMiCo software developments by dtec.bw - Digitalization and Technology Research Center of the Bundeswehr (project MaST). dtec.bw is funded by the European Union - NextGenerationEU. We further thank for software sustainability support by the Deutsche Forschungsgemeinschaft (DFG, German Research Foundation), project SuMo (project no. 443188038).

References

1. Anderson, J.D.: Computational Fluid Dynamics: The Basics with Applications. McGraw-Hill Education (2005)
2. Barsky, S., Delgado-Buscalioni, R., Coveney, P.V.: Comparison of molecular dynamics with hybrid continuum-molecular dynamics for a single tethered polymer in a solvent. J. Chem. Phys. **121**(5), 2403–2411 (2004)
3. Chourdakis, G., Davis, K., Rodenberg, B., Schulte, M., Simonis, F., Uekermann, B., Abrams, G., Bungartz, H.-J., Cheung Yau, L., Desai, I., Eder, K., Hertrich, R., Lindner, F., Rusch, A., Sashko, D., Schneider, D., Totounferoush, A., Volland, D., Vollmer, P., Koseomur, O.Z.: preCICE v2: a sustainable and user-friendly coupling library [version 2; peer review: 2 approved]. Open Res. Eur. **2**(51) (2022)
4. Chourdakis, G., Schneider, D., Uekermann, B.: Openfoam-precice: coupling openfoam with external solvers for multi-physics simulations. OpenFOAM®Journal **3**, 1–25 (2023)
5. The OpenFOAM Foundation. OpenFOAM, 2023. [Online; accessed 02 March 2023]
6. Gratl, F.A., Seckler, S., Tchipev, N., Bungartz, H.-J., Neumann, P.: Autopas: auto-tuning for particle simulations. In: 2019 IEEE International Parallel and Distributed Processing Symposium Workshops (IPDPSW), Rio de Janeiro (2019). IEEE
7. Grinberg, L., Insley, J.A., Fedosov, D.A., Morozov, V., Papka, M.E., Karniadakis, G.E.: Tightly coupled atomistic-continuum simulations of brain blood flow on petaflop supercomputers. Comput. Sci. Eng. **14**(6), 58–67 (2012)
8. Jafari, V., Wittmer, N., Neumann, P.: Massively parallel molecular-continuum flow simulation with error control and dynamic ensemble handling. In: International Conference on High Performance Computing in Asia-Pacific Region, HPCAsia2022, pp. 52–60, New York, NY, USA (2022). Association for Computing Machinery
9. Jarmatz, P., Maurer, F., Wittenberg, H., Neumann, P.: Mamico: non-local means and pod filtering with flexible data-flow for two-way coupled molecular-continuum hpc flow simulation. J. Comput. Sci. **61** (2022)

10. Jarmatz, P., Wittenberg, H., Jafari, V., Das Sharma, A., Maurer, F., Wittmer, N., Neuman, P.: Mamico 2.0: an enhanced open-source framework for high-performance molecular-continuum flow simulation. SoftwareX **20** (2022)
11. Kalweit, M., Drikakis, D.: Coupling strategies for hybrid molecular-continuum simulation methods. Proc. Inst. Mech. Eng., Part C: J. Mech. Eng. Sci. **222**(5), 797–806 (2008)
12. Karniadakis, G., Beskok, A., Aluru, N.: Microflows and Nanoflows – Fundamentals and Simulation. Springer (2005)
13. Limbach, H.J., Arnold, A., Mann, B.A., Holm, C.: Espresso-an extensible simulation package for research on soft matter systems. Comput. Phys. Commun. **174**(9), 704–727 (2006)
14. Mohamed, K.M., Mohamad, A.A.: A review of the development of hybrid atomistic-continuum methods for dense fluids. Microfluid. Nanofluidics **8**, 283–302 (2009)
15. Neumann, P., Bian, X.: MaMiCo: transient multi-instance molecular-continuum flow simulation on supercomputers. Comput. Phys. Commun. **220**, 390–402 (2017)
16. Neumann, P., Flohr, H., Jarmatz, P., Tchipev, N., Bungartz, H.-J.: MaMiCo: software design for parallel molecular-continuum flow simulations. Comput. Phys. Commun. **200**, 324–335 (2016)
17. Niethammer, C., Becker, S., Bernreuther, M., Buchholz, M., Eckhardt, W., Heinecke, A., Werth, S., Bungartz, H.-J., Glass, C.W., Hasse, H., Vrabec, J., Horsch, M.: ls1 mardyn: the massively parallel molecular dynamics code for large systems. J. Chem. Theory Comput. **10**(10), 4455–4464 (2014)
18. Rapaport, D.C.: The Art of Molecular Dynamics Simulation. Cambridge University Press (2004)
19. Ritos, K., Borg, M.K., Lockerby, D.A., Emerson, D.R., Reese, J.M.: Hybrid molecular-continuum simulations of water flow through carbon nanotube membranes of realistic thickness. Microfluid. Nanofluidics **19**, 997–1010 (2015)
20. Seckler, S., Gratl, F., Heinen, M., Vrabec, J., Bungartz, H.-J., Neumann, P.: Autopas in ls1 mardyn: massively parallel particle simulations with node-level auto-tuning. J. Comput. Sci. **50**, 101296 (2021)
21. Germann, T.C., Kadau, K.: Trillion-atom molecular dynamics becomes a reality. Int. J. Mod. Phys. C **19**(09), 1315–1319 (2008)
22. Tchipev, N., Seckler, S., Heinen, M., Vrabec, J., Gratl, F., Horsch, M., Bernreuther, M., Glass, C.W., Niethammer, C., Hammer, N., Krischok, B., Resch, M., Kranzlmüller, D., Hasse, H., Bungartz, H.-J., Neumann, P.: TweTriS: twenty trillion-atom simulation. Int. J. High Perform. Comput. Appl. **33**(5), 838–854 (2019)
23. Thompson, A.P., Aktulga, H.M., Berger, R., Bolintineanu, D.S., Brown, W.M., Crozier, P.S., in 't Veld, P.J., Kohlmeyer, A., Moore, S.G., Nguyen, T.D., Shan, R., Stevens, M.J., Tranchida, J., Trott, C., Plimpton, S.J.: LAMMPS - a flexible simulation tool for particle-based materials modeling at the atomic, meso, and continuum scales. Comput. Phys. Commun. **271**, 108171 (2022)
24. Viot, L., Piel, Y., Neumann, Y.: From desktop to supercomputer: computational fluid dynamics augmented by molecular dynamics using MaMiCo and preCICE. In: Proceedings of ISC HPC 2023 Workshops (Second Combined Workshop on Interactive and Urgent High-Performance Computing) (2023). Accepted
25. Wittenberg, H., Neumann, P.: Transient two-way molecular-continuum coupling with OpenFOAM and MaMiCo: a sensitivity study. Computation **9**(12), 128 (2021)

MIX
Papier aus verantwortungsvollen Quellen
Paper from responsible sources
FSC® C105338
FSC
www.fsc.org

If you have any concerns about our products,
you can contact us on
ProductSafety@springernature.com

In case Publisher is established outside the EU,
the EU authorized representative is:
Springer Nature Customer Service Center GmbH
Europaplatz 3, 69115 Heidelberg, Germany

Printed by Libri Plureos GmbH
in Hamburg, Germany